W9-ANB-581

Laboratory Manual for

Starr's *Biology: Concepts and Applications* and

Starr & Taggart's *Biology: The Unity and Diversity of Life*

James W. Perry & David Morton
FROSTBURG STATE UNIVERSITY

Contributions to Animal Diversity
by Ronald E. Barry, Jr.

and Evolutionary Agents
by James H. Howard

Wadsworth Publishing Company   Belmont, California   A Division of Wadsworth, Inc.

Publisher: Jack Carey
Editorial Assistant: Kristin Milotich
Production Editor: Gary Mcdonald
Managing Designer: Andrew H. Ogus
Print Buyer: Karen Hunt
Art Editors: Donna Kalal, Marta Kongsle,
  Roberta Broyer
Permissions Editor: Jeanne Bosschart
Copy Editor: George Dyke
Technical Illustrators: John Waller, Judy Waller,
  Joan Olson, Jill Turney, Carole Lawson
Cover: Gary Head
Cover Photo Credits: Wolf: © Jim Brandenburg/
  Minden Pictures; Rain Forest: © Thomas D. Mangelsen
Compositor: Thompson Type
Color Separator: Rainbow Graphic Arts
Printer: Von Hoffmann Press

I⟨T⟩P ™

International Thomson Publishing
The trademark ITP is used under license.

11 12 13 14 15 16 – 05 04 03 02 01 00 99

ISBN 0-534-21069-4

Contents

Preface

Greetings from the authors! We're happy that you are examining our work. We believe you'll find the information below an invaluable introduction to this laboratory manual.

Audience

The manual is designed for students at the college entry level and assumes that the student has had **no previous college biology or chemistry**. Depending upon the choice of exercises (or portions of exercises) selected, the manual is **appropriate for majors and nonmajors alike**. We've found the exercises suitable for preparation for our majors in the biological sciences (biology, wildlife-fisheries, environmental planning, and preprofessional programs); at the same time, they **provide the foundational knowledge** necessary for nonscience majors to make informed decisions about biological questions in their everyday life.

The Lab Manual's **terminology conforms to that of Starr and Taggart's** *Biology* **and to Starr's** *Biology: Concepts and Applications*, although you'll find that the **exercises support virtually any biology text** used in an introductory course.

What's Important to Us

In preparing this lab manual we paid particular attention to **pedagogy, clarity of procedures and terminology, illustrations, and practicality of materials** used.

Pedagogy

The exercises have been written so that the conscientious student can accomplish the objectives of each exercise with minimal input from an instructor. The **procedure sections** of the exercises are **more detailed** and step-by-step than in other manuals. Instructions follow a natural progression of thought so the instructor need not conduct every movement.

We strived to make **each portion of the exercise part of a continuous flow of thought**. Thus, we do not wait until the post-lab questions to ask students to record conclusions when it is more appropriate to do so within the body of the procedure.

Terms required to accomplish objectives are **bold-faced**. Scientific names and precautionary statements, or those needing emphasis, are in *italics*.

The use of scientific names has been deemphasized when it is not relevant to understanding the subject. Generally, however, these names do appear in parentheses, since the labels on many prepared microscope slides bear only the scientific name.

Format

Each exercise includes:

1. **Objectives:** a list of desired outcomes.
2. **Introduction:** to pique student interest, indicate relevance, and provide background.
3. **Materials:** a list for each portion of the exercise so that a student can quickly gather the necessary supplies. Materials are listed "Per student," "Per pair," "Per group," and "Per lab room."
4. **Procedures:** including safety notes, illustrations of apparatus, figures to be labeled, drawings to be made, tables for recording data, graphs to be drawn, and questions that lead to conclusions to be drawn. The procedures are listed in easy-to-follow numbered steps.
5. **Pre-lab Questions:** ten multiple-choice questions that the student should be able to answer after reading the exercise but prior to entering the laboratory.
6. **Post-lab Questions** that draw upon knowledge gained from *doing* the exercise and that the student should be able to answer *after* finishing the exercise. These **post-lab questions assess both recall** (preparing students for the lab practicals) **and understanding**.

Practical Post-lab Questions

Virtually all courses use laboratory *practical* examinations. We explain to our students the difference between lecture-type questions, in which they need to read and provide an answer based upon the written word, and practical-type questions, in which a response is dependent upon *observation*.

We believe the **post-lab questions** should **draw upon the knowledge gained by observation**. Consequently, we've incorporated illustrations into the post-lab questions. These illustrations typically are similar, *but not identical,* to those in the procedures. Thus, they assess the student's ability to use knowledge gained during the exercise in a new situation.

Flexible Quiz Options

Each exercise has a set of pre-lab questions. We have found, through nearly thirty years of experience, that students left to their own initiative typically come to laboratory unprepared to do the exercise. Few read the exercise beforehand. One answer to this is to incorporate some sort of preexercise activity. At the same time, we recognize that grading a large number of lab papers each week may put an unreasonable burden on instructors. Consequently, we decided on a multiple-choice format, which is easy to grade but still accomplishes the pedagogical goal.

In our own course, we have students take a pre-lab quiz consisting of the questions in their lab manual, in scrambled order to discourage memorization. These scrambled quizzes are reproduced in our *Instructor's Manual*. Our quiz takes literally two minutes of lab time and counts as a portion of the lab grade, thus rewarding students for preparation.

Other instructors have told us they use the pre-lab questions to assess learning *after* the exercise has been completed. We encourage you to be creative with the manual; do what you like best.

Lab Length and Exercise Options

We realize there is wide variation in the amount of time each instructor devotes to laboratory activities. To provide maximum **flexibility** for the instructor, the procedure portions of the exercises are divided by major headings. Once the introduction has been studied, **portions of the procedures can be deleted or put on as demonstrations without sacrificing the pedagogy of the exercise as a whole**.

It's our experience that if the lecture section has covered the topic prior to the lab, students find the exercise much more relevant and understandable. We strive to create this situation in our course, and thus no time is spent on a lecture-style introduction in the lab itself before the exercise begins. We therefore have two full hours for real scientific investigation and need delete very little material to complete the exercise in the time alotted.

Optional, Inquiry Experiments

We all realize that the best way to learn science is to *do* science. Thus, in this manual you will find reference to **numerous optional experiments** that are **inquiry oriented** and are extensions based upon an activity just completed. Each follows the cognitive techniques whose foundations are laid in the first exercise, "Scientific Method." These optional experiments are included in the Instructor's Manual and may be copied to provide the opportunity for students to perform their own investigations.

Illustrations

Perhaps our new illustrations, more than anything else, will be most quickly noticed as you thumb through the manual. Color illustrations are pedagogically sound because color increases both the utility of the illustrations and the comfort level of the students. We've included everything a student needs visually to accomplish the objectives of each exercise. There is **no longer a need for students to purchase supplemental pictorial atlases;** thus, **students should save money.**

Most illustrations of microscopic specimens are labeled to provide orientation and clarity. A few are unlabeled but provided with leaders to which students can attach labels. In other cases, more can be gained by requesting the student to do simple drawings. Space has been included in the manual for these, with boxes for drawings of macroscopic specimens and circles for microscopic specimens.

Materials

The equipment and supplies used in the exercises are readily available from biological and laboratory supply houses. Many can be collected from nature. We've attempted to keep instrumentation as simple and inexpensive as possible.

As anyone who lives in a temperate climate knows, it may be necessary to adjust the sequence of the laboratory exercises to accommodate seasonal availability of certain materials. However, we've provided alternatives, including the use of preserved specimens, wherever possible, to avoid this problem.

Instructor's Manual

There is no need to worry, "Where am I going to get *that?*" or, "How do I prepare *this?*" Our *Instructor's Manual* includes:

- procedures for preparing reagents, materials, and equipment
- scheduling information for materials needing advance preparation
- approximate quantities of material needed
- answers for pre-lab questions
- answers for post-lab questions
- answers to Mendelian genetics problems (Exercise 11)
- tear-out sheets of pre-lab questions, in scrambled order from those in the lab manual, for those who wish to duplicate them for quizzes
- vendors and item numbers for supplies
- text and preparation guide for optional experiments/ activities

James W. Perry
(301) 689-4173

David Morton
(301) 689-4355

Department of Biology
Frostburg State University
Frostburg, Maryland 21532-1099

And in the End . . .

We would like to express our special thanks to Gary Mcdonald and Jack Carey and other individuals at Wadsworth involved in this project: Andrew Ogus, Donna Kalal, Marta Kongsle, Roberta Broyer, and Karen Hunt.

There are very few things in life that are perfect. We don't suppose that this lab manual is one of them. We think your students will enjoy the exercises. Perhaps you and they will find places where rephrasing will make the activity better. Please, call or send us a note expressing your opinions and any ideas you wish to share; encourage your students to do likewise.

Reviewers

Our thanks to the following colleagues for their assistance during the extensive review and class-testing processes.

To the Student

Welcome! You are about to embark on a journey through the cosmos of life. You will learn things about yourself and your surroundings that will broaden and enrich your life. You will have the opportunity to marvel at the microscopic world, to be fascinated by the cellular events occurring in your body at this very moment, and to gain an appreciation for the environment, including the marvelous diversity of the plant and animal world.

We can offer a number of suggestions to make your collegiate experience in biology a pleasant one. The first step toward that goal has been taken by us; we have written a laboratory guide that is "user friendly." You will be able to hear the authors speaking with you as though we were there to share your experience. Both of us share a personal belief that the more we make you feel comfortable with us, the more likely you are to share our enthusiasm for biology. It would be naive for us to suppose that each and every one of you will be biology majors at graduation. But one thing we all must realize is that we are citizens of "spaceship Earth." The fate of our spaceship is largely in your hands because you are the decision makers of the future. As has been so aptly stated, "We inherited the earth from our parents and grandparents, but we are only the caretakers for our children and grandchildren."

As caretakers we need to be informed about the world about us. That's why we enroll in colleges and universities with the hope of gaining a liberal education. In doing so, we establish a basis on which to make educated decisions about the future of the planet. Each exercise in this manual contains a lesson in life that is of a more global nature than the surroundings of your biology laboratory.

In order to enhance your biology education, take the initiative to put yourself at the best possible advantage: Don't miss class; read your text assignment routinely; read the laboratory exercise before you come to the lab.

Each exercise in the manual is organized in the same way:

1. **Objectives** tell exactly what you should learn from the exercise. If you wish to know what will be on the exam, consult the objectives for each exercise.

2. The **introduction** provides you with background information for the exercise and is intended to stimulate your interest.

3. The **materials** list for each portion of the exercise allows you to determine at a glance whether you have all the necessary supplies needed to do the activity.

4. The **procedure** for each section, in easy-to-follow step-by-step fashion, describes the activity. Within the procedure, spaces are provided to make required drawings. Questions, with space for answers, are posed asking you to draw conclusions about an activity you are engaged in. You'll find a lot of illustrations, most of which are labeled and others which are not but have leaders for you to attach labels. The terms to be used as labels are found in the procedure and in a list accompanying the illustration. Sometimes we believe it best for you to make a simple drawing, and have inserted boxes or circles for your sketches. Where appropriate, tables and graphs are present for recording your data.

5. **Pre-lab questions** can be answered easily by simply reading the exercise. They're meant to "set the stage" for the lab period by emphasizing some of the more salient points.

6. **Post-lab questions** are intended to be done after the laboratory is completed. Some are straightforward interpretations of what you have done, while others require additional thought and perhaps some research in your textbook. In fact, some have no "right" or "wrong" answer at all!

In our experience, students are much too reluctant to ask questions for fear of appearing stupid. Remember, there is no such thing as a stupid question. *Speak up!* Think of yourselves as "basic learners" and your instructors as "advanced learners." Interact and ask questions so that you and your instructors can further your/their respective educations.

Laboratory Supplies and Procedures

Materials and Supplies Kept in the Lab at All Times

The materials listed below will be always available in the lab room. Familiarize yourself with their location prior to beginning the exercises.

- compound light microscopes
- dissection microscopes
- glass microscope slides
- coverslips
- lens paper
- tissue wipes
- plastic 15-cm rulers
- dissecting needles
- razor blades
- assorted glassware-cleaning brushes
- detergent for washing glassware
- distilled water
- hand soap
- paper toweling
- safety equipment (see separate list)

Laboratory Safety

None of the exercises in this manual are inherently dangerous. Some of the chemicals are corrosive (causing burns to the skin), others are poisonous if ingested or inhaled in large amounts. Contact with your eyes by otherwise innocuous substances may result in permanent eye injury; remember, once your sight is lost, it's probably lost forever. **Locate the safety items described below and then study the list of basic safety rules.**

1. *Eyewash bottle or eye bath*
   Should any substance be splashed in your eyes, wash them thoroughly.

2. *Fire extinguisher*
   Read the directions for use of the fire extinguisher.

3. *Fire blanket*
   Should someone's clothing catch fire, wrap the blanket around the individual and roll the person on the floor to smother the flames.

4. *First-aid kit*
   Minor injuries such as small cuts can be treated effectively in the lab. Open the first-aid kit to determine its contents.

5. *Safety goggles*
   Eye protection should be worn during the more experimental exercises.

Safety Rules

1. Do not eat, drink, or smoke in the laboratory.

2. Wash your hands with soap and warm water before leaving the laboratory.

3. When heating a test tube, point the mouth of the tube away from yourself and other people.

4. Always wear shoes in the laboratory.

5. Keep extra books and clothing in designated places so that your work area is as uncluttered as possible.

6. If you have long hair, tie it back when in the laboratory.

7. Read labels carefully before removing substances from a container. **Never return a substance to a container.**

8. Discard used chemicals and materials into appropriately labeled containers. Certain chemicals should not be washed down the sink; these will be indicated by your instructor.

> **CAUTION**
> **Report all accidents and spills to your instructor immediately!**

Instructions for Washing Laboratory Glassware

1. Place contents to be discarded in proper waste container as described in exercise.

2. Rinse glassware with tap water.

3. Add a small amount of glassware cleaning detergent.

4. Scrub with an appropriately sized brush.

5. Rinse with tap water until detergent disappears.

6. Rinse three times with distilled water (dH_2O).

7. Allow to dry in inverted position on drying rack (if available).

When glassware is clean, dH_2O sheets off rather than remaining on the surface in droplets.

EXERCISE 1
Scientific Method

OBJECTIVES

After completing this exercise you will be able to:

1. define *scientific method, mechanist, vitalist, cause and effect, teleology, induction, deduction, experimental group, control group, independent variable, controlled variables, dependent variable, correlation, theory, principle, bioassay;*

2. explain the nature of scientific knowledge;

3. describe the basic steps of the scientific method;

4. state the purpose of an experiment;

5. explain the difference between *cause and effect* and *correlation;*

6. describe the design of a typical research article in biology.

INTRODUCTION

Biology is an empirical science in that it relies on experience gained by observing and testing nature. To expand and augment our biological senses—sight, hearing, smell, taste, and so on, humans have developed many sophisticated tools. Some of these tools are used to quantify (measure) observations that could otherwise be described only by words. Other tools expand our existing senses, such as when the light microscope allows us to view objects that are invisible to the unaided eye. Still other tools translate phenomena that cannot be sensed into forms that can, such as when the viewing screen of an electron microscope turns electrons into visible light.

To appreciate biology or, for that matter, the nature of any body of scientific knowledge, one must first understand how that knowledge is gathered. For this reason, before studying measurement in Exercise 2 and microscopes in Exercise 3, we will first examine the scientific method.

The **scientific method** is a process that empirically tests possible answers to questions about nature in ways that can be duplicated or verified. Questions are generated from careful observations of nature. Answers supported by the results of tests are added to the body of scientific knowledge and contribute to the concepts presented in your textbook and other science books. Although these concepts are as up-to-date as possible, they are considered open to further questions and modifications.

MATERIALS

Per student:
• a typical research article in biology

Per student group:
• a set of five named, related organisms

PROCEDURE

A. Natural Philosophy

The roots of the scientific method can be found in ancient Greek philosophy. The natural philosophy of Aristotle and his colleagues was mechanistic rather than vitalistic. A **mechanist** believes that living things, along with the rest of the universe, are governed only by natural forces, while a **vitalist** believes that the universe is at least partially governed by supernatural powers. Mechanists look for interrelationships between the structures and functions of living things, and the processes that shape them. Their explanations of nature deal in **cause and effect**—the idea that one thing is the result of another thing. In contrast, vitalists often use teleological explanations of natural events. **Teleology** not only states that nature possesses a supernatural design but also that it has purpose.

Although most modern scientists are mechanists, it is often difficult to keep teleology out of science. For example, the statement, "The digestive tract processes food to provide the fuel needed to sustain the body," is a teleological statement because it suggests that the digestive tract processes food for the purpose of producing fuel for the body.

Some people claim that the AIDS (acquired immune deficiency syndrome) epidemic is a result of the wrath of a supernatural being directed at sexual promiscuity. Is this statement mechanistic or vitalistic?

_____

Others claim that AIDS may develop after infection by a specific retrovirus. Is this statement mechanistic or vitalistic?

_____

Here's a more difficult example. A fertilized egg develops so that it will become an adult. Is this statement mechanistic or vitalistic?

_____

Explain why you chose this answer. _____

_____

_____

Table 1-1 Natural Philosophy Observations	
Name of Organism	**Observations**
_____	_____
	_____
	_____
	_____
_____	_____
	_____
	_____
_____	_____
	_____
	_____
_____	_____
	_____
	_____
_____	_____

Aristotle and his colleagues developed three rules to examine the laws of nature. They are:

RULE 1 The careful collection of observations about some aspect of nature.

RULE 2 The careful examination of these observations as to their similarities and differences.

RULE 3 The production of a principle or generalization about the aspect of nature being studied.

Now let's see how good you are at natural philosophy.

1. Carefully observe your group's set of five related organisms. In table 1-1, list your observations of each organism, keeping in mind that you will be asked next to pose questions about the similarities and differences among them.

2. In table 1-2, carefully list the characteristics of these organisms that you observe to be similar and those that you observe to be different.

3. Now write a statement about a characteristic these organisms share in common or one that varies among them. _____

_____

_____

Of course, this example is biased because your instructor picked the five organisms you observed and because the sample size was very small. The ancient Greek philosophers would have examined many more organisms before they would have either decided a group of organisms was related or described their identifying characteristics. This is why you are asked to produce a statement rather than a principle.

The major defect of natural philosophy was that it accepted the idea of *absolute truth*. This suppressed the testing of principles once they had been produced. Aristotle's belief in spontaneous generation, the principle that some life can arise from nonliving things (for example, maggots from spoiled meat), survived over two thousand years of controversy before finally being discredited by Louis Pasteur in 1860.

Rejection of the idea of absolute truth and the testing of principles either by experimentation or by further pertinent observation are the essence of the modern scientific method.

B. Modern Scientific Method

1. Observation. As with natural philosophy, the scientific method starts with *careful observation*. An investigator may make observations from nature or from

Table 1-2 Analysis of Natural Philosophy Observations	
Similarities	**Differences**
_____	_____
_____	_____
_____	_____
_____	_____
_____	_____

the written words of other investigators, which are published in books or research articles in scientific journals and are available in the storehouse of human knowledge, the library.

2. Question. The second step of the scientific method is to *ask a question* about these observations. The quality of this question will depend on how carefully the observations were made and analyzed.

> **N O T E**
>
> **The purpose of the italicized statements such as the one that follows is to give an example of how each part of the scientific method is applied in practice.**

A question that comes from the research interests of one of the authors is, "Does excess iron in the diet result in a decreased growth rate?"

Your observations of the five organisms must have provoked a number of questions. For example, "What is the function of a characteristic shared by all five?" or "What is the function of a characteristic unique to one of the organisms?" Write a specific question about the statement you made in section A.3.

_____

_____

_____

3. Research Hypothesis. The third step of the scientific method is to *make a research hypothesis*. This is a general statement of the answer to the question and is derived by inductive reasoning. **Induction** is a logical process by which all pertinent observations are combined and considered before producing a general statement.

Here is a research hypothesis for the example question above: "Excess iron in the diet of mammals will decrease growth rate."

Write a research hypothesis answering your question.

_____

_____

4. Prediction. The fourth step of the scientific method is to *formulate a prediction*. If the research hypothesis is accurate, then a prediction based on the research hypothesis should also be accurate. This is deductive or "if-then" reasoning. **Deduction** is a logical process by which a prediction is produced from a general statement.

A prediction deduced from the example research hypothesis is, "If mammals are fed excess iron in their diets, then they will have a decreased growth rate."

Now, make a prediction based on your research hypothesis.

_____

_____

5. Experiment or Pertinent Observations. The fifth step of the scientific method is to *design an experiment* or to *make pertinent observations* to test the prediction made from the research hypothesis. One begins with *statements of a null hypothesis (H_0) and an alternative hypothesis (H_a)*. The H_0 and H_a each describe opposite results for the test. After analysis of the actual test results, H_0 will be accepted or rejected. The H_a is accepted only if H_0 is rejected. An H_0 is usually phrased negatively.

The following are hypotheses based on the example prediction.

H_0: There is no observable difference in growth between weanling rats (rats that have just finished nursing) fed food with high and average amounts of iron.

H_a: There is an observable difference in growth between weanling rats fed food with high and average amounts of iron.

Write null and alternative hypotheses based on your prediction.

H_0: _____

_____

H_a: _____

_____

In any test there are three kinds of variables. The **independent variable** is the condition or event under study. The **dependent variable** is the condition or event that may change due to the independent variable. The **controlled variables** are all the other conditions and events, which the investigators attempt to keep the same.

In an experiment of classical design, the individuals under study are divided into two groups: an **experimental group** that is subject to the independent variable and a **control group** that is not.

In an experiment to test the example H_0, weanling rats would be separated at random into two groups and raised on a balanced diet except for the iron content. The experimental group would receive a high iron diet, and the control group would receive an average iron diet. The independent variable is the increase in the iron content of the diet; controlled variables are the same room temperature, water availability, and housing conditions; and the dependent variable is the amount of growth.

Sometimes the best test of a hypothesis is not an actual experiment but pertinent observations. One of the most important principles in biology, Darwin's theory of natural selection, was developed by this nonexperimental approach.

For example, testing the null hypothesis that not all of the three main groups of mammals bear live young would require observation of reproduction in as many mammals as possible from each of the three groups.

Variables also can be described for the nonexperimental approach.

For this example, the controlled variable is the requirement that all of the observed organisms be mammals; the independent variable is the classification of each of the observed mammals into one of the three main groups of mammals; and the dependent variable is the method of reproduction — for example, live bearing or egg laying — for each mammal observed. In fact, members of one of the three main groups of mammals, the monotremes, do lay eggs. This group has only two members, the duckbilled platypus and the spiny anteater.

If your hypotheses are best suited to an experimental approach, complete section **a** below. However, if your hypotheses are best suited to a nonexperimental approach, skip **a** and complete section **b**.

a. Design an experiment to test your hypotheses.

independent variable — _____

_____

dependent variable — _____

_____

controlled variables — _____

_____

experimental group — _____

_____

control group — _____

_____

b. Describe what pertinent observations you would need to make to test your hypotheses.

_____

_____

_____

Describe the variables for your nonexperimental approach.

controlled variable — _____

_____

independent variable — _____

_____

dependent variable — _____

_____

6. Conclusion. To *make a conclusion* — the last step in one cycle of the scientific method — you use the results of the experiment or pertinent observations to test H_0. If your prediction does not occur, then you must accept H_0. If your prediction does occur, it indicates that you may reject H_0 and accept H_a. However, you can never completely accept or reject a hypothesis; all you can do is state the probability that one is correct or incorrect. To quantify this probability, scientists use a branch of mathematics called *statistical analysis*.

In the example experiment on the effect of excess iron in the diet on growth, H_0 will be accepted if the amount of growth in the two groups is the same and rejected if there are significant differences.

Even if the growth of the experimental rats is significantly less than that of the control group, this does not necessarily mean that it was caused by the high iron content of their food. It could be a coincidence or the effect of some unforeseen and thus uncontrolled variable. For this reason, the results of experiments and observations must be *repeatable* by the same and other investigators.

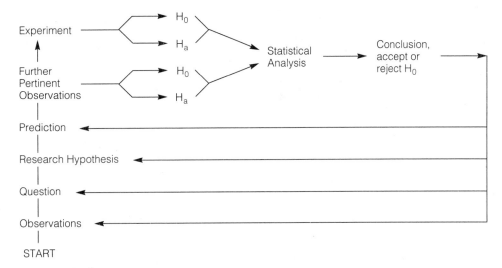

H_0

H_a

H_0

H_a

Experiment

Further
Pertinent
Observations

Statistical
Analysis

Conclusion,
accept or
reject H_0

Prediction

Research Hypothesis

Question

Observations

START

Figure 1-1 Scheme of the scientific method. (After Glase, 1975.)

Even if the results are repeatable, it still does not necessarily mean that the high iron in the diet directly caused the lower growth rate. *Cause and effect,* especially in biology, is rarely proven in experiments. We cannot say that the high iron content of the food caused the decreased growth rate. We can, however, say that the two are correlated. A **correlation** is a relationship between the independent and the dependent variables.

One cycle of the modern scientific method is summarized in figure 1-1. Quite often the results of an experiment or a pertinent observation require a modification of the research hypothesis and/or the prediction and, therefore, a new experiment. Even if the conclusion is clear-cut, it usually creates several new questions. It is typical for one cycle of the modern scientific method to lead to several others. For biologists, the modern scientific method is a "round trip" repeated many times in their never-ending search for knowledge about life.

C. The Research Article

The account of one or several related cycles of the scientific method is usually initially reported in depth in a research article published in a scientific journal. The goal of the scientific community is to be cooperative as well as competitive. Research articles both share knowledge and provide enough information so that the results of experiments or pertinent observations described by those articles may be repeated.

Check the design of a typical research article in biology and list the titles and functions of its various sections.

1. for example, Abstract — summary of paper

2. _____

3. _____

4. _____

5. _____

6. _____

In which section do you find the question being asked or the hypothesis?

_____

Where would you look to find the details necessary to repeat this experiment or observation?

_____

Not all scientific journals have the same format. For example, *Science* presents its research articles in narrative form, and many of the details of the scientific method are tacitly understood (for example, there is usually no statement of the null or alternate hypotheses). However, adherence to the modern scientific method is expected, and the scientific community understands that it is as important to expose mistakes as it is to praise new knowledge.

D. Theories and Principles

When exhaustive experiments and observations consistently support an important hypothesis, it is accepted as a **theory.** A theory that stands the test of time may be elevated to the status of a **principle.** Theories and principles are always considered when new hypotheses are formulated. However, like hypotheses, theories and principles may be modified or even discarded in the light of new knowledge. Biology, like life itself, is not static but is constantly changing in time.

O P T I O N A L

Experiment: A Bioassay

Your instructor may provide you with a simple experiment.

_____ 1. The natural philosophy of Aristotle and his colleagues was (a) mechanistic, (b) vitalistic, (c) teleological, (d) b and c.

_____ 2. A person who believes that the universe is at least partially controlled by supernatural powers can best be described as a(n) (a) teleologist, (b) vitalist, (c) empiricist, (d) mechanist.

_____ 3. The major defect in the natural philosophy of the ancient Greeks was a belief in (a) vitalism, (b) teleology, (c) cause and effect, (d) absolute truth.

_____ 4. The first step of the scientific method is to (a) ask a question, (b) make a research hypothesis, (c) observe carefully, (d) formulate a prediction.

_____ 5. Which series of letters puts the first four steps of the scientific method listed in question 4 in the correct order? (a) a,b,c,d, (b) a,b,d,c, (c) c,a,b,d, (d) d,c,a,b.

_____ 6. The alternative hypothesis (a) is the same as the null hypothesis, (b) is the opposite of the null hypothesis, (c) is usually phrased negatively, (d) a and c.

_____ 7. The variable(s) that investigators try to keep the same for both the experimental and the control groups is (are) (a) independent, (b) controlled, (c) dependent, (d) a and c.

_____ 8. The variable(s) that is (are) _always_ different between the experimental and the control groups is (are) (a) independent, (b) controlled, (c) dependent, (d) a and c.

_____ 9. The results of an experiment (a) don't have to be repeatable, (b) should be repeatable by the investigator, (c) should be repeatable by other investigators, (d) b and c.

_____ 10. The detailed report of an experiment is usually published in a (a) newspaper, (b) book, (c) scientific journal, (d) magazine.

EXERCISE 1

Scientific Method

POST-LAB QUESTIONS

1. How does the modern scientific method differ from the natural philosophy of the ancient Greeks?

2. Describe the six steps of one full cycle of the scientific method.

   a.

   b.

   c.

   d.

   e.

   f.

3. What is tested by an experiment?

4. Within the framework of an experiment, what is(are) the:
   a. independent variable

   b. dependent variable

   c. controlled variables

5. Is the statement, "In most biology experiments, the relationship between the independent and the dependent variable can best be described as cause and effect," true or false? Explain your answer.

6. What is the function of research articles in scientific journals?

7. Compare the nature of religious and scientific knowledge in terms of whether they are mechanistic or vitalistic.

8. Compare the nature of religious and scientific knowledge in terms of the idea of absolute truth.

9. Do you think the differences between religious and scientific knowledge make it difficult to debate points of conflict between them? Explain your answer.

Measurement

OBJECTIVES

After completing this exercise you will be able to:

1. define *qualitative observation, quantitative observation, meniscus, density, thermometer;*

2. recognize graduated cylinders, beakers, Erlenmeyer flasks, different types of pipets, and a triple beam balance;

3. explain the concepts of length, volume, and mass in metric units;

4. measure and estimate length, volume, and mass in metric units;

5. explain the concept of temperature;

6. measure and estimate temperature in degrees Celsius.

INTRODUCTION

We examine our world in two ways, qualitatively and quantitatively. A **qualitative observation** of something describes a characteristic important in understanding what that thing is; it is not a numerical observation. For example, "He is tall" is a qualitative observation of one of the authors. A **quantitative observation** of the same characteristic would involve a measurement or count. For example, "His height is 6 feet, 4 inches or 1 meter, 93 centimeters" is a quantitative observation.

One of the requirements of the scientific method is that results be repeatable. As numerical results are more precise than verbal descriptions, scientific observations are usually made as quantitative as possible.

Logically, units in the ideal system of measurement should be easy to convert from one to another (for example, inches to feet or centimeters to meters) and from one related measurement to another (for example, length to area, and area to volume). The metric system meets these requirements and is used by the majority of citizens and countries in the world. Universally, it is preferred by science educators and researchers. In most nonmetric countries, governments have launched programs to hasten the conversion to metrics. Any country that fails to do so could be at a serious economic and scientific disadvantage. In fact, the U.S. Department of Defense adopted the metric system in 1957, and all cars made in the USA have metric components.

Table 2-1 Prefixes for Metric System Units	
Prefix of Unit (Symbol)	**Part of Reference Unit**
nano (n)	$1/1,000,000,000 = 0.000000001 = 10^{-9}$
micro (μ)	$1/1,000,000 = 0.000001 = 10^{-6}$
milli (m)	$1/1,000 = 0.001 = 10^{-3}$
centi (c)	$1/100 = 0.01 = 10^{-2}$
kilo (k)	$1000 = 10^{3}$

The metric reference units are the **meter** for length, the **liter** for volume, the **gram** for mass, and the **degree Celsius** for temperature. Regardless of the type of measurement, the same prefixes are used to designate the relationship of a unit to the reference unit. Table 2-1 lists the prefixes we will introduce in this and subsequent exercises.

As you can see, the metric system is a decimal system of measurement. Metric units are 10, 100, 1,000 and sometimes 1,000,000 or more times larger or smaller than the reference unit. Thus, it is easy to convert from one measurement to another either by multiplying or dividing by 10 or a multiple of 10:

$$\begin{array}{ccccc} \times 1000 & \times 1000 & \times 10 & \times 100 \\ \text{nanounit} \underset{\div 1000}{\overset{}{\rightleftarrows}} \text{microunit} \underset{\div 1000}{\overset{}{\rightleftarrows}} \text{milliunit} \underset{\div 10}{\overset{}{\rightleftarrows}} \text{centiunit} \underset{\div 100}{\overset{}{\rightleftarrows}} \text{unit} \end{array}$$

In this exercise we examine the metric system and compare it to the American Standard system of measurement (feet, quarts, pounds, and so on).

MATERIALS

Per student pair:

- 30-centimeter ruler with metric and American (English) Standard units on opposite edges
- 250-mL beaker made of heat-proof glass
- 250-mL Erlenmeyer flask
- 3 graduated cylinders: 10-mL, 25-mL, 100-mL
- 1-quart jar or bottle marked with a fill line
- one-piece plastic dropping pipet (not graduated) or Pasteur pipet and bulb
- graduated pipet and safety bulb or filling device
- 1-pound brick of coffee
- ceramic coffee mug
- 1-gallon milk bottle

- metric tape measure
- 1-liter measuring cup
- thermometer with both Celsius (°C) and Fahrenheit (°F) scales (−20° to 110°C)
- hot plate
- 3 boiling chips
- thermometer holder

Per student group:
- a triple beam balance

Per lab room:
- source of distilled water
- metric bathroom scale
- source of ice

PROCEDURE

A. Length

Length is the measurement of a line, end to end. The standard unit is the *meter,* and the most commonly used related units of length are:

$$1,000 \text{ millimeters (mm)} = 1 \text{ meter (m)}$$
$$100 \text{ centimeters (cm)} = 1 \text{ m}$$
$$1,000 \text{ m} = 1 \text{ kilometer (km)}$$

For orientation purposes, the yolk of a chicken egg is about 3 cm in diameter. Since the differences between these metric units are based on 10 or multiples of 10, it is fairly easy to convert a measurement in one unit to another.

1. For example, if you wanted to convert 1.7 km to centimeters, the first step would be to determine how many centimeters there are in 1 km. Remember, like units may be cancelled.

$$\frac{100 \text{ cm}}{1 \text{ m}} \times \frac{1,000 \text{ m}}{1 \text{ km}} = \frac{100,000 \text{ cm}}{1 \text{ km}}$$

The second step is to multiply the number by this fraction.

$$\frac{1.7 \text{ km}}{1} \times \frac{100,000 \text{ cm}}{1 \text{ km}} = 170,000 \text{ cm}$$

The last calculation can also be done quickly by shifting the decimal point five places to the right.

$$1.\underline{700000} \text{ km} = 170,000.0 \text{ cm}$$

Alternatively, conversion of 1.7 km to centimeters can be done by adding exponents.

$$\frac{1.7 \text{ km}}{1} \times 10^2 \frac{\text{cm}}{\text{m}} \times 10^3 \frac{\text{m}}{\text{km}}$$
$$= 1.7 \times 10^5 \text{ cm} = 170,000 \text{ cm}$$

Using the most comfortable method, calculate how many millimeters there are in 4.8 m.

_____ mm

Now let's convert 17 mm to meters.

step 1 $\dfrac{1 \text{ m}}{100 \text{ cm}} \times \dfrac{1 \text{ cm}}{10 \text{ mm}} = \dfrac{1 \text{ m}}{1,000 \text{ mm}}$

step 2 $\dfrac{17 \text{ mm}}{1} \times \dfrac{1 \text{ m}}{1,000 \text{ mm}} = 0.017 \text{ m}$

or shift the decimal point three places to the left

$\underline{0017}.0 \text{ mm} = 0.017 \text{ m}$

or add exponents

$$\frac{17 \text{ mm}}{1} \times 10^{-2} \frac{\text{m}}{\text{cm}} \times 10^{-1} \frac{\text{cm}}{\text{mm}}$$
$$= 17 \times 10^{-3} \text{ m} = 0.017 \text{ m}$$

Calculate how many kilometers there are in 16 cm.

_____ km

2. Precisely measure the length of this page in centimeters to the nearest tenth of a centimeter with the metric edge of a ruler. Note that the space between each centimeter is divided by nine lines into 10 millimeters.

The page is _____ cm long.

Calculate the length of this page in millimeters, meters, and kilometers.

_____ mm   _____ m   _____ km

Now repeat the above measurement using the American Standard edge of the ruler. Measure the length of this page in inches to the nearest eighth of an inch.

_____ in

Convert your answer to feet and yards.

_____ ft   _____ yd

It is much easier to convert *units of length* in the metric system than in the American Standard system.

B. Volume

Volume is the space a given object occupies. The standard unit of volume is the *liter* (L), and the most commonly used subunit, the *milliliter* (mL). There are 1,000 mL in 1 liter. A chicken egg has a volume of about 60 mL.

The volume of a box is the height multiplied by the width multiplied by the depth. The amount of water contained in a cube with sides 1 cm long is 1 cubic centimeter (cc), which for all practical purposes equals 1 mL (fig. 2-1).

1. How many milliliters are there in 1.7 liters?

_____ mL

How many liters are there in 1.7 mL?

_____ L

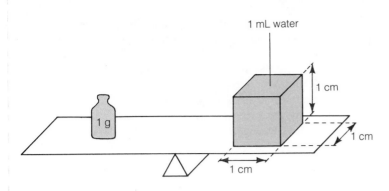

Figure 2-1 Illustration of the relationship between the units of length, volume, and mass in the metric system.

2. Use the illustrations in figure 2-2 to recognize **graduated cylinders, beakers, Erlenmeyer flasks,** and the different types of **pipets.** Some of these objects may be made of glass; some may be plastic. Some will be calibrated in milliliters and liters; some may not be.

3. Pour some water into a 100-mL graduated cylinder and observe the boundary between fluid and air, the **meniscus.** Due to surface tension the meniscus is curved, not flat. The high surface tension of water is due to its cohesive and adhesive or "sticky" properties. Draw the meniscus in the plain cylinder outlined in figure 2-3. The correct reading of the volume is at the *lowest* point of the meniscus.

4. Using the 100-mL graduated cylinder, pour water into a 1-quart jar or bottle. About how many milliliters of water are needed to fill the vessel up to the line?

_____ mL

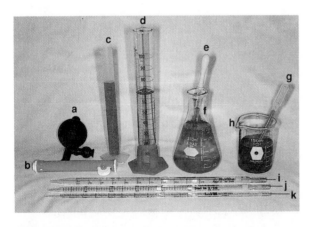

Figure 2-2 Apparatus commonly used to measure volume: (**a**) pipet safety bulb, (**b**) pipet filling device, (**c**) plastic graduated cylinder, (**d**) glass graduated cylinder, (**e**) Pasteur pipet and bulb, (**f**) Erlenmeyer flask, (**g**) plastic dropping pipet, (**h**) beaker, (**i** to **k**) graduated pipets. (Photo by D. Morton and J. W. Perry.)

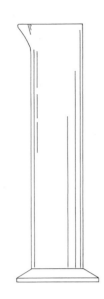

Figure 2-3 Draw a meniscus in this plain cylinder.

5. Pipets are used to transfer small volumes from one vessel to another. Some pipets are not graduated (for example, Pasteur pipets and most one-piece plastic dropping pipets). Other pipets are graduated.

Fill a 250-mL Erlenmeyer flask with distilled water. Use a plastic dropping pipet or Pasteur pipet with a bulb to withdraw some water. To find out how many drops there are in 1 mL, count the number of drops needed to fill a 10-mL graduated cylinder to the 1-mL mark.

There are _____ drops/mL.

C. Mass

Mass is the quantity of matter in a given object. The standard unit is the **kilogram** (kg), and other commonly used units are the *milligram* (mg) and *gram* (g). There are 1,000,000 mg in 1 kg and 1,000 g in 1 kg. A chicken egg has a mass of about 60 g.

1. How many milligrams are there in 1 g?

_____ mg

Convert 1.7 g to milligrams and kilograms.

_____ mg _____ kg

2. Our 1-cc cube in section B, if filled with 1 mL of water, would have a mass of 1 g (fig. 2-1). The mass of other materials depends on their **density** (water is defined as having a density of 1).

$$density = \frac{mass}{volume}$$

Approximately how many liters are present in 1 cubic meter of water? As each of the sides of 1 cubic meter are 100 centimeters in length, it is easy to calculate the number of cubic centimeters (that is,

100 cm × 100 cm × 100 cm = 1,000,000 cc). Now just change cc to mL and convert 1,000,000 mL to liters.

_____ L

In kg, what is its mass?

_____ kg

In your "mind's eye" contemplate calculating how many pounds there are in a cubic yard of water. It is easier to convert *between different units* of the metric system than those of the American Standard system.

3. Determine the mass of an unknown volume of water.* Mass may be measured with a **triple beam balance,** which gets its name from its three beams (fig. 2-4). A movable mass hangs from each beam.

Slide all of the movable masses to zero. The middle and back masses each click into the leftmost notch, and the front mass is moved to the far left. Clear the pan of all objects and make sure it is clean. The balance marks should line up, indicating that the beam is level and that the pan is empty. If the balance marks do not line up, rotate the zero adjust knob until they do.

Place a 250-mL beaker on the pan. The right side of the beam should rise. Slide the mass on the middle beam until it clicks into the notch at the 100-g mark. If the right end of the beam tilts down below the stationary balance mark, you have added too much mass. Move the mass back a notch. If the right end remains tilted up, additional mass is needed. Keep adding 100-g increments until the beam tilts down; then move the mass back one notch. Repeat this procedure on the

_____

*Modified from C. M. Wynn and G. A. Joppich, *Laboratory Experiments for Chemistry: A Basic Introduction,* 3rd ed. Wadsworth, 1984.

back beam, adding 10 g at a time until the beam tilts down, and then backing up one notch. Next, slide the front movable mass until the balance marks line up. The sum of the masses indicated on the three beams gives the mass of the beaker. The space between the numbered gram markings on the front beam is divided by nine unnumbered lines into ten sections, each representing 0.1 g. The mass of the beaker to the nearest tenth of a gram is

_____ g.

Add an unknown amount of water and repeat the above procedure. The mass of the water will equal the combined masses of the beaker and water minus that of the beaker alone.

mass of the beaker plus the water _____ g

minus the mass of the beaker   _____ g

equals the mass of the water   _____ g

Now measure the volume of the water in milliliters with a graduated cylinder. What was the volume?

_____ mL

You may have wondered why we have avoided the term *weight* in the above discussion. That is because mass is a quantity of matter, while weight depends on the gravitational field in which the matter is located. Thus, if you were on the moon you would weigh less, but your mass would be the same as on earth. Although it is technically incorrect, mass and weight are often used interchangeably.

4. Using the triple beam balance, determine the mass of, or weigh, a pound of coffee in grams.

_____ g

D. Estimation of Measurements

Now that you have had some experience using metric units, let's try estimating the measurements of some everyday items. Also, you may consult the metric/American Standard conversion table in Appendix 1 at the end of the lab manual.

1. Estimate the length of your index finger in centimeters.

_____ cm

2. Estimate your lab partner's height in meters.

_____ m

3. How many milliliters will it take to fill a ceramic coffee mug?

_____ mL

4. How many liters will it take to fill a 1-gallon milk bottle?

_____ L

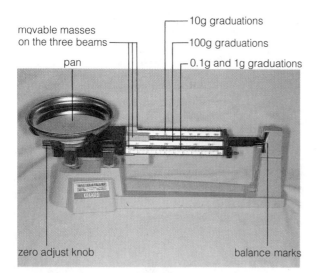

movable masses on the three beams
pan
10g graduations
100g graduations
0.1g and 1g graduations
zero adjust knob
balance marks

Figure 2-4 Triple beam balance. (Photo by D. Morton and J. W. Perry.)

Table 2-2 Differences Between Estimates and Measurements

Number	Estimate	Measurement	Estimate − Measurement
1	_____ cm	_____ cm	_____ cm
2	_____ m	_____ m	_____ m
3	_____ mL	_____ mL	_____ mL
4	_____ L	_____ L	_____ L
5	_____ g	_____ g	_____ g
6	_____ kg	_____ kg	_____ kg

Table 2-3 Measure of Everyday Packed Items

Item	American Standard	Metric
_____	_____	_____
_____	_____	_____
_____	_____	_____
_____	_____	_____
_____	_____	_____
_____	_____	_____
_____	_____	_____
_____	_____	_____
_____	_____	_____
_____	_____	_____

5. Estimate the weight of some small personal item (for example, loose change) in grams.

_____ g

6. Estimate your or your lab partner's weight in kilograms.

_____ kg

7. Transfer your estimates to table 2-2. Then check each of your results using either a ruler, metric tape measure, 100-mL graduated cylinder, 1-liter measuring cup, triple beam balance, or metric bathroom scale, recording your measurements in table 2-2. Complete table 2-2 by calculating the difference between each estimate and measurement.

8. Today, many packaged items have the volume or weight listed in both American Standard and metric units. Before your next lab period find and list ten such items in table 2-3.

E. Temperature

The degree of hot or cold of an object is temperature. More specifically, it is the average kinetic energy of molecules. Heat always flows from high to low temperatures. This is why hot objects left at room temperature always cool to the surrounding or ambient temperature, while cold objects warm up. Consequently, to keep a heater hot and the inside of a refrigerator cold requires energy. **Thermometers** are instruments used to measure temperature.

> **CAUTION**
> Mercury is toxic. If a thermometer breaks, immediately inform your instructor.

1. Using a thermometer with both *Celsius* (°C) and *Fahrenheit* (°F) scales, measure room temperature and the temperature of cold and hot running tap water.

	°C	°F
room temperature	_____	_____
cold running tap water	_____	_____
hot running tap water	_____	_____

2. Fill a 250-mL beaker with ice about three-fourths full and add cold tap water to just below the ice. Wait for three minutes and measure the temperature.

	°C	°F
ice water	_____	_____

Remove the thermometer and discard the ice water into the sink.

3. Fill the beaker with warm tap water to about three-fourths full and add three boiling chips. Use a thermometer holder to clip the thermometer onto the rim of the beaker so that the bulb of the thermometer is halfway into the water. Boil the water in the beaker by placing it on a hot plate.

CAUTION

Your instructor will give you specific instructions on how to set up the equipment in your lab for boiling water.

After the water boils, record its temperature.

	°C	°F
boiling water	_____	_____

Turn off the hot plate and let the water and beaker cool to below 50°C before pouring the water into the sink.

4. To convert Celsius degrees to Fahrenheit degrees, multiply by 9/5 and add 32. Is 4°C the temperature of a hot or cool day?

_____

What temperature is this in degrees Fahrenheit?

_____ °F

5. To convert Fahrenheit degrees to Celsius degrees, subtract 32 and multiply by 5/9. What is body temperature, 98.6°F, in degrees Celsius?

_____ °C

6. In summary, the formulas for these temperature conversions are:

$$°F = (°C) \left(\frac{9}{5}\right) + 32 \quad °C = (°F - 32)\left(\frac{5}{9}\right)$$

PRE-LAB QUESTIONS

____ 1. The observation that a person is tall is (a) qualitative, (b) quantitative, (c) numerical, (d) b and c.

____ 2. Length is the measurement of (a) a line, end to end, (b) the space a given object occupies, (c) the quantity of matter present in an object, (d) the degree of hot or cold of an object.

____ 3. Volume is the measurement of (a) a line, end to end, (b) the space a given object occupies, (c) the quantity of matter present in an object, (d) the degree of hot or cold of an object.

____ 4. Mass is the measurement of (a) a line, end to end, (b) the space a given object occupies, (c) the quantity of matter present in an object, (d) the degree of hot or cold of an object.

____ 5. A millicurie, a unit of radioactivity, is (a) a tenth of a curie, (b) a hundredth of a curie, (c) a thousandth of a curie, (d) a millionth of a curie.

____ 6. A kilowatt, a unit of electrical power, is (a) ten watts, (b) a hundred watts, (c) a thousand watts, (d) a million watts.

____ 7. If your mass were 70 kilograms (kg) on the earth, how much would your mass be on the moon? (a) less than 70 kg, (b) more than 70 kg, (c) 70 kg, (d) none of the above.

____ 8. The metric system is the measurement system of choice for (a) science educators, (b) science researchers, (c) the citizens of most countries in the world, (d) all of the above.

____ 9. Above zero degrees, the actual number of degrees Celsius for any given temperature is _____ the number of degrees Fahrenheit. (a) higher than, (b) lower than, (c) the same as, (d) a or b.

____ 10. A thermometer measures (a) the degree of hot or cold, (b) temperature, (c) a and b, (d) none of the above.

E X E R C I S E   2

Measurement

P O S T - L A B   Q U E S T I O N S

1. Convert 1.24 m to millimeters, centimeters, and kilometers.

   _____ mm

   _____ cm

   _____ km

2. Observe the following carefully and read the volume.

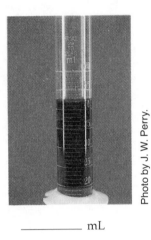

Photo by J. W. Perry.

   _____ mL

3. Construct a conversion table for mass. Construct it so that if you wish to convert a measurement from one unit to another, you multiply it by the number at the intersection of the original unit and the new unit.

Original Units	New Unit		
	mg	g	kg
mg	1	_____	_____
g	_____	1	_____
kg	_____	_____	1

4. One °F is _____ (larger, smaller) than 1°C.

5. The gravity on the planet Uranus is many times that of Earth. What would happen to the mass and weight of a visiting astronaut compared to the earth values of these measures?

6. How are length, area, and volume related in terms of the three dimensions of space?

7. How is it possible for objects of the same volume to have a different mass?

8. What is the importance of measurement to science?

9. If you were to choose between the metric and American Standard systems of measurement for future generations, which one would you choose? Take into consideration their ease of use and degree of standardization with the rest of the world.

EXERCISE 3
Microscopy

OBJECTIVES

After completing this exercise you will be able to:

1. define *magnification, resolving power, contrast, field of view, parfocal, parcentral, depth of field, working distance;*

2. describe how to care for a compound light microscope;

3. recognize and give the function of the parts of a compound light microscope;

4. accurately align a compound light microscope;

5. correctly use a compound light microscope;

6. use the microscopic units of length: the micrometer, nanometer, and Ångstrom;

7. make a wet mount;

8. correctly use a dissecting microscope;

9. describe the usefulness of the phase-contrast, transmission electron, and scanning electron microscopes;

10. use your skills to enjoy a fascinating world unavailable to the unaided eye.

INTRODUCTION

A microscope is an instrument that contains at least one *lens* and is used to view a specimen, or the detail in a specimen, that cannot be seen with the unaided eye. The lenses of light microscopes are made of transparent glass. A lens focuses the light rays emanating from a specimen to produce an image of that specimen.

Microscopy has three basic concepts: magnification, resolving power, and contrast. **Magnification** is the factor by which the image of a specimen is enlarged. **Resolving power** is the degree to which two adjacent points in a specimen are seen as separate; it depends on the preservation of detail in the image during the magnifying process. **Contrast** is how well details stand out against the background of an image.

Why can't you simply bring a specimen closer to the eye to see its detail, much as you might do to read the fine print in a contract? The lens of the eye focuses an image of what you view onto the light-sensing surface of the eye, the retina. Unfortunately, the normal eye lens cannot focus on an object closer than about 10 cm. At this distance you can see two specimen details separated by 0.1 mm. Because most cells are between 0.01 mm and 0.1 mm in diameter, they cannot be seen without a microscope.

I. The Compound Light Microscope

The compound light microscope has at least two lenses, an **ocular,** which you look into, and an **objective,** which scans the specimen. The basic function of a compound microscope is to bring the specimen very close to the eye and help the eye lens focus its image on the retina. It actually does this by producing a series of magnified images. Ultimately, the greater the proportion of the retina covered by the final image of the specimen, the greater its magnification. Magnification without detail is empty, and with a light microscope, the maximum useful magnification is about 1,000 times the diameter of the specimen ($1000 \times$). Above this value, detail is missing.

To see detail, there has to be contrast. Dyes are usually added to sections of biological specimens to increase contrast.

Like automobiles, there are many models of compound light microscopes and, for a given instrument, many possible accessories. Typical examples are illustrated in figure 3-1, and one is diagrammed in figure 3-2. If your microscope is significantly different from figure 3-2, your instructor will distribute an unlabeled diagram. If the instructor assigns to you a specific microscope for your use in the laboratory, record its identification code here.

My microscope is ——————.

The following activities will emphasize the practical aspects of microscopy. Additional reading on the basic optics of microscopy is provided in the Instructor's Manual.

Before removing your microscope from the cabinet read section A.

MATERIALS

Per student:

- compound light microscope
- lens paper
- lint-free cloth (optional)
- unlabeled diagram of the compound light microscope model used in your course (optional)
- prepared slide with a whole mount of stained diatoms
- prepared slide with Wright-stained smear of mammalian blood
- prepared slide with mounted letter *e*
- index card

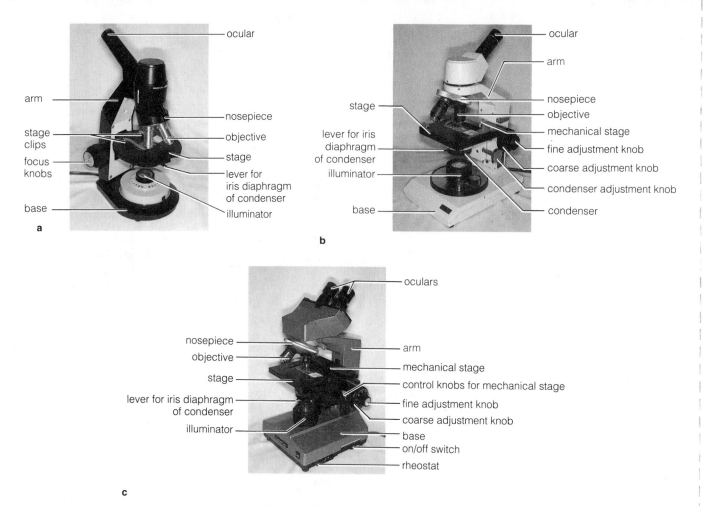

Figure 3-1 Compound microscopes. (Photos by D. Morton and J. W. Perry.)

• prepared slide with crossed colored threads coded for thread order
• prepared slide with unstained fibers
• 15-cm plastic ruler
• directions on how to calibrate and use an ocular micrometer (optional)

Per student group (4):
• bottle of lens-cleaning solution (optional)
• dropper bottle of immersion oil (optional)

Per lab room:
• labeled chart of a compound light microscope

PROCEDURE

A. Care of a Compound Light Microscope

1. To carry a microscope to and from your lab bench, grasp the **arm** with your dominant hand and support the **base** with the other hand, always keeping the microscope upright. *Do not try to carry anything else at the same time.* Label the arm and the base on figure 3-2 or the diagram given to you by your instructor.

> **CAUTION**
> **Never wipe a glass lens with anything other than lens paper.**

2. Remove the dust cover and clean the exposed parts of the optical system. First, blow off any loose dust that may be on the ocular and then gently brush off any remaining dust with a piece of lens paper.

If the part is still dirty, breathe on the lens and gently polish it with a rotary motion, using a fresh piece of lens paper. If the part is still dirty, *and with your instructor's approval,* clean the lens with a piece of lens paper moistened with lens-cleaning solution.

3. Always remember that your microscope is a precision instrument. Never force any of its moving parts.

4. It is just as difficult to see clearly through a dirty slide as through a dirty microscope. Clean dirty slides with a lint-free cloth or with lens paper before using.

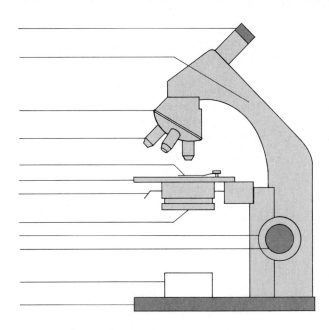

Figure 3-2 Compound light microscope.

Labels: ocular, arm, base, illuminator, condenser, lever for iris diaphragm of condenser, stage, stage clip, coarse adjustment knob, fine adjustment knob, nosepiece

5. At the end of an exercise, make sure the last slide has been removed from the stage and *rotate the nosepiece so that the low-power objective is in the light path.* If your instrument focuses by moving the body tube, turn the coarse adjustment so that it is racked all the way down. If your microscope has an electric cord, neatly fold it up on itself and tie it with a plastic strap or rubber band. Otherwise, wind the cord around the base of the arm of the microscope.

6. Replace the dust cover before returning your microscope to the cabinet.

B. Parts of the Compound Light Microscope

Now that you know how to care for your microscope, remove the instrument assigned to you from the cabinet and place it on your lab bench. Use figure 3-1 and the chart on the wall of your lab room to identify the various parts of your microscope. Read each step below and manipulate the parts *only where indicated.* Before you start, make sure the shortest objective is in the light path.

1. *Light source.* The compound microscope uses transmitted light to illuminate a transparent specimen usually mounted on a glass slide. Newer microscopes have a built-in illuminator. Locate the illuminator, the *off/on switch,* and perhaps also a *rheostat,* which is used to vary the intensity of the light. On some models the switch and the rheostat are combined. Turn on the light source. Look through the ocular. If the illuminator has a rheostat, adjust the intensity so that the light is not too bright.

Label the illuminator in figure 3-2 or the diagram given to you by your instructor.

Older microscopes may have a *double-sided mirror* to reflect light through the specimen. Angle the flat surface of the mirror so that the light path is directed through the hole in the stage (the platform where the specimen sits). Use the curved surface of the mirror only when illumination is inadequate.

2. *Condenser.* For maximum resolving power, a **condenser** — with a *condenser lens* and *iris diaphragm* — focuses the light source on the specimen so that each of its points is evenly illuminated. The **lever for the iris diaphragm of the condenser** is used to open and close the condenser. Establish if there is a *condenser adjustment knob* to set the height of the condenser. *Do not turn the knob; you will learn how to use it later.*

There may be a *filter holder* under the condenser with a blue or frosted glass disk. Many manufacturers of microscopes believe that blue light is more pleasing to the eye because, when used with an incandescent bulb, it produces a color balance similar to daylight conditions. Also, theoretically at least, blue light gives better resolving power because of its shorter wavelength. The frosted glass disk scatters light and may be useful in producing even illumination at low magnifications.

Label the condenser and lever for the iris diaphragm of the condenser on figure 3-2 or the diagram given to you by your instructor.

Simpler models have a revolving disk with a series of holes of different sizes to regulate the illumination of the specimen. As long as the specimen has adequate contrast, use the smallest hole that gives maximum illumination. The correct hole is usually marked to match it with the objective or with the total magnification in use at that time.

3. *Stage.* A specimen mounted on a glass slide is held in place on the **stage** by either a pair of **stage clips** or a *mechanical stage* so that the specimen is suspended over a central hole (fig. 3-1).

Label the stage and stage clips (or mechanical stage) on figure 3-2 or the diagram given to you by your instructor.

If your microscope has stage clips, place a prepared slide of stained diatoms under their free ends. *Never remove the stage clips, because they make it easier to move a slide in small increments.*

If your microscope has a mechanical stage, position a prepared slide of stained diatoms on the stage by releasing the tension on the spring-loaded movable arm. There are two knobs to the right or left of the stage: one to move the specimen forward and backward, the other to move it laterally.

On most mechanical stages, each direction has a *vernier scale* so that you can easily locate interesting fields again and again. A vernier scale consists of two scales running side by side, a long one in millimeters and a short one, 9 mm in length and divided into ten equal subdivisions.

Figure 3-3 A vernier scale. (Photo by D. Morton and J. W. Perry.)

To take a reading, note the whole number on the long scale coinciding with or just below the zero line of the short scale. If the whole number of the long scale and the zero of the short scale coincide, the first place after the decimal point is zero. Otherwise, the first place after the decimal point is the value of the line on the short scale that coincides (or nearly coincides) with one of the next nine lines after the whole number on the long scale. For example, the correct reading of the vernier scale in figure 3-3 is 19.6 mm.

4. *Focusing knobs.* The **coarse focus adjustment knob** is for use with low- and medium-power objectives, while the **fine focus adjustment knob** is for critical focusing, especially with high-dry and oil-immersion objectives. On some microscopes, you move the body tube of the instrument up and down to focus the specimen; on others you move the stage.

Modern microscopes also have a *preset focus lock,* which stops the stage at a particular height. After setting this lock, you can lower the stage with the coarse focus knob, to facilitate changing of the specimen, and then raise it to focusing height without fear of colliding the specimen against the objective. There may also be a *focus tension adjustment knob,* usually located inside of the left-hand coarse focus knob.

Turn the coarse focus knob. Do you turn the knob toward you or away from you to bring the slide and objective closer together?

_____

Label the coarse and fine focus adjustment knobs on figure 3-2 or the diagram given to you by your instructor.

5. *Objectives.* Most microscopes have several objectives mounted on a revolving **nosepiece.** The magnifying power of each objective is labeled on its side. Usually included are these objectives: a *low-power* or scanning (4×), a *medium-power* (10×), a *high-dry* (about 40×), and perhaps an *oil-immersion objective* (about 100×).

The other number often labeled on the side of nosepiece objectives is the **numerical aperture** (N.A.). The larger the numerical aperture, the greater the resolving power and useful magnification.

Record the magnifying power and N.A. of the objectives on your microscope in table 3-1. If your instrument does not have a particular objective, indicate that it is not present (NP).

Objectives are **parfocal.** That is, once an objective has been focused, you can rotate to another one and the image will remain in coarse focus, requiring only slight movement of the fine focus knob. Objectives are also **parcentral,** meaning that the center of the field of view remains about the same for each objective. The **field of view** is the circle of light you see when looking into the microscope.

Objectives have more or less different lengths, with the lower-power objectives being shorter than the higher-power ones. That is, the working distance of objectives decreases with magnification. **Working distance** is the space between the objective lens and the slide. Therefore, the higher the power of the objective in use, the closer the objective is to the slide—and the more careful you must be.

Label the nosepiece and objective on figure 3-2 or the diagram given to you by your instructor.

6. *Ocular.* Oculars are generally 10×. Since each objective has a different magnifying power, the total magnification is calculated by multiplying the magnifying power of the ocular by that of the objective in use. What is the total magnification of the ocular and high-dry objective on your microscope?

_____ ×

Your microscope will have one or two oculars mounted on a *monocular* or *binocular head,* respectively. There may be a pointer mounted in an ocular so that you can easily show a specimen detail to the instructor or another student. For a monocular microscope, it is best to use your dominant eye to look down the ocular, keeping your other eye open.

To determine your dominant eye:

a. Look at a small object on the far wall of your room with both eyes open.

b. Form the thumb and index finger of one hand into a circle and place this circle in your line of sight, at arm's length, so that it surrounds the object.

c. Close your right eye. If the object shifts out of the circle to your left, your right eye is probably dominant. If the object remains in the circle, your left eye is probably dominant.

d. Repeat a and b. This time close your left eye. If the object shifts to the right, your left eye is dominant. If the object remains within the circle, your right eye is dominant. The more pronounced the shift, the greater the dominance. If there is no shift, neither eye is dominant.

Table 3-1 Objectives Present on Assigned Compound Light Microscope		
Objective	**Magnifying Power**	**N.A.**
low-power	_____	_____
medium-power	_____	_____
high-dry	_____	_____
oil-immersion	_____	_____

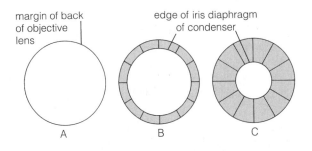

Figure 3-4 Correct setting for condenser iris diaphragm: Drawing B is correct. In A you cannot see the edge of the iris diaphragm. In C the diaphragm has been closed too much.

Binocular microscopes need to be adjusted to the distance between your pupils (*interpupillary distance*) and for any difference in power (*diopter*) between the lenses of each eye. These adjustments are described in section C.

Models with *zoom objectives* usually have the total magnification printed on the *zoom control*. A few models have only one zoom objective.

C. Aligning a Compound Light Microscope with In-base Illumination and a Condenser with an Iris Diaphragm

Aligning your microscope properly will not only help you see specimen detail clearly but will also protect your eyes from strain.

1. Rotate the nosepiece until the medium-power objective is in the light path. Open the iris diaphragm.

2. If it is not already there, place the prepared slide of stained diatoms on the stage; center and carefully focus on it. *Skip steps 3 and 4 if your microscope is monocular. Skip step 5 if your microscope does not have a control to adjust the height of the condenser.*

3. If your microscope is binocular, adjust the interpupillary distance. Hold a different ocular tube with each hand and, while looking at the specimen, pull the tubes apart or push them together until you see one field of view. After making this adjustment, read and record the number off the scale.

My interpupillary distance is _____ .

From now on you can set the interpupillary distance at this number.

4. Now compensate for any difference in diopter between the lenses of each eye:

a. If there is one diopter adjustment ring around the left ocular tube, cover the left eye with an index card and focus your microscope using the fine focus knob. Now uncover the left eye and cover the right one. Use the diopter adjustment ring to bring the specimen into focus.

b. If both ocular tubes have a diopter adjustment ring, set the left one to the same number as the interpupillary distance, cover the right eye with an index card, and focus on the specimen. Then uncover the right eye and cover the left one. Use the diopter adjustment ring on the right ocular tube to bring the specimen into focus.

5. Place a sharp point (pencil, dissecting needle, or some similar object) on top of the illuminator and bring the silhouette into sharp focus by adjusting the height of the condenser.

6. Use the lever to set the iris diaphragm of the condenser:

a. If the ocular on your microscope is removable (and with the permission of the instructor), carefully slide it out and put the ocular open-end down on a piece of lens paper in a safe place. Then, while looking down the ocular tube, adjust the iris diaphragm until the edge of the aperture lies just inside the margin of the back lens element of the objective (fig. 3-4). Replace the ocular.

b. If the ocular cannot be removed, close the condenser diaphragm and then open it until there is no further increase in brightness. Now close it again, stopping when you see the brightness begin to diminish.

7. If your microscope has a rheostat, adjust the illumination to a level that lets you see specimen detail and that is comfortable for your eyes. To maintain the same illumination at higher magnifications, you will have to increase its intensity.

8. For best results, repeat steps 6 and 7 each time you use a different objective.

D. Using Different Magnifications

It is safest to observe a specimen on a slide first with low power and then, step by step, with higher-power objectives. This way you can avoid colliding the objective against the slide or vice versa. Also, it is easier to use a lower-power objective to locate a specific specimen detail. This is why the low-power objective is sometimes called the scanning objective. Since the magnification is in diameters, the area of the field of view decreases dramatically with increasing magnifi-

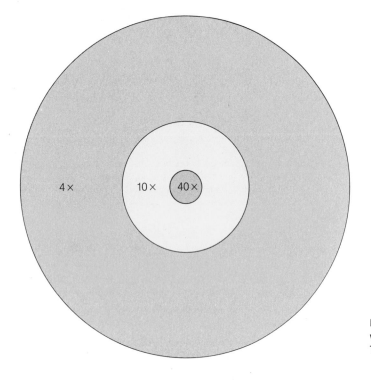

Figure 3-5 Illustration of the decreasing area of the field of view when a 4×, 10×, and 40× objective is used with a 10× ocular. The actual area of each circle has been enlarged 10×.

cation (fig. 3-5). It is just as easy to lose a specimen detail at higher magnifications; but once a specimen detail is lost, it is always easier to find it again if you switch to a lower-power objective.

Now follow these steps to use each objective:

1. Rotate the low-power objective into the light path.

2. If it is not already there, place a prepared slide of stained diatoms on the stage, securing it with either the stage clips or the movable arm of the mechanical stage.

3. Look through the ocular. Bring the diatoms into focus using the *coarse focus knob*. Adjust the illumination as described in steps 6 and 7 of section C. At this magnification the diatoms appear small. Center a diatom by moving the slide.

4. Rotate the nosepiece so that the medium-power objective is in the light path. Adjust the illumination. Focus the diatom.

5. Rotate the nosepiece so that the high-dry objective is in the light path. Adjust the illumination. Focus the diatom using the *fine focus adjustment knob*.

OPTIONAL

6. To use the oil-immersion objective, replace the prepared slide of diatoms with one with a smear of mammalian blood and repeat steps 1 through 5 above. Center and focus on a white blood cell. (See fig. 34-4b on page 459.) Most of the cells are red blood cells stained a light pink. A few cells are white blood cells with prominent blue-stained nuclei.

To use an oil-immersion objective:

a. Rotate the nosepiece so that the light path is midway between the high-dry and oil-immersion objectives.

b. Place a small drop of immersion oil onto the coverslip, using the circle of light above the specimen as a reference point.

c. Rotate the nosepiece so that the oil-immersion objective is in the oil.

d. Adjust the illumination and focus the white blood cell with the fine focus knob.

e. After examining the white blood cell, rotate the oil-immersion objective out of the light path. Carefully wipe the oil from the oil-immersion objective with lens paper.

f. Remove the slide from the stage and wipe the oil from the coverslip with lens paper.

Immersion oil, because it has optical properties similar to glass, increases resolving power and useful magnification.

E. Orientation of the Image

1. If you have not already done so, remove the prepared slide of diatoms and replace it with a prepared slide with the letter *e*. With the medium-power objective in the light path, position the slide with the specimen (the letter *e*) right-side up on the stage, center the specimen in the field of view, and carefully bring it into focus.

2. Draw in figure 3-6 the image of the *e* as you see it through the ocular. Record the total magnification used in the line at the end of the legend.

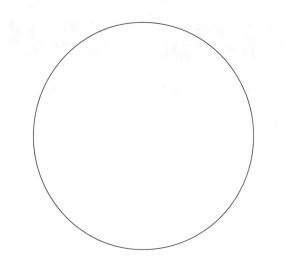

Figure 3-6 Drawing of the letter *e* as seen through the ocular (_____ ×).

Is the image right-side up or upside down compared to the specimen?

_____

Compared to the specimen, is the image backward as well as upside down?

(yes or no) _____

In summary, the image is *inverted* with respect to the specimen.

3. Move the specimen to the right while watching it through the microscope. In which direction does the image move?

_____

4. Move the specimen away from you. In which direction does the image move?

_____

5. Remove and put away the slide.

F. Depth of Field

The **depth of field** is the distance through which you can move the specimen and still have it remain in focus. Remember, the working distance—the space between the objective lens and the coverslip—decreases with increasing magnifying power. Therefore, the higher the power of the objective in use, the closer the objective is to the slide—and the more careful you must be.

1. Obtain a prepared slide of three crossed colored threads. *This exercise requires care, since you are probably not yet adept at focusing on a specimen.* Once you have the threads in focus (using first the low-power objective and then the medium-power objective), you need only use the fine focus knob to focus with the high-dry objective. After switching to the high-dry objective, try rotating the fine focus knob one-half turn away from you and then a full turn toward you. If you have not found the plane of focus, next try one and one-half turns away from you and two full turns toward you, and so on. If you work deliberately, you will find the plane of focus and will not crack the coverslip.

How many threads are in focus using the

low-power objective? _____

medium-power objective? _____

high-dry objective? _____

With which objective is it easiest to focus a specimen?

_____

At which magnification is it most difficult to focus a specimen?

_____

2. Specimens have depth. Utilizing the high-dry objective, determine which of the three threads is closest to the slide. (Each slide label has a code on it. When you believe that you have discovered which thread is closest to the slide, check with your instructor to find out if you are correct.)
Which thread is on the bottom?

_____

Which of the three threads is closest to the objective?

_____

Focusing carefully with the fine focus knob, move from the bottom to the upper thread. Did you move the knob away from you or toward you?

_____

3. Remove and put away the slide.

G. Using the Iris Diaphragm or Revolving Disk with Different-diameter Holes

1. Place a specimen of unstained fibers on the stage. Locate and focus on these fibers using the medium-power objective. Make sure the condenser and iris diaphragm, or revolving disk, are correctly set.

2. Close the iris diaphragm or move the revolving disk to smaller holes. Does this procedure increase or decrease contrast?

_____

Although this procedure is useful when viewing specimens with low contrast, it should be used as a last resort because resolving power is also decreased.

3. Remove and put away the slide.

H. Units of Measurement

The basic metric unit of length at the light-microscopic level is the micrometer (μm). Even smaller units, the nanometer (nm) and Ångstrom (Å), are often used at the electron-microscopic level.

1,000 μm = 1 mm
1,000 nm = 1 μm
10,000 Å = 1 μm

How many nanometers are there in 1 mm?

_____ nm

How many Ångstroms are there in 1 nm?

_____ Å

I. Determining the Diameter of the Field of View

1. Rotate the low-power objective into the light path. What is the total magnification?

_____ ×

2. Place a transparent 15-cm ruler on the stage.

3. What is the diameter of the field of view?

_____ mm

4. Repeat step 3 with the medium-power objective in the light path. The total magnification is

_____ ×, and the diameter of the field of view is

_____ mm.

5. Use the following formula to estimate the diameter of the field of view when the high-dry objective is in the light path.

$$\frac{\text{total magnification} \times \text{mm counted} \atop \text{using low-power} \times \text{with that} \atop \text{objective} \quad \text{objective}}{\text{total magnification of} \atop \text{high-dry objective}} = \underline{\quad} \text{mm}$$

Once it has been calculated, convert this value to micrometers.

_____ μm

6. Complete table 3-2.

Table 3-2 Diameter of Field of View

Objective	Magnifying Power	Diameter of Field of View
low-power	_____	_____
medium-power	_____	_____
high-dry	_____	_____
oil-immersion	_____	_____

7. You will use this information to estimate size by observing the percentage of the diameter of the field of view taken up by the specimen or part of the specimen. Using the medium-power objective, estimate the percentage of the diameter of the field of view covered by the letter *e*. The approximate diameter of the letter *e* is:

$$\frac{\text{percent} \times \text{diameter of field of view (mm)}}{100\%} = \underline{\quad} \text{mm}$$

OPTIONAL

8. If your microscope is equipped with an ocular micrometer, your instructor may provide directions on how to accurately measure specimen details.

II. How to Make a Wet Mount

In the midseventeenth century, Robert Hooke used a microscope to discover tiny, empty compartments in thin shavings of cork. He named them *cells*. Repeating this historic observation is a good way to learn how to prepare a wet mount.

MATERIALS

Per student:

• compound microscope, lens paper, a bottle of lens-cleaning solution (optional), a lint-free cloth (optional)
• cork
• razor blade
• glass microscope slide
• glass coverslip
• dissecting needle

Per student group (4):

• dropper bottle of distilled water

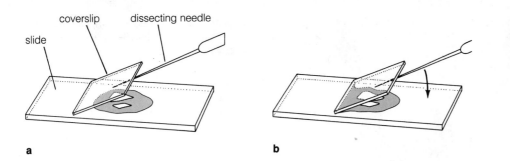

Figure 3-7 How to make a wet mount.

PROCEDURE

1. Carefully use a razor blade to cut a number of *very thin shavings* from a cork stopper. Place them on a glass microscope slide.

2. Gently add a drop of distilled water.

3. Place one end of a glass coverslip to the right or left of the specimen so that the rest of the slip is held at a 45° angle over the specimen (fig. 3-7a).

4. Slowly lower the coverslip with a dissecting needle so as not to trap air bubbles (fig. 3-7b).

5. Observe the wet mount, first at low magnification and then with higher power. Air may be trapped either in the cork or as free bubbles (fig. 3-8). Trapped air will appear dark and refractive around its edges. This effect is due to sharply bending rays of light.

Draw what you see in figure 3-9. Note the total magnification used to make the drawing.

6. Clean and dispose of the slide and coverslip as indicated by your instructor.

III. Dissection Microscope

Dissecting microscopes (fig. 3-10) have a large working distance between the specimen and the objective lens. They are especially useful in viewing larger specimens (including thicker slide-mounted specimens) and in manipulating the specimen (when dissection is required, for example).

The large working distance also allows for illumination of the specimen from above (reflected light) as well as from below (transmitted light). Reflected light shows up surface features on the specimen better than transmitted light does.

MATERIALS

Per student group:

• dissection microscope

• specimens appropriate for viewing with the dissecting microscope (for example, a prepared slide with a whole mount of a small organism, bread mold an insect mounted on a pin stuck in a cork, a small flower)

Figure 3-8 Free air bubble (250 ×). (Photo by J. W. Perry.)

Figure 3-9 Drawing of the microscopic structure in a cork shaving (_____ ×).

PROCEDURE

1. View under a dissecting microscope one or more of the specimens provided by your instructor. What is the magnification range of this microscope?

_____ × to _____ ×

2. Is the image of the specimen inverted as in the compound microscope?

_____

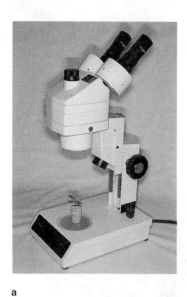

a

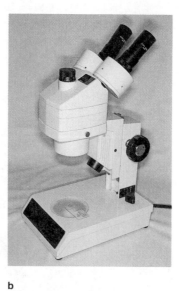

b

Figure 3-10 Dissecting microscope (**a**) reflected light, (**b**) transmitted light. (Photos by D. Morton and J. W. Perry.)

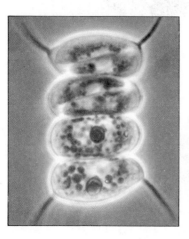

a Light micrograph (phase contrast)

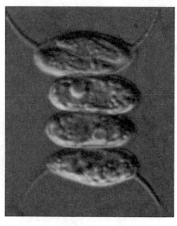

b Light micrograph (Nomarski process)

Figure 3-11 Comparison of how different types of microscopes reveal detail in cells of the green alga *Scenedesmus.* (Photos courtesy J. Pickett-Heaps.)

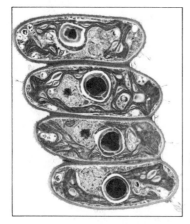

c Transmission electron micrograph, thin section

d Scanning electron micrograph

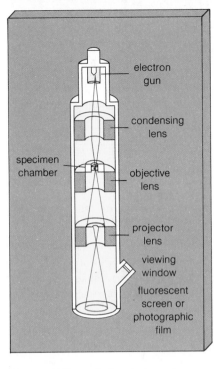

Figure 3-12 Transmission electron microscope. (After Starr and Taggart, 1989.)

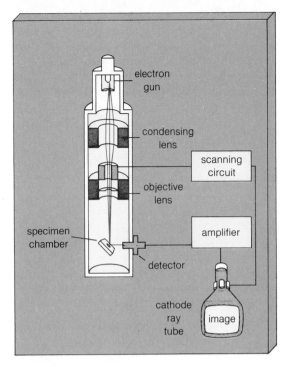

Figure 3-13 Scanning electron microscope. (After Starr and Taggart, 1989.)

3. Describe the type of illumination used by your dissecting microscope. Is there a choice?

_____

_____

_____

_____

IV. Other Microscopes

In future exercises you will have the opportunity to examine pictures taken with other types of microscopes (fig. 3-11). Some will be of living cells taken with a phase-contrast microscope or similar instrument, including those using the *Nomarski process*. Others will be of very thin-sectioned specimens taken with a transmission electron microscope. Still others will be of surfaces produced by signals from the scanning electron microscope.

As the name suggests, a **phase-contrast microscope** manipulates phase differences between the light passing through different parts of a specimen and its surrounding medium, turning them into intensity differences. This process produces contrast in the image. If the same specimen were viewed with a conventional compound microscope, the overall intensity of the image would be the same, and detail could not be seen. This technique is especially useful for observing living specimens.

The **transmission electron microscope** (TEM) is not unlike an upside-down compound light microscope (fig. 3-12). Because electrons have a much shorter wavelength than visible light, the use of electrons results in much greater resolving power. Therefore, details not visible under the light microscope or in *photomicrographs* (pictures taken through a light microscope) can be seen at magnifications greater than $1,000\times$. In fact, magnifications up to $1,000,000\times$ or more are possible.

A photomicrograph can be taken in color or black and white, but a *transmission electron micrograph* is always in black and white. Sections prepared for the TEM are stained with heavy-metal ions to render them electron-dense. In a transmission electron micrograph, the darker areas are more electron-dense than the lighter areas.

The **scanning electron microscope** (SEM) (fig. 3-13) scans the surface of a specimen with a narrow beam of electrons. The surface of the specimen usually has been previously coated with a thin layer of gold. The energy in the scanning beam causes the gold to emit additional electrons, which are picked up by a detector. Much like a television set constructing a picture from signals received from an antenna or the cable, an SEM produces a picture of the surface of the specimen on a cathode-ray tube.

The resolving power and maximum useful magnification of the SEM are less than those of the TEM but higher than those of light microscopes. The SEM is used to investigate the fine structure of surfaces, and, at magnifications comparable to the light-microscopic level, the SEM gives a three-dimensional view of sectioned specimens. The lighter areas in a *scanning electron micrograph* correspond to the emission of higher numbers of electrons from that part of the specimen.

Electron microscopes differ from light microscopes in three major ways: (1) Electrons are used instead of light, (2) magnetic lenses replace glass lenses, and (3) the electron path has to be maintained in a high vacuum.

PRE-LAB QUESTIONS

_____ 1. Magnification (a) is the factor by which the image of a specimen is enlarged, (b) is how well details stand out against the background in the image, (c) is the degree to which two adjacent points in a specimen are seen as separate in the image, (d) focuses the light rays emanating from a specimen to produce an image of that specimen.

_____ 2. Resolving power (a) is the factor by which the image of a specimen is enlarged, (b) is how well details stand out against the background in the image, (c) is the degree to which two adjacent points in a specimen are seen as separate in the image, (d) focuses the light rays emanating from a specimen to produce an image of that specimen.

_____ 3. A lens (a) is the factor by which the image of a specimen is enlarged, (b) is how well details stand out against the background in the image, (c) is the degree to which two adjacent points in a specimen are seen as separate in the image, (d) focuses the light rays emanating from a specimen to produce an image of that specimen.

_____ 4. Contrast (a) is the factor by which the image of a specimen is enlarged, (b) is how well details stand out against the background in the image, (c) is the degree to which two adjacent points in a specimen are seen as separate in the image, (d) focuses the light rays emanating from a specimen to produce an image of that specimen.

_____ 5. The maximum useful magnification for a light microscope is about (a) $100\times$, (b) $1,000\times$, (c) $10,000\times$, (d) $100,000\times$.

_____ 6. The two image-forming lenses of a compound light microscope are the (a) condenser and objective, (b) condenser and ocular, (c) objective and ocular, (d) condenser and eye.

_____ 7. Dyes are usually added to sections of biological specimens to increase (a) resolving power, (b) magnification, (c) contrast, (d) all of the above.

_____ 8. If the magnification of the two image-forming lenses are both $10\times$, the total magnification of the image will be (a) $1\times$, (b) $10\times$, (c) $100\times$, (d) $1,000\times$.

_____ 9. The distance through which a microscopic specimen can be moved and still have it remain in focus is called the (a) field of view, (b) working distance, (c) depth of focus, (d) magnification.

_____ 10. Electron microscopes differ from light microscopes in that (a) electrons are used instead of light, (b) magnetic lenses replace glass lenses, (c) the electron path has to be maintained in a high vacuum, (d) a, b, and c.

EXERCISE 3

Microscopy

POST-LAB QUESTIONS

1. What is the function of each of the following parts of a compound light microscope?

   a. condenser lens

   b. iris diaphragm

   c. objective

   d. ocular

2. In order, list the lenses in the light path between a specimen viewed with the compound light microscope and its image on the retina of the eye.

3. What happens to contrast and resolving power when the aperture of the condenser (that is, the size of the hole through which light passes before it reaches the specimen) of a compound light microscope is decreased?

4. What happens to the field of view in a compound light microscope when the total magnification is increased?

5. Describe the importance of the following concepts to microscopy.
   a. magnification

   b. resolving power

   c. contrast

6. What type of microscope would you use to examine the following:
   a. the living surface of a finger

   b. a prepared slide of a section of a finger

   c. the insides of a single cell in a thin section of part of a finger

   d. bacteria on a single cell on the surface of a finger

7. Which photomicrograph of unstained cotton fibers was taken with the iris diaphragm closed? _____

a

Photo by J. W. Perry.

b

Photo by J. W. Perry.

8. List how you would care for and put away your compound light microscope at the end of the laboratory period.

EXERCISE 4
Structure and Function of Living Cells

OBJECTIVES

After completing this exercise you will be able to:

1. define *cell, cell theory, prokaryotic, eukaryotic, nucleus, cytomembrane system, organelle,* karyon, *multinucleate, ctyoplasmic streaming, colloid, sol, gel, envelope;*

2. list the structural features shared by all cells;

3. describe the similarities and differences between prokaryotic and eukaryotic cells;

4. identify the cell parts described within this exercise;

5. state the function for each cell part;

6. distinguish between plant and animal cells;

7. recognize the structures presented in **boldface** in the procedure section.

INTRODUCTION

Structurally and functionally, all life has one common feature: All living organisms are composed of **cells.** The development of this concept began with Robert Hooke's seventeenth-century observation that slices of cork were made up of small units. He called these units "cells" because their structure reminded him of the small cubicles that monks lived in. Over the next 100 years the **cell theory** emerged. This theory has three principles: (1) All organisms are composed of one or more cells; (2) the cell is the basic *living* unit of organization; and (3) all cells arise from preexisting cells.

Although cells vary in organization, size, and function, all share three structural features: (1) All possess a **plasma membrane** defining the boundary of the living material; (2) all contain a region of **DNA** (deoxyribonucleic acid), which stores genetic information; and (3) all contain **cytoplasm,** everything inside the plasma membrane that is not part of the DNA region.

With respect to internal organization, there are two basic types of cells, **prokaryotic** and **eukaryotic.** Study table 4-1, comparing the more important differences between prokaryotic and eukaryotic cells.

The Greek word *karyon* means "kernel," referring to the nucleus. Thus, *prokaryotic* means "before a nucleus," while *eukaryotic* indicates the presence of a "true nucleus."

Prokaryotic cells typical of modern bacteria and cyanobacteria are believed to be similar to the first cells, which arose on earth 3.5 billion years ago. Eukaryotic cells probably evolved from prokaryotes.

This exercise will familiarize you with the basics of cell structure and the function of prokaryotes (prokaryotic cells) and eukaryotes (eukaryotic cells).

Table 4-1 Comparison of Prokaryotic and Eukaryotic Cells

Characteristics	Cell Type	
	Prokaryotic	Eukaryotic
Genetic material	Located within cytoplasm, not bounded by a special membrane	Located in **nucleus,** a double membrane-bounded compartment within the cytoplasm
	Consists of a single molecule of DNA	Numerous molecules of DNA combined with protein
		Organized into chromosomes
Cytoplasmic structures	Small ribosomes	Large ribosomes
	Photosynthetic membranes arising from the plasma membrane (in some representatives only)	**Cytomembrane system,** a system of connected membrane structures
		Organelles, membrane-bounded compartments specialized to perform specific functions
Kingdoms represented	Monera	Protista Fungi Plantae Animalia

I. Prokaryotic Cells

MATERIALS

Per student:

• dissecting needle

• compound microscope

• microscope slide

• coverslip

Per student pair:

• distilled water (dH_2O) in dropping bottle

Per student group (table):

• culture of a cyanobacterium (either *Anabaena* or *Oscillatoria*)

Per lab room:

• 3 bacterium-containing nutrient agar plates (demonstration)

• 3 demonstration slides of bacteria (coccus, bacillus, spirillum)

PROCEDURE

Observe the culture plate containing bacteria growing on the surface of a nutrient medium. Can you see the individual cells with your naked eye?

_____

Observe the microscopic preparations of bacteria on *demonstration* next to the culture plate. The three slides represent the three different shapes of bacteria. Which objective lenses are being used to view the bacteria?

_____

Would you say bacteria are large or small organisms?

_____

Can you discern any detail within the cytoplasm?

_____

In the space provided in figure 4-1, make a sketch of what you see through the microscope. Record the magnification you are using to view the bacteria in the blank provided at the end of the legend. Next to your sketch record the approximate size of the bacterial cells. (Return to page 26 of Exercise 3 if you've forgotten how to estimate the size of an object being viewed through the microscope.)

Examine the electron micrograph of the bacterium *Escherichia coli* (fig. 4-2). The cell is bound by the **cell wall,** a structure chemically distinct from the wall of plant cells but serving the same primary function. The **plasma membrane** is tightly appressed to (lying flat against) the internal surface of the cell wall and is

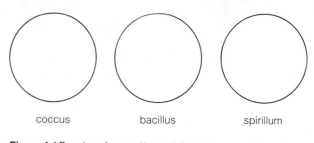

coccus   bacillus   spirillum

Figure 4-1 Drawing of several bacterial cells (_____ ×). Approximate size = _____ μm.

difficult to distinguish. Look for two components of the **cytoplasm: Ribosomes,** electron-dense particles (they appear black) that give the cytoplasm its granular appearance; and a relatively electron-transparent region (appears light) containing fine threads of DNA.*

Another type of prokaryotic cell is exemplified by cyanobacteria, such as *Oscillatoria* and *Anabaena.* Cyanobacteria (sometimes called blue-green algae) are commonly found in water and damp soils. They obtain their nutrition by converting the sun's energy through photosynthesis.

With a dissecting needle, remove a few filaments from the cyanobacterial culture, placing them in a drop of water on a clean microscope slide. Place a coverslip over the material and examine it first with the low-power objective and then using the high-dry objective (or oil-immersion objective, if your microscope is so equipped). Make a sketch in figure 4-3 of the cells you see at high power. Estimate the size of a *single* cyanobacterial cell and record the magnification you used in making your drawing.

Now examine the electron micrograph of *Anabaena* (fig. 4-4), identifying *cell wall, cytoplasm,* and *ribosomes.* The cyanobacteria also possess membranes that function in photosynthesis. Identify these **photosynthetic membranes,** which look like tiny threads within the cytoplasm.

Look at the figure legend for figures 4-2 and 4-4. Judging by the magnification of each electron micrograph, which cell is larger, the bacterium *E. coli* or the cyanobacterium *Anabaena*?

_____

Because the electron micrograph of *Anabaena* is of relatively low magnification, the plasma membrane is not obvious, but if you could see it, it would be found just under the cell wall.

_____

*Recently, new techniques of transmission electron microscopy have demonstrated that the region shown on figure 4-2 — sometimes called a *nucleoid* — is an artifact due to older chemical preservation methods. It now seems that the DNA is scattered within the cytoplasm, rather than aggregated.

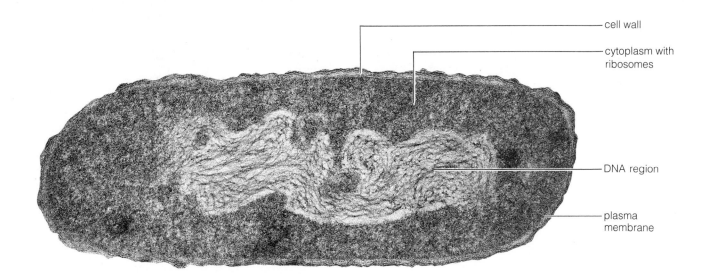

cell wall

cytoplasm with ribosomes

DNA region

plasma membrane

Figure 4-2 Electron micrograph of the bacterium *Escherichia coli* (55,000 ×). (Photo courtesy of G. Cohen-Bazire.)

Figure 4-3 Drawing of several prokaryotic cells of a cyanobacterium (_____ ×). Approximate size = _____ μm.

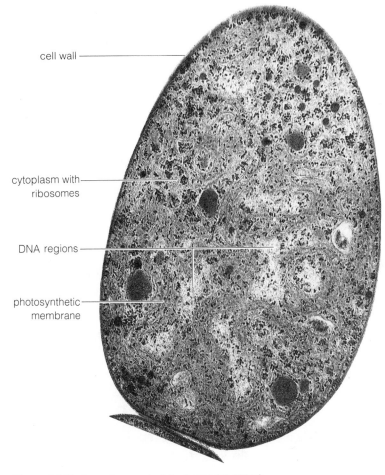

cell wall

cytoplasm with ribosomes

DNA regions

photosynthetic membrane

Figure 4-4 Electron micrograph of *Anabaena* (13,000 ×). (Photo courtesy R. D. Warmbrodt.)

II. Eukaryotic Cells

MATERIALS

Per student:

- textbook
- toothpick
- microscope slide
- coverslip
- culture of *Physarum polycephalum*
- compound microscope
- forceps
- dissecting needle

Per student pair:

- methylene blue in dropping bottle
- safranin O in dropping bottle
- distilled water (dH$_2$O) in dropping bottle

Per student group (table):

- *Elodea* in water-containing culture dish
- onion bulb
- tissue paper
- alcohol-containing disposal jar

Per lab room:

- model of animal cell
- model of plant cell

PROCEDURE

A. Animal Cells Observed with the Light Microscope

1. *Human cheek cells.* Using the broad end of a clean toothpick, gently scrape the inside of your cheek. Stir the scrapings into a drop of distilled water on a clean microscope slide and add a coverslip.

> **CAUTION**
> Dispose of used toothpick in the jar containing alcohol.

Because the cells are almost transparent, decrease the amount of light entering the objective lens to increase the contrast. (See section G, Exercise 3, page 25.) Find the cells using the low-power objective of your microscope; then switch to the high-dry objective for detailed study. Find the **nucleus,** a centrally located spherical body within the **cytoplasm** of each cell.

Now stain your cheek cells with a dilute solution of methylene blue, a dye that stains the nucleus darker than the surrounding cytoplasm. To stain your slide, follow the directions illustrated in figure 4-5.

First add a drop of the stain to one edge of the coverslip. Then draw the stain under the coverslip by

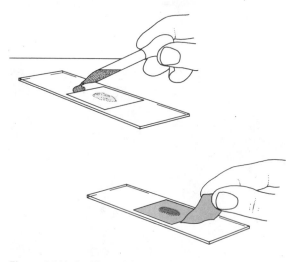

Figure 4-5 Method for staining specimen under coverslip of microscope slide.

touching a piece of tissue paper to the *opposite* side of the coverslip. It is not necessary to remove the coverslip.

In figure 4-6, make a sketch of the cheek cells, labeling *cytoplasm, nucleus,* and the location of the *plasma membrane.* (A light microscope cannot resolve the plasma membrane, but the boundary between the cytoplasm and the external medium indicates its location.) Many of the cells will be folded or wrinkled due to their thin, flexible nature. Estimate and record in your sketch the size of the cells. (The method for estimating the size is found in step 7 on page 26.)

2. *Physarum polycephalum.* Although, strictly speaking, the slime mold *Physarum* is not considered an animal (depending upon the authority, it's considered a protist or a fungus), its cellular structure is somewhat typical of animal cells. *Physarum* is a unicellular organism, so it contains all the metabolic machinery for independent existence.

Place a plain microscope slide on the stage of your compound microscope. This will serve as a platform on which you can place a culture dish. Now obtain a petri dish culture of *Physarum*, remove the lid, and place it on the platform. Observe initially with the low-power objective and then with the medium-power objective. Place a coverslip over part of the organism before rotating the high-dry objective into place. This prevents the agar from getting on the lens.

Physarum is **multinucleate**, meaning that more than one nucleus occurs within the cytoplasm. Unfortunately, the nuclei are tiny; you will not be able to distinguish them from other granules in the cytoplasm. The outer boundary of the cytoplasm is the *plasma membrane.* Locate the boundary. Once again, the resolving power of your microscope is not sufficient to allow you to actually view the membrane.

Watch the cytoplasm of the organism move. This intracellular motion is known as **cytoplasmic streaming.** Although not visible with the light microscope

Labels: cytoplasm, nucleus, plasma membrane

Figure 4-6 Drawing of human cheek cells (_____ ×).
Approximate size = _____ μm.

Labels: cytoplasm (gel), cytoplasm (sol), plasma membrane

Figure 4-7 Drawing of a portion of *Physarum* (_____ ×).

without using special techniques, contractile proteins called *microfilaments* are believed responsible for cytoplasmic streaming.

As in all cells, the cytoplasm of *Physarum* is a semisolid **colloid,** a state of matter in which the particles are too large to be dissolved but too small to settle out. Note that the outer portion of the cytoplasm appears solid; this is the **gel** state of the cytoplasm. Notice that the granules closer to the interior are in motion within a fluid; this portion of the cytoplasm is in the **sol** state. Movement of the organism occurs as the sol-state cytoplasm at the advancing tip pushes against the plasma membrane, causing the region to swell outward. The sol-state cytoplasm flows into the region, becoming converted to the gel state along the margins.

In figure 4-7 make a labeled sketch of the portion of *Physarum* that you have been observing.

OPTIONAL
Experiment: Cytoplasmic Streaming

Your instructor may provide you with an experiment about cytoplasmic streaming.

B. Animal Cells as Observed with the Electron Microscope

Studies with the electron microscope have yielded a wealth of information on the structure of eukaryotic cells. Structures too small to be seen with the light microscope have been identified. These include many **organelles,** structures in the cytoplasm that have been separated ("compartmentalized") by enclosure in membranes. Examples of organelles are the nucleus, mitochondria, endoplasmic reticulum, and Golgi bodies. Although the cells in each of the five kingdoms

have some peculiarities unique to that kingdom, electron microscopy has revealed that all cells are fundamentally similar.

Study figure 4-8, a three-dimensional representation of an animal cell. With the aid of figure 4-8, identify the parts on the model of the animal cell that is on *demonstration*.

Figure 4-9 is an electron micrograph (EM) of an animal cell (kingdom Animalia). Study the electron micrograph and identify, with the aid of figure 4-8 and any electron micrographs in your textbook, each structure listed below.

Pay particular attention to the membranes surrounding the nucleus and mitochondria. Note that these two are each bounded by *two* membranes, which are commonly referred to collectively as an **envelope.**

Using your textbook as a reference, list the function for each of the following cellular components.

1. plasma membrane _____

_____

2. cytoplasm _____

_____

3. nucleus (the plural is *nuclei*) _____

_____

   a. nuclear envelope _____

   b. nuclear pores _____

4. chromatin _____

_____

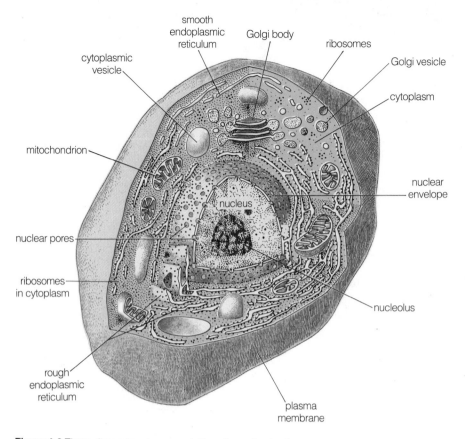

Figure 4-8 Three-dimensional representation of an animal cell as seen with the electron microscope. (After Wolfe, 1985.)

5. nucleolus (the plural is *nucleoli*) _____

_____

6. endoplasmic reticulum (ER)

   a. rough ER _____

   b. smooth ER _____

7. Golgi body _____

8. mitochondrion (the plural is *mitochondria*) _____

_____

C. Plant Cells Seen with the Light Microscope

1. *Elodea leaf cells.* Young leaves at the growing tip of *Elodea* are particularly well suited for studying cell structure because these leaves are only a few cell layers thick.

With a forceps, remove a single young leaf, mount it on a slide in a drop of distilled water, and cover with a coverslip. Examine the leaf first with the low-power objective. Then concentrate your study on several cells using the high-dry objective. Refer to figure 4.10.

You will probably be struck by the abundance of green bodies in the cytoplasm. These are the **chloroplasts,** organelles that function in photosynthesis and that are typical of green plants. You may see numerous

dark lines running parallel to the long axis of the leaf. These are the air-containing **intercellular spaces.** The **cell wall,** a structure distinguishing plant from animal cells, may be visible as a clear area surrounding the cytoplasm.

After the cells have warmed a bit, you should see **cytoplasmic streaming** taking place. Movement of the chloroplasts along the cell wall is the most obvious evidence of cytoplasmic streaming. Microfilaments are thought to be responsible for this intracellular motion.

Remember that you are looking at a three-dimensional object. In the middle portion of the cell, is the large, clear **central vacuole,** which can take up from 50% to 90% of the cell interior. Because the vacuole in *Elodea* is transparent, it cannot be seen with the light microscope.

The chloroplasts occur in the cytoplasm surrounding the vacuole, so they will appear to be in different locations, depending on where you focus in the cell. If your focus is the upper or lower surface, the chloroplasts will appear to be scattered throughout the cell. But if you focus in the center of the cell (by raising or lowering the objective with the fine focus knob), you will see the chloroplasts in a thin layer of cytoplasm along the wall.

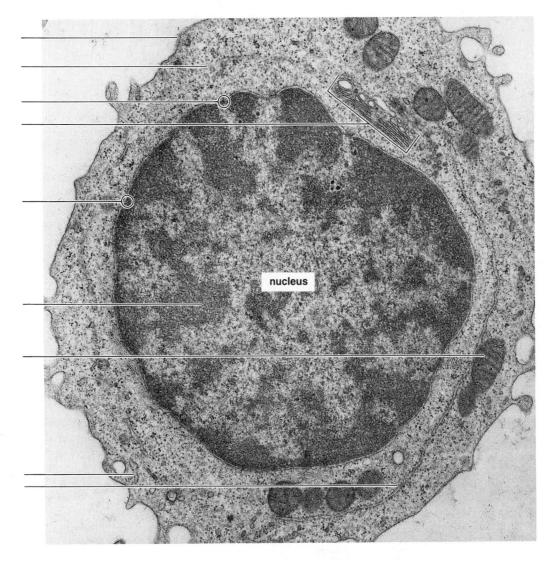

nucleus

Labels: plasma membrane, cytoplasm, nuclear envelope, nuclear pore, chromatin, rough ER, smooth ER, Golgi body, mitochondrion

Figure 4-9 Electron micrograph of an animal cell (2,000 ×). (Photo courtesy W. R. Hargreaves.)

Locate the **nucleus** within the cytoplasm. It will appear as a clear or slightly amber body that is slightly larger than the chloroplasts. (You may need to examine several cells to find a clearly defined nucleus.)

How would you describe the three-dimensional shape of the *Elodea* leaf cell?

_____

_____

_____

The shape of the chloroplasts and nucleus?

_____

_____

_____

chloroplasts (surrounding a nucleus)

cell wall   nucleus   central vacuole

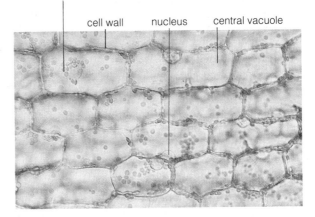

Figure 4-10 *Elodea* cells (300 ×). (Photo by J. W. Perry.)

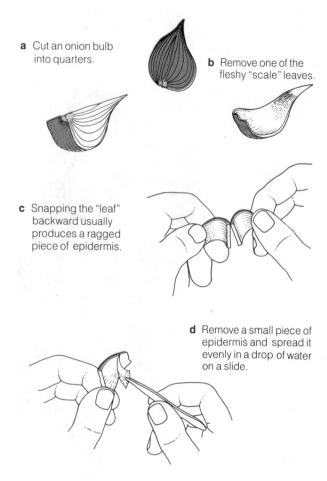

a Cut an onion bulb into quarters.

b Remove one of the fleshy "scale" leaves.

c Snapping the "leaf" backward usually produces a ragged piece of epidermis.

d Remove a small piece of epidermis and spread it evenly in a drop of water on a slide.

e Gently lower a coverslip to prevent trapping air bubbles. Examine with your microscope. Add more water to the edge of the coverslip with an eye dropper if the slide begins to dry.

Figure 4-11 Method for obtaining onion scale cells. (From Peter Abramoff and Robert G. Thomson, *Laboratory Outlines in Biology III*. Copyright © 1962, 1963 Peter Abramoff and Robert G. Thomson. Copyright © 1964, 1966, 1972, 1982 W. H. Freeman and Company. Used by permission.)

Now add a drop of safranin stain to make the cell wall more obvious. Add the stain the same way you stained your cheek cells with methylene blue (fig. 4-5).

2. *Onion scale cells.* Make a wet mount of a colorless scale of an onion bulb, using the technique described in figure 4-11. The *inner* face of the scale is easiest to remove, as shown in figure 4-11d.

Observe your preparation with your microscope, focusing first with the low-power objective. Continue your study, switching to the medium-power and finally the high-dry objective. Refer to figure 4-12.

Identify the **cell wall** and **cytoplasm.** The **nucleus** should be a prominent sphere within the cytoplasm. Examine the nucleus more carefully at high magnification. Within it find one or more **nucleoli** (the singular is *nucleolus*). Nucleoli are rich in a nucleic acid known as RNA (ribonucleic acid), while the rest of the nucleus is largely DNA (deoxyribonucleic acid), the genetic material.

Numerous **oil droplets** should be visible in the form of granule-like bodies within the cytoplasm. These oil droplets are a form of stored food material. You may be surprised to learn that onion scales are actually leaves! Which cellular components present in *Elodea* leaf cells are absent in onion leaf cells?

_____

If you are using the pigmented tissue from a red onion, you should see a purple pigment located in the vacuole. In this case the cell wall appears as a bright line.

D. Plant Cells as Seen with the Electron Microscope

The electron microscope has made obvious some of the unique features of plant cells. Study figure 4-13, a three-dimensional representation of a typical plant cell.

With the aid of figure 4-13, identify the structures present on the model of a plant cell that is on *demonstration*.

Now examine figure 4-14, a transmission electron micrograph from a corn leaf. Label all of the structures listed below. Notice that the chloroplast has an envelope, just as do the nucleus and mitochondria. With the help of figure 4-13 and any transmission electron micrographs in your textbook, list the function of each.

1. cell wall _____

_____

2. chloroplast _____

_____

3. vacuole _____

_____

4. vacuolar membrane _____

_____

5. plasma membrane _____

_____

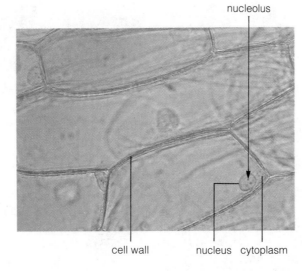

Figure 4-12 Onion bulb leaf cells (75×). (Photo by J. W. Perry.)

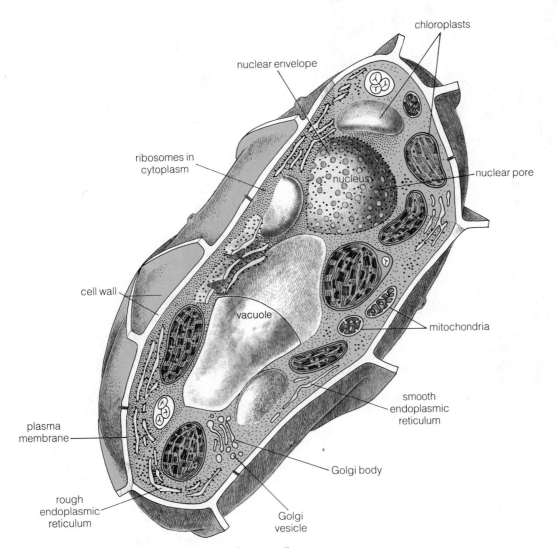

Figure 4-13 Three-dimensional representation of a plant cell as seen with the electron microscope. (After Wolfe, 1985.)

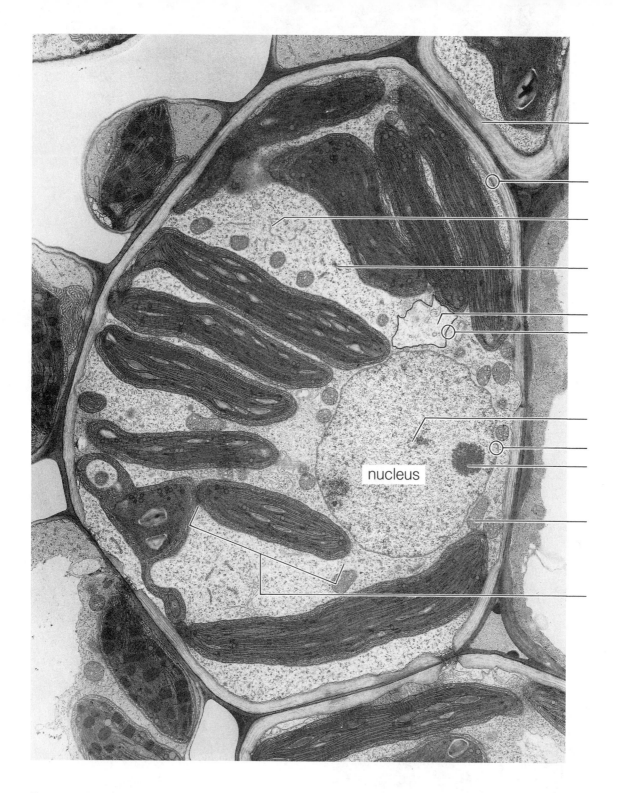

nucleus

Figure 4-14 Electron micrograph of a corn leaf cell (3,000 ×). (Courtesy R. F. Evert and M. A. Walsh.)

Labels: cell wall, chloroplast, vacuole, vacuolar membrane, plasma membrane, nuclear envelope, chromatin, nucleolus, endoplasmic reticulum (ER), Golgi body, mitochondrion

6. cytoplasm _____

_____

7. nucleus _____

    a. nuclear envelope _____

    _____

    b. nuclear pore _____

    _____

    c. chromatin _____

    _____

    d. nucleolus _____

    _____

8. endoplasmic reticulum (ER)

    a. rough ER _____

    b. smooth ER _____

9. Golgi body _____

10. mitochondrion _____

_____

PRE-LAB QUESTIONS

____ 1. The person responsible for first using the term *cell* was (a) Darwin, (b) Leeuwenhoek, (c) Hooke, (d) Watson.

____ 2. All cells contain (a) a nucleus, plasma membrane, and cytoplasm, (b) a cell wall, nucleus, and cytoplasm, (c) DNA, plasma membrane, and cytoplasm, (d) mitochondria, plasma membrane, and cytoplasm.

____ 3. Prokaryotic cells *lack* (a) DNA, (b) a true nucleus, (c) a cell wall, (d) none of the above.

____ 4. The word *eukaryotic* refers specifically to a cell containing (a) photosynthetic membranes, (b) a true nucleus, (c) a cell wall, (d) none of the above.

____ 5. A bacterium is an example of a (an) (a) prokaryotic cell, (b) eukaryotic cell, (c) plant cell, (d) all of the above.

____ 6. Methylene blue (a) is used to kill cells that are moving too quickly to observe; (b) renders cells nontoxic; (c) is a portion of the electromagnetic spectrum used by green plant cells; (d) is a biological stain used to increase contrast of cellular constituents.

____ 7. Components typical of plant cells but not of animal cells are (a) nuclei, (b) cell walls, (c) mitochondria, (d) ribosomes.

____ 8. A central vacuole (a) is found only in plant cells, (b) may take up between 50% and 90% of the cell's interior, (c) both of the above, (d) none of the above.

____ 9. The intercellular spaces between plant cells (a) contain air, (b) are responsible for cytoplasmic streaming, (c) are nonexistent, (d) contain chloroplasts.

____ 10. An envelope (a) surrounds the nucleus, (b) surrounds mitochondria, (c) consists of two membranes, (d) all of the above.

Name _____   Section Number _____

EXERCISE 4

Structure and Function of Living Cells

POST-LAB QUESTIONS

1. a. What structural differences did you observe between prokaryotic and eukaryotic cells?

   b. Are the cells in this electron micrograph below prokaryotic or eukaryotic?

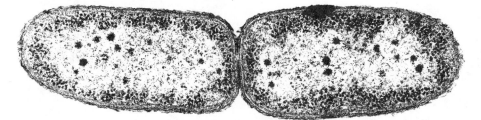

(Photos by J. J. Cardamone, Jr., University of Pittsburgh/BPS.)

2. Did all living cells that you saw in lab contain mitochondria?

3. Did all living plant cells you observed in lab contain chloroplasts?

4. Describe a major distinction between plant and animal cells.

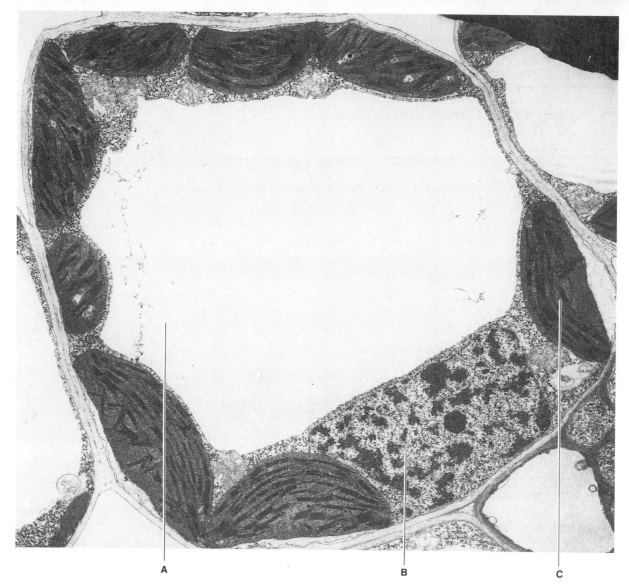

(11,300×). (Photo by M. C. Ledbetter.)

5. Observe the electron micrograph above.

   a. Is the cell prokaryotic or eukaryotic?

   b. Identify the labeled structures.

     A. _____   C. _____

     B. _____

6. How does the size of the bacterial cells you observed compare with that of your human cheek cells and the *Elodea* cells? Make a generalization about the size of prokaryotic and eukaryotic cells.

7. Look at the photomicrograph below taken with a technique that gives a three-dimensional impression. Identify the structures labeled A, B, and C.

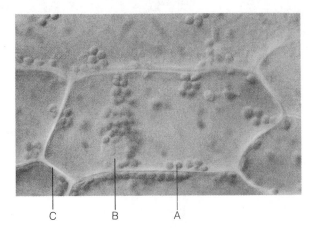

A. _____

B. _____

C. _____

(750 ×). (Photo by J. W. Perry.)

8. a. Is the electron micrograph below a plant or an animal cell?

   b. Identify structures labeled A and B.

   A. _____

   B. _____

   c. What are the numerous "wavy lines" within the cell?

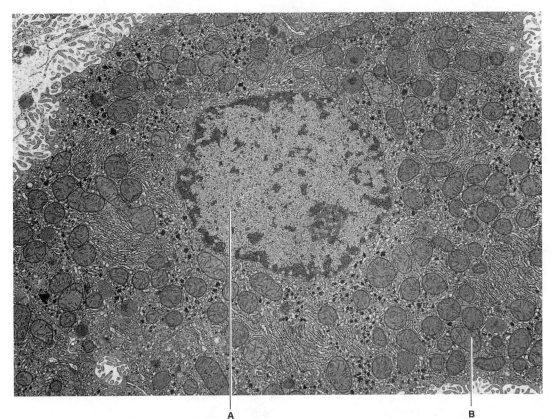

(15,000 ×). (Photo by G. L. Decker.)

9. Below is a high-magnification photomicrograph of an organism you observed in this exercise. Are the cells in this chain prokaryotic or eukaryotic?

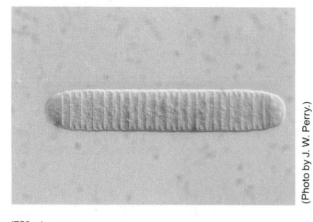

(Photo by J. W. Perry.)

(750×).

10. What structure(s) found in plant cells are primarily responsible for cellular support?

Diffusion, Osmosis, and the Functional Significance of Biological Membranes

OBJECTIVES

After completing this exercise you will be able to:

1. define *solvent, solute, solution, differentially permeable, diffusion, osmosis, concentration gradient, equilibrium, turgid, plasmolyzed, plasmolysis, turgor pressure, tonicity, hypertonic, isotonic, hypotonic;*

2. describe the structure of cellular membranes;

3. distinguish between diffusion and osmosis;

4. determine the effects of concentration and temperature on diffusion;

5. describe the effects of hypertonic, isotonic, and hypotonic solutions on red blood cells and *Elodea* leaf cells.

INTRODUCTION

Water is a great environment. Life is believed to have originated in the water. Without it, life as we know it would cease to exist. If, as is suspected, life in our solar system is unique to earth, it is probably because ours is the only planet known to possess free water on its surface.

Living cells are made up of 75% to 85% water. Virtually all substances entering and leaving cells are dissolved in water, making it the **solvent** most important for life processes. The substances dissolved in water are called **solutes** and include such substances as salts and sugars. The combination of a solvent and dissolved solute is a **solution.**

The cytoplasm of living cells contains numerous solutes, like sugars and salts. These are in solution.

All cells possess membranes composed of a phospholipid bilayer that contains embedded and surface proteins. Membranes are boundaries that solutes must cross to reach the cellular site where they will be utilized for the processes of life. These membranes regulate the passage of substances into and out of the cell. They are **differentially permeable,** allowing some substances to move easily while completely excluding others.

Although there are several methods by which solutes may enter cells, the most common is diffusion. Once within the cell, solutes move through the cytoplasm by diffusion, sometimes assisted by cytoplasmic streaming. Simply stated, **diffusion** is the movement of solute molecules from a region of high concentration to one of lower concentration. Diffusion occurs without requiring the expenditure of cellular energy.

So far, we've described the movement only of solute molecules across membranes. However, water (the solvent) also moves across the membrane. The movement of *water* across differentially permeable membranes is **osmosis.** Think of osmosis as a special form of diffusion, one occurring from a region of higher *water* concentration to one of lower *water* concentration.

The difference in concentration of like molecules in two regions is called a **concentration gradient.** Diffusion and osmosis take place *down* concentration gradients. Over time, the concentration of solvent and solute molecules becomes equally distributed, the gradient ceasing to exist. At this point the system is said to be at **equilibrium.**

Molecules are always in motion, even at equilibrium. Thus, solvent and solute molecules continue to move because of randomly colliding molecules. However, at equilibrium there is no *net change* in the concentration.

This exercise will introduce you to the principles of diffusion and osmosis.

> **NOTE**
>
> Start parts II and III before doing any other portion of this exercise.

I. Rate of Diffusion of Solutes

Solutes move within a cell's cytoplasm largely because of diffusion. However, the rate of diffusion (the distance diffused in a given amount of time) is affected by factors such as temperature and the size of the solute molecules. Recall from Exercise 4 that cytoplasm is a colloid. The following experiment demonstrates the effects of these two factors within a common colloid, gelatin (the substance of Jell-O®), to simulate cytoplasm.

MATERIALS

Per student:
• metric ruler

Per student group (table):
• 1 set of 3 screw-cap test tubes half-filled with 5% gelatin, to which the following dyes have been added: potassium dichromate, aniline blue, Janus green; labeled with each dye and marked "5°C"
• 1 set of 3 screw-cap test tubes as above but marked "Room Temperature"

Per lab room:
• 5°C refrigerator

PROCEDURE

Two sets of three screw-cap test tubes have been half-filled with 5% gelatin. One mL of a dye has been added to each test tube. On the line below, record the time at which your instructor tells you the experiment was started.

_____

Set 1 is in a 5°C refrigerator; set 2 is at room temperature.

1. Remove set 1 from the refrigerator and compare the distance the dye has diffused in corresponding tubes of each set.

> **CAUTION**
> Be certain the cap to each tube is tight!

2. Invert and hold each tube vertically in front of a white sheet of paper. Use a metric ruler to measure how far each dye has diffused from the gelatin's surface. Record this distance in table 5-1.

3. Determine the *rate* of diffusion for each dye by using the formula:

> **rate = distance/elapsed time (hours)**

Time experiment ended: _____

Time experiment started: _____

Elapsed time: _____ hours

Which of the solutes diffused the slowest (regardless of temperature)?

_____

Which diffused the fastest?

_____

What effect did temperature have on the rate of diffusion?

_____

Make a conclusion about diffusion of a solute in a gel, relating the rate of diffusion to the molecular weight of the solute.

_____

_____

> **NOTE**
> Return set 1 to the refrigerator.

II. Osmosis

Osmosis occurs when different concentrations of water are separated by a differentially permeable membrane. One example of a differentially permeable membrane within a living cell is the plasma membrane. This experiment demonstrates osmosis by using dialysis membrane, a differentially permeable cellulose sheet that permits the passage of water but obstructs passage of larger molecules. If you could examine the membrane with a scanning electron microscope, you would see that it is porous. Thus molecules larger than the pores cannot pass through the membrane.

Table 5-1 Effect of Temperature on Diffusion Rate of Various Solutes					
	Set 1 (5°C)			Set 2 (Room Temp.)	
Solute (dye)	Distance (mm)	Rate	Distance (mm)	Rate	
Potassium dichromate (MW* = 294)					
Janus green (MW = 511)					
Aniline blue (MW = 738)					

*MW = molecular weight, a reflection of the mass of a substance. (MW is determined by adding the atomic weights of all elements comprising a compound.)

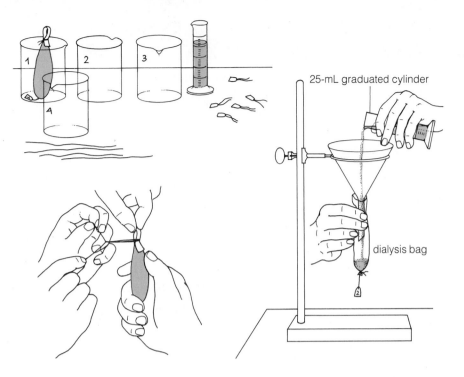

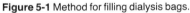

Figure 5-1 Method for filling dialysis bags.

MATERIALS

Per student group (4):

• 4 15-cm lengths of dialysis tubing, soaking in dH_2O

• 8 10-cm pieces of string

• ring stand and funnel apparatus (fig. 5-1)

• 25-mL graduated cylinder

• 4 small string tags

• china marker

• 4 400 mL beakers

Per student group (table):

• dishpan half-filled with dH_2O

• paper toweling

• balance

Per lab room:

• source of dH_2O (at each sink)

• sucrose solutions 15% and 30%

• scissor (at each sink)

PROCEDURE

Work in groups of four for this experiment.

1. Obtain four sections of dialysis tubing, each 15 cm long, that have been presoaked in distilled water. Recall that the dialysis tubing is permeable to water molecules but not to sucrose.

2. Fold over one end of each tube and tie it tightly with string.

3. Attach a string tag to the tied end of each bag and number them from 1 through 4.

4. Slip the open end of the bag over the stem of a funnel (fig. 5-1). Using a graduated cylinder to measure volume, fill the bags as follows:

> **NOTE**
> Be sure to rinse the cylinder if it has been used for measuring sucrose.

Bag 1. 10 mL of distilled water

Bag 2. 10 mL of 15% sucrose

Bag 3. 10 mL of 30% sucrose

Bag 4. 10 mL of distilled water

5. As each bag is filled, force out excess air by squeezing the bottom end of the tube.

6. Fold the end of the bag and tie it securely with another piece of string.

7. Rinse each filled bag in the dishpan containing distilled water (dH_2O); gently blot off the excess water with paper toweling.

8. Weigh each bag to the nearest 0.5 g.

9. Record the weights in the column marked "0 min" of table 5-2.

Table 5-2 Change in Weight as a Consequence of Osmosis

No.	Bag Contents/ Beaker Contents	Bag Weight (grams)					Weight Change (grams)
		0 Min	20 Min	40 Min	60 Min	80 Min	
1	distilled water/ distilled water						
2	15% sucrose/ distilled water						
3	30% sucrose/ distilled water						
4	distilled water/ 30% sucrose						

10. Number four 400-mL beakers with a china marker.

11. Add 200 mL of dH_2O to beakers 1 through 3.

12. Add 200 mL of 30% sucrose solution to beaker 4.

13. Place bags 1 through 3 in the correspondingly numbered beakers.

14. Place bag 4 in the beaker containing 30% sucrose.

15. After 20 minutes, remove each bag from its beaker, blot off the excess fluid, and weigh each bag.

16. Record the weight of each bag in table 5-2.

17. Return the bags to their respective beakers immediately after weighing.

18. Repeat steps 15 to 17 at 40, 60, and 80 minutes from time zero.

At the end of the experiment, take the bags to the sink, cut them open, pour the contents down the drain and discard the bags in the wastebasket. Pour the contents of the beakers down the drain and wash them according to the instructions given on page x.

Make a *qualitative* statement about what you observed.

_____

_____

_____

Was the direction of *net* movement of water in bags 2 to 4 into or out of the bags?

_____

Which bag gained the most weight? Why?

_____

_____

III. Differential Permeability of Membranes

Dialysis tubing is a differentially permeable material that provides a means to demonstrate the movement of substances through cellular membranes.

MATERIALS

Per student group (4):

- 1 25-cm length of dialysis tubing, soaking in dH_2O
- 2 10-cm pieces of string
- bottle of 1% soluble starch in 1% sodium sulfate (Na_2SO_4)
- dishpan half-filled with dH_2O
- 400-mL graduated beaker
- ring stand and funnel apparatus (fig. 5-1)
- bottle of 1% albumin in 1% sodium chloride (NaCl)
- 8 test tubes
- test tube rack
- china marker
- 25-mL graduated cylinder
- iodine (I_2KI) solution in dropping bottle
- 2% barium chloride ($BaCl_2$) in dropping bottle
- 2% silver nitrate ($AgNO_3$) in dropping bottle
- biuret reagent in dropping bottle
- albustix reagent strips (optional)
- scissors

Per lab room:

- series of 4 test tubes in test tube rack demonstrating positive tests for starch, sulfate ion, chloride ion, protein

PROCEDURE

Work in groups of four for this experiment.

1. Obtain a 25-cm section of dialysis tubing that has been soaked in distilled water (dH_2O).

Table 5-3 Results of Tests for Substances in Beaker		
	At Start of Experiment	After 75 Minutes
Starch	−	
Sulfate ion	−	
Chloride ion	+	
Albumin	+	

Table 5-4 Results of Tests for Substances in Dialysis Bag		
Contents of Dialysis Bag: (+) = Presence, (−) = Absence		
	At Start of Experiment	After 75 Minutes
Starch	+	
Sulfate ion	+	
Chloride ion	−	
Albumin	−	

2. Fold over one end of the tubing and tie it securely with string to form a leakproof bag (fig. 5-1).

3. Slip the open end of the bag over the stem of a funnel and fill the bag approximately half full with 25 mL of a solution of 1% soluble starch in 1% sodium sulfate (Na_2SO_4).

4. Remove the bag from the funnel; fold and tie the open end of the bag.

5. Rinse the tied bag in a dishpan partially filled with dH_2O.

6. Pour 200 mL of a solution of 1% albumin (a protein) in 1% sodium chloride (NaCl) into a 400-mL beaker.

7. Place the bag into the fluid in the beaker.

8. Record the time in this space:

_____

9. With a china marker, label eight test tubes, numbering them 1 through 8.

10. Seventy-five minutes after the start of the experiment, pour 20 mL of the *beaker contents* into a *clean* 25-mL graduated cylinder.

11. Decant (pour out) 5 mL from the graduated cylinder into each of the first four test tubes.

12. Perform the following tests, recording the results in table 5-3. Your instructor will have a series of test tubes showing positive tests for starch, sulfate and chloride ions, and proteins. You should compare your results with the known positives.

a. *Test for starch.* Add several drops of iodine solution (I_2KI) from the dropper bottle to test tube 1. If starch is present, the solution will turn blue-black.

b. *Test for sulfate ion.* Add several drops of 2% barium chloride ($BaCl_2$) from the dropper bottle to test tube 2. If sulfate ions (SO_4^-) are present, a white precipitate of barium sulfate ($BaSO_4$) will form.

c. *Test for chloride ion.* Add several drops of 2% silver nitrate ($AgNO_3$) from the dropper bottle to test tube 3. A milky-white precipitate of silver chloride (AgCl) indicates the presence of chloride ions (Cl^-).

d. *Test for protein.* Add several drops of biuret reagent from the dropper bottle to test tube 4. If protein is present, the solution will change from blue to pinkish-violet. The more intense the violet hue, the greater the quantity of the protein.

An alternative method for determining the presence of protein is the use of albustix reagent strips. Presence of protein is indicated by green or blue-green coloration of the paper.

13. Wash the graduated cylinder, using the technique described on page x.

14. Thoroughly rinse the bag in the dishpan of dH_2O.

15. Using a scissors, cut the bag open and empty the contents into the 25-mL graduated cylinder.

16. Decant 5-mL samples into each of the four remaining test tubes.

17. Perform the tests for starch, sulfate ions, chloride ions, and protein on tubes 5 through 8, respectively.

18. Record the results of this series of tests in table 5-4.

To which substances was the dialysis tubing permeable?

_____

What physical property of the dialysis tubing might explain its differential permeability?

_____

19. Discard contents of test tubes and beaker down sink drain. Wash glassware by technique described on page x.

20. Discard dialysis tubing in wastebasket.

IV. Plasmolysis in Plant Cells

Plant cells are surrounded by a rigid cell wall, composed primarily of the glucose polymer, cellulose. Recall from Exercise 4 that many plant cells have a large central vacuole surrounded by the vacuolar membrane. The vacuolar membrane is differentially permeable. Normally the solute concentration within the cell's central vacuole is greater than that of the external environment. Consequently, water moves into the cell, creating **turgor pressure,** which presses the cytoplasm against the cell wall. Such cells are said to be **turgid.** Many nonwoody plants (like beans and peas) rely on turgor pressure to maintain their rigidity and erect stance.

This experiment demonstrates the effects of external solute concentration on the structure of plant cells.

MATERIALS

Per student:

• forceps

• 2 microscope slides

• 2 coverslips

• compound microscope

Per student group (table):

• *Elodea* in tap water

• 2 dropping bottles of distilled water (dH$_2$O)

• 2 dropping bottles of 20% sodium chloride (NaCl)

PROCEDURE

1. With a forceps, remove two young leaves from the tip of an *Elodea* plant.

2. Mount one leaf in a drop of distilled water on a microscope slide and the other in 20% NaCl solution on a second microscope slide.

3. Place coverslips over both leaves.

4. Observe the leaf in distilled water with the compound microscope. Focus first with the medium-power objective and then switch to the high-dry objective.

5. Label the photomicrograph of turgid cells (fig. 5-2).

6. Now observe the leaf mounted in 20% NaCl solution. After several minutes the cell will have lost water, causing it to become **plasmolyzed.** (This process is called **plasmolysis.**) Label the plasmolyzed cells shown in figure 5-3.

Tonicity is a description of one solution's solute concentration compared to that of another solution. A solution containing a lower concentration of solute molecules than another is **hypotonic** *relative to the second.* Solutions containing equal concentrations of solute are **isotonic** to each other, while one containing a greater concentration of solute relative to a second is **hypertonic.**

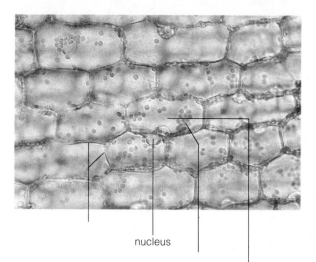

Labels: cell wall, chloroplasts in cytoplasm, central vacuole

Figure 5-2 Turgid *Elodea* cells (300 ×). (Photo by J. W. Perry.)

Were the contents of the vacuole in the *Elodea* leaf in distilled water hypotonic, isotonic, or hypertonic compared to the distilled water?

_____

Was the 20% sodium chloride solution hypertonic, isotonic, or hypotonic relative to the cytoplasm?

_____

If a hypotonic and a hypertonic solution are separated by a differentially permeable membrane, which direction will the water move?

_____

Name two differentially permeable membranes that are present within the *Elodea* cells and that were involved in the plasmolysis process.

1. _____

2. _____

V. Osmotic Changes in Red Blood Cells

Animal cells lack the rigid cell wall of a plant. The external boundary of an animal cell is the differentially permeable plasma membrane. Consequently, an animal cell increases in size as water enters the cell. However, since the plasma membrane is relatively fragile, it ruptures when too much water enters the cell. This is because of excessive pressure pushing out against the membrane. Conversely, if water moves out of the cell, it will become plasmolyzed and look spiny.

In this experiment, you will use red blood cells to demonstrate the effects of osmosis in animal cells.

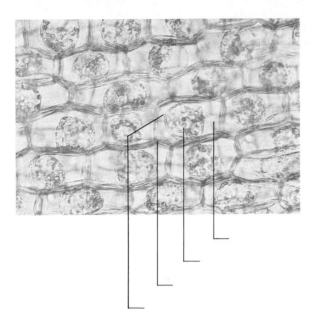

Labels: cell wall, chloroplasts in cytoplasm, plasma membrane, space (between cell wall and plasma membrane)

Figure 5-3 Plasmolyzed *Elodea* cells (300 ×). (Photo by J. W. Perry.)

MATERIALS

Per student:

• compound microscope

Per student group (4):

• 3 clean screw-cap test tubes

• test tube rack

• metric ruler

• china marker

• bottle of 0.9% sodium chloride (NaCl)

• bottle of 10% NaCl

• bottle of distilled water (dH₂O)

• 3 disposable plastic pipets

• 3 clean microscope slides

• 3 coverslips

Per student group (table):

• bottle of sheep blood (in ice bath)

Per lab room:

• source of distilled water

PROCEDURE

Work in groups of four for this experiment, but do the microscopic observations individually.

1. Observe the scanning electron micrographs in figure 5-4. Figure 5-4a illustrates the normal appearance

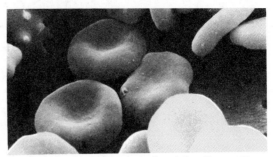

a Red blood cells in an isotonic solution ("normal")

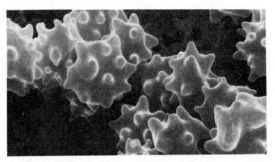

b Red blood cells in a hypertonic solution ("crenate")

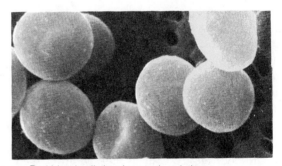

c Red blood cells in a hypotonic solution

Figure 5-4 Scanning electron micrographs of red blood cells. (Photos from M. Sheetz, R. Painter, and S. Singer. Reproduced from *The Journal of Cell Biology*, 1976, 70: 193, by copyright permission of the Rockefeller University Press and M. Sheetz.)

of red blood cells. They are described as biconcave disks; that is, they are circular in outline with a depression in the center of both surfaces. Cells in an isotonic solution would appear like those in figure 5-4a.

Figure 5-4b shows cells that have been plasmolyzed. (In the case of red blood cells, plasmolysis is given a special term, *crenation;* the blood cell is said to be *crenate.*)

Figure 5-4c represents cells that have taken in water but have not yet burst. (Burst red blood cells are said to be *hemolyzed,* and of course they can't be seen.) Note their swollen, spherical appearance.

Table 5-5 Effect of Salt Solutions on Red Blood Cells

Tube No.	Contents of Tube	Print Visible?	Microscopic Appearance of Cells	Tonicity of External Solution*
1				
2				
3				

*With respect to that inside the red blood cell at the start of the experiment.

2. Obtain three clean screw-cap test tubes.

3. Lay test tubes 1 and 2 against a metric ruler and mark lines indicating 5 cm *from the bottom of each tube.*

4. Fill each tube as follows:

Tube 1: 5 cm of 0.9% sodium chloride (NaCl)
        5 drops of sheep blood

Tube 2: 5 cm of 10% NaCl
        5 drops of sheep blood

5. Lay test tube 3 against a metric ruler and mark lines indicating 0.5 cm and 5 cm *from the bottom of the tube.*

6. Fill tube 3 to the 0.5-cm mark with 0.9% NaCl, and to the 5-cm mark with distilled water. Then add 5 drops of sheep blood. Enter the contents of each tube in the appropriate column of table 5-5.

7. Replace the caps and mix the contents of each tube by inverting several times (fig. 5-5a).

8. Hold each tube flat against the printed page of your lab manual (fig. 5-5b). *Only if the blood cells are hemolyzed should you be able to read the print.*

9. Record your observations in the column marked "Print Visible?" of table 5-5.

10. Number three clean microscope slides.

11. With three *separate* disposable pipets, remove a small amount of blood from each of the three tubes. Place 1 drop of blood from tube 1 on slide 1, 1 drop from tube 2 on slide 2, and 1 drop from tube 3 on slide 3.

12. Cover each drop of blood with a coverslip.

13. Observe the three slides with your compound microscope, focusing first with the medium-power objective and finally with the high-dry objective. (Hemolyzed cells are virtually unrecognizable; all that remains are membranous "ghosts," which are difficult to see with the microscope.)

14. In figure 5-6 make a sketch of the cells from each tube. Label the sketches, indicating whether the cells are normal, plasmolyzed (crenate), or hemolyzed.

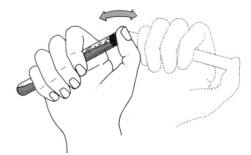

Figure 5-5 Method for studying effects of different solute concentrations on red blood cells. (After Abramoff and Thomson, 1982.)

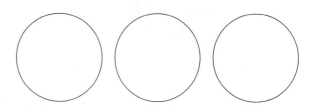

Figure 5-6 Microscopic appearance of red blood cells in different solute concentrations (_____ ×).

15. Record the microscopic appearance in table 5-5.

16. Record in table 5-5 the tonicity of the sodium chloride solutions you added to the test tubes.

Why do red blood cells burst when put in a hypotonic solution whereas *Elodea* leaf cells do not?

_____

_____

OPTIONAL
VI. Experiment: Determining the Concentration of Solutes in Cells

Your instructor may provide you with an experiment about determining solute concentration.

OPTIONAL
VII. Experiment: An Investigation of Active Transport

Your instructor may provide you with an experiment about active transport.

PRE-LAB QUESTIONS

_____ 1. If one were to identify the most important compound for sustenance of life, it would probably be (a) salt, (b) $BaCl_2$, (c) water, (d) I_2KI.

_____ 2. A solvent is (a) the substance in which solutes are dissolved, (b) a salt or sugar, (c) one component of a biological membrane, (d) differentially permeable.

_____ 3. Diffusion (a) is an energy-requiring process, (b) is the movement of molecules from a region of higher concentration to one of lower concentration, (c) occurs only across differentially permeable membranes, (d) none of the above.

_____ 4. Cellular membranes (a) consist of a phospholipid bilayer containing embedded proteins, (b) control the movement of substances into and out of cells, (c) are differentially permeable, (d) all of the above.

_____ 5. An example of a solute would be (a) Janus green B, (b) water, (c) sucrose, (d) a and c above.

_____ 6. Dialysis membrane is (a) differentially permeable, (b) used in these experiments to simulate cellular membranes, (c) permeable to water but not to sucrose, (d) all of the above.

_____ 7. Specifically, osmosis (a) requires the expenditure of energy, (b) is diffusion of water from one region to another, (c) is diffusion of water across a differentially permeable membrane, (d) none of the above.

_____ 8. Which of the following reagents does _not_ fit with the substance being tested for? (a) biuret reagent–protein, (b) $BaCl_2$–starch, (c) $AgNO_3$–chloride ion, (d) albustix–protein.

_____ 9. When the cytoplasm of a plant cell is pressed against the cell wall, the cell is said to be (a) turgid, (b) plasmolyzed, (c) hemolyzed, (d) crenate.

_____ 10. If one solution contains 10% NaCl and another contains 30% NaCl, the 30% solution is _____ with respect to the 10% solution. (a) isotonic, (b) hypotonic, (c) hypertonic, (d) plasmolyzed.

EXERCISE 5

Diffusion, Osmosis, and the Functional Significance of Biological Membranes

POST-LAB QUESTIONS

1. How does diffusion differ from osmosis?

2. If a 10% sugar solution is separated from a 20% sugar solution by a differentially permeable membrane, in which direction will there be a net movement of water?

3. Based upon your observations in this exercise, would you expect dialysis membrane to be permeable to sucrose? Why?

4. Suppose you were having a party. You wish to serve crisp celery, but your celery has become limp, and the stores are closed. What might you do to make the celery crisp (turgid) again?

5. Plant fertilizer consists of numerous different solutes. A small dose of fertilizer may enhance plant growth, but overfertilization can kill the plant. Why might overfertilization have this effect?

6. A human lost at sea without fresh drinking water is effectively lost in an osmotic desert. Why would drinking salt water be harmful?

7. Suppose you wanted to dissolve a solute in water. Without shaking or swirling the solution, what might you do to increase the rate at which the solute would go into solution? Relate your answer to your method's effect on the motion of the molecules.

8. What does the word *lysis* mean?

   (Now does the name of the disinfectant Lysol make sense?)

9. Why don't plant cells undergo osmotic lysis?

10.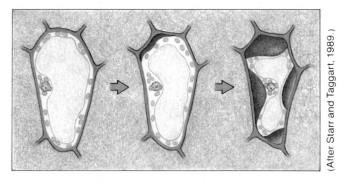

(After Starr and Taggart, 1989.)

   a. The above represents a plant cell that has been placed in a solution. What *process* is taking place as one follows the arrows? What is happening at the cellular level when a wilted plant is watered and begins to recover from the wilt?

   b. Is the solution in which the cells have been placed hypotonic, isotonic, or hypertonic relative to the cytoplasm?

Enzymes: Catalysts of Life

OBJECTIVES

After completing this exercise you will be able to:

1. define *catalyst, enzyme, activation energy, enzyme-substrate complex, substrate, product, active site, denaturation, cofactor;*

2. explain how an enzyme operates;

3. recognize benzoquinone as a brown substance formed in damaged plant tissue;

4. indicate the substrates for the enzyme catechol oxidase;

5. describe the effect of temperature on the rate of chemical reactions in general and on enzymatically controlled reactions in particular;

6. describe the effect that an atypical pH may have on enzyme action;

7. indicate how a cofactor might operate and identify a cofactor for catechol oxidase.

INTRODUCTION

Life without enzymes is unimaginable. The energy required by your muscles simply to open your laboratory manual would take years to accumulate without enzymes. Due to the presence of enzymes, the myriad chemical reactions occurring in your cells at this very moment are being completed in a fraction of a second rather than the years or even decades that would be otherwise required.

Enzymes are proteins that function as biological catalysts. A **catalyst** is a substance that lowers the amount of energy necessary for a chemical reaction to proceed. You might think of this so-called **activation energy** as a hump to be negotiated. Enzymes decrease the size of the hump, in effect turning a mountain into a molehill (fig. 6-1).

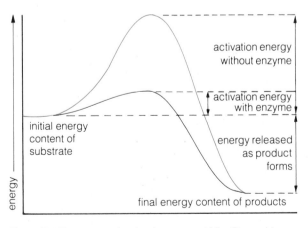

Figure 6-1 Enzymes and activation energy. (After Starr and Taggart, 1989.)

By lowering the activation energy, an enzyme affects the *rate* at which reaction occurs. Enzyme-boosted reactions may proceed from 100 thousand to 10 million times faster than they would without the enzyme.

In an enzyme-catalyzed reaction, the reactant (the substance being acted upon) is called the **substrate.** Substrate molecules combine with enzyme molecules to form a temporary **enzyme-substrate complex. Products** are formed, and the enzyme molecule is released unchanged. Thus, the enzyme is not used up in the process and is capable of catalyzing the same reaction again and again. This can be summarized as follows:

$$\text{substrate} \xrightarrow{\text{enzyme}} \begin{array}{c}\text{enzyme-}\\\text{substrate}\\\text{complex}\end{array} \rightarrow \text{products} + \text{enzyme}$$

Although thousands of enzymes are present within cells, we will examine only one, catechol oxidase (also known as tyrosinase), to demonstrate the effects of several factors that influence enzyme action. These factors include:

1. temperature;

2. pH (hydrogen-ion concentration of the environment);

3. specificity (how discriminating the enzyme is in catalyzing different potential substrates); and

4. cofactor necessity (the need for a metallic ion for enzyme activity).

I. Formation and Detection of Benzoquinone

Catechol oxidase is an enzyme that catalyzes the production of benzoquinone and water from catechol:

$$\begin{array}{c}\text{catechol} + \frac{1}{2}O_2 \\ \text{(substrate)}\end{array} \xrightarrow[\text{potato extract}]{\begin{array}{c}\text{catechol oxidase}\\\text{(enzyme in}\end{array}} \begin{array}{c}\text{benzoquinone} + H_2O\\\text{(product)}\end{array}$$

This is an oxidation reaction, with catechol and oxygen as the substrates. Hence, the enzyme gets its preferred name, *catechol oxidase.* (This suffix *-ase* is a tipoff that the substance is an enzyme.)

Catechol and catechol oxidase are present in the cells of many plants, although in undamaged tissue they are separated in different compartments of the cells. Injury causes mixing of the substrate and enzyme, producing benzoquinone, a brown substance.

You've probably noticed the brown coloration of a damaged apple or the blackening of an injured potato tuber. Benzoquinone inhibits the growth of certain microorganisms that cause rot.

In this experiment you will form the product, benzoquinone, and establish a color intensity scale to be used in subsequent experiments.

MATERIALS

Per student:

• disposable plastic gloves

Per student group (4):

• 3 test tubes

• test tube rack

• metric ruler

• ice bath with wash bottle of potato extract containing catechol oxidase

• wash bottle containing 1% catechol solution

• bottle of distilled water (dH$_2$O)

• china marker

Per lab room:

• 40°C waterbath

• vortex mixer (optional)

CAUTION

Some of the chemicals (catechol, hydroquinone) used in these experiments may be hazardous to your health if they are ingested or taken in through your skin. Wear disposable plastic gloves for all experiments.

PROCEDURE

Work in groups of four for all experiments in this exercise.

1. With a china marker, label three test tubes I$_1$, I$_2$, and I$_3$. Place your initials on each test tube for later identification.

2. Lay the test tubes against a metric ruler and mark lines on the tubes corresponding to 1 cm and 2 cm *from the bottom* of each tube.

3. Fill each tube as follows:

Tube I$_1$:

• 1 cm of potato extract containing catechol oxidase

• 1 cm of 1% catechol solution

Tube I$_2$:

• 1 cm of potato extract containing catechol oxidase

• 1 cm of dH$_2$O

Table 6-1 Formation and Detection of Benzoquinone

Time (minutes)	Tube I$_1$ Potato Extract and Catechol	Tube I$_2$ Potato Extract and Water	Tube I$_3$ Catechol and Water
0			
5			
10			
15			

NOTE

Be certain to return potato extract to ice bath IMMEDIATELY after use in this and subsequent experiments.

Tube I$_3$:

• 1 cm of 1% catechol solution

• 1 cm of dH$_2$O

4. Shake all tubes (using a vortex mixer if available).

5. Record the color of the solution in each tube in the "Time 0" spaces of table 6-1.

6. Place the tubes in a 40°C waterbath.

7. At 5-minute intervals over the next 15 minutes, check the tubes and record the color of the solutions in table 6-1.

NOTE

If your laboratory has a spectrophotometer, use it to determine color changes that take place during the experiment. Follow directions for use as supplied by your instructor.

8. Remove the tubes from the waterbath and *save them for comparison with results of other experiments you will perform.*

What is the brown-colored substance that appeared in test tube I$_1$?

_____

What was the substrate for the reaction that occurred in tube I$_1$?

_____

What was the product of the reaction in tube I$_1$?

_____

What substances lacking in tubes I$_2$ and I$_3$ account for the absence of the brown-colored substance?

I$_2$ _____ I$_3$ _____

Table 6-2 Color Intensity Scale

Intensity	Tube No.	Color of Product
0	I_2, I_3	
5	I_1	

What is the purpose of having tubes I_2 and I_3?

_____

After 15 minutes, the catechol should be completely oxidized. The color of the product in tube I_1 will be considered to be a "5" on a color intensity scale of 0 to 5, while the color of the substance in tubes I_2 and I_3 will be considered to be "0." You will use this scale to make comparisons in experiments II to V. Fill in table 6-2.

Keep the contents in tubes I_1, I_2, and I_3, and refer to them in making comparisons in subsequent experiments.

II. Enzyme Specificity

Generally, enzymes are substrate-specific, acting on one particular substrate or a small number of structurally similar substrates. This specificity is due to the three-dimensional structure of the enzyme. In order for the enzyme-substrate complex to form, the structure of the substrate must complement very closely that of the **active site** of the enzyme. The active site is a special region of the enzyme to which the substrate binds. The active site has a small amount of moldability, so that the active site and substrate become fully complementary to each other, as illustrated in figure 6-2.

This experiment demonstrates the ability of the enzyme catechol oxidase to catalyze the oxidation of two different but structurally similar substrates: catechol and hydroquinone.

Examine the chemical structure of each compound:

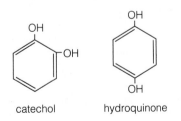

catechol   hydroquinone

You need not memorize these structural formulas, but do notice that both are ring structures with two hydroxyl (—OH) groups attached.

Keep this in mind as you do the next experiment, in which you will determine how specific (discriminating) catechol oxidase is for particular substrates.

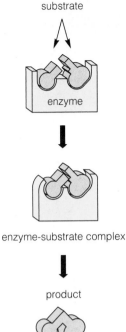

Figure 6-2 Induced-fit model of enzyme action. (After Starr and Taggart, 1989.)

MATERIALS

Per student group (4):
- 3 test tubes
- test tube rack
- metric ruler
- china marker
- wash bottle containing 1% catechol
- wash bottle containing 1% hydroquinone
- ice bath with wash bottle of potato extract containing catechol oxidase

Per lab room:
- 40°C waterbath
- vortex mixer (optional)

PROCEDURE

1. With a china marker, label three clean test tubes II_1, II_2, and II_3. Include your initials for identification.

2. Lay the test tubes against a metric ruler and mark lines indicating 1 cm and 2 cm _from the bottom_ of each test tube.

Table 6-3 Specificity of Catechol Oxidase

Time (minutes)	Relative Color Intensity on a Scale of 0 to 5		
	Tube II_1: with Catechol	Tube II_2: with Hydroquinone	Tube II_3: with dH_2O
0			
5			
10			

3. Fill each tube as follows:

Tube II_1:

• 1 cm of potato extract containing catechol oxidase

• 1 cm of 1% catechol

Tube II_2:

• 1 cm of potato extract containing catechol oxidase

• 1 cm of 1% hydroquinone

Tube II_3:

• 1 cm of potato extract containing catechol oxidase

• 1 cm of distilled water (dH_2O)

4. Gently shake the test tubes to mix the contents.

5. Compare the color intensity of the solution in each test tube *with the standards produced in experiment I* and record at Time 0 in table 6-3.

6. Place the test tubes in a 40°C waterbath.

7. Examine the test tubes after 5 and 10 minutes, recording the color intensity (scale 0 to 5) of the contents of each in table 6-3.

Upon which substrate does catechol oxidase work best, forming the most benzoquinone in the shortest amount of time?

_____

Based upon your knowledge of the structure of the two substrates, what apparently determines the specificity of catechol oxidase?

_____

_____

Why was tube II_3 included in this experiment?

_____

_____

III. Effect of Temperature on Enzyme Activity

The rate at which chemical reactions take place is largely determined by the temperature of the environment. *Generally, for every 10°C rise in temperature, the reaction rate doubles.* Within a rather narrow range, this is true for enzymatic reactions also. However, because enzymes are proteins, excessive temperature alters their structure, destroying their ability to function. When an enzyme's structure is changed sufficiently to destroy its function, the enzyme is said to be **denatured.** Most enzymatically controlled reactions have an *optimum* temperature and pH, that is, one temperature and pH where activity is maximized.

MATERIALS

Per study group (4):

• 6 test tubes

• test tube rack

• metric ruler

• china marker

• wash bottle containing 1% catechol

• ice bath with wash bottle of potato extract containing catechol oxidase

• 3 400-mL graduated beakers

• heat-resistant glove

• Celsius thermometer

Per student group (table):

• 2 hot plates *or* burner, tripod support, wire gauze, and matches or striker

• boiling chips

Per lab room:

• source of room-temperature water

• 40°C waterbath

• 60°C waterbath

• 80°C waterbath

• vortex mixer (optional)

PROCEDURE

1. Half fill one 400-mL beaker with tap water. Add a few boiling chips and turn on the hotplate to the highest temperature setting, *or,* if your lab is equipped with burners, light the burner. Bring the water to a boil and then turn the heat down so that the water just continues to boil.

2. Put 150 mL of tap water into a second beaker and then add ice to the water.

3. Half fill a third beaker with water from the source at room temperature.

4. With a china marker, label six test tubes III_1 through III_6. Include your initials for identification.

5. Lay the test tubes against a metric ruler and mark off lines indicating 1 cm and 2 cm *from the bottom* of each tube.

6. Fill each tube to the 1-cm mark with potato extract containing catechol oxidase.

7a. Place tube III_1 in the 400-mL beaker of ice water. Measure the water temperature and record here:

_____ °C.

b. Place tube III_2 in the 400-mL beaker containing room-temperature water. Room temperature is

_____ °C.

c. Place tube III_3 in the 40°C waterbath.

d. Place tube III_4 in the 60°C waterbath.

e. Place tube III_5 in the 80°C waterbath.

f. Place tube III_6 in the 400-mL beaker containing boiling water. The boiling water is at

_____ °C.

8. Allow the test tubes to remain at the various temperatures for five minutes.

9. Remove the tubes and add catechol to the 2-cm line on each. Agitate the tubes (with a vortex mixer if available) to mix the contents.

> **CAUTION**
> Wear a heat-resistant glove when handling heated glassware.

10. Record in table 6-4 the relative color intensity (scale 0 to 5) of the solution in each tube, using the standard established in experiment I. Return each tube to its respective temperature bath immediately after recording.

11. Shake frequently (by hand) all tubes over the next 15 minutes, recording in table 6-4 the relative color intensity at 5, 10, and 15 minutes after adding catechol.

12. Plot the data from table 6-4 *for the 10-minute reading* on the graph in figure 6-3.

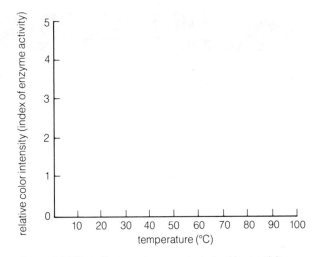

Figure 6-3 Effect of temperature on catechol oxidase activity.

What is the temperature *range* over which catechol oxidase is active?

_____

What is the *optimum* temperature for activity of this enzyme?

_____

What happens to enzyme activity at very high temperatures?

_____

_____

IV. Effect of pH on Enzyme Activity

Another factor influencing the rate of enzyme catalysis is the hydrogen-ion concentration (pH) of the solution. pH, like temperature, affects the three-dimensional shape of enzymes, thus regulating their function. Most enzymes operate best when the pH of the solution is near neutrality (pH 7). Others, however,

Table 6-4 Effect of Temperature on Enzyme Action						
	Relative Color Intensity on a Scale of 0 to 5					
Time (minutes)	Tube III_1 _____°C	Tube III_2 _____°C	Tube III_3 40°C	Tube III_4 60°C	Tube III_5 80°C	Tube III_6 _____°C
0						
5						
10						
15						

Table 6-5 Effect of pH on Enzyme Activity

Time (minutes)	Relative Color Intensity on a Scale of 0 to 5						
	Tube IV_1 pH 2	Tube IV_2 pH 4	Tube IV_3 pH 6	Tube IV_4 pH 7	Tube IV_5 pH 8	Tube IV_6 pH 10	Tube IV_7 pH 12
0							
5							
10							
15							

have pH optima in the acidic or basic range, corresponding to the environment in which they normally are found.

This experiment will allow you to determine the pH optimum of catechol oxidase.

MATERIALS

Per student group (4):

• 7 test tubes

• test tube rack

• metric ruler

• china marker

• wash bottle containing 1% catechol

• ice bath with wash bottle of potato extract containing catechol oxidase

Per lab room:

• 40°C waterbath

• phosphate buffer series, pH 2–12 (2, 4, 6, 7, 8, 10, 12)

• vortex mixer (optional)

PROCEDURE

1. With a china marker, label seven test tubes IV_1 through IV_7. Include your initials for identification.

2. Lay the test tubes against a metric ruler and mark lines indicating 4 cm, 5 cm, and 6 cm *from the bottom* of each tube.

3. Take your test tubes to the location of the phosphate buffer series and fill each tube according to the following directions:

Tube Number	Fill to the 4-cm Mark with Buffer of
1	pH 2
2	pH 4
3	pH 6
4	pH 7
5	pH 8
6	pH 10
7	pH 12

4. Return to your work area and add 1 cm of potato extract containing catechol oxidase to each of the seven tubes (thus bringing the total volume of each to the 5-cm mark). Agitate the tubes by hand.

5. Add 1% catechol to each of the seven tubes, bringing the total volume to the 6-cm mark. Agitate the contents of the tubes, using a vortex mixer if available.

6. Record in table 6-5 at Time 0 the relative color intensity of each tube *immediately after adding the 1% catechol.*

7. Place the tubes in the 40°C waterbath.

8. Agitate the tubes frequently over the next 15 minutes. At 5-minute intervals, record in table 6-5 the relative color intensity of each tube.

9. Plot the data from table 6-5 *for your 10-minute reading* on the graph in figure 6-4.

What is the *range* of pH over which catechol oxidase catalyzes catechol to benzoquinone?

_____

What is the *optimum* pH for activity of catechol oxidase?

_____

V. Necessity of Cofactors for Enzyme Activity

Some enzymatic reactions occur only when the proper *cofactors* are present. **Cofactors** are non-protein organic molecules and metal ions that are part of the

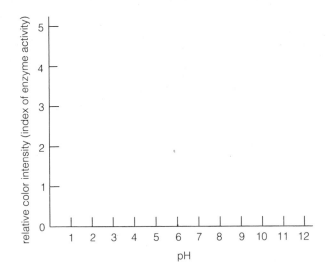

Figure 6-4 Effect of pH on catechol oxidase activity.

structure of the active site, making possible the formation of the enzyme-substrate complex.

In this experiment we will use phenylthiourea (PTU), which binds strongly to copper, to remove copper ions. Thus, we will be able to determine if copper is a cofactor necessary for producing benzoquinone from catechol.

MATERIALS

Per student group (4):

• 2 test tubes

• test tube rack

• metric ruler

• china marker

• ice bath with wash bottle of potato extract containing catechol oxidase

• wash bottle containing 1% catechol solution

• bottle of distilled water (dH₂O)

• china marker

• scoopula (small spoon)

• phenylthiourea crystals in small screw-cap bottle

Per lab room:

• 40°C waterbath

• vortex mixer (optional)

• bottle of 95% ethanol (at each sink)

• tissues (at each sink)

PROCEDURE

1. With a china marker, label two test tubes V_1 and V_2. Include your initials for identification.

2. Lay the test tube against a metric ruler and mark lines indicating 1 cm and 2 cm *from the bottom* of each tube.

Table 6-6 Is Copper a Cofactor for Catechol Oxidase?		
Time (minutes)	Relative Color Intensity on a Scale of 0 to 5	
	Tube V_1: With PTU	Tube V_2: Without PTU
0		
10		

3. Add potato extract containing catechol oxidase to the 1-cm mark of each test tube.

4. Using a scoopula, add five crystals of phenylthiourea (PTU) to tube V_1. Do not add anything to tube V_2.

CAUTION
PTU is poisonous.

5. Agitate frequently by hand the contents of both test tubes during the next five minutes.

6. Add 1% catechol to the 2-cm mark of both test tubes and agitate the contents of the tubes (with a vortex mixer if available). Record at Time 0 in table 6-6 the relative color intensities (scale of 0 to 5).

7. Place the tubes in a 40°C waterbath. Agitate the tubes several times during the next 10 minutes.

8. Remove the tubes from the waterbath and compare their relative color intensities. Record your observations in table 6-6.

Did benzoquinone form in tube V_1?

_____

In tube V_2?

_____

From this experiment, what can you conclude about the necessity for copper for catechol oxidase activity?

_____

What substance used in this experiment contained copper?

_____

NOTE
After completing all experiments, take your dirty glassware to the sink and wash it following directions on page x. Use 95% ethanol to remove the china marker. Invert the test tubes in the test tube racks so they drain. Tidy up your work area, making certain all materials used in this exercise are there for the next class.

_____ 1. Enzymes are (a) biological catalysts, (b) agents that speed up cellular reactions, (c) proteins, (d) all of the above.

_____ 2. Enzymes function by (a) being consumed (used up) in the reaction, (b) lowering the activation energy of a reaction, (c) combining with otherwise toxic substances in the cell, (d) adding heat to the cell to speed up the reaction.

_____ 3. The substance that an enzyme combines with is (a) another enzyme, (b) a cofactor, (c) a coenzyme, (d) the substrate.

_____ 4. Enzyme specificity refers to the (a) need for cofactors for the function of some enzymes, (b) fact that enzymes catalyze one particular substrate or a small number of structurally similar substrates, (c) effect of temperature on enzyme activity, (d) effect of pH on enzyme activity.

_____ 5. For every 10°C rise in temperature, the rate of most chemical reactions will (a) double, (b) triple, (c) increase by 100 times, (d) stop.

_____ 6. When an enzyme becomes denatured, it (a) increases in effectiveness, (b) loses its requirement for a cofactor, (c) forms an enzyme-substrate complex, (d) loses its function.

_____ 7. An enzyme may lose its ability to function because of (a) excessively high temperatures, (b) a change in its three-dimensional structure, (c) a large change in the pH of the environment, (d) all of the above.

_____ 8. pH is a measure of (a) an enzyme's effectiveness, (b) enzyme concentration, (c) the hydrogen-ion concentration, (d) none of the above.

_____ 9. Catechol oxidase (a) is an enzyme found in potatoes, (b) catalyzes the production of catechol, (c) has as its substrate benzoquinone, (d) is a substance that encourages the growth of microorganisms.

_____ 10. The relative color intensity used in the experiments of this exercise (a) is a consequence of production of benzoquinone; (b) is an index of enzyme activity; (c) may differ depending on the pH, temperature, or presence of cofactors, respectively; (d) all of the above.

E X E R C I S E   6

Enzymes: Catalysts of Life

P O S T - L A B   Q U E S T I O N S

1. Eggs may contain bacteria such as *Salmonella*. Considering what you've learned in this exercise, explain how cooking eggs makes them safe to eat.

2. Explain what happens to catechol oxidase when the pH is on either side of the optimum.

3. As you demonstrated in this experiment, high temperatures inactivate catechol oxidase. How is it that some bacteria live in the hot springs of Yellowstone Park at temperatures as high as 73°C?

4. What would you expect the pH optimum to be for an enzyme secreted into your stomach?

5. Why do you think high fevers alter cellular functions?

6. Some surgical procedures involve lowering a patient's body temperature during periods when blood flow must be restricted. What effect might this have on enzyme-controlled cellular metabolism?

7. At one time it was believed that individuals who had been submerged under water for longer than several minutes could not be resuscitated. Recently this has been shown to be false, especially if the person was in cold water. Explain why cold-water "drowning" victims might survive prolonged periods under water.

8. Is it necessary to have one enzyme molecule for every substrate molecule that needs to be catalyzed? Why or why not?

9. Explain the difference between *substrate* and *active site*.

10. Name the substrate for the following enzymes:

a. urease _____

b. sucrase _____

c. DNA polymerase _____

d. ATPase _____

Photosynthesis: Capture of Light Energy

OBJECTIVES

After completing this exercise you will be able to:

1. define *photosynthesis, autotroph, heterotroph, chlorophyll, chromatogram, absorption spectrum, carotenoid;*

2. describe the role of carbon dioxide in photosynthesis;

3. determine the effect of white and colored light on the rate of photosynthesis;

4. determine the wavelengths absorbed by pigments;

5. identify the pigments in spinach chloroplast extract;

6. identify the carbohydrate produced in geranium leaves during photosynthesis;

7. identify the structures composing the chloroplast and indicate the function of each structure in photosynthesis.

INTRODUCTION

Photosynthesis, the process by which light energy converts inorganic compounds to organic substances with the subsequent release of elemental oxygen, may very well be the most important biological event sustaining life. Without it, most living things would starve, and atmospheric oxygen would become depleted to a level incapable of supporting animal life. Ultimately, the source of light energy is the sun, although on a small scale we may substitute artificial light.

Nutritionally, two types of organisms exist in our world, autotrophs and heterotrophs. **Autotrophs** (*auto* means self, *troph* means feeding) synthesize organic molecules (carbohydrates) from inorganic carbon dioxide. The vast majority of autotrophs are photosynthetic organisms with which you are familiar — plants, as well as some protistans. These organisms use light energy to produce carbohydrates. (A few bacteria are able to produce their organic carbon compounds chemosynthetically, that is, using chemical energy.)

By contrast, **heterotrophs** must rely directly or indirectly on autotrophs for their nutritional carbon and metabolic energy. Heterotrophs include animals, fungi, many protistans, and most bacteria.

In both autotrophs and heterotrophs, carbohydrates originally produced by photosynthesis are broken down by *cellular respiration* (Exercise 8), releasing the energy captured from the sun for metabolic needs.

The photosynthetic reaction can be conveniently summarized by the equation

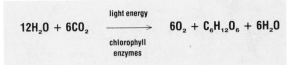

$$2H_2O + CO_2 \xrightarrow[\substack{\text{chlorophyll} \\ \text{enzymes}}]{\text{light energy}} O_2 + (CH_2O) + H_2O$$
(water) (carbon dioxide)   (oxygen) (carbohydrate) (water)

If the carbohydrate produced is glucose ($C_6H_{12}O_6$), the reaction becomes

$$12H_2O + 6CO_2 \xrightarrow[\substack{\text{chlorophyll} \\ \text{enzymes}}]{\text{light energy}} 6O_2 + C_6H_{12}O_6 + 6H_2O$$

Although glucose is often produced during photosynthesis, it is usually converted to another transport or storage compound unless it is to be used immediately for carbohydrate metabolism. In plants and many protistans, the most common storage carbohydrate is *starch*, a compound made up of numerous glucose units linked together. Starch is designated by the chemical formula $(C_6H_{12}O_6)_n$, where *n* indicates a large number. Most plants transport carbohydrate as sucrose.

The following experiments will acquaint you with the principles of photosynthesis.

I. Absorption of Light by Chloroplast Extract

We tend to think of sunlight as being white. However, as you will see in this experiment, white light consists of a continuum of wavelengths. If we see light of just one wavelength, that light will appear colored.

When light hits a pigmented surface, some of the wavelengths are absorbed and others are reflected or transmitted. This experiment demonstrates *which* wavelengths are absorbed, transmitted, or reflected by *particular* pigments, among them the photosynthetic pigment **chlorophyll**.

MATERIALS

Per lab room:

• several spectroscope setups (fig. 7-1a)

• sets of colored pencils (violet, blue, green, yellow, orange, red)

• colored filters (blue, green, red)

• small test tube containing pigment extract

Per student:

• hand-held spectroscope (optional; fig. 7-1b)

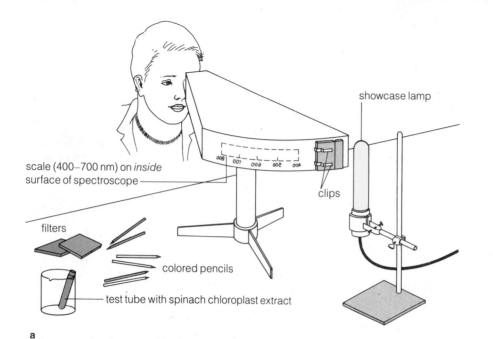

scale (400–700 nm) on *inside* surface of spectroscope

showcase lamp

clips

filters

colored pencils

test tube with spinach chloroplast extract

a

b

Figure 7-1 Use of a spectroscope. (**a**) Table-mounted. (After Abramoff and Thomson, 1982.) (**b**) Hand-held. (Photo by J. W. Perry.)

PROCEDURE

Work alone for this experiment.

One means of separating light into its component parts is by viewing the light through a spectroscope. The spectroscope contains a prism that causes the formation of a spectrum of colors. A nanometer scale is imposed upon the spectrum to indicate the wavelength of each component of white light.

1. Observe the spectrum of white light given off by an incandescent bulb (showcase lamp) through the spectroscope (see fig. 7-1). With the colored pencils provided, record the positions of the colors violet, blue, green, yellow, orange, and red on the scale of figure 7-2.

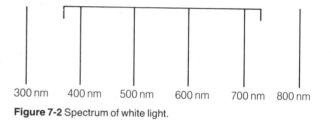

300 nm   400 nm   500 nm   600 nm   700 nm   800 nm

Figure 7-2 Spectrum of white light.

2. Observe the spectrum produced by the three colored filters using the spectroscope.

Which color(s) is (are) absorbed when a red filter is placed between the light and the prism?

_____

When a blue filter is used?

_____

A green filter?

_____

Make a general statement concerning the color of a pigment (filter) and the absorption of light by that pigment.

_____

_____

3. Now obtain a small test tube containing spinach chloroplast pigment extract and place it between the light source and the spectroscope. By adjusting the height of the tube so that the upper portion of the light passes through the pigment extract and the lower portion is white light, you can compare the absorption spectrum of the pigment extract with the spectrum of white light. An **absorption spectrum** is a spectrum of light waves absorbed by a particular pigment.

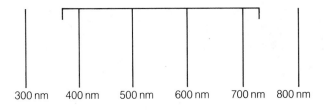

300 nm 400 nm 500 nm 600 nm 700 nm 800 nm

Figure 7-3 Transmission spectrum of chloroplast extract.

By contrast, the wavelengths that pass through the pigment extract and are visible in the spectroscope make up the *transmission spectrum* of the pigment.

4. Using the colored pencils, record the transmission spectrum of the chloroplast extract on the scale in figure 7-3.

How does the absorption spectrum of the chloroplast extract compare with the absorption spectrum of the green filter?

_____

How might you explain the difference in absorption by the green filter and by the chloroplast pigment extract? (You might want to do Part III before answering this question.)

_____

_____

II. Role of Light and Carbon Dioxide in Photosynthesis: Leaf Disk Assay*

MATERIALS

Per student:
• 2 10-mL syringes
• china marker
• soda straw punch
• several actively growing young plants
• 50-mL beaker
• Roscolux® light-filter cylinders or material

Per lab bench:
• bottle of sodium bicarbonate (NaHCO₃) solution
• transparent adhesive tape (optional)

Per lab room:
• source of distilled water (dH₂O)
• light bank
• heat absorber (optional)

_____

*Adapted with permission from Steucek, G. L., R. J. Hill, and Class/Summer 1982, 1985. *Am. Biol. Teacher* 471:96–99.

PROCEDURE

Work alone for this experiment.

Your instructor will assign an experimental light variable (white light or colored Roscolux® filter) to different students. Thus, some will use no filter, while others will use red, green, or blue filters.

> **NOTE**
> **Read through the entire set of directions before starting the experiment.**

1. Obtain two clean 10-mL syringes. With a china marker, number them "1" and "2" and add your initials. The syringes will be used as miniature photosynthesis chambers.

2. Separate the plungers from the bodies of the syringes.

3. Using the punch made from a small-diameter soda straw, cut 10 leaf disks from young, actively growing leaves by supporting the leaf with your index finger while pressing and using a twisting motion of the straw (fig. 7-4a).

4. Blow the 10 disks into the body of Chamber No. 1 (fig. 7-4b) and replace the plunger. Be careful not to damage the leaf disks as you depress the plunger.

5. Obtain about 20 mL of distilled water (dH₂O) in a clean 50-mL beaker. Insert the tip of Chamber No. 1 into the dH₂O and withdraw about 8 mL (8 cc) into the chamber. The leaf disks should be floating at this time (fig. 7-4c).

6. Hold the chamber tip-upward and, by depressing the plunger, expel the air from the chamber, as a health-care provider would before administering an injection.

7. Seal the tip of the chamber using your index finger and pull back on the plunger to create a partial vacuum within the chamber (fig. 7-4d). You will notice considerable resistance to the withdrawal of the plunger if you have a good seal. You should see air bubbles coming from the edge of the leaf disks.

8. *Simultaneously,* release your index finger and the plunger (fig. 7-4e).

9. Repeat steps 7 and 8 until all leaf disks sink. It may help to lightly rap the chamber on the edge of your desk or hand to remove air bubbles sticking to the edge of the leaf disks.

10. Repeat steps 3–9 with 10 more disks and Chamber No. 2, but rather than using dH₂O in step 5, use sodium bicarbonate (NaHCO₃) from the stock bottle. (Pour about 20 mL into the 50-mL beaker.) NaHCO₃ is a source of carbon dioxide (CO₂).

11. If you have been assigned a colored filter as a light variable, place that filter around both chambers. (If the colored filter cylinder has not been provided, simply wrap the filter cellophane around the body of the chamber and tape with transparent adhesive tape.)

a Using the straw, cut leaf disks from leaves of five or six day-old plants.

b Remove cap from the tip of the syringe. Pull the plunger out of the syringe. Blow the leaf disks out of the straw into the syringe. Replace the plunger.

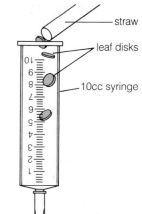

straw

leaf disks

10cc syringe

c Draw 8ml of NaHCO₃ solution into the syringe. Invert syringe as shown, tip-end up. Gently push the plunger to remove all the air.

d Put your finger over the syringe tip and pull the plunger. This will create a vacuum, which will pull the air and oxygen from the leaf disks.

e Tip the end of the syringe down so that the leaf disks are in the solution. Release plunger; remove your finger. Turn syringe back up and tap the side repeatedly until all of the disks sink.

f Place the syringe narrow-end up about 5cm from the light-bank lights. Tap the syringe with your finger or on edge of desk to dislodge the floating disks.

light bank

Figure 7-4 Setting up photosynthesis chambers.

12. Set both chambers, tips up, 5 cm from the fluorescent light bank (fig. 7-4f); or, if using an incandescent floodlamp, set both chambers immediately behind the heat absorber (transparent container filled with water). Be certain the heat absorber is 30 cm from the floodlamp.

As photosynthesis proceeds, oxygen accumulates in the intercellular spaces of a plant. This floating leaf disk assay for photosynthesis utilizes the *rate* at which oxygen is produced or consumed as a measure of the rate at which photosynthesis is occurring as a whole. In this experiment, you have first *removed* that oxygen

	My Chamber 1 ($NaHCO_3$)			My Chamber 2 (dH_2O)			Class Chamber 1 ($NaHCO_3$)			Class Chamber 2 (dH_2O)	
ET*	NDF**		%***	NDF		%	ET	NDF		%	NDF	%
2							2					
4							4					
6							6					
8							8					
10							10					
12							12					
14							14					
16							16					
18							18					
20							20					

* = Elapsed time (minutes)
** = Number disks floating
*** = NDF/10 $\times$ 100%

by vacuum infiltration, replacing it with a liquid. Thus, although leaf disks initially float, after infiltration they sink to the bottom of the chamber.

As photosynthesis takes place within the leaf disks, they rise once again.

13. Based upon the observations you made in part I and the general equation for photosynthesis in the introduction of this exercise, make a research hypothesis concerning the response you expect in both chambers.

Hypothesis: _____

_____

14. At 2-minute intervals for 20 minutes, invert the chambers to agitate the leaf disks and then immediately return them under the lights to their previous orientation (tip upward). Count the total number of disks that are floating every two minutes. The time required for a leaf disk to float is an index of the net rate of photosynthesis (oxygen produced by photosynthesis minus oxygen consumed by aerobic respiration). Record your data in table 7-1.

15. At the end of the 20-minute experiment, pool your data with all other students who used the same light variable as you and calculate the average number of leaves floating at each 2-minute interval. As a scientist, why would you do this rather than rely on your experiment alone for data and drawing conclusions?

_____

16. Your instructor has reproduced a graph similar to that in figure 7-5 on the chalkboard or overhead projector. One student representing each light variable should plot the data points on the graph with appropriately colored chalk or markers, using an "●" for Chamber 1 (dH_2O) and a "+" for Chamber 2 ($NaHCO_3$).

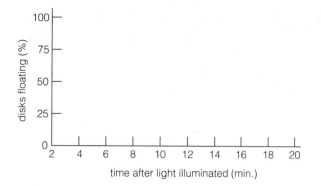

Figure 7-5 Floating leaf disk assay for photosynthesis.

17. Using colored pencils (black for white light), reproduce the class's aggregate data in figure 7-5.

Was your hypothesis confirmed? _____

Explain the differences in results between those disks in dH$_2$O and NaHCO$_3$.

_____

_____

Explain the differences in results with different colors of light illuminating the leaf disks.

_____

_____

III. Separation of Photosynthetic Pigments by Paper Chromatography

Paper chromatography allows substances to be separated from one another on the basis of their physical characteristics. A chloroplast pigment extract has been prepared for you by soaking spinach leaves in cold acetone and ethanol. Although the extract appears green, other pigments present may be masked by the chlorophyll. In this experiment, you will use paper chromatography to separate any pigments present. Separation occurs due to the solubility of the pigment in the chromatography solvent and the affinity of the pigments for absorption to the paper surface. The finished product, showing separated pigments, is called a **chromatogram.**

MATERIALS

Per student:

• chromatography paper, 3-cm-by-15-cm sheet
• metric ruler

Per student pair:

• chloroplast pigment extract in foil-wrapped dropping bottle
• chromatography chamber containing solvent
• colored pencils (green, blue-green, yellow, orange)

PROCEDURE

1. Obtain a 3-cm-by-15-cm sheet of chromatography paper. *Touch only the edges of the paper,* because oil from your fingers may interfere with development of the chromatogram.

2. Using a ruler, make a *pencil* line (do *not* use ink) about 2 cm from the bottom of the paper.

3. Load the paper by applying a droplet of the chloroplast pigment extract near the center of the pencil line. Allow the pigment spot to dry for about thirty seconds. Several applications of extract on the same spot are necessary to get enough pigment for a good chromatogram. Be certain to allow the pigment to air-dry between applications.

4. Insert your "loaded" chromatography paper, spot-side down, into a chromatography chamber—a bottle containing a solvent consisting of 10% acetone in petroleum ether. The level of the solvent should cover the bottom of the strip but no portion of the pigment spot. Seal the chromatography chamber and allow the solvent to rise on the paper. (Two chromatograms can be inserted in a single bottle, but attempt to keep each separate.)

> **CAUTION**
> Avoid inhaling the solvent vapors. Keep the chamber tightly capped whenever possible.

5. Watch the separation take place over the next ten minutes. When the solvent is within about 1 cm of the top of the paper, the separation is complete. Remove the strip, close the chromatography chamber, and allow the chromatogram to dry.

6. Using colored pencils, record the results as a sketch in figure 7-6, showing the relative position of the colors along the paper. Beginning nearest the original pigment spot, identify and label the yellow-green pigment **chlorophyll b.** Moving upward, find the blue-green **chlorophyll a,** two yellow-orange **xanthophylls** in the middle, and an orange **carotene** at the top. Xanthophylls and carotenes belong to the class of pigments called **carotenoids.**

pigment extract

pencil line

Labels: chlorophyll b, chlorophyll a, xanthophyll, carotene

Figure 7-6 Chloroplast pigment chromatogram.

You may preserve your chromatogram for future reference by keeping it in a dark place (for example, between the pages of your textbook). Light causes the chromatogram to fade.

What pigments are contained within the chloroplasts of spinach leaves?

_____

What common "vegetable" is particularly high in carotenes?

_____

(Have you ever heard of a medical condition called "carotenosis"? If not, go to the library to look up this term in a medical dictionary.)

IV. Relationship Between Light and Photosynthetic Products

As indicated by the overall formula of photosynthesis, one end product is a carbohydrate (CH_2O). But a number of different carbohydrates have the empirical formula CH_2O. In this experiment, you will perform a test to determine the specific carbohydrate stored in photosynthesizing geranium leaves.

MATERIALS

Per student group (4):
- china marker
- 4 400-mL graduated beakers
- hot plate
- heat-resistant glove
- 2 20-by-150-mm test tubes
- 25-mL graduated cylinder
- iodine (I_2KI) solution in dropping bottle
- bottle of starch solution
- bottle of 95% ethanol
- forceps
- 2 petri dishes
- I_2KI solution in foil-wrapped stock bottle

Per lab room:
- source of distilled water (dH_2O)
- light-grown geranium plant or leaves of geranium plant with "masks"
- dark-grown geranium plant or leaves of geranium plant with "masks" (kept in dark place)

PROCEDURE

Work in groups of four.

1. With a china marker, label two 400-mL beakers, one "L" (for leaf kept in light), the other "D" (for leaf placed in dark). Add 150 mL of tap water to each beaker, set them on a hotplate, and turn on the hotplate to the highest setting. Allow the water to come to a boil. (You will use the boiling water a bit later.)

2. Obtain two clean 20-by-150-mm test tubes. With a graduated cylinder, measure out 5 mL of distilled water, add it to one test tube, and then add 3 drops of iodine (I_2KI) solution.

3. Measure out 5 mL of starch solution, add it to the second test tube, and then add 3 drops of I_2KI.

4. Record the results of this test in table 7-2.

What does the blue-black color indicate?

_____

_____

Table 7-2 Reaction of I_2KI with Water and Starch	
Solution	**Color After Addition of I_2KI**
Water	
Starch	

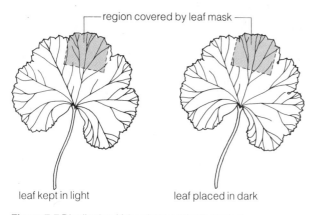

region covered by leaf mask

leaf kept in light   leaf placed in dark

Figure 7-7 Distribution of the photosynthetic product _____. (fill in this blank)

5. Select a leaf from each of two geranium plants. One plant has been growing in bright light for several hours; the other has been kept in the dark for a day or more. Both leaves should have had an area of the lamina (blade) masked by an opaque design.

6. Remove the opaque cover. Place each leaf for about thirty seconds in the appropriately labeled beaker of boiling water. Using a heat-resistant glove, remove the beaker from the hotplate and pour the water down the drain.

7. Pour about 100 mL of 95% ethanol into each of the two beakers, place on the hotplate, and bring the alcohol to a gentle boil.

> **CAUTION**
>
> Ethanol is highly flammable. Use only electric hotplates, never open flame.

8. When the pigments have been extracted (one to two minutes), remove the leaves from the alcohol with forceps and place them in two appropriately labeled petri dishes containing water.

9. After 60 seconds, gently pour the water off into one of the beakers and flood the leaves with I_2KI solution *from the larger stock bottle.*

10. After several minutes, pour the I_2KI from the beakers into the sink, rinse the leaves in cold water, and observe the pattern of staining. Show the distribution of the stain by shading in and labeling figure 7-7. In the blank provided in the legend for figure 7-7, record the *substance* that I_2KI stains.

What does the blue-black coloration of the leaf indicate?

_____

_____

_____

Why did the masked area fail to stain?

_____

_____

V. Structure of the Chloroplast

Work individually.

The chloroplast is the organelle concerned with photosynthesis. Study figure 7-8, an artist's conception of the three-dimensional structure of a chloroplast.

Like the mitochondrion and the nucleus, the chloroplast is surrounded by two membranes. Within the **stroma** (semifluid matrix), identify the **thylakoid disks** stacked into **grana** (a single stack is a **granum**). The chloroplast pigment molecules are located on the surface of the thylakoid disks. It is within the interior of the disks where hydrogen ion buildup occurs. As these ions are expelled back into the stroma, ATP is formed. Within the stroma, the ATP is used to generate organic compounds. These compounds may be converted to carbohydrates, lipids, and amino acids from carbon dioxide, water, and other raw materials.

Now examine figure 7-9, a high-magnification electron micrograph of a chloroplast. With the aid of figure 7-8, label the electron micrograph.

If the plant is killed and fixed for electron microscopy after being exposed to strong light, the chloroplasts will contain **starch grains.** Note the large starch grain present in this chloroplast. (Starch grains appear as ellipsoidal white structures in electron micrographs.)

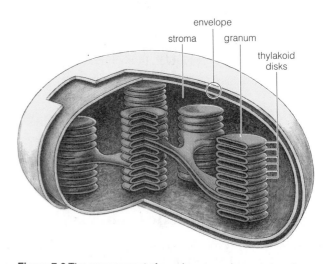

envelope

stroma   granum

thylakoid disks

Figure 7-8 The arrangement of membranes and compartments inside a chloroplast. (After Wolfe, 1985.)

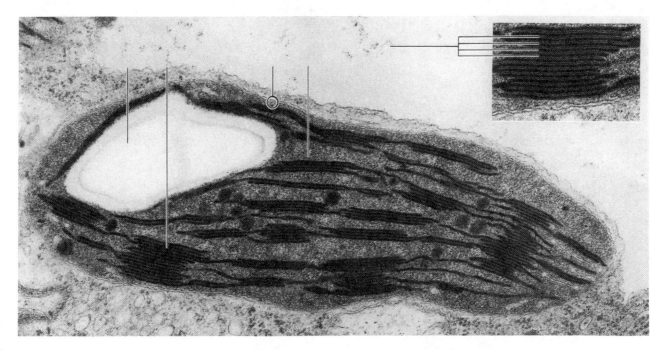

Labels: chloroplast membrane, thylakoid disks, granum, stroma, starch

Figure 7-9 Electron micrograph of chloroplast (10,000 ×). Inset: a single granum (20,000 ×). (Photo courtesy R. R. Duke.)

O P T I O N A L

VI. Experiment: An Investigation of Photosynthesis and Assimilate Transport

Your instructor may provide you with an additional experiment about photosynthesis and carbohydrate transport.

PRE-LAB QUESTIONS

_____ 1. The raw materials used for photosynthesis include (a) O_2, (b) $C_6H_{12}O_6$, (c) $CO_2 + H_2O$, (d) CH_2O.

_____ 2. A device useful for viewing the spectrum of light is a (a) spectroscope, (b) volumeter, (c) chromatogram, (d) chloroplast.

_____ 3. Which of the following is *not* true of the floating leaf disk assay for photosynthesis? (a) it is a direct means for measuring the amount of carbohydrate produced during photosynthesis; (b) it utilizes oxygen production as an indication of photosynthesis; (c) the number of floating leaves is an indication of the rate of photosynthesis; (d) none of the above.

_____ 4. A paper chromatogram is useful for (a) measuring the amount of photosynthesis, (b) determining the amount of gas evolved during photosynthesis, (c) separating pigments based upon their physical characteristics, (d) determining the distribution of chlorophyll in a leaf.

_____ 5. Which of the following pigments would you find in a geranium leaf? (a) chlorophyll, xanthophyll, phycobilins; (b) chlorophyll a, chlorophyll b, carotenoids; (c) phycocyanin, xanthophyll, fucoxanthin; (d) carotenoids, chlorophylls, phycoerythrin.

_____ 6. Which reagent would you use to determine the distribution of the carbohydrate stored in leaves? (a) starch, (b) Benedict's solution, (c) chlorophyll, (d) I_2KI.

_____ 7. An example of a heterotrophic organism is (a) a plant, (b) a geranium, (c) a human, (d) none of the above.

_____ 8. Organisms capable of producing their own food are known as (a) autotrophs, (b) heterotrophs, (c) omnivores, (d) herbivores.

_____ 9. Grana are (a) the same as starch grains, (b) the site of ATP production within chloroplasts, (c) part of the outer chloroplast membrane, (d) contained within mitochondria and nuclei.

_____ 10. The ultimate source of energy trapped during photosynthesis is (a) CO_2, (b) H_2O, (c) O_2, (d) sunlight.

PHOTOSYNTHESIS: CAPTURE OF LIGHT ENERGY

EXERCISE 7

Photosynthesis: Capture of Light Energy

POST-LAB QUESTIONS

1. Would you illuminate your house plants with a green light bulb? Why or why not?

2. Examine the figure below, showing the location of starch in two geranium leaves treated in much the same way as you did in Part IV of your experiments. Explain the results you see.

(Photo by H. M. Clarke.)

3. Explain the statement: "Without autotrophic organisms, heterotrophic life would cease to exist."

4. Why do you suppose that a chloroplast kept in darkness for some time prior to being fixed for electron microscopy does not contain starch?

5. With the results of the preceding experiments in mind, what might you do to increase the vigor of your house plants?

6. Below is a photograph taken through a spectroscope. What color was the pigment extract used to produce this spectrum? What color(s) did this extract absorb?

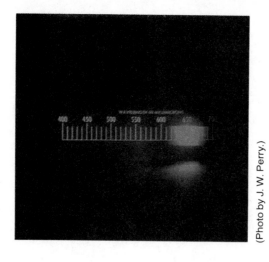

(Photo by J. W. Perry.)

7. Examine the electron micrograph below.

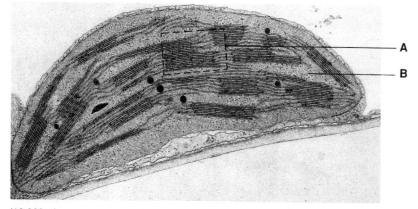

(18,900 ×).

(Photo by W. P. Wergin, courtesy E. H. Newcomb.)

a. What is this structure?

b. Identify the stack of membranes labeled A.

c. Identify the region labeled B.

d. Would the production of organic compounds during the light-independent reactions occur in region B or on the membranes labeled A?

e. Would you expect the plant in which this structure was found to have been illuminated with strong light immediately before it was prepared for electron microscopy? Why or why not?

8. Examine the photograph below of the leaf of *Coleus*. Describe an experiment that would allow you to determine if the deep purple portion of the leaf is photosynthesizing.

(Photo by J. W. Perry.)

9. Examine the photograph below of a water plant that has been illuminated inside a cylinder containing a CO_2-rich solution. Identify the chemical composition of the numerous bubbles that you see.

(Photo by J. W. Perry.)

10. The reason for the extinction of the large dinosaurs that once roamed our planet has long been a subject of conjecture. Recently, evidence has been found of the impact of a large meteor with the Earth at about the time of this mass extinction. We believe the amount of dust and debris put into the atmosphere upon impact was enormous. Utilizing your knowledge of photosynthesis, speculate as to why the dinosaurs became extinct.

EXERCISE 8
Respiration: Energy Conversion

OBJECTIVES

After completing this exercise you will be able to:

1. define *metabolism, reaction, metabolic pathway, respiration, ATP, phosphorylation, exergonic reaction, endergonic reaction, glycolysis;*

2. give the overall balanced equations for aerobic respiration and alcoholic fermentation;

3. distinguish between the products and efficiency of aerobic respiration and fermentation;

4. identify the structures and list the functions of each part of a mitochondrion;

5. explain the relationship between photosynthesis and respiration.

INTRODUCTION

The first law of thermodynamics states that energy can neither be created nor destroyed, only converted from one form to another. Because all living organisms have a constant energy requirement, they have mechanisms to gather, store, and use energy. Collectively, these mechanisms are called **metabolism.** A specific metabolic step is a **reaction,** and a sequence of such reactions is a **metabolic pathway.**

During Exercise 7, we investigated the metabolic pathways by which green plants capture light energy and use it to make carbohydrates such as glucose. Carbohydrates are temporary energy stores. The process by which energy stored in carbohydrates is released to the cell is **respiration.**

The energy needed for living processes is stored in the chemical bonds holding carbohydrate atoms together. However, the cell cannot *directly* use the chemical bond energy of carbohydrates. Rather, the energy must be converted by a metabolic pathway to form **adenosine triphosphate (ATP),** the so-called *universal energy currency* of the cell. The bond energy of carbohydrates is transferred to ATP during **phosphorylation,** the addition of a phosphate group to adenosine *di*phosphate (ADP). When the bond holding this new phosphate group is broken during respiration, the energy released is available for a great variety of cellular reactions. Thus, the needs of a cell are linked by the energy(ATP)-releasing **exergonic reactions** of respiration to the energy(ATP)-requiring **endergonic reactions.** This last group of reactions is important for maintaining or synthesizing cellular structures, or for doing cellular work.

Both autotrophs and heterotrophs undergo respiration. Autotrophs utilize the carbohydrates they have produced to build new cells and maintain cellular machinery. Heterotrophic organisms may obtain materials for respiration two ways: by digesting plant material or by digesting the tissues of animals that have previously digested plants.

Regardless of the type of organism or environmental conditions, the first series of metabolic steps occurring during respiration is **glycolysis.** The word *glycolysis* should be a tip-off concerning what happens during this process. The Greek word *glykos* means "sweet," referring to sugar, while *lysis* means "loosening." During glycolysis the 6-carbon sugar glucose, $C_6H_{12}O_6$, is split into two 3-carbon pyruvate molecules. This universal event occurs within the cytoplasm of all living cells, whether bacteria, protistans, fungi, plants, or animals, including humans. The net energy yield from glycolysis is two ATP per molecule of glucose.

After glycolysis, the fate of the pyruvate formed depends upon the specific organism and/or environmental conditions. Consequently, we can consider that there are four alternative pathways through which the products of glycolysis may proceed: (1) **aerobic respiration,** an oxygen-dependent pathway common in most organisms; (2) **anaerobic electron transport** (sometimes also called *anaerobic respiration*), a pathway utilized by some bacteria; (3) **alcoholic fermentation,** an ethanol-producing process occurring in some yeasts; and (4) **lactate fermentation,** a pathway taken by some bacteria as well as animal cells that normally rely on aerobic respiration, but that are subjected to oxygen-deficient conditions. (During strenuous activity, your skeletal muscles may switch to lactate fermentation for energy.)

Perhaps the most important aspect to remember about these four processes is that aerobic respiration is by far the most energy-efficient. **Efficiency** refers to the amount of energy captured in the form of ATP relative to the amount available within the bonds of the carbohydrate.

For aerobic respiration, the general equation is

$$C_6H_{12}O_6 + 6O_2 \xrightarrow{\text{enzymes}} 6CO_2 + 6H_2O + 36ATP^*$$

glucose · · · oxygen · · · carbon dioxide · · · water · · · chemical energy

If glucose is broken down completely to CO_2 and H_2O, about 686,000 calories of energy are released. Each ATP molecule represents about 7,500 calories of usable

---

*Depending upon the tissue, as many as thirty-eight ATP may be found.

energy. The 36 ATP represent 270,000 calories of energy (36 × 7,500 calories). Thus, aerobic respiration is about 39% efficient [(270,000/686,000) × 100%].

By contrast, fermentation and anaerobic electron transport yield only 2 ATP. Thus, these processes are only about 2% efficient [(2 × 7,500/686,000) × 100%]. Obviously, breaking down carbohydrates by aerobic respiration gives a bigger payback than the other means.

I. Aerobic Respiration

The fate of pyruvate molecules produced by glycolysis depends upon the organism and environmental conditions. If oxygen is abundant and the organism normally undergoes aerobic respiration, pyruvate is further metabolized by a cyclic (circular) pathway known as the *Krebs cycle,* which generates a small amount of ATP and releases CO_2. For the most part, the Krebs cycle functions to reduce (donate electrons to) special electron carriers. These electron carriers eventually become oxidized (lose electrons) during **electron transport phosphorylation,** during which large amounts of ATP are produced. The Krebs cycle and electron transport phosphorylation occur in the mitochondrion.

Why must oxygen be present? Whenever one substance is oxidized (loses electrons), another must be reduced (accept, or gain, those electrons). The final electron acceptor of electron transport phosphorylation is oxygen. Tagging along with the electrons as they pass through the electron transport process are protons (H^+). When the electrons and protons are captured by oxygen, water (H_2O) is formed:

$$2H^+ + 2e^- + 1/2O_2 \rightarrow H_2O$$

In the following experiments we examine aerobic respiration in two sets of seeds.

A. Carbon Dioxide Production

Seeds contain stored food material in the form of some carbohydrate. When a seed germinates, the carbohydrate is broken down, liberating energy (ATP) needed for growth of the enclosed embryo into a seedling.

Two days ago, one set of dry pea seeds was soaked in water to start the germination process. Another set was not soaked. This experiment will compare carbon dioxide production between germinating pea seeds, germinating pea seeds that have been boiled, and ungerminated (dry) pea seeds.

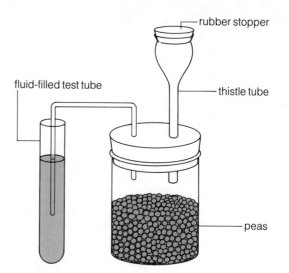

Figure 8-1 Respiration bottle apparatus.

MATERIALS

Per student group (4):
- 600-mL beaker
- hot plate *or* burner, wire gauze, tripod, and matches
- heat-resistant glove
- 3 respiration bottle apparatus (fig. 8-1)
- china marker
- phenol red solution

Per lab room:
- germinating pea seeds
- ungerminated (dry) pea seeds

PROCEDURE

Work in groups of four.

1. Place about 250 mL of tap water in a 600-mL beaker, put the beaker on a heat source, and bring the water to a boil.

2. Obtain three respiration bottle setups (fig. 8-1). With a china marker, label one "Germ" for germinating pea seeds, the second "Germ-Boil" for those you will boil, and the third "Ungerm" for ungerminated seeds.

3. From the class supply, obtain and put enough germinating pea seeds into the two appropriately labeled respiration bottles to fill them approximately halfway. Fill the third bottle half full with ungerminated (dry) pea seeds.

4. Dump the germinating peas from the "Germ-Boil" bottle into the boiling water bath; continue to boil for 5 minutes. After 5 minutes, turn off the heat source, put on a heat-resistant glove, and remove the water bath. Pour the water off into the sink and cool the boiled peas by pouring cold water into the beaker. Allow 5 minutes for the peas to cool; then pour off the water. Now replace the peas into the "Germ-Boil" respiration bottle.

Table 8-1 CO$_2$ Evolution by Pea Seeds

Pea Seeds	Indicator color (phenol red)	Conclusion (CO$_2$ present or absent)
Germinating-unboiled	_____	_____
Germinating-boiled	_____	_____
Ungerminated	_____	_____

5. Fit the rubber stopper with attached glass tubes into the respiration bottles. Add enough water to the test tube to cover the end of the glass tubing that comes out of the respiration bottle. (This keeps gases from escaping from the respiration bottle.)

6. Insert rubber stoppers into the thistle tubes.

7. Set the three bottles aside for the next 1¼ hours and do the other experiments in this exercise.

Now start the next series of experiments while you allow this one to proceed.

8. After 1¼ hours, pour the water in each test tube into the sink and replace it with an equal volume of dilute phenol red solution. Phenol red solution, which should appear pinkish in the stock bottle, will be used to test for the presence of carbon dioxide (CO$_2$) within the respiration bottles. If CO$_2$ is bubbled through water, carbonic acid (H$_2$CO$_3$) forms, as shown by the following equation:

$$CO_2 + H_2O \rightarrow H_2CO_3$$

Phenol red solution is mostly water. When the phenol red solution is basic (pH $>$ 7), it is pink; when it is acidic (pH $<$ 7), the solution is yellow. The phenol red solution in the stock bottle is

_____ (color);

therefore, the stock solution is

_____ (acidic/basic).

9. Put several hundred mL of tap water in the 600-mL beaker.

10. Remove the stopper plugging the top of the thistle tube and *slowly* pour water from the beaker into each thistle tube. The water will force out gases present within the bottles. If CO$_2$ is present, the phenol red will become yellow.

11. Record your observations in table 8-1.

Which set of seeds was undergoing respiration?

_____

What happened during boiling that caused the results you found? (Hint: enzymes.)

_____

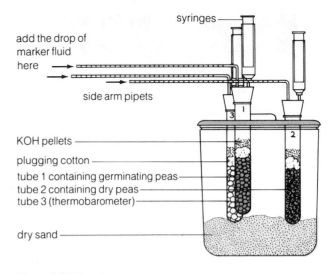

Figure 8-2 Volumeter.

B. Oxygen Consumption

One set of pea seeds has been soaked in water for the past 48 hours to initiate germination. In this experiment you will measure the respiratory rate of germinating and ungerminated seeds as determined by oxygen consumption.

MATERIALS

Per student group (4):
- volumeter (fig. 8-2)
- china marker
- 80 germinating pea seeds
- 80 ungerminated (dry) pea seeds
- glass beads
- nonabsorbent cotton
- metric ruler
- bottle of potassium hydroxide (KOH) pellets
- 1/4 teaspoon measure
- marker fluid in dropping bottle

PROCEDURE

Work in groups of four.

1. Obtain a volumeter set up as in figure 8-2. Skip to step 6 if your instructor has already assembled the volumeters as described by steps 2 through 5.

2. Remove the test tubes from the volumeter. With a china marker, number the tubes and then fill as follows:

Tube 1: 80 germinating (soaked) pea seeds

Tube 2: 80 ungerminated (dry) pea seeds plus enough glass beads to bring the total volume equal to that of Tube 1

Table 8-2 Respiratory Rate as Measured by Oxygen Consumption

Times (minutes)	Thermobarometer			Tube 1: Germinating Peas			Tube 2: Dry Peas		
	Reading	Total Change in Volume		Reading	Total Change in Volume	Total Oxygen Consumption	Reading	Total Change in Volume	Total Oxygen Consumption
0									
5									
10									
15									
20									
25									
30									
35									
40									
45									
50									
55									
60									

Tube 3: Enough glass beads to equal the volume of Tube 1. This tube serves as a thermobarometer and is used to correct experimental reading to account for changes in temperature and barometric pressure taking place during the experiment.

3. Pack cotton *loosely* into each tube to a thickness of about 1.5 cm above the peas/beads.

4. Measure out 1 cubic centimeter (cm^3) (about 1/4 teaspoon) of potassium hydroxide (KOH) pellets and pour them atop the cotton.

> **CAUTION**
>
> **Potassium hydroxide can cause burns. Do not get any on your skin or clothing. If you do, wash immediately with copious amounts of water.**

KOH absorbs carbon dioxide (CO$_2$) given off during aerobic respiration. Since the volumeter measures change in gas volume, any gas *given off* during respiration must be removed from the tube so an accurate measure of O$_2$ consumption can be made.

5. Insert the stopper-syringe assembly in place.

6. Add a small drop of marker fluid to each side arm pipet by touching the dropper to the end of each. The drop should be taken into the side arm by capillary action. Gently withdraw the plunger of each syringe and adjust the position of the drop so it is between 0.80 and 0.90 cm^3 on the scale of the graduated pipet.

7. Adjust each side arm pipet so it is parallel to the table top. Wait 5 minutes before starting data collection.

8. At time 0, record in table 8-2 the position of the marker droplet within each pipet. Record readings for each tube at 5-minute intervals for the next 60 minutes. (If respiration is rapid and the marker moves too near the end of the scale, *carefully* depress the syringe to readjust its position. The new readings are then added to the old readings as the data are being recorded.)

9. To determine total change in volume of gas within each tube, subtract each subsequent reading from the first (time 0).

10. At the end of the experiment, correct for any volume changes caused by changes in temperature or barometric pressure by using the reading obtained from the thermobarometer. If the thermobarometric marker moves *toward* the test tube (decrease in volume), *subtract* the volume change from the last total oxygen consumption measurement of the pea-containing test tubes. If the marker droplet moves *away* from the test tube (increase in volume), *add* the volume change to the last total oxygen consumption measurement for each pea-containing test tube. In figure 8-3, graph the consumption of oxygen over time.

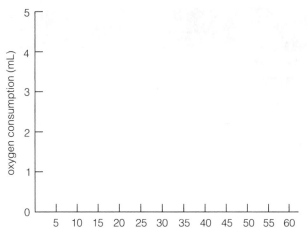

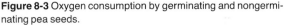

Figure 8-3 Oxygen consumption by germinating and nongerminating pea seeds.

Use a "+" for data points of germinating peas, a "●" for dry peas.

How do the respiratory rates for germinating and nongerminating seeds compare?

_____

_____

How do you account for this difference?

_____

_____

It takes 820 cm³ of oxygen to completely oxidize 1 gram of glucose. How much glucose are the germinating peas consuming per hour?

_____

II. Fermentation

Despite relatively low energy yield, fermentation provides sufficient energy for certain organisms to survive. Alcoholic fermentation by yeast is the basis for the brewing industry. It's been said that yeast and alcoholic fermentation have made Milwaukee famous.

The chemical equation for this process is:

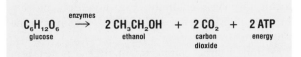

Starch (amylose), a common storage carbohydrate in plants, is a polymer consisting of a chain of repeating glucose ($C_6H_{12}O_6$) units. The polymer has the chemical formula $(C_6H_{12}O_6)_n$,* where n represents a

---

*A number of carbohydrates share this same chemical formula but differ slightly in the arrangement of their atoms. These carbohydrates are called *structural isomers*.

large number. Starch is broken down by the enzyme amylase into individual glucose units. To summarize:

$$(C_6H_{12}O_6)_n \xrightarrow{\text{amylase}} C_6H_{12}O_6 + C_6H_{12}O_6 + C_6H_{12}O_6 + \ldots$$

This experiment demonstrates the action of yeast cells on carbohydrates.

MATERIALS

Per student group (4):
• china marker
• 3 50-mL beakers
• 25-mL graduated cylinder
• bottle of 10% glucose
• bottle of 1% starch
• 0.5% amylase in bottle fitted with graduated pipet
• 3 glass stirring rods
• 1/4 teaspoon measure (optional)
• 3 fermentation tubes
• 15-cm metric ruler

Per lab room:
• 0.5-gram pieces of fresh yeast cake
• scale and weighing paper (optional)
• 37°C incubator

PROCEDURE

Work in groups of four.

1. Using a china marker, number three 50-mL beakers.

2. With a *clean* 25-mL graduated cylinder, measure out and pour 15 mL† of the following solutions into each beaker:

> **NOTE**
> **Wash the graduated cylinder between solutions.**

Beaker 1: 15 mL of 10% glucose
Beaker 2: 15 mL of 1% starch
Beaker 3: 15 mL of 1% starch; then, using the graduated pipet to measure, add 5 mL of 0.5% amylase.

3. Wait five minutes and then to each beaker add a 0.5-gram piece of fresh cake yeast. Stir with *separate* glass stirring rods.

---

†The amount of fluid needed to fill the fermentation tube depends upon its size. Your instructor may indicate the required volume.

Table 8-3 Evolution of Gas by Yeast Cells

Tube	Solution	Distance from tip of tube to fluid level (mm)			Volume* of gas evolved (mm³)
		20 min	40 min	60 min	
1	10% glucose + yeast				
2	1% starch + yeast				
3	1% starch + yeast + amylase				

*To calculate the volume of gas evolved, use the following equation: $V = \pi r^2 h$, where π = 3.14, r = radius of tail of fermentation tube ($r = \frac{1}{2}d$), h = distance from top of tail to level of solution.

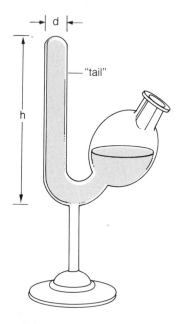

Figure 8-4 Fermentation tube.

4. When each is thoroughly mixed, pour the contents into three correspondingly numbered fermentation tubes (fig. 8-4). Cover the opening of the fermentation tube with your thumb and invert each fermentation tube so that the "tail" portion is filled with the solution.

5. Place the tubes in a 37°C incubator.

6. At intervals of 20, 40, and 60 minutes after the start of the experiment, remove the tubes and, using a metric ruler, measure the distance from the tip of the tail to the fluid level. Record your results in table 8-3. Calculate the volume of gas evolved using the formula at the bottom of table 8-3.* (If time is short, do your calculations later.)

What gas accumulates in the tail portion of the fermentation tube?

---

*Some fermentation tubes are graduated, thus making unnecessary the use of the formula in determining the volume of gas evolved.

III. Ultrastructure of the Mitochondrion

1. Study figure 8-5, an artist's interpretation of the three-dimensional structure of a mitochondrion, the respiratory organelle of all living eukaryotic cells. The mitochondrion has frequently been referred to as the "powerhouse of the cell," because most of the cell's chemical energy (ATP) is produced here.

2. Now observe figure 8-6, a high-magnification electron micrograph of a mitochondrion. Identify and label the **outer membrane** separating the organelle from the cytoplasm.

3. Note the presence of an inner membrane, folded into fingerlike projections. Each projection is called a **crista** (the plural is *cristae*). The folding of the inner membrane greatly increases the surface area upon which many of the chemical reactions of aerobic respiration take place. Label the crista.

4. Identify and label the **outer compartment,** the space between the inner and outer membranes. The outer compartment serves as a reservoir for hydrogen ions.

5. Finally, identify and label the **inner compartment** (sometimes called the *matrix*), the interior of the mitochondrion.

Now that you know the structure of the mitochondrion, you can visualize the events that take place in the production of chemical energy needed for life. Glycolysis, the first step in cellular respiration, takes place in the cytoplasm. Pyruvate and NADH formed during glycolysis enter the mitochondrion, moving through both the outer and inner membranes to the inner compartment.

Within the inner compartment, the pyruvate is broken down in the Krebs cycle, forming $FADH_2$ and NADH. A small amount of ATP also is produced during these reactions.

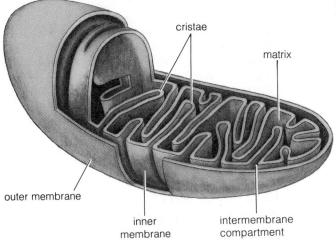

Figure 8-5 The membranes and compartments of a mitochondrion. (After Wolfe, 1985.)

cristae

matrix

outer membrane

inner membrane

intermembrane compartment

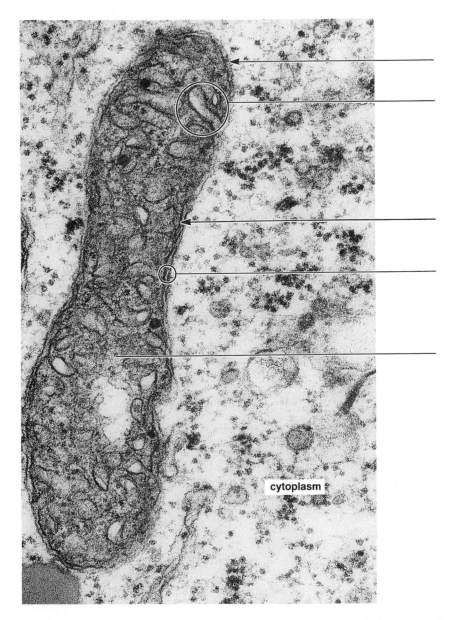

cytoplasm

Figure 8-6 Transmission electron micrograph of a mitochondrion (23,250 ×). (Photo courtesy S. E. Eichhorn.)

Labels: outer membrane, inner membrane, crista, intermembrane compartment, matrix

RESPIRATION: ENERGY CONVERSION

Electron transport phosphorylation occurs as electrons and hydrogen ions are stripped from NADH and FADH$_2$. (Note that NAD and FAD are abbreviations for coenzymes, but the H and H$_2$ attached to these coenzymes indicate the presence of attached hydrogens.) These hydrogen ions move through channel proteins embedded in the inner membrane and accumulate in the outer compartment. They flow back into the inner compartment through an enzyme called *ATP synthase*, which is embedded in the inner membrane. Energy associated with the flow drives the coupling of ADP and inorganic phosphate to form ATP. Most of the ATP generation during aerobic respiration occurs during electron transport phosphorylation.

OPTIONAL
IV. Experiment: Effect of Temperature on Goldfish Respiration Rate

Your instructor may provide you with an experiment to observe and determine the respiration rate of an animal.

PRE-LAB QUESTIONS

_____ 1. A metabolic pathway is a (a) single, specific reaction that starts with one compound and ends up with another; (b) sequence of chemical reactions that are part of the metabolic process; (c) series of events that occurs only in autotrophs; (d) all of the above.

_____ 2. The "universal energy currency" of the cell is (a) glucose, (b) C$_6$H$_{12}$O$_6$, (c) ATP, (d) H$_2$O.

_____ 3. The most efficient type of respiration is (a) fermentation, (b) anaerobic electron transport, (c) aerobic respiration, (d) lactate production.

_____ 4. When muscle cells are subjected to periods of strenuous activity, during which oxygen is not replaced within the cells as fast as it's used, the muscle cells switch from (a) lactate fermentation to aerobic respiration, (b) lactate fermentation to alcoholic fermentation, (c) alcoholic fermentation to lactate fermentation, (d) aerobic respiration to lactate fermentation.

_____ 5. The purpose of the thermobarometer in a volumeter is to (a) judge the amount of O$_2$ evolved during respiration, (b) determine the volume changes as a result of respiration, (c) indicate oxygen consumption by germinating pea seeds, (d) indicate volume changes resulting from changes in temperature or barometric pressure.

_____ 6. Phenol red is used in the experiments as (a) an O$_2$ indicator, (b) a CO$_2$ indicator, (c) a sugar indicator, (d) an enzyme.

_____ 7. If the pH of a phenol red solution is 2, (a) phenol red will be pink, (b) the solution is acidic, (c) phenol red will be yellow, (d) b and c above.

_____ 8. Which of the following enzymes breaks down starch into glucose? (a) kinase, (b) maltase, (c) fructase, (d) amylase.

_____ 9. Oxygen is necessary for life because (a) photosynthesis depends upon it, (b) it serves as the final electron acceptor during aerobic respiration, (c) it is necessary for glycolysis, (d) all of the above.

_____ 10. Yeast cells produce (a) ATP, (b) ethanol, (c) CO$_2$, (d) all of the above.

EXERCISE 8

Respiration: Energy Conversion

POST-LAB QUESTIONS

1. During aerobic respiration, glucose ($C_6H_{12}O_6$) is broken down to form several end products. Which end products contain:

   a. the carbon atoms from glucose? _____

   b. the hydrogen atoms from glucose? _____

   c. the oxygen atoms from glucose? _____

   d. the energy stored in the glucose molecules? _____

2. Compare aerobic respiration, anaerobic electron transport, and fermentation in terms of:

   a. efficiency of obtaining energy from glucose

   b. end products

3. Besides the product formed in the alcoholic fermentation experiment, what other products might be formed by other types of fermentation?

4. How much ATP is derived when one molecule of glucose goes through glycolysis?

   _____

   What does this tell you about the source of ATP obtained during fermentation and anaerobic electron transport pathways?

5. Examine the electron micrograph to the right of an organelle found in all living eukaryotic cells.

   a. Identify this organelle.

   b. What portions of aerobic respiration occur in region **b**?

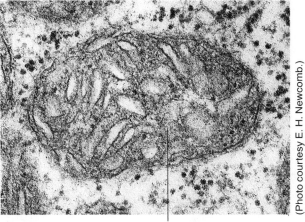

(20,000 ×).   **b**

(Photo courtesy E. H. Newcomb.)

c. What substance is produced as hydrogen ions cross from the space between the inner and outer membranes into region **b**?

d. What portion of cellular respiration takes place *outside* of this organelle in the cytoplasm?

6. How would you explain this statement: "The ultimate source of our energy is the sun"?

7. Oxygen is used during aerobic respiration. What biological process is the source of the oxygen?

8. Bread is made by mixing flour, water, sugar, and yeast to form a dense dough. Why does the dough rise? What gas is responsible for the holes in bread?

9. The first law of thermodynamics seems to conflict with what we know about ourselves. For example, after strenuous exercise we run out of "energy." We must eat to replenish our energy stores. Where has that energy gone? What form has it taken?

10. As a plant grows, not all of the carbohydrates produced by photosynthesis are stored as starch, nor are they all respired to produce ATP. What cellular constituents would be synthesized by some of the carbohydrate produced?

Mitosis and Cytokinesis: Nuclear and Cytoplasmic Division

OBJECTIVES

After completing this exercise you will be able to:

1. define *fertilization, zygote, DNA, chromosome, prokaryotic fission, mitosis, cytokinesis, nucleoprotein, sister chromatid, centromere, meristem, derivative;*

2. identify the stages of the cell cycle;

3. distinguish between mitosis and cytokinesis as they take place in animal and plant cells;

4. identify the structures involved in nuclear and cell division (those in **boldface**) and describe the role each plays.

INTRODUCTION

"All cells arise from preexisting cells." This is one tenet of the cell theory. It is easy to understand this concept when thinking of a single-celled *Amoeba* or bacterium. Each cell divides to give rise to two entirely new individuals. But it is enormously fascinating that each of us began life as *one* single cell and developed into an astonishingly complex animal. This cell has *all* the hereditary information we'll ever get.

In higher plants and animals, **fertilization,** the fusion of egg and sperm nuclei, produces a single-celled **zygote.** The zygote divides into two cells, these two into four, and so on to produce a multicellular organism. During cell division each new cell receives a complete set of hereditary information and an assortment of cytoplasmic components.

Recall from Exercise 4 that there are two basic cell types, prokaryotic and eukaryotic. The genetic material of both consists of **DNA (deoxyribonucleic acid).** In prokaryotes, the DNA molecule is organized into a single **chromosome.** Prior to cell division, the chromosome duplicates. Then the cell undergoes **prokaryotic fission,** the splitting of a preexisting cell into two, with each new cell receiving a full complement of the genetic material.

In eukaryotes, the process of cell division is more complex, primarily because of the much more complex nature of the hereditary material. Here the chromosomes consist of DNA and proteins complexed together within the nucleus. Cell division is preceded by duplication of the chromosomes and usually involves two processes: **mitosis** (nuclear division) and **cytokinesis** (cytoplasmic division). Whereas mitosis results in the production of two nuclei, both containing identical chromosomes, cytokinesis ensures that each new cell contains all the metabolic machinery necessary for sustenance of life. We will restrict further discussion within this exercise to a consideration of what occurs in eukaryotic cells only.

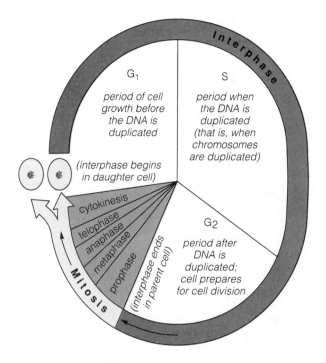

Figure 9-1 The eukaryotic cell cycle.

Dividing cells pass through a regular sequence of events called the cell cycle (fig. 9-1). Notice that the majority of the time is spent in interphase and that actual nuclear division—mitosis—is but a brief portion of the cycle.

Interphase is broken into three parts (fig. 9-1): the G_1 period, during which cytoplasmic growth takes place; the S period, when the DNA is duplicated; and the G_2 period, when structures directly involved in mitosis are synthesized.

Unfortunately, because of the apparent relative inactivity that early microscopists observed, interphase was given the misnomer "resting stage." In fact, we know now that interphase is anything but a resting period. The cell is producing new DNA, assembling proteins from amino acids, and synthesizing or breaking down carbohydrates. In short, interphase is a very busy time in the life of a cell.

I. Chromosomal Structure

During much of a cell's life, each DNA-protein complex, the **nucleoprotein,** is extended as a thin strand within the nucleus. In this form it is called **chromatin.**

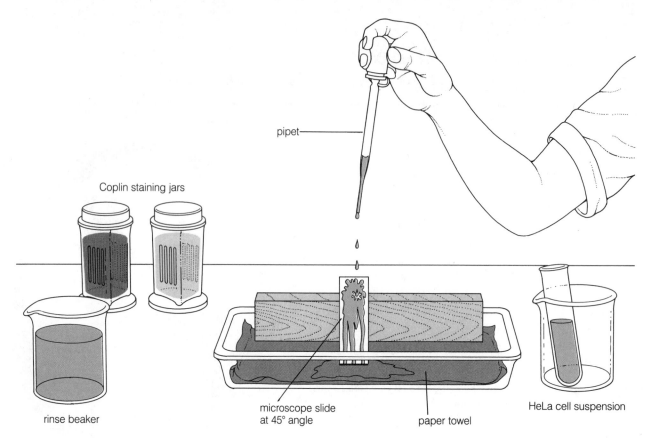

pipet

Coplin staining jars

rinse beaker

microscope slide
at 45° angle

paper towel

HeLa cell suspension

Figure 9-2 Applying HeLa cells to slide.

Prior to the onset of nuclear division, the genetic material duplicates itself, and the chromatin condenses. These two identical condensed nucleoproteins are called **sister chromatids** and are attached at the **centromere.** The centromere gives the appearance of dividing each chromatid into two "arms." Collectively, the two attached sister chromatids are referred to as a *duplicated chromosome.* Before looking at actual chromosomes, let's examine the background of the materials we'll use.

In 1951 Henrietta Lacks died from cervical cancer. Prior to her death, some of the cancerous (tumor) cells were removed from her body and grown in culture. The cells were allowed to divide repeatedly in this artificial culture environment and resulted in formation of what is known as a *cell line.* These HeLa cells (the abbreviation coming from *He*nrietta *La*cks) live on today and are used for research on cell and tumor growth. They're also useful for showing chromosomal structure in the biology laboratory.

Normal human body cells contain 23 pairs of chromosomes (46 total chromosomes). Human cells of tumor origin (such as the HeLa cells) often produce greater chromosome numbers when grown in culture. Cells containing up to 200 chromosomes have been observed, although 50 to 70 per cell are most frequently found. This portion of the exercise will allow you to observe the chromosomes of the descendant tumor cells of Henrietta Lacks.

NOTE

There is no danger from this exercise. The tumor-producing properties are nontransmissible.

MATERIALS

Per student:

• clean microscope slide prechilled in cold 40% methanol

• coverslip

• disposable pipet

• paper toweling

• compound microscope (with oil-immersion objective preferable)

Per student pair:

• metric ruler

• tube of HeLa cells

• dropper bottle containing Permount mounting medium

Per lab room:

• Coplin staining jars containing stains 1 and 2

• 1-L beaker containing distilled water (dH$_2$O)

PROCEDURE

Refer to figure 9-2.

1. Place a paper towel on your work surface.

2. Remove a clean glass microscope slide from the cold methanol and lean it against a surface at a 45° angle, with one short edge resting on the paper towel.

3. Most of the cells are at the bottom of the culture tube. Gently resuspend the cells by inserting your pipet and squeezing the bulb. This expels air from the pipet, which disperses the cells. Remove a small cell sample with the pipet. Holding the pipet about 18–36 cm above the slide, allow 8–10 drops of the cell suspension to "splat" onto the upper edge of the wet slide and to tumble down the slide.

> **NOTE**
>
> **The slide must be wet when "splatting" takes place.**

4. Allow the cells to *air dry completely.*

You will now stain the slide three times with both stains, for *one second each time.* The time in the stain is critical. Count "one thousand, one;" this will be one second.

5. Go to the location of the staining solutions and dip the slide in stain 1 for one second. Withdraw the slide and then dip twice more. Drain the slide of the excess stain, blotting the bottom edge of the slide on the paper toweling before proceeding.

6. Immediately dip the slide in stain 2 for one second. (Repeat twice more.) Drain and blot the excess stain as before.

7. Rinse the slide in distilled water (dH$_2$O) by swishing it gently back and forth in the beaker.

8. Allow the slides to *air dry completely.*

9. Place 2 drops of Permount mounting medium on the slide in the region of the stained cells. Using the technique illustrated in figure 3-7 (page 27), place a coverslip on the slide. (NOTE: Of course, you *do not* add the drop of water shown in figure 3-7.) Squeeze out excess Permount by applying gentle pressure to the coverslip with the blunt eraser end of a pencil.

> **CAUTION**
>
> **Allow Permount to dry for an hour before attempting to view it using an oil-immersion objective.**

10. Place the slide on the microscope stage and observe your chromosome spread, focusing first with the medium-power objective and then with the high-dry objective. Locate cells that appear to have burst and have the chromosomes spread out. The number of good spreads will be low; so careful observation of many cells is necessary.

Labels: chromosome, sister chromatid, centromere

Figure 9-3 Drawing of human chromosomes from HeLa cells (_____×).

> **NOTE**
>
> **If your microscope has an oil-immersion objective, proceed to step 11. If not, skip to step 12.**

11. Once a good spread has been located, rotate the high-dry objective out of the light path and place a drop of immersion oil on that spot. Rotate the oil-immersion objective into the light path.

12. Observe the structure of the chromosomes, identifying sister chromatids and centromeres. Draw several of the chromosomes in figure 9-3, labeling the required parts.

13. Select ten cells in which the chromosomes are visible and count the number of chromosomes you observe in each cell, inserting your results in table 9-1. When you have counted the chromosomes in ten individual cells, calculate the average number of chromosomes per cell.

Now examine figure 9-4, an electron micrograph of a human chromosome. Label the two chromatids, centromere, and the duplicated chromosome.

> **NOTE**
>
> **Use illustrations in your textbook to aid you in the following study.**

| Table 9-1 Chromosome Numbers in HeLa Cells | | | | | | | | | | | |
Cell No: 1	2	3	4	5	6	7	8	9	10	Average

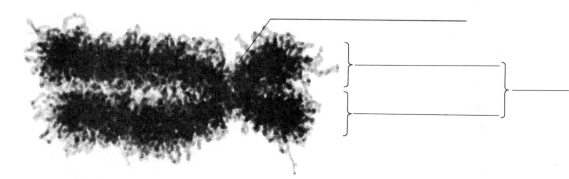

Figure 9-4 Electron micrograph of a chromosome (26,200 ×). (Photo by E. J. DuPraw.)

Labels: chromatid (2), centromere, duplicated chromosome

II. The Cell Cycle in Plant Cells: Onion Roots

Nuclear and cell divisions in plants are, for the most part, localized in specialized regions called meristems. **Meristems** are regions of active growth.

A meristem contains cells that have the capability to divide repeatedly. Each division results in two cells. One of these, the **derivative,** eventually differentiates (becomes specialized for a particular function), generally losing its ability to divide again.* The other cell, however, remains meristematic and eventually divides again. This process, summarized by figure 9-5, accounts for the unlimited or prolonged growth of vegetative (nonreproductive) plant meristems.

Plants have two types of meristems: apical and lateral. Apical meristems are found at the tips of plant organs (shoots and roots) and increase length. Lateral meristems, located beneath the bark of woody plants, increase girth.

MATERIALS

Per student:

• prepared slide of onion (*Allium*) root tip mitosis

• compound microscope

---

*In some instances, recently formed derivatives can also divide, increasing the number of derivatives.

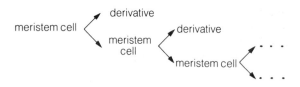

Figure 9-5 Cell division in plant meristems.

PROCEDURE

Obtain a prepared slide of a longitudinal section of an *Allium* (onion) root tip. This slide has been prepared from the terminal several millimeters of an actively growing root. It was "fixed" (killed) by chemicals to preserve the cellular structure and stained with dyes that have an affinity for the structures involved in nuclear division.

Focus first with the low-power objective of your compound microscope to get an overall impression of the root's morphology.

Concentrate your study in the region about 1 mm behind the actual tip. This region is the apical meristem of the root (fig. 9-6).

A. Interphase and Mitosis

1. Interphase. Use the medium-power objective to scan the apical meristem. Note that most of the nuclei are in interphase.

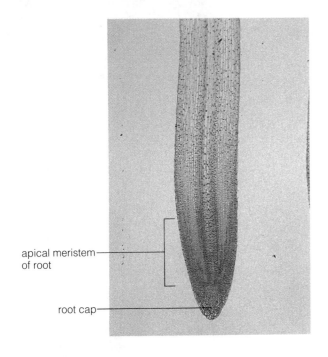

apical meristem
of root

root cap

Figure 9-6 Root tip, l.s. (20 ×). (Photo by J. W. Perry.)

Switch to the high-dry objective, focusing on a single interphase cell. Note the distinct **nucleus,** with one or more **nucleoli,** and the **chromatin** dispersed within the bounds of the **nuclear envelope.** Label these features in cell 1 of figure 9-7.

2. Mitosis.

a. *Prophase.* During **prophase** the chromatin condenses, rendering the duplicated chromosomes visible as threadlike structures. At the same time, microtubules outside the nucleus are beginning to assemble into **spindle fibers.** Collectively, the spindle fibers make up the **spindle,** a three-dimensional structure widest in the middle and tapering to a point at the two **poles** (opposite ends of the cell). *You will not see the spindle during prophase.*

Find a nucleus in prophase. Draw and label a prophase nucleus in cell 2 of figure 9-7.

The transition from prophase to metaphase is marked by the fragmentation and disappearance of the nuclear envelope. At about the same time the nucleoli disappear.

b. *Metaphase.* When the nuclear envelope is no longer distinct, the cell is in **metaphase.** Identify a metaphase cell by locating a cell with the duplicated chromosomes, each consisting of two **sister chromatids,** lined up midway between the two poles. This imaginary midline is called the **spindle equator.** (You will not be able to distinguish the chromatids.) The spindle has moved into the space the nucleus once occupied. The microtubules have become attached to the chromosomes at the **kinetochores,** groups of proteins that form the outer faces of the centromeres. Find a cell in metaphase. Label cell number 3 of figure 9-7.

c. *Anaphase.* During **anaphase** sister chromatids of each chromosome separate, each chromatid moving toward an opposite pole.

Find an early anaphase cell, recognizable by the slightly separated chromatids. Notice that the chromatids begin separating at the centromere. The last point of contact before separation is complete is at the ends of the "arms" of each chromatid. Although incompletely understood, the mechanism of chromatid separation is based upon action of the spindle-fiber microtubules. Once separated, each chromatid is referred to as an individual chromosome. Note that now the chromosome consists of a *single* chromatid.

Find a later anaphase cell and draw it in cell 4 of figure 9-7.

d. *Telophase.* When the chromosomes (formerly sister chromatids) arrive at opposite poles, the cell is in **telophase.** The spindle disorganizes. The chromosomes expand again, and a nuclear envelope re-forms around each newly formed daughter nucleus.

Find a telophase cell and label individual chromosomes, nuclei, and nuclear envelopes on cell 5 of figure 9-7.

B. Cytokinesis in Onion Cells

Cytokinesis, division of the cytoplasm, usually follows mitosis. In fact, it often overlaps with telophase. Find a cell undergoing cytokinesis in the onion root tip. In plants, cytokinesis takes place by **cell plate formation** (fig. 9-8). During this process Golgi body-derived vesicles migrate to the spindle equator, where they fuse. Their contents contribute to the formation of a new cell wall, and their membranes make up the new plasma membrane. In most plants, cell plate formation starts in the *middle* of the cell.

1. Examine figure 9-8, an electron micrograph showing cell plate formation. Note the microtubules that are part of the spindle apparatus.

2. Find a cell undergoing cytokinesis. With your light microscope, the developing **cell plate** appears as a line running horizontally between the two newly formed nuclei. Return to cell 5 of figure 9-7 and label the developing cell plate.

Recently divided cells are often easy to distinguish by their square, boxy appearance. Find two recently divided **daughter cells;** then draw and label their contents in cell 6 of figure 9-7. Include cytoplasm, nuclei, nucleoli, nuclear envelopes, and chromatin. What is the difference between chromatin and chromosomes?

_____

_____

_____

Following cytokinesis, the cell undergoes a period of growth and enlargement, during which time the nucleus is in interphase. Interphase may be followed by another mitosis and cytokinesis, or in some cells interphase may persist for the rest of a cell's life.

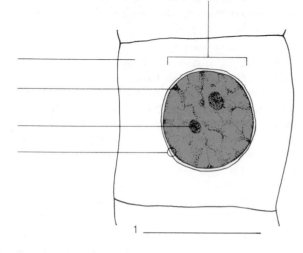

1 _____

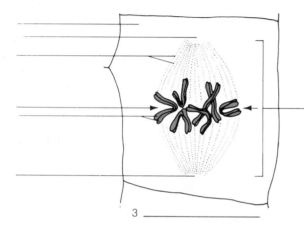

3 _____

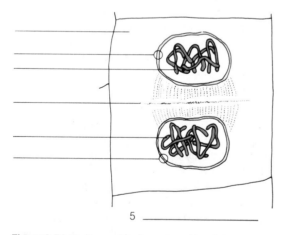

5 _____

Figure 9-7 Interphase, mitosis, and cytokinesis in onion root tip cells. (After H. Clark, 1937.)

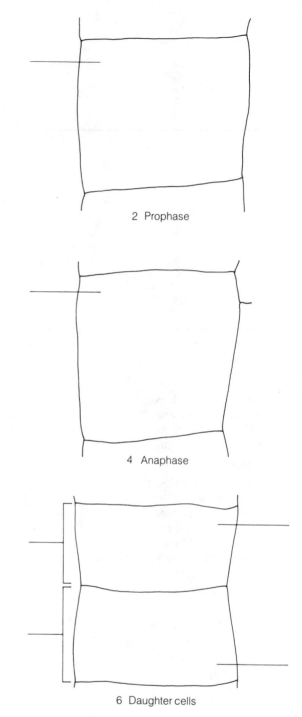

2 Prophase

4 Anaphase

6 Daughter cells

Labels: interphase, cytoplasm, nucleus, nucleolus, chromatin, nuclear envelope, metaphase, spindle fibers, spindle, pole, spindle equator (between arrows), sister chromatids, telophase and cell plate formation, chromosome, cell plate, daughter cell (Note: Some terms are used more than once.)

III. The Cell Cycle in Animal Cells: Whitefish Blastula

Fertilization of an ovum by a sperm produces a zygote. In animal cells, the zygote undergoes a special type of cell division *(cleavage)* in which no increase in cytoplasm occurs between divisions. A ball of cells called a blastula is produced by cleavage. Within the blastula, repeated nuclear and cytoplasmic divisions take place; consequently, the whitefish blastula is an excellent example in which to observe the cell cycle of an animal.

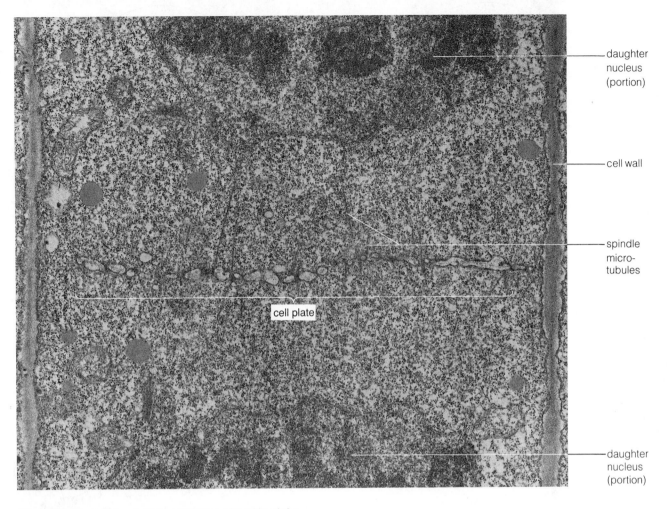

— daughter nucleus (portion)

— cell wall

— spindle micro-tubules

cell plate

— daughter nucleus (portion)

Figure 9-8 Transmission electron micrograph of cytokinesis by cell plate formation in a plant cell (2,000×). (Photo by W. P. Wergin, courtesy E.H. Newcomb.)

Note a difference between plants and animals: Whereas plants have meristems where divisions continually take place, animals do not have specialized regions to which mitosis and cytokinesis are limited. Indeed, divisions occur continually throughout many tissues of an animal's body, replacing worn-out or damaged cells.

With several important exceptions, mitosis in animals is remarkably like that in plants. These exceptions will be pointed out as we go through the cell cycle.

MATERIALS

Per student:

• prepared slide of whitefish blastula mitosis
• compound microscope

PROCEDURE

Obtain a slide labeled "whitefish blastula." Scan it with the low-power objective and then at medium power. This slide has numerous sections of a blastula.

Select one section (fig. 9-9) and then switch to the high-dry objective for detailed observation.

As you examine the slides, draw the cells to show the correct sequence of events in the cell cycle of whitefish blastula.

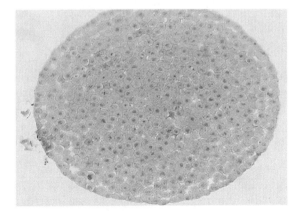

Figure 9-9 Section of blastula (75×). (Photo by J. W. Perry.)

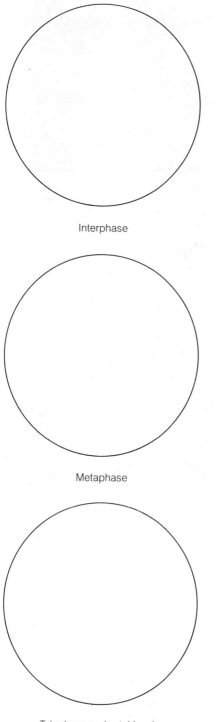

Interphase

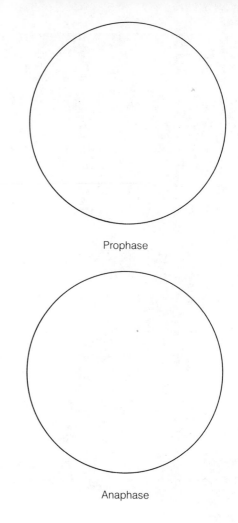

Prophase

Metaphase

Anaphase

Telophase and cytokinesis

Figure 9-10 Drawings of cell cycle stages in whitefish blastula.

Labels: cytoplasm, nucleus, plasma membrane, spindle, chromosomes, spindle equator, sister chromatids, daughter nuclei, chromatin, furrow (Note: Some terms are used more than once.)

A. Interphase and Mitosis

1. Interphase. Locate a cell in **interphase.** As you observed in the onion root tip, note the presence of the nucleus and chromatin within it. Note also the absence of a cell wall.

Draw an interphase cell above the word "Interphase" in figure 9-10 and label cytoplasm, nucleus, and plasma membrane.

2. Mitosis.

a. *Prophase.* The first obvious difference between mitosis in plants and animals is found in **prophase.** Unlike the onion cells, those of whitefish contain **centrioles** (fig. 9-11). As seen with the electron microscope, centrioles are barrel-shaped structures consisting of nine radially arranged triplets of microtubules.

One pair of centrioles was present in the cytoplasm in the G$_1$ stage of interphase. These centrioles duplicated during the S stage of interphase. Subsequently, one new and one old centriole migrated to each pole.

Although the centrioles are too small to be resolved with your light microscope, you can see a starburst pattern of spindle fibers that appear to radiate from the centrioles. Other microtubules extend between the centrioles, forming the **spindle** (fig. 9-12). The chromosomes become visible as the chromatin condenses.

Find a prophase cell, identifying the spindle and starburst cluster of fibers about the centriole.

Draw the prophase cell in the proper location on figure 9-10. Label spindle, chromosomes, cytoplasm, and the position of the plasma membrane.

b. *Metaphase.* As was the case in plant cells, during **metaphase** the spindle fiber microtubules become attached to the **kinetechore** of each centromere region, and the duplicated chromosomes (each consisting of two **sister chromatids**) line up on the **spindle equator.** Locate a metaphase cell.

Draw the metaphase cell in the proper location on figure 9-10. Label the chromosomes on the spindle equator, spindle, and plasma membrane.

c. *Anaphase.* Again similar to that observed in plant cells, **anaphase** begins with the separation of sister chromatids into individual (daughter) chromosomes. Observe a blastula cell in anaphase.

Draw the anaphase cell in the proper location on figure 9-10. Label the separating sister chromatids, spindle, cytoplasm, and plasma membrane.

d. *Telophase.* **Telophase** is characterized by the arrival of the individual (daughter) chromosomes at the poles. A nuclear envelope forms around each daughter nucleus. Find a telophase cell. Is the spindle still visible?

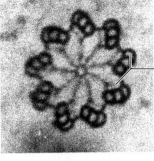

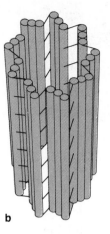

triplet of microtubules

a

b

Figure 9-11 (a) Transmission electron micrograph (122,000 ×); **(b)** artist's drawing of centriole. (Photo courtesy I. R. Gibbons.)

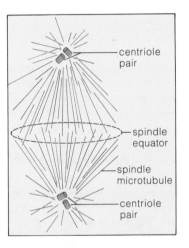

centriole pair

spindle equator

spindle microtubule

centriole pair

Figure 9-12 Spindle apparatus in animal cell. (From Starr and Taggart, 1989.)

Is there any evidence of a nuclear envelope forming around the chromosomes?

Draw the telophase cell. Label daughter nuclei, chromatin, cytoplasm, and plasma membrane.

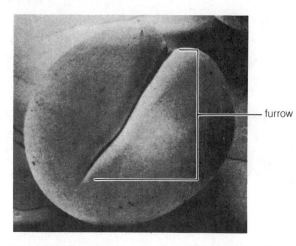

furrow

Figure 9-13 Cytokinesis in an animal cell. The scanning electron micrograph shows the cleavage furrow caused by the contraction of a microfilament ring just beneath the plasma membrane. (From H. Beans and R. G. Kessel, *American Scientist* 64: 279–290.)

B. Cytokinesis in Animal Cells

A second major distinction between cell division in plants and animals occurs during cytoplasmic division. Cell plates are absent in animal cells. Instead, cytokinesis takes place by **furrowing.**

To visualize how furrowing takes place, imagine wrapping a string around a balloon and slowly tightening the string until the balloon has been pinched in two. In life, the animal cell is pinched in two, forming two discrete cytoplasmic entities, each with a single nucleus. Figure 9-13 illustrates the cleavage furrow in a frog zygote.

Find a cell in the blastula undergoing cytokinesis. The telophase cell that you drew in figure 9-10 may also show an early stage of cytokinesis. Label the cleavage furrow if it does.

IV. Chromosome Squashes

You can make your own chromosome squash preparation quite simply. Cytologists and taxonomists do this routinely to count chromosomes. Observing whole sets of chromosomes is useful for studying chromosomal abnormalities and for determining if two organisms are different species.

MATERIALS

Per student:
- onion or daffodil root tips
- sharp razor blade
- 2 dissecting needles
- microscope slide and coverslip
- compound microscope

Per student pair:
- acetocarmine stain in dropping bottle
- iron alum in dropping bottle
- burner and matches

PROCEDURE

We will use onion or daffodil root tips that have been fixed, preserved, and softened to make squashes.

1. Obtain a single root and place it on a clean microscope slide. Notice that the terminal 2 mm or so is opaque white. This is the apical meristem.

2. With a sharp razor blade, separate the apical meristem from the rest of the root. Discard all *but* this meristem region.

3. Add a drop of acetocarmine stain and tease the tissue apart with dissecting needles.

4. Add a drop of iron alum. The iron intensifies the staining of the chromosomes.

5. Place a coverslip over the root tip. Spread the cells out by gently pressing down on the coverslip with your finger or a pencil eraser. Gently heat the slide over a flame.

> **BE CAREFUL**
> **You don't want to cause the fluid to boil away!**

6. Examine the preparation with your light microscope. Identify all stages of the cell cycle that have been described above.

What do you notice about the shape of the cells after this preparation?

_____

_____

_____

____ 1. Reproduction in prokaryotes occurs primarily through the process known as (a) mitosis, (b) cytokinesis, (c) furrowing, (d) fission.

____ 2. The genetic material (DNA) of eukaryotes is organized into (a) centrioles, (b) spindles, (c) chromosomes, (d) microtubules.

____ 3. The process of cytoplasmic division is known as (a) meiosis, (b) cytokinesis, (c) mitosis, (d) fission.

____ 4. The product of nucleoprotein duplication is (a) two chromatids, (b) two nuclei, (c) two daughter cells, (d) two spindles.

____ 5. The correct sequence of stages in *mitosis* is (a) interphase, prophase, metaphase, anaphase, telophase; (b) prophase, metaphase, anaphase, telophase; (c) metaphase, anaphase, prophase, telophase; (d) prophase, telophase, anaphase, interphase.

____ 6. During prophase, duplicated chromosomes (a) consist of chromatids, (b) contain centromeres, (c) consist of nucleoproteins, (d) all of the above.

____ 7. During the S period of interphase (a) cell growth takes place, (b) nothing occurs because this is a resting period, (c) chromosomes divide, (d) synthesis (or replication) of the nucleoproteins takes place.

____ 8. Chromatids separate during (a) prophase, (b) telophase, (c) cytokinesis, (d) anaphase.

____ 9. Cell plate formation (a) occurs in plant cells but not in animal cells, (b) begins during telophase, (c) is a result of fusion of Golgi vesicles, (d) all of the above.

____ 10. Centrioles and a starburst cluster of spindle fibers would be found in (a) both plant and animal cells, (b) only plant cells, (c) only animal cells, (d) none of the above.

EXERCISE 9

Mitosis and Cytokinesis: Nuclear and Cytoplasmic Division

POST-LAB QUESTIONS

1. If the chromosome number of a typical onion root tip cell is 16 before mitosis, what is the chromosome number of each newly formed nucleus after nuclear division has taken place?

2. Why must the DNA be duplicated during the S phase of the cell cycle, prior to mitosis?

3. In plants, what name is given to a region where mitosis occurs most frequently?

4. Distinguish among interphase, mitosis, and cytokinesis.

5. Distinguish between the structure of a duplicated chromosome before mitosis and the chromosome produced by separation of two chromatids during mitosis.

6. The cells in the photomicrographs below have been stained to show microtubules comprising the spindle apparatus. Identify the stage of mitosis in each and label the region indicated on (**b**).

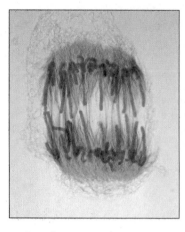

a stage ? _____

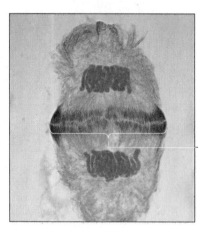

—region ? _____

b stage ? _____

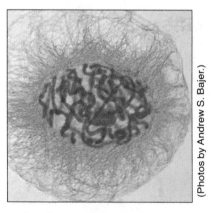

(Photos by Andrew S. Bajer.)

c stage ? _____

7. Observe photomicrographs (**a**) and (**b**) below. Is (**a**) from a plant or an animal?

Note the double nature of the blue "threads." Each individual component of the doublet is called a _____ . Is (**b**) from a plant or an animal?

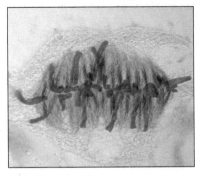

a plant or animal? _____

structure? _____

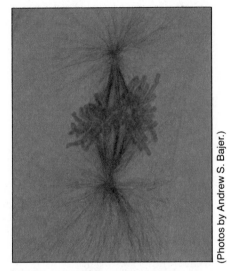

(Photos by Andrew S. Bajer.)

b plant or animal? _____

8. What would happen if a cell underwent mitosis but not cytokinesis?

9. Name two features of animal cell mitosis and cytokinesis you can use to distinguish these processes from those occurring in plant cells.

a.

b.

10. Why do you suppose cytokinesis generally occurs in the cell's midplane?

Meiosis: Basis of Sexual Reproduction

OBJECTIVES

After completing this exercise you will be able to:

1. define *meiosis, homologue (homologous chromosome), diploid, haploid, gene, gene pair, allele, gamete, ovum, sperm, gametic meiosis, fertilization, sporic meiosis, locus, synapsis, zygote;*

2. indicate the differences and similarities between meiosis and mitosis;

3. describe the basic differences between the life cycles of higher plants and higher animals;

4. describe the process of meiosis, recognizing the events that occur during each stage;

5. describe the significance of crossing over, independent assortment, and segregation;

6. identify the meiotic products in male and female animals.

INTRODUCTION

Like mitosis, meiosis is a process of nuclear division. During mitosis, the number of chromosomes within the daughter nuclei remains the same as was present in the parental nucleus. In meiosis, however, the genetic complement is halved, resulting in daughter nuclei containing only one-half the number of chromosomes as the parental nucleus. Thus, while mitosis is sometimes referred to as an *equational division,* meiosis is often called *reduction division.* Moreover, while mitosis is completed after a single nuclear division, two divisions, called meiosis I and meiosis II, occur during meiosis. Table 10-1 summarizes the differences between mitosis and meiosis.

In the body cells of most eukaryotes, chromosomes exist in pairs called **homologues** (homologous chromosomes): that is, there are two chromosomes that are physically similar and contain genetic information for the same traits. To visualize this, press your palms together, lining up your fingers. Each "finger pair" represents one pair of homologues.

When both homologues are in the *same* nucleus, the nucleus is **diploid** (2n); when only one of the homologues is present, the nucleus is **haploid** (n). If the parental nucleus normally contains the diploid (2n) chromosome number before meiosis, all four daughter nuclei contain the haploid (n) number at the completion of meiosis.

The reduction in chromosome number is the basis for sexual reproduction. In animals, the cells containing the daughter nuclei produced by meiosis are called **gametes: ova** (singular is *ovum*) if the parent is female, **sperm** cells if male. As you probably know, gametes

Table 10-1 Comparison of Mitosis and Meiosis

Mitosis	Meiosis
Equational division: amount of genetic material remains constant	Reduction division: amount of genetic material is halved
Completed in one division	Requires two divisions for completion
Produces two genetically identical nuclei	Produces two to four genetically different nuclei
Generally produces cells not directly involved in sexual reproduction	Produces cells for sexual reproduction

are produced in the gonads—ovaries and testes, respectively. In fact, this is the *only* place where meiosis occurs in higher animals. The simple diagram in figure 10-1 illustrates the life cycle of a higher animal.

Note where meiosis has occurred—during gamete production. Consequently, this is called **gametic meiosis.** During **fertilization** (the fusion of a sperm nucleus with an ovum nucleus), the diploid chromosome number is restored as the two haploid gamete nuclei fuse.

What about plants? Do plants have sex? Indeed they do. However, the plant life cycle is a bit more complex than that of animals. Plants of a single species have two completely different body forms. The primary function of one is production of gametes. This plant is called a *gametophyte* ("gamete-producing plant") and is haploid. Because the entire plant is haploid, gametes are produced in specialized organs (gametangia) by mitosis. The other body form is diploid and is called a *sporophyte.* This diploid sporophyte has specialized organs (sporangia) where meiosis occurs, producing haploid meiospores (hence the name *sporophyte,* "spore-producing plant"). When spores germinate, they grow into gametophytes.

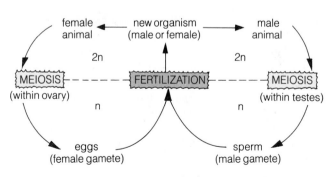

Figure 10-1 Life cycle of a higher animal.

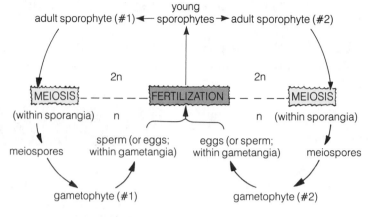

Figure 10-2 Life cycle of a plant.

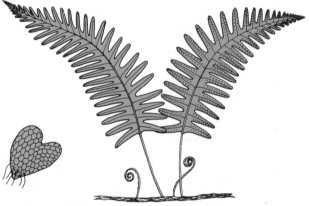

gametophyte (2x)   sporophyte (0.1x)

Figure 10-3 Gametophyte and sporophyte of the same fern plant.

Examine figure 10-3, which illustrates the gametophyte and sporophyte of a fern plant. Remember, the gametophyte and sporophyte are different, free-living stages of the *same* species of fern. Look at figure 10-2, a diagram of a typical plant life cycle. Again, note the consequence of meiosis. In plants it results in the production of meiospores. Hence, this type of meiosis is called **sporic meiosis.**

You should understand an important concept from these diagrams: *Meiosis always reduces the chromosome number. The diploid chromosome number is eventually restored during fertilization.*

Understanding meiosis is an absolute necessity for understanding the patterns of inheritance in Mendelian genetics. Gregor Mendel, an Austrian monk, spent years deciphering the complexity of simple genetics. Although he knew nothing of genes and chromosomes, he noted certain patterns of inheritance and formulated three principles, now known as Mendel's principles of recombination, segregation, and independent assortment. The following activities will demonstrate the events of meiosis and the genetic basis for Mendel's principles.

I. Demonstration of Meiosis Using Pop Beads

MATERIALS

Per student pair:

• 8 chains of simulated chromosomes consisting of pop beads with magnetic centromeres
• marking pens
• 8 pieces of string, each 40 cm long
• meiotic diagram cards similar to those illustrated within this exercise
• colored pencils

Per student group (table):

• bottle of 95% ethanol to remove marking ink
• tissues

PROCEDURE

Work in pairs.

Within the nucleus of an organism each chromosome bears **genes,** which are units of inheritance. Genes may exist in two or more alternative forms called **alleles.** Thus each homologue bears *genes* for the same traits; these are called **gene pairs.** However, the homologues may or may not have the same *alleles.* An example will help here.

Suppose the trait in question is flower color and that a flower has only two possible colors, red or white (fig. 10-4a,b). The gene is coding (providing the information) for flower color. Now there are two homologues in the same nucleus, so each bears the gene for flower color. *But,* on one homologue, the *allele* might code for red flowers, while the allele on the other homologue might code for white flowers (fig. 10-4c). There are two other possibilities. The alleles on *both* homologues might be coding for red flowers (fig. 10-4d), or they *both* might be coding for white flowers. (These three possibilities are mutually exclusive.)

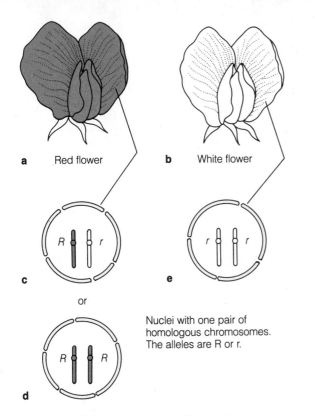

Figure 10-4 Chromosomal control of flower color.

A. Meiosis Without Crossing Over

Obtain four chains of pop beads (see fig. 10-5). The beads on two chains should be one color while those of the other two chains should be another color. All beads *within* a chain should be the *same* color.

The chains of beads represent chromatids of duplicated homologous chromosomes (homologues), each bead a gene, and the magnet the centromere.

We start by assuming that these chromosomes represent the diploid condition. The two colors represent the origin of the chromosomes: One homologue

(color) came from the male parent, and the other homologue (color) came from the female parent.

You have four chains of beads because chromosome replication occurred during the S stage of interphase (fig. 9-1), prior to the onset of meiosis. During chromosome replication, the gene pairs also duplicate. Thus alleles on sister chromatids are identical.

How many sister chromatids are there in a duplicated chromosome?

_____

How many chromosomes are represented by four sister chromatids?

_____

What is the diploid number of the starting (parental) nucleus? (Hint: Count the number of homologues to obtain the diploid number.)

_____

As mentioned previously, genes may exist in two or more alternative forms, called alleles. The location of an allele on a chromosome is its **locus** (plural: *loci*). Using the marking pen, mark two loci on each chromatid with letters to indicate alleles for a common trait. For example, suppose the homologous chromosomes code for two traits, skin pigmentation and the presence of attached earlobes in humans. Let the capital letter *A* represent the allele for normal pigmentation, lower case *a* the allele for albinism (the absence of skin pigmentation); let *F* represent free earlobes and *f* attached earlobes.

A suggested sequence is illustrated in figure 10-5.

Obtain a meiotic diagram card that appears similar to figure 10-6. Manipulate your model chromosomes through the stages of meiosis described below, locating the chromosomes in the correct diagram circles (representing nuclei) as you go along. Reference to the drawing in figure 10-6 will be made in the proper steps. *Do not draw on the meiotic diagram cards.*

1. Interphase. During interphase the nuclear envelope is intact, and the chromosomes are randomly distributed throughout the nucleoplasm (semifluid substance within the nucleus). Both duplicated chro-

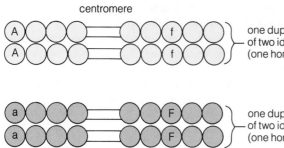

Figure 10-5 One pair of homologous pop-bead chromosomes.

MEIOSIS: BASIS OF SEXUAL REPRODUCTION

parental nucleus

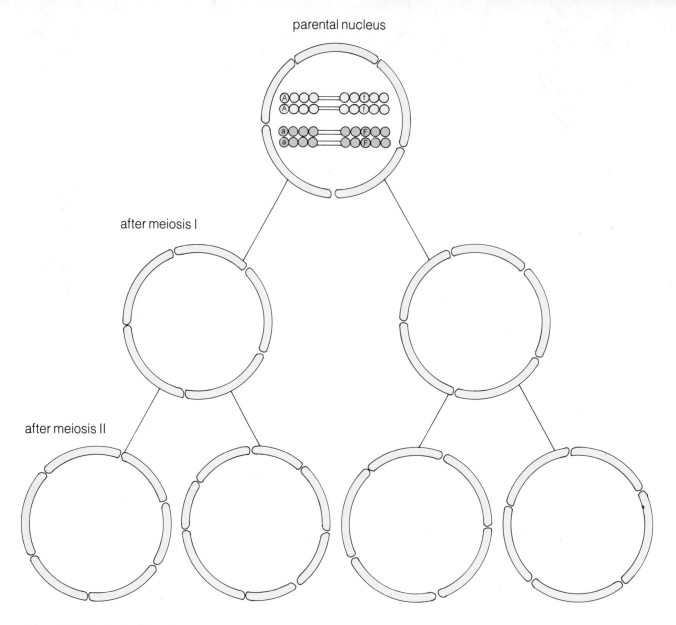

after meiosis I

after meiosis II

Figure 10-6 Meiosis without crossing over.

mosomes (four chromatids) should be in the parental nucleus, indicating that DNA duplication has taken place. The sister chromatids of each homologue should be attached by their magnetic centromeres, but the two homologues should be separate. Your model nucleus contains a diploid number (2n) = 2.

The pop bead chromosomes should appear during interphase in the parental nucleus as shown in figure 10-6. Be sure to mark the location of the alleles. Use different pencil colors to keep the homologues separate.

2. Meiosis I.

a. *Prophase I.* During the first prophase the parental nucleus contains two duplicated homologous chromosomes, each made up of two sister chromatids joined at their centromeres. The homologues pair with each other. This pairing is called **synapsis.** Slide the two homologues together.

Twist the chromatids about one another to simulate synapsis.

The nuclear envelope disorganizes at the end of prophase I.

b. *Metaphase I.* Homologous chromosomes now move toward the spindle equator, the centromeres of each homologue coming to lie *on either side of the equator.* Spindle fibers, consisting of aggregations of microtubules, attach to the centromeres. One homologue becomes attached to microtubules extending from one pole, and the other homologue becomes attached to microtubules extending from the opposite spindle pole.

To simulate the spindle fibers, attach one piece of string to each centromere. Then lay the free ends of one pair of strings toward one spindle pole and the ends of the other pair toward the opposite pole.

c. *Anaphase I.* During anaphase I, the homologous chromosomes separate, one homologue moving toward one pole, the other toward the opposite pole. The movement of the chromosomes is apparently the result of shortening of some spindle fibers and lengthening of others. Each homologue is still in the duplicated form, consisting of two sister chromatids.

Pull the two strings of one homologous pair toward its spindle pole and the other toward the opposite spindle pole, separating the homologues from one another.

d. *Telophase I.* Continue pulling the string spindle fibers until each homologue is now at its respective pole. The first meiotic division is now complete. You should have two nuclei, each containing a single chromosome consisting of two sister chromatids.

Draw your pop-bead chromosomes as they appear after meiosis I on the two nuclei labeled "after meiosis I" of figure 10-6. Depending on the organism involved, an interphase (interkinesis) and cytokinesis may precede the second meiotic division, or each nucleus may proceed directly into meiosis II.

It is important to note here that DNA synthesis *does not* occur following telophase I (between meiosis I and meiosis II).

Before meiosis II the spindle is rearranged into two spindles, one for each nucleus.

3. Meiosis II.

a. *Prophase II.* At the beginning of the second meiotic division, the sister chromatids are still attached by their centromeres. During prophase II, the nuclear envelope disorganizes, and the chromatin recondenses.

b. *Metaphase II.* Within each nucleus, the duplicated chromosome aligns with the equator, the centromeres lying *on the equator.* Spindle fiber microtubules attach the centromeres of each chromatid to opposite spindle poles.

Your string spindle fibers should be positioned just as they were during prophase I. Note that each nucleus contains only *one* duplicated chromosome consisting of *two* sister chromatids.

c. *Anaphase II.* The sister chromatids separate, moving to opposite poles. Pull on the string until the two sister chromatids separate. After the sister chromatids separate, each is an individual (not duplicated) daughter chromosome.

d. *Telophase II.* Continue pulling on the string spindle fibers until the two daughter chromosomes are at opposite poles. The nuclear envelope re-forms around each chromosome. Four daughter nuclei now exist. Note that each nucleus contains one individual chromosome (formerly a chromatid) originally present within the parental nucleus.

Draw your pop-bead chromosomes as they appear after meiosis II in the "gamete nuclei" of figure 10-6.

Your diagram should indicate the genetic (chromatid) complement *before* meiosis and *after* each meiotic division, *not* the stages of each division.

Remember that meiosis takes place in both male and female organisms. If the parental nucleus was from a male, what is the gamete called? (See fig. 10-1.)

_____

If female?

_____

Is the parental nucleus diploid or haploid?

_____

Are the nuclei produced after the *first* meiotic division diploid or haploid?

_____

Are the nuclei of the gametes diploid or haploid?

_____

If you answered the above questions correctly, you might logically ask, "If the chromosome number of the gametes is the same as that produced after the first meiotic division, why bother to have two separate divisions? After all, the genes present are the same in both gametes and first-division nuclei."

There are two answers to this apparent paradox. The first, and perhaps the most obvious, is that the second meiotic division ensures that a *single* chromatid (non-duplicated chromosome) is contained within each gamete. After gametes fuse, producing a zygote, the genetic material duplicates prior to the zygote's undergoing mitosis. If gametes contained two chromatids, the zygote would have four, and duplication prior to zygote division would produce eight, twice as many as the organism should have. If DNA duplication within the zygote were not necessary for the onset of mitosis, this problem would not exist. Alas, DNA synthesis apparently is a necessity to initiate mitosis.

You can discover the second answer for yourself by continuing on with the exercise, for although you have simulated meiosis, you have done so without showing what happens in *real* life. That's the next step. . . .

B. Meiosis with Crossing Over

A very important event that results in a reshuffling of alleles on the chromatids occurs during prophase I. Recall that synapsis results in pairing of the homologues. During synapsis, the chromatids break, and portions of chromatids bearing genes for the same characteristic (but perhaps *different* alleles) are exchanged between *non-sister* chromatids. This event is called **crossing over,** and it results in recombination (shuffling) of alleles. Look again at figure 10-5. Distinguish between sister and non-sister chromatids.

To simulate crossing over, break three beads from the five-bead arms of two non-sister chromatids, ex-

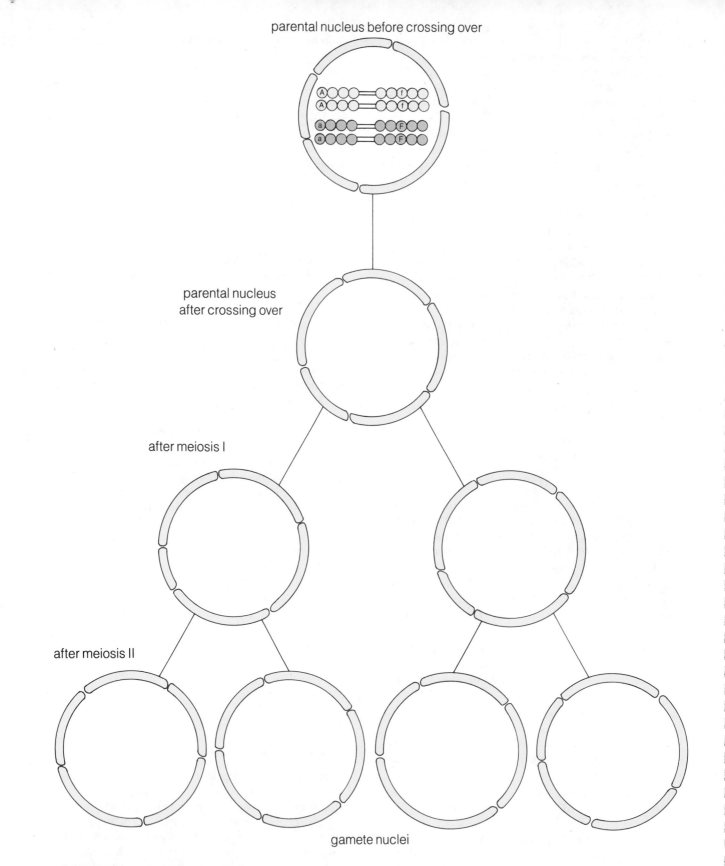

parental nucleus before crossing over

parental nucleus after crossing over

after meiosis I

after meiosis II

gamete nuclei

Figure 10-7 Meiosis after crossing over.

changing bead color between the two arms. During actual crossing over the chromosomes may break anywhere within the arms.

Crossing over is virtually a universal event in meiosis.

Manipulate your model chromosomes through meiosis I and II again, this time watching what happens to the distribution of the alleles as a consequence of the crossing over. Fill in figure 10-7 as you did before, but this time, show the effects of crossing over. Again, use different colors in making your sketches.

Is the distribution of alleles present in the gamete nuclei after crossing over the same as that which was present without crossing over?

_____

Is the distribution of alleles present in the gamete nuclei after crossing over the same as that in the nuclei after the first meiotic division?

_____

Crossing over provides for genetic recombination, resulting in increased variety. How many different *types* of daughter chromosomes are present in the gamete nuclei without crossing over (fig. 10-6)?

_____

How many different types are present with crossing over (fig. 10-7)?

_____

We think you would agree that a greater number of *types* of daughter chromosomes indicates greater *variety*. Now, does crossing over result in increased or decreased variety?

_____

Recall that the parental nucleus contained a pair of homologues, each homologue consisting of two sister chromatids. Because sister chromatids are identical in all respects, they have the same alleles of a gene (see fig. 10-5). As your models showed, the alleles on non-sister chromatids may not (or may) be identical; they bear the same gene but may have different alleles.

What is the difference between a gene and an allele?

_____

Let's look at a single set of alleles that are on your model chromosomes, say, the alleles for pigmentation, *A* and *a*. Both alleles were present in the parental nucleus. How many are present in the gametes?

_____

This illustrates Mendel's first principle, segregation. *Segregation* means that during gamete formation, the alleles are separated (segregated) from each other and end up in different gametes.

C. Demonstrating Independent Assortment

You have just demonstrated meiosis in which only one pair of homologues was present (n = 1, 2n = 2). Now obtain another set of model chromosomes (four more chains with magnetic centromeres). These two chromosomes should be distinct from the original set. The easiest way to accomplish this is to make the chains different colors, different lengths, and/or with different numbers of beads on the arms on either side of the centromeres.

Let's assign a gene to our second set of homologues. Suppose this gene codes for the production of an enzyme necessary for metabolism. On one homologue (consisting of two chromatids), mark the letter *P*, representing the allele causing production of the enzyme. On the other homologue, let *p* represent the allele that interferes with normal enzyme production. (For now, it is not important to remember these traits; they're real situations used simply as examples.)

We now have a parental nucleus where there are two sets of homologous chromosomes (four homologues). Here the diploid number (2n) is 4. Count the number of duplicated chromosomes. How many are there?

_____

This is the 2n number.

You know that meiosis is reduction division, so you can predict the number of individual chromosomes (the haploid number) each gamete will have after meiosis II. Do so.

_____

Manipulate your model chromosomes through meiosis I and II. Simulate crossing over with the original set of models. Keep this in mind: Crossing over and recombination occur between non-sister chromatids of homologous chromosomes, but *not* between *nonhomologous* chromosomes.

Fill in figure 10-8, showing the outcome of meiosis in a nucleus with two sets of homologues.

How many individual chromosomes does each gamete contain?

_____

Are the gametes the same genetically or different from each other?

_____

Go through meiosis again, searching for different possibilities in chromosome distribution that would make the gametes different.

Does the distribution of the alleles for production of the enzyme to different gametes on the second set of homologues have any bearing on the distribution of the alleles on the first set (alleles for skin pigmentation and earlobe condition)?

_____

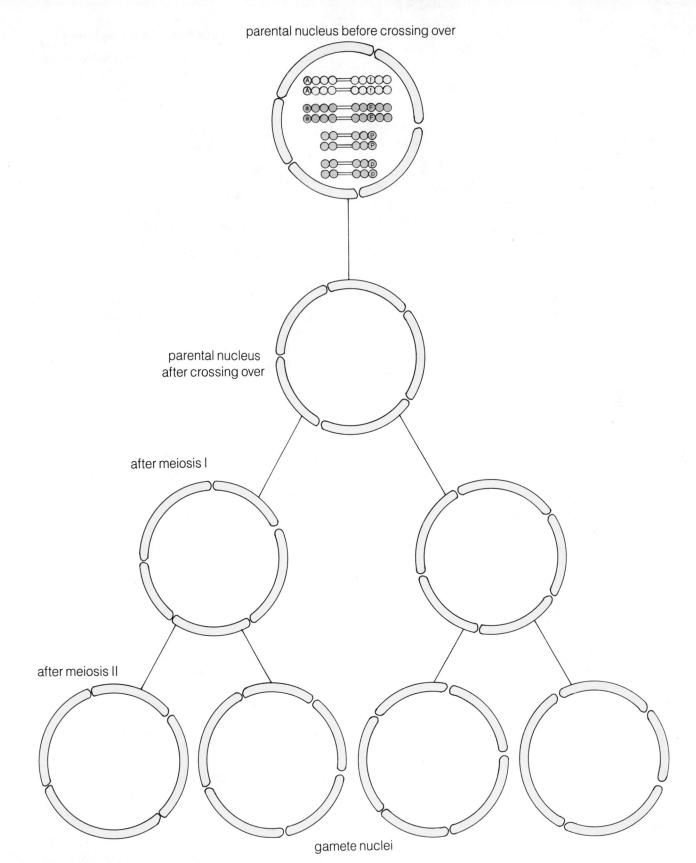

Figure 10-8 Meiosis in a nucleus where 2n = 4.

This distribution demonstrates the principle of independent assortment, which states that segregation of alleles into gametes is independent of the segregation of alleles for other traits, *as long as the genes are on different sets of homologous chromosomes.* Genes that are on different (nonhomologous) chromosomes are said to be **nonlinked.** By contrast, genes for different traits that are on the same chromosome are **linked.**

Because the genes for enzyme production and those for skin pigmentation and earlobe attachment are on different homologous chromosomes, these genes are

_____

while the genes for skin pigmentation and earlobe attachment are

_____

because they are on the same chromosome.

In reality, most organisms have many more than two sets of chromosomes. Humans have 23 pairs (2n = 46), while some plants literally have hundreds!

A thorough understanding of meiosis is necessary to understand genetics. With this basis you will find doing problems involving Mendelian genetics easy and fun. Without an understanding of meiosis, Mendelian genetics will be hopelessly confusing.

> **NOTE**
> **Remove marking ink from pop beads with 95% ethanol and tissues.**

II. Meiosis in Animal and Plant Cells

Now that you have a conceptual understanding of meiosis, let's see the actual divisions as they occur in living organisms.

In animals, as mentioned previously, meiosis results in the production of gametes—ova in females and sperm in males.

MATERIALS

Per lab room:
- set of demonstration slides of meiosis in grasshopper testes and lily anther
- set of models illustrating meiosis and fertilization in roundworm

Per student pair:
- scissors
- tape or glue

PROCEDURE

A. Meiosis in Male Animals

In male animals meiosis occurs in the testes.

1. Examine figure 10-9a. A diploid reproductive cell, the *spermatogonium,* first enlarges into a *primary spermatocyte.* The primary spermatocyte undergoes meiosis I to form two haploid *secondary spermatocytes.* After meiosis II, four haploid *spermatids* are produced, which develop flagella during differentiation into four sperm cells. This process is called *spermatogenesis.*

2. Examine the demonstration slide of spermatogenesis in grasshopper testes.

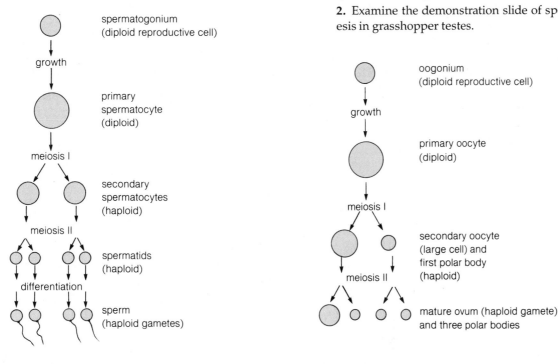

a spermatogenesis

b oogenesis

Figure 10-9 Gametogenesis in animals. (From Starr, 1991.)

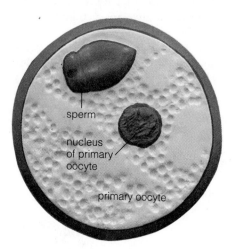

sperm

nucleus
of primary
oocyte

primary oocyte

Sperm entrance

Labels: primary oocyte, sperm,
homologous chromosomes,
sister chromatids

Prophase I

Labels: primary oocyte, homologous
chromosomes, centrioles, spindle, sister
chromatids, sperm nucleus

Late Metaphase (Early Anaphase I)

Labels: primary oocyte, homologous
chromosomes, sister chromatids, spindle

Later Anaphase I (Telophase I)

Labels: first polar body, four chromatids,
secondary oocyte, sperm nucleus

First polar body formation

Labels: first polar body, centrioles, four
chromatids, spindle, sperm nucleus

Later Metaphase II (Early Anaphase II)

Labels: first polar body, unduplicated
chromosomes, mature ovum,
sperm nucleus

Telophase II and cytokinesis

Labels: sperm nucleus, ovum nucleus,
zygote

Fertilization

Figure 10-10 Animal meiosis, ovum
formation, and fertilization.

B. Meiosis in Female Animals

In the ovaries of female animals, *ova* (eggs) are produced by meiosis during the process called oogenesis (fig. 10-9b). Unlike spermatogenesis, only one of the meiotic products becomes a gamete.

1. Examine figure 10-9b. The diploid reproductive cell, called an *oogonium*, grows into a *primary oocyte*. The primary oocyte undergoes meiosis I, one product being the *secondary oocyte*, the other a *polar body*. Notice the difference in size of the secondary oocyte and the polar body. This is because the secondary oocyte ends up with nearly all of the cytoplasm after meiosis I. Following meiosis II, only the secondary oocyte becomes a mature, haploid *ovum;* depending on the species, the polar body may or may not undergo meiosis II. In any case, the polar bodies are extremely small and do not function as gametes.

2. Observe the models illustrating oogenesis in the demonstration series. These models represent the events as they occur in the roundworm, an organism that has only two pairs of homologous chromosomes (2n = 4).

3. As you study the models and read the description for each stage, cut out the photographs of the models (page 131) and tape or glue them in the proper sequence in figure 10-10. Label each stage in your correctly sequenced photographs.

a. Meiosis in the primary oocyte does not begin until a *sperm* penetrates the cytoplasm. In this model, note that the oogonium's nucleus is intact. A photograph of this stage has been inserted in figure 10-10 to get you started.

b. *Prophase I.* The second model represents prophase I. Note that the nuclear envelope has disorganized. Each dark dot represents a chromatid. How many chromatids are there?

_____

How many chromosomes does this represent?

_____

c. *Late Metaphase I (or Early Anaphase I).* (The third model represents a transition between metaphase I and anaphase I.) During metaphase I, the *homologous chromosomes* become located on either side of the spindle equator. The *spindle* is distinct, the component fibers seemingly attached to the *centrioles,* here represented by two small dots. The homologous chromosomes are beginning to separate. Note the *sperm nucleus* within the cytoplasm of the primary oocyte.

d. *Later Anaphase I (or Early Telophase I).* The homologous chromosomes move toward opposite spindle poles. Remember, each homologous chromosome consists of two sister chromatids. The sperm nucleus remains "lying in wait."

e. *Formation of the first polar body.* Cytokinesis takes place, separating the homologous chromosomes. One set of homologues resides in a small cell with relatively little cytoplasm. This is the first polar body.

Two non-homologous chromosomes (four chromatids) remain in the larger cell, which is now called the *secondary oocyte.* A nuclear envelope does not form about these chromosomes, so essentially the secondary oocyte is in prophase II.

f. *Late Metaphase II (or Early Anaphase II).* Now the sister chromatids of the chromosomes within the secondary oocyte line up on the spindle equator. (The models show them on opposite sides of the equator.) A new spindle with centrioles is present as the sperm nucleus remains in wait. In the roundworm, the polar body does not undergo meiosis II.

g. *Telophase II and cytokinesis.* A thin line represents cytokinesis occurring to form the second polar body. How many unduplicated chromosomes (formerly sister chromatids) does the mature haploid *ovum* contain?

_____

(The models do not show formation of the second polar body.)

h. *Fertilization.* The final model represents fertilization, the fusion of the ovum nucleus with the sperm nucleus. With fertilization, the large ovum becomes the first diploid cell, the zygote. How many chromosomes does the zygote contain?

_____

C. Meiosis in Plants

For the sake of brevity, we will examine meiosis in the male reproductive structure of flowering plants only. Recall from our earlier discussion that meiosis in plants results in meiospore production, not directly into gametes. The details of the life cycle of flowering plants will be considered in Exercise 21.

1. Examine the demonstration slides of meiosis beginning with the diploid *microsporocytes*. Microsporocytes are the cells within a flower that undergo meiosis to produce haploid *microspores*. Eventually these microspores develop into pollen grains, which in turn produce sperm.

2. As you examine the slides, cut out the photomicrographs on pages 133 and 135 and arrange them on figure 10-11 to depict the meiotic events leading to microspore formation.

a. *Interphase.* During interphase the *nucleus* of each diploid *microsporocyte* is distinct, containing granular appearing chromatin. The cells are compactly arranged.

b. *Early Prophase I.* Now the chromatin has begun to condense into discrete *chromosomes*, which have the appearance of fine threads within the nucleus.

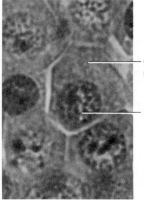

diploid
microsporocyte

nucleus

Interphase

Labels: nucleus, chromosomes
Early Prophase I

Label: chromosomes
Mid-Prophase I

Label: chromosomes
Late Prophase I

Labels: spindle equator, spindle, spindle
fibers, pole
Metaphase I

Label: spindle fibers
Early Anaphase I

Labels: pole, homologous chromosomes
Later Anaphase I

Labels: pole, homologous chromosomes,
spindle fibers
Telophase I

Label: cell plate
Cytokinesis I

Labels: nuclei, cell wall, daughter cells
Interkinesis

Labels: daughter cells, nuclei
Prophase II

Labels: chromosomes, spindle equator
Metaphase II

Labels: sister chromatids (unduplicated
chromosomes)
Anaphase II

Labels: cell plate, nuclei
Telophase II & Cytokinesis

Figure 10-11 Meiosis and microsporogenesis in the anther.

c. *Mid-Prophase I.* Additional condensation of the *chromosomes* has taken place. Pairing of homologous chromosomes is taking place.

d. *Late Prophase I.* The chromosomes have condensed into short, rather fat structures. Synapsis and crossing over are taking place. Note that the nuclear envelope has disorganized.

e. *Metaphase I.* The homologous chromosomes lie in the region of the *spindle equator*. The *spindle,* composed of *spindle fibers,* can be discerned as fine lines running toward the *poles.* (Note the absence of centrioles in plant cells.)

f. *Early Anaphase I.* Separation of homologous chromosomes is beginning to take place.

g. *Later Anaphase I.* Homologous chromosomes have nearly reached the opposite poles. Reduction division has occurred.

h. *Telophase I.* The homologous chromosomes have aggregated at opposite poles. The spindle remains visible.

i. *Cytokinesis I.* The *cell plate* is forming in the midplane of the cell. Spindle fibers, which are aggregations of microtubules, are visible running perpendicularly through the cell plate. The microtubules are directing the movement of Golgi vesicles, which contain the materials that form the cell plate.

A nuclear envelope has re-formed about the chromosomes, resulting in a well-defined nucleus in each *daughter cell*.

j. *Interkinesis.* In these plant cells, a short stage exists between meiosis I and II. Distinct nuclei are apparent in the two daughter cells. A cell wall has formed across the entirety of the midplane.

k. *Prophase II.* The chromosomes in each nucleus of the two daughter cells condense again into distinct, threadlike bodies. As was the case at the end of prophase I, the nuclear envelope disorganizes.

l. *Metaphase II.* Chromosomes consisting of sister chromatids line up on the spindle equator in both cells. (The photomicrograph shows the very early stages of separation of the chromatids.)

m. *Anaphase II.* The sister chromatids (now more appropriately considered *unduplicated chromosomes*) are being drawn to their respective poles in each cell.

Before anaphase II begins, sister chromatids are attached to each other along their length. Shortening of the spindle fibers, which are attached to the chromatids at kinetochores within their centromeres, causes separation of the chromatids, beginning in the region of the centromere. This causes a V-shaped configuration of the chromosomes.

n. *Telophase II and cytokinesis.* Nuclear envelopes are now re-forming around each of the four sets of chromosomes. Cell plate formation is occurring perpendicular to the cell wall that was formed after telophase I.

After cell wall formation is complete, the four haploid cells (microspores) will separate. Subsequently, each will develop into a pollen grain inside which sperm cells will be formed.

PRE-LAB QUESTIONS

_____ 1. In meiosis the number of chromosomes _____, while in mitosis, it _____. (a) is halved, is doubled, (b) is halved, remains the same, (c) is doubled, is halved, (d) remains the same, is halved.

_____ 2. The term "2n" means that (a) the diploid chromosome number is present, (b) the haploid chromosome number is present, (c) within a single nucleus chromosomes exist in homologous pairs, (d) a and c.

_____ 3. In higher animals, meiosis results in the production of (a) egg cells (ova), (b) gametes, (c) sperm cells, (d) all of the above.

_____ 4. Recombination of alleles on non-sister chromatids occurs during (a) anaphase I, (b) meiosis II, (c) telophase II, (d) crossing over.

_____ 5. Alternative forms of genes are called (a) homologues, (b) locus, (c) loci, (d) alleles.

_____ 6. If both homologous chromosomes of each pair exist in the same nucleus, that nucleus is (a) diploid, (b) unable to undergo meiosis, (c) haploid, (d) none of the above.

_____ 7. DNA duplication occurs during (a) interphase, (b) prophase I, (c) prophase II, (d) interkinesis.

_____ 8. A daughter chromosome (a) is formed during anaphase II, (b) is the same as a homologous chromosome, (c) is the result of separation of chromatids, (d) a and c.

_____ 9. Humans (a) don't undergo meiosis, (b) have forty-six chromosomes, (c) produce gametes by mitosis, (d) all of the above.

_____ 10. Gametogenesis in female animals results in (a) four sperm, (b) one gamete and three polar bodies, (c) four functional ova, (d) a haploid ovum and three diploid polar bodies.

E X E R C I S E 1 0

Meiosis: Basis of Sexual Reproduction

P O S T - L A B Q U E S T I O N S

1. If a cell of an organism had forty-six chromosomes before meiosis, how many chromosomes would exist in each nucleus after meiosis?

2. Suppose one sister chromatid of a chromosome has the allele *H*. What allele will the other sister chromatid have? (Assume crossing over has not taken place.)

3. Suppose that two alleles on one homologous chromosome are *A* and *B,* and the other homologous chromosome's alleles are *a* and *b*. How many different genetic types of gametes would be produced *without* crossing over?

   _____

   What are those types?

   If crossing over were to occur, how many different genetic types of gametes could occur?

   _____

   List them.

4. List two differences between mitosis and meiosis.

   1. _____

   2. _____

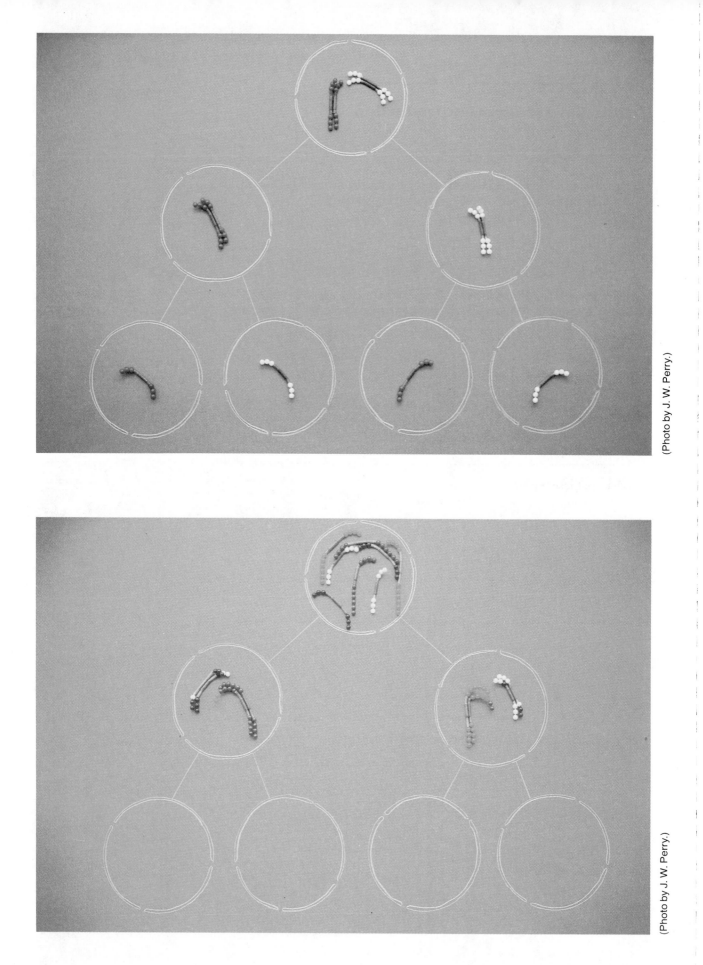

(Photo by J. W. Perry.)

(Photo by J. W. Perry.)

EXERCISE 10

5. Observe the meiotic diagram illustrated at the top of the facing page. Describe completely what's wrong with the diagram.

6. Observe the meiotic diagram illustrated at the bottom of the facing page.

   a. What is the diploid (2n) number of the parental nucleus?

   b. What *event* has occurred that is illustrated within the nuclei after meiosis I?

   c. Using colored pencils, complete the meiotic diagram.

7. From a genetic viewpoint, of what significance is fertilization?

8. In animals, meiosis results directly in gamete production, while in plants meiospores are produced. Where do the gametes come from in the life cycle of a plant?

9. What basic difference exists between the life cycles of higher plants and higher animals?

10. How would you argue that sporic meiosis is the basis for sexual reproduction in plants, even though the *direct* result is a meiospore rather than a gamete?

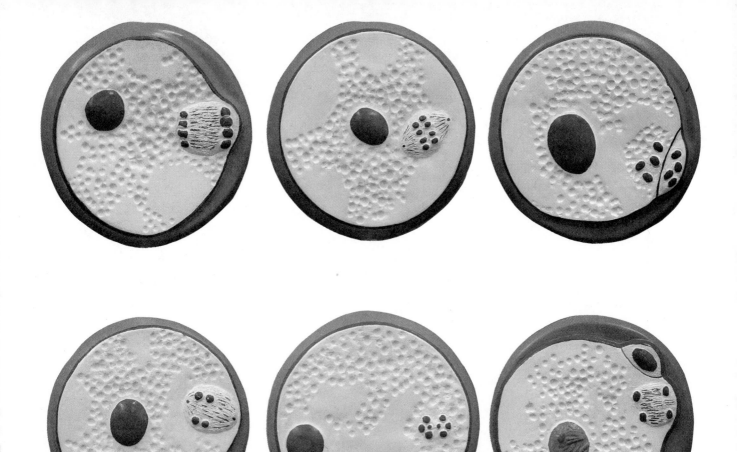

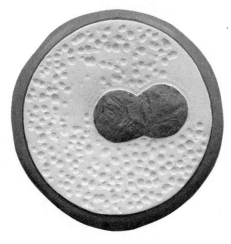

Photos for Figure 10-10 Meiosis in female roundworm. Cut the photographs from this page and arrange them in the proper sequence in figure 10-10. (Photos by J. W. Perry.)

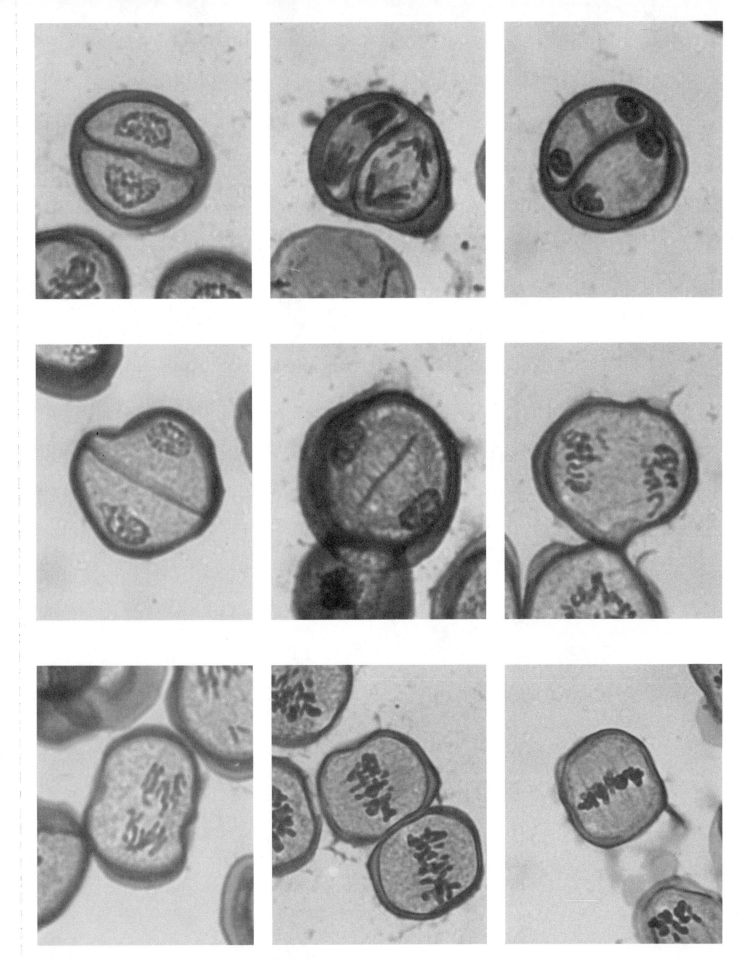

Photos for Figure 10-11 Meiosis in flowering plants. Cut from this page and arrange in proper sequence in figure 10-11. (Photos by J. W. Perry.) *Continues.*

133

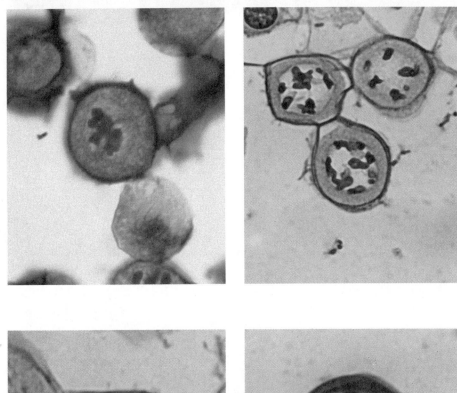

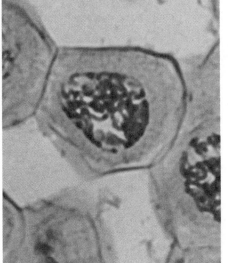

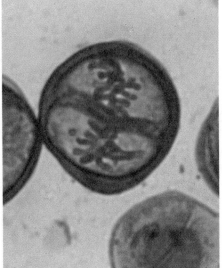

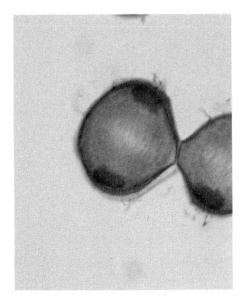

Photos for Figure 10-11 *Continued*

EXERCISE 11
Mendelian Genetics

OBJECTIVES

After completing this exercise you will be able to:

1. define *true-breeding, hybrid, monohybrid cross, diploid, haploid, genotype, phenotype, dominant, recessive, complete dominance, homozygous, heterozygous, incomplete dominance, codominance, sex-linked, dihybrid cross, probability, multiple alleles, chi-square test;*

2. solve problems illustrating monohybrid and dihybrid crosses, including those with incomplete dominance, codominance, sex-linkage, and problems involving multiple alleles;

3. Use a chi-square test to determine if observed results are consistent with expected results;

4. determine your phenotype for traits used in this exercise and give your probable genotype for these traits.

INTRODUCTION

In 1866 an Austrian monk, Gregor Mendel, presented the results of painstaking experiments on the inheritance of the garden pea. Those results were heard, but probably not understood, by Mendel's audience. Now, more than a century later, Mendel's work seems elementary to modern-day geneticists, but its importance cannot be overstated. The principles generated by Mendel's pioneering experimentation are the foundation for genetic counseling so important today to families with health disorders having a genetic basis. It's also the framework for the modern research that is making inroads in treating diseases previously believed to be incurable. In this era of genetic engineering—the incorporation of foreign DNA into chromosomes of unrelated species—it is easy to lose sight of the basics of the process that makes it all possible.

Recent advances in molecular genetics have resulted in the production of insulin and human growth hormone by genetic engineering techniques. Cancer patients are being treated with cells that have been removed from their own bodies, genetically altered to enhance their tumor-destroying capacity, and then reinserted in the hope that microscopic tumors escaping the surgeon's scalpel may be destroyed.

This newfound technology has not been without controversy, however. Release into the environment of genetically engineered microorganisms that may make crops resistant to disease-causing organisms (or even capable of withstanding temperatures that normally would freeze plants) has met with strong opposition.

In the future, *you* may be called upon to help make decisions about issues like these. To make an educated judgment, you must understand the basics, just as Mendel did. The genetics problems in this exercise should start you well on your way.

MATERIALS

Per student group (table):
- genetic corn ears illustrating monohybrid and dihybrid crosses

Optional:
- pop beads used in Exercise 10
- marking pen
- bottle of 70% ethanol
- simulated chromosomes, consisting of pop beads with magnetic centromeres, and meiotic diagram cards (see p. 116, Exercise 10)
- hand lens

I. Monohybrid Problems with Complete Dominance

Garden peas have both male and female parts in the same flower and are able to self-fertilize. For his experiments, Mendel chose parental plants that were **true-breeding,** meaning that all self-fertilized offspring displayed the same form of a trait as their parent. For example, if a true-breeding purple-flowered plant is allowed to self-fertilize, all of the offspring will have purple flowers.

When parents that are true-breeding for *different* forms of a trait are crossed—for example, purple flowers and white flowers—the offspring are called **hybrids.** When only one trait is being studied, the cross is called a **monohybrid cross.**

We'll look first at monohybrid problems.

1. Most organisms are diploid; that is, they contain homologous chromosomes with genes for the same traits. The location of a gene on a chromosome is its *locus* (plural: *loci*). Two genes at homologous loci are called a *gene pair.* Chromosomes have numerous genes, as illustrated in figure 11-1. Genes may exist in

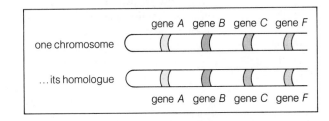

Figure 11-1 Arrangement of genes on homologous chromosomes.

different forms, called *alleles*. Let's consider one gene pair at the *F* locus. There are three possibilities for the allelic makeup at the *F* locus:

Both alleles are *FF*.

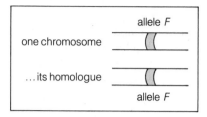

Both alleles are *ff*.

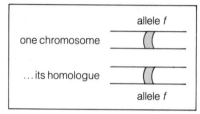

One allele is *F*, and the other is *f*.

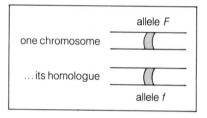

Gametes, on the other hand, are haploid, containing only one of the two homologues, and thus only one of the two alleles for a specific trait.

The **genotype** of an organism is its genetic constitution, that is, the alleles present.

For each of the following diploid genotypes, indicate the possible genotypes of the gametes.

Diploid genotype	Gamete genotype
FF	____
ff	____
Ff	____ , ____

If you don't understand the process that gives rise to the gamete genotypes, manipulate the pop-bead models that you used in the meiosis exercise. Using a marking pen, label one bead of each chromosome and go through the meiotic divisions that give rise to the gametes.

It is imperative that you understand meiosis before attempting to do genetics problems.

2. During fertilization, two gamete nuclei fuse, and the diploid condition is restored. Give the diploid genotype produced by fusion of the following gamete genotypes.

Gamete genotype	× Gamete genotype	→ Diploid genotype
f	*f*	____
F	*f*	____
F	*F*	____

3. Now let's attach some meaning to genotypes. As you see from the previous problems, the genotype is an expression of the actual genetic makeup of the organism. The **phenotype** is the observable result of the genotype, that is, what the organism looks like because of its genotype. (Although phenotype is determined primarily by genotype, in many instances environmental factors can modify phenotype.)

Human earlobes are either attached or free (fig. 11-2). This trait is determined by a single gene consisting of two alleles, *F* and *f*. An individual whose genotype is *FF* or *Ff* has free earlobes. This is the **dominant** condition. Note that the presence of one *or* two *F* alleles results in the dominant phenotype, free earlobes. The allele *F* is said to be dominant over its allelic partner, *f*. The **recessive** phenotype, attached earlobes, occurs only when the genotype is *ff*. In the case of **complete dominance**, the dominant allele completely masks the expression or affect of the recessive allele.

When both alleles in a nucleus are identical, the nucleus is **homozygous**. Those having both dominant alleles are homozygous dominant.

When both recessives are present in the same nucleus, the individual is said to be *homozygous recessive* for the trait.

Suppose a man has the genotype *FF*. What is the genotype of his gamete (sperm) nuclei?

_____

Suppose a woman has attached earlobes. What is her genotype?

_____

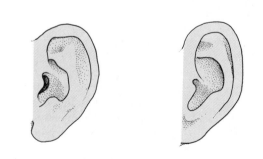

Figure 11-2 Free and attached earlobes in humans.

What allele(s) do her gametes (ova) carry?

_____

Suppose these two individuals produce a child. Show the genotype of the child by doing the cross:

sperm genotype   x   ovum genotype

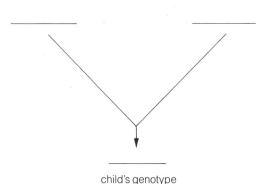

child's genotype

When both the dominant and recessive alleles are present within a single nucleus, the individual is **heterozygous** for that trait.

What is the phenotype of the child? (That is, does this child have attached or free earlobes?)

_____

4. In garden peas, purple flowers are dominant over white flowers. Let P represent the allele for purple flowers, p the allele for white flowers.

a. What is the phenotype (color) of the flowers with the following genotypes:

Genotype	Phenotype
PP	_____
pp	_____
Pp	_____

NOTE

Always be sure to distinguish clearly between upper- and lowercase letters.

A white-flowered garden pea is crossed with a homozygous dominant purple-flowered plant.

b. What is the genotype of the gametes of the white-flowered plant?

_____

c. What is the genotype of the gametes of the purple-flowered plant?

_____

d. What is the genotype of the plant produced by the cross?

_____

e. What is the phenotype of the plant produced by the cross?

_____

A convenient method of performing the mechanics of a cross is to use a Punnett square. The circles along the top and side of the Punnett square represent the gamete nuclei. Insert the proper letters indicating the genotypes of the gamete nuclei for the above cross in the circles and then fill in the Punnett square.

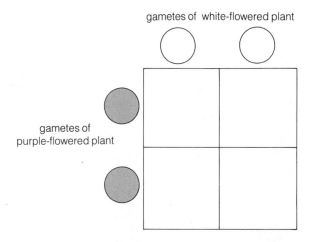

A heterozygous plant is crossed with a white-flowered plant. Fill in the Punnett square and give the genotypes and phenotypes of the offspring.

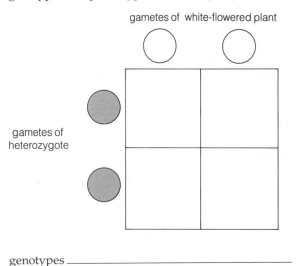

genotypes _____

phenotypes _____

(Draw a line from the genotype to its respective phenotype.)

For the remaining problems, you may wish to draw your own Punnett squares on a separate sheet of paper.

5. In mice, black fur (B) is dominant over brown fur (b). Breeding a brown mouse and a homozygous black mouse produces all black offspring.

a. What is the genotype of *the gametes* produced by the brown-furred parent?

_____

b. What is the genotype of the brown-furred parent?

_____

c. What is the genotype of the black-furred parent?

_____

d. What is the genotype of the black-furred offspring?

_____

By convention, P stands for the parental generation. The offspring are called the *first filial generation,* abbreviated F_1. If these F_1 offspring are crossed, their offspring are called the *second filial generation,* designated F_2. Note the following diagram.

e. If two of the F_1 mice are bred with one another, what will the phenotype of the F_2 be, and in what proportion?

phenotype _____

proportion _____

6. The presence of horns on Hereford cattle is controlled by a single gene. The hornless (H) condition is dominant over the horned (h) condition. A hornless cow was crossed repeatedly with the same horned bull. The following results were obtained in the F_1 offspring:

    8 hornless cattle
    7 horned cattle

What are the genotypes of the parents?

cow _____

bull _____

7. In fruit flies, red eyes (R) are dominant over purple eyes (r). Two red-eyed fruit flies were crossed, producing the following offspring:

    76 red-eyed flies
    24 purple-eyed flies

a. What is the approximate ratio of red-eyed to purple-eyed flies?

_____

b. Based upon your experience with previous problems, what two genotypes give rise to this ratio?

_____

c. What are the genotypes of the parents?

_____

d. What is the genotypic ratio of the F_1?

_____

e. What is the phenotypic ratio of the F_1?

_____

II. Monohybrid Problems with Incomplete Dominance

8. Petunia flower color is governed by two alleles, but neither allele is truly dominant over the other. Petunias with the genotype R^1R^1 are red-flowered, those that are heterozygous (R^1R^2) are pink, and those with the R^2R^2 genotype are white. This is an example of **incomplete dominance.** (Note that superscripts are used rather than upper- and lowercase letters to describe the alleles.)

a. If a white-flowered plant is crossed with a red-flowered petunia, what is the genotypic ratio of the F_1?

_____

b. What is the phenotypic ratio of the F_1?

_____

c. If two of the F_1 offspring were crossed, what phenotypes would appear in the F_2?

_____

d. What would be the genotypic ratio in the F_2 generation?

_____

III. Monohybrid Problems Illustrating Codominance

9. Another type of monohybrid inheritance involves the expression of *both* phenotypes in the heterozygous situation. This is called **codominance.**

One of the well-known examples of codominance occurs in the coat color of Shorthorn cattle. Those with reddish-gray coats are heterozygous (RR'), and result from a mating between a red (RR) Shorthorn and one that's white ($R'R'$). Roan cattle do not have roan-colored hairs, as would be expected with incomplete dominance, but rather appear roan as a result of both red and white hairs being on the same animal. Thus the roan coloration is not a consequence of blending of pigments within each hair. Because the *R* and *R'* alleles are *both* fully expressed in the heterozygote, they are codominant.

a. If a roan Shorthorn cow is mated with a white bull, what will be the genotypic and phenotypic ratios in the F_1 generation?

genotypic ratio _____

phenotypic ratio _____

b. List the parental genotypes of crosses that could produce at least some:

white offspring _____

roan offspring _____

IV. Monohybrid, Sex-linked Problems

10. In humans, as well as in other primates, sex is determined by special sex chromosomes. An individual containing two X chromosomes is a female, while an individual possessing an X and a Y chromosome is a male. (Rare exceptions of XY females and XX males have recently been discovered.)

I am a male/female (circle one).

a. What sex chromosomes do you have?

_____

b. In terms of sex chromosomes, what type of gametes (ova) does a female produce?

_____

c. What are the possible sex chromosomes in a male's sperm cells?

_____

d. The gametes of which parent will determine the sex of the offspring?

_____

11. The sex chromosomes bear alleles for traits, just like the other chromosomes in our bodies. Genes that occur on the sex chromosomes are said to be **sex-linked.** More specifically, the genes present on the X chromosome are said to be X-linked. There are many more genes present on the X chromosome than are found on the Y chromosome. Nonetheless, those genes found on the Y chromosome are said to be Y-linked.

The Y chromosome is smaller than its homologue, the X chromosome. Consequently, some of the loci present on the X chromosome are absent on the Y chromosome.

In humans, color vision is X-linked; the gene for color vision is located on the X chromosome but is absent from the Y chromosome. Figure 11-3 illustrates

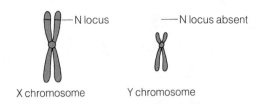

Figure 11-3 Diagrammatic representation of a sex-linked trait, *N*.

the appearance of duplicated sex chromosomes, each consisting of two sister chromatids.

In figure 11-4 sketch the appearance and distribution of the sex chromosomes as they would appear in gametes after meiosis.

Normal color vision (X^N) is dominant over color blindness (X^n). Suppose a color-blind man fathers children of a woman with the genotype $X^N X^N$.

a. What is the genotype of the father?

_____

b. What proportion of daughters would be color-blind?

_____

c. What proportion of sons would be color-blind?

_____

12. One of the daughters from the above problem marries a color-blind man.

a. What proportion of their sons will be color-blind? (Another way to think of this is to ask what the *chances* are that their sons will be color-blind.)

_____

b. Explain how a color-blind daughter might result from this couple.

_____

_____

V. An Observable Monohybrid Cross

Examine the monohybrid genetic corn demonstration. This illustrates a monohybrid cross between plants producing purple kernels and plants producing yellow kernels. Note that all the first-generation kernels (F_1) are purple, while the second-generation ear (F_2) has both purple kernels and yellow kernels. Count the purple kernels and then the yellow kernels. _____ *purple:* _____ *yellow.* When reduced to the lowest common denominator, is this ratio closest to 1:1, 2:1, 3:1, or 4:1? _____. This is called the *phenotypic ratio.*

13. A corncob represents the products of multiple instances of sexual reproduction. Each kernel represents a single instance; fertilization of one egg by one sperm produced *each* kernel. Thus each kernel represents a different cross.

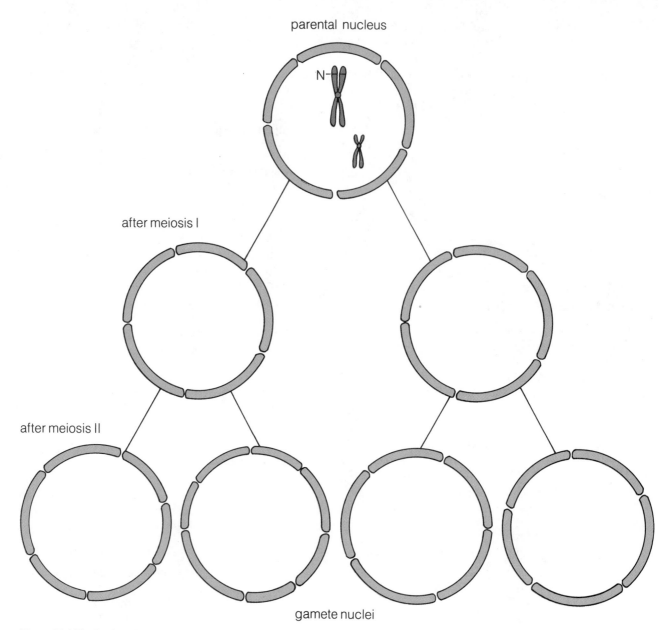

parental nucleus

after meiosis I

after meiosis II

gamete nuclei

Figure 11-4 Distribution of sex chromosomes after meiosis.

a. What genotypes produce a purple phenotype?

_____

b. Which allele is dominant?

_____

c. What is the genotype of the yellow kernels on the F$_2$ ear?

_____

d. Suppose you were given an ear with purple kernels. How could you determine its genotype with a single cross?

_____

_____

VI. Dihybrid Problems

All the problems so far have involved the inheritance of only one trait; that is, they were monohybrid problems. We will now examine cases in which two traits are involved: **dihybrid problems.**

> **NOTE**
>
> **We will assume that the genes for these traits are carried on different (nonhomologous) chromosomes.**

Examine the demonstration of dihybrid inheritance in corn. Notice that not only are the kernels two different colors (one trait), but they are also differently shaped (second trait). Kernels with starchy endosperm (the carbohydrate-storing tissue) are smooth, while those with sweet endosperm are shriveled. Notice that all *four* possible phenotypic combinations of color and shape are present in the F_2 generation.

14. In humans, a pigment in the front part of the eye masks a blue layer at the back of the iris. The dominant allele *P* causes production of this pigment. Those who are homozygous recessive (*pp*) lack the pigment, and the back of the iris shows through, resulting in blue eyes. (Other genes determine the color of the pigment, but in this problem we'll consider only the presence or absence of *any* pigment at the front of the eye.)

Dimpled chins (*D* = allele for dimpling) are dominant over undimpled chins (*d* = allele for lack of dimple).

a. List all possible genotypes for an individual with pigmented iris and dimpled chin.

_____

b. List the possible genotypes for an individual with pigmented iris but lacking a dimpled chin.

_____

c. List the possible genotypes of a blue-eyed, dimple-chinned individual.

_____

d. List the possible genotypes of a blue-eyed individual lacking a dimpled chin.

_____

15. Suppose an individual is heterozygous for both traits (eye pigmentation and chin form).

a. What is the genotype of such an individual?

_____

b. What are the possible genotypes of that individual's gametes?

_____

If determining the answer for the last question was difficult, recall from Exercise 10 that the principle of independent assortment states that genes on different (nonhomologous) chromosomes are separated out independently of one another during meiosis. That is, the occurrence of an allele for eye pigmentation in a gamete has no bearing on which allele for chin form will occur in that same gamete.

There is a useful convention for determining possible gamete genotypes produced during meiosis from a given parental genotype. Using the genotype *PpDd* as an example, here's the method:

Follow the four arrows to determine the four gamete genotypes.

c. Suppose two individuals heterozygous for both eye pigmentation and chin form have children. What are the possible genotypes of their children?

_____

You can set up a Punnett square to do dihybrid problems just as you did with monohybrid problems. However, depending upon the parental genotypes, the square may have as many as sixteen boxes, rather than just four. Insert the possible genotypes of the gametes from one parent in the top circles and the gamete genotypes of the other parent in the circles to the left of the box.

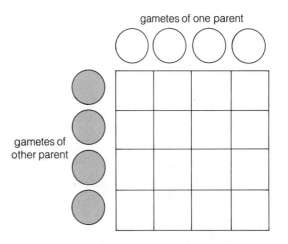

Possible genotypes of children produced by two parents heterozygous for both eye pigmentation and chin form are

_____

d. What is the ratio of the genotypes?

_____

e. What is the phenotypic ratio?

_____

16. You would probably agree that it is unlikely that a family will have sixteen children. In fact, one of the most useful facets of problems such as these is that they allow you to *predict* what the chances are for a phenotype occurring. Genetics is really a matter of **probability,** the likelihood of the occurrence of any particular outcome.

To take a simple example, consider that the probability of coming up with heads in a single toss of a coin is one chance in two, or 1/2.

Now apply this example to the question of the probability of having a certain genotype. Look at your Punnett square in problem 15. The probability of having a genotype is the sum of all occurrences of that genotype. For example, the genotype *PPDD* occurs in one of the sixteen boxes. The probability of having the genotype *PPDD* is 1/16.

a. What is the probability of an individual from the above problem having the genotype:

ppDD _____

PpDd _____

PPDd _____

To extend this idea, let's consider the probability of flipping heads twice in a row with our coin. The chance of flipping heads the first time is 1/2. The same is true for the second flip. The chance (probability) that we will flip heads twice in a row is $1/2 \times 1/2 = 1/4$. The probability that we could flip heads three times in a row is $1/2 \times 1/2 \times 1/2 = 1/8$.

b. Returning to eye color and chin form, state the probability that three children born to these parents will have the genotype *ppdd*.

_____

c. What is the probability that three children born to these parents will have dimpled chins and pigmented eyes?

_____

d. What is the genotype of the F_1 generation when the father is homozygous for both pigmented eyes and dimpled chin, but the mother has blue eyes and no dimple?

_____

e. What is the phenotype of the individual(s) you determined in letter *d.* above?

_____

17. A pigment-eyed, dimple-chinned man marries a blue-eyed woman without a dimpled chin. Their firstborn child is blue-eyed and has a dimpled chin.

_____

a. What are the possible genotypes of the father?

_____

b. What is the genotype of the mother?

_____

c. What alleles may have been carried by the father's sperm?

_____

18. Suppose a dimple-chinned, blue-eyed man whose father lacked a dimple marries a woman who is homozygous recessive for both traits.

a. What would be the expected genotypic ratio of children produced in this marriage?

_____

b. What would be the expected phenotypic ratio?

_____

19. In his original work on the genetics of garden peas, Mendel found that yellow seed color (*YY*, *Yy*) was dominant over green seeds (*yy*) and that round seed shape (*RR*, *Rr*) was dominant over shrunken seeds (*rr*). Mendel crossed pure-breeding (homozygous) yellow, round-seeded plants with green, shrunken-seeded plants.

a. What would be the genotype and phenotype of the F_1 produced from such a cross?

_____

b. If the F_1 plants are crossed, what would be the expected phenotypic ratio of the F_2?

_____

VII. Multiple Alleles

20. The major blood groups in humans are determined by **multiple alleles,** that is, there are *more than* two possible alleles, any one of which can occupy a locus.

In this ABO blood group system, a single gene can exist in any of three allelic forms: I^A, I^B, or i. The alleles *A* and *B* code for production of antigen A and antigen B (two proteins) on the surface of red blood cells. Alleles *A* and *B* are codominant, while allele *i* is recessive.

Four blood groups (phenotypes) are possible from combinations of these alleles (table 11-1).

a. Is it possible for a child with blood type O to be produced by two AB parents?

_____

Explain.

_____

_____

_____

Table 11-1 The ABO Blood Groups

Blood Type	Antigens Present	Antibody Present	Genotype
O	neither A nor B	A and B	ii
A	A	B	$I^A I^A$ or $I^A i$
B	B	A	$I^B I^B$ or $I^B i$
AB	AB	neither A nor B	$I^A I^B$

b. In a case of disputed paternity, the child is type O, the mother type A. Could an individual of the following blood types be the father?

O _____

A _____

B _____

AB _____

VIII. Do Your Results Fit the Expected? Chi-square Analysis of Data

You have found that in monohybrid crosses, the expected phenotypic ratio of offspring produced by a cross between two heterozygous individuals is 3:1. Similarly, in dihybrid situations, the expected ratio would be 9:3:3:1. These are *expected* ratios. Stated in the manner described in the exercise on the scientific method (Exercise 1), it is your *hypothesis* that the ratio will be 3:1 or 9:3:3:1.

By now, you are probably wondering how this ratio can be interpreted if there are not exact multiples of four in the case of monohybrid problems, or of sixteen in the case of dihybrids. If the results don't fit exactly as expected, could the differences be due to chance variation? For example, if you crossed two heterozygous purple garden peas (question 4) and counted sixty-four purple and seventeen white offspring, is the deviation from the expected 3:1 ratio a result of chance or of some other factor?

You can determine whether experimentally obtained data are a satisfactory approximation of the expected data—or, stated another way, whether your hypothesis is correct—using a **chi-square test.**

The formula for chi square (χ^2) is

$$\chi^2 = \Sigma \frac{(O - E)^2}{E}$$

where

O = the *observed* number of individuals,

E = the *expected* number of individuals,

and

Σ = the sum of all values of $(O - E)^2/E$

for the various categories of phenotypes.

Let's see how we use this with our garden-pea example. Suppose eighty-one flowers are counted in a cross. Our hypothesis (expectation) is that three-fourths of them will be purple:

$3/4 \times 81 = 60.75$

Similarly, we expect one-fourth to be white:

$1/4 \times 81 = 20.25$

Examine table 11-2, noting how these values are used.

$$\chi^2 = \Sigma [0.174 + 0.522]$$
$$= 0.696$$

Now, how do we interpret the χ^2 value we found? Suppose that the expected and observed values were identical. Thus, χ^2 would equal zero. You might guess that a number very close to zero would indicate close agreement between observed and expected and that a large χ^2 would suggest that "something unusual" was taking place. But almost always there will be small deviations between observed and expected because of chance, *even when the hypothesis being tested is correct.*

When does the χ^2 value indicate that chance alone cannot explain this deviation? Statisticians have agreed that if the likelihood of deviation is more than one in twenty (or 5% = 0.05), then the disagreement between observed and expected cannot be explained by chance. Here we must consult a table of χ^2 values to make our decision (table 11-3).

In the example above, the χ^2 value is 0.696. Since this was a monohybrid problem with only two categories of possible outcomes (purple or white flowers) the number of degrees of freedom (n in the left hand column of table 11-3) is one. We read across the table until we come to 0.05 and find the number 3.841. Because 0.696, our calculated χ^2 value, is less than 3.841, it is likely that the variation in the observed and expected is the result of chance, and that our hypothesized outcome is correct. A value *greater than* 3.841, however, would indicate that chance alone could not explain the deviation between observed and expected, and we would reject our hypothesis.

Table 11-2 Summary of Calculations of Chi Square for Garden Peas						
Phenotype	Genotype	O	E	(O − E)	(O − E)²	(O − E)²/E
Purple	P_	64	60.75	3.25	10.56	0.174
White	pp	17	20.25	− 3.25	10.56	0.522
Total		81	81	0		0.696

Table 11-3 Distribution of χ^2

n	Probability													
	.99	.98	.95	.90	.80	.70	.50	.30	.20	.10	.05	.02	.01	.001
1	.00016	.00063	.00393	.0158	.0642	.148	.455	1.074	1.642	2.706	3.841	5.412	6.635	10.827
2	.0201	.0404	.103	.211	.446	.713	1.386	2.408	3.219	4.605	5.991	7.824	9.210	13.815
3	.115	.185	.352	.584	1.005	1.424	2.366	3.665	4.642	6.251	7.815	9.837	11.345	16.268
4	.297	.429	.711	1.064	1.649	2.195	3.357	4.878	5.989	7.779	9.488	11.668	13.277	18.465
5	.554	.752	1.145	1.610	2.343	3.000	4.351	6.064	7.289	9.236	11.070	13.388	15.086	20.517
6	.872	1.134	1.635	2.204	3.070	3.828	5.348	7.231	8.558	10.645	12.592	15.033	16.812	22.457
7	1.239	1.564	2.167	2.833	3.822	4.671	6.346	8.383	9.803	12.017	14.067	16.622	18.475	24.322
8	1.646	2.032	2.733	3.490	4.594	5.527	7.344	9.524	11.030	13.362	15.507	18.168	20.090	26.125
9	2.088	2.532	3.325	4.168	5.380	6.393	8.343	10.656	12.242	14.684	16.919	19.679	21.666	27.877
10	2.558	3.059	3.940	4.865	6.179	7.267	9.342	11.781	13.442	15.987	18.307	21.161	23.209	29.588

The term *degrees of freedom* requires further explanation. The number of degrees of freedom is always one *less* than the number of categories of possible outcomes. Thus, you were dealing with a dihybrid problem with a ratio of 9:3:3:1 (four possible phenotypes), n would equal three.

Now let's put this test into practice.

21. In fruit flies, red eyes (R) are dominant over white eyes (r). A student performs a cross between a heterozygous red-eyed fly and a white-eyed fly. The student counts the offspring and finds sixty-five red-eyed flies and forty-nine white-eyed flies.

What is the expected phenotypic ratio of this cross?

_____

Using a χ^2 test, determine if the deviation between the observed and the expected is probably the result of chance.

$\chi^2 = $ _____

Conclusion _____

_____

22. Once again in fruit flies, gray body (G) is dominant over ebony body (g). A red-eyed, gray-bodied fly known to be heterozygous for both traits is mated with a white-eyed fly that is heterozygous for body color.

What is the expected phenotypic ratio for this mating?

_____

The observed offspring consisted of fifteen white-eyed, ebony-bodied flies; thirty-one white-eyed, gray-bodied flies; twelve red-eyed, ebony-bodied flies; and thirty-eight red-eyed, gray-bodied flies.

What is the χ^2 value for this cross?

_____

Is it likely that the observed results "fit" the expected?

The chi-square technique is not restricted to monohybrid or dihybrid problems; it may be used to evaluate any cross in which an expected ratio can be determined.

IX. Some Readily Observable Human Traits

In the preceding problems, we examined several human traits that are fairly simple and that follow the Mendelian pattern of inheritance. Most of our traits are much more complex, involving many genes or interactions between genes. As an example, hair color is determined by at least four genes, each one coding for the production of melanin, a brown pigment. Because the effect of these genes is cumulative, hair color can range from blond (little melanin) to very dark brown (much melanin).

Clearly, human traits are most interesting to humans. A number of traits listed below exhibit Mendelian inheritance. For each, examine your phenotype and fill in table 11-4. List your possible genotype(s) for each trait. When convenient, examine your parents' phenotypes and attempt to determine your actual genotype.

1. *Mid-digital hair (fig. 11-5a).* Examine the joint of your fingers for the presence of hair, the dominant condition (MM, Mm). Complete absence of hair is due to the homozygous-recessive condition (mm). You may need a hand lens to determine your phenotype. Even the slightest amount of hair indicates the dominant condition.

2. *Tongue rolling (fig. 11-5b).* The ability to roll one's tongue is due to a dominant allele, T. The homozygous-recessive condition, t, results in inability to roll one's tongue.

3. *Widow's peak (fig. 11-5c).* Widow's peak describes a distinct downward point in the frontal hairline and is due to the dominant allele, W. The recessive allele, w, results in a continuous hairline. (Omit study of this trait if baldness is affecting the hairline.)

Table 11-4 Summary of My Mendelian Traits

Trait	My Phenotype	My Possible Genotype(s)	Mom's Phenotype	Mom's Possible Genotype	Dad's Phenotype	Dad's Possible Genotype	My Possible or Probable Genotype
Mid-digital hair							
Tongue rolling							
Widow's peak							
Earlobe attachment							
Hitchhiker's thumb							
Relative finger length							

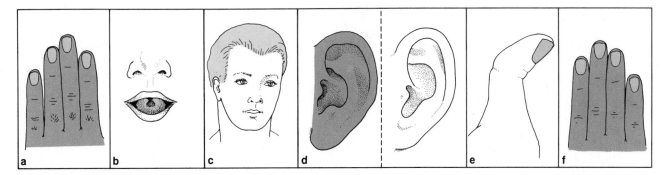

Figure 11-5 Some readily observable human Mendelian traits.

4. *Earlobe attachment (fig. 11-5d).* Most individuals have free earlobes (*FF, Ff*). Homozygous recessives (*ff*) have earlobes attached directly to the head.

5. *Hitchhiker's thumb (fig. 11-5e).* Although considerable variation exists in this trait, we will consider those individuals who *cannot* extend their thumbs backward to approximately 45° to be carrying the dominant allele, *H*. Homozygous-recessive persons (*hh*) can bend their thumbs at least 45°, if not farther.

6. *Relative finger length (fig. 11-5f).* An interesting sex-influenced (*not sex-linked*) trait relates to the relative lengths of the index and ring finger. In males, the allele for a short index finger (*S*) is dominant. In females, it is recessive. In rare cases each hand may be different. If one or both index fingers are greater than or equal to the length of the ring finger, the recessive genotype is present in males, and the dominant present in females.

X. Experiment: Inheritance in a Fungus

Your instructor may provide you with an optional experiment about inheritance of spore color in the fungus *Sordaria*.

PRE-LAB QUESTIONS

_____ 1. In a monohybrid cross (a) only one trait is being considered, (b) the parents are always homozygous, (c) the parents are always heterozygous, (d) no hybrid is produced.

_____ 2. The genetic makeup of an organism is its (a) phenotype, (b) genotype, (c) locus, (d) gamete.

_____ 3. An allele whose expression is completely masked by the expression or effect of its allelic partner is (a) incompletely dominant, (b) homozygous, (c) dominant, (d) recessive.

_____ 4. The physical appearance of an organism, resulting from interactions of its genetic makeup and its environment, is (a) phenotype, (b) hybrid vigor, (c) dominance, (d) genotype.

_____ 5. An organism that is heterozygous for a trait is (a) haploid, (b) homozygous, (c) diploid, (d) all of the above.

_____ 6. Codominance occurs when (a) the phenotype for both (or all) alleles is expressed, (b) the individual is heterozygous, (c) the organism is homozygous recessive, (d) a and b above.

_____ 7. A gene located only on the female (X) chromosome having no allelic partner on the Y chromosome would be (a) incompletely dominant, (b) codominant, (c) sex-linked, (d) heterozygous.

_____ 8. The sex chromosome determining maleness is (a) the Y chromosome, (b) the X chromosome, (c) sex-linked, (d) heterozygous.

_____ 9. A nucleus containing only one of the two homologues is (a) sex-linked, (b) an improbable event, (c) diploid, (d) haploid.

_____ 10. An example of a trait controlled by multiple alleles is (a) baldness in males, (b) color blindness, (c) the ABO blood groups, (d) blue Andalusian chickens.

E X E R C I S E 1 1

Mendelian Genetics

P O S T - L A B Q U E S T I O N S

1. Explain the implications of the principle of independent assortment as it applies to distribution of alleles in gametes.

2. What does it mean to say that certain traits are sex-linked?

3. Distinguish between incomplete dominance and codominance.

4. Define the term *multiple alleles*.

5. Suppose you have two traits controlled by genes on separate chromosomes. If sexual reproduction occurs between two heterozygous parents, what will the genotypic ratio of all possible gametes be?

6. What is the probability that parents will bear five sons and no daughters?

7. How does probability differ from actuality?

8. Studies have suggested (although not proved) that whether you are right- or left-handed may be hereditary. Homozygous-dominant (RR) people are strongly right-handed and are not easily influenced to change preferences. Homozygous-recessive individuals are strongly left-handed. Heterozygous individuals are more variable. They are potentially ambidextrous but are easily influenced by environment or training.

   a. Would you characterize handedness as an example of complete dominance, incomplete dominance, or codominance?

   b. Would it be possible for a left-handed person to be heterozygous?

   c. Would it be possible for two left-handed parents to have a right-handed child? Explain.

9. For this problem, assume that one allele is completely dominant over the other.

   a. Suppose two individuals heterozygous for a *single* trait have children. What is the expected phenotypic ratio of the offspring?

   b. If two individuals heterozygous for *two* traits have children, what would be the expected phenotypic ratio of the offspring?

   c. Remember that the gene for each trait is located at a locus, a physical region on the chromosome. Suppose that crossing two individuals heterozygous for two traits resulted in the same phenotypic ratio as for a single trait. Are the genes for these two traits on separate chromosomes or on the same chromosome? Explain your answer.

10. Explain the usefulness of the chi-square test.

Nucleic Acids: Blueprints for Life

OBJECTIVES

After completing this exercise you will be able to:

1. define *DNA, RNA, purine, pyrimidine, principle of base pairing, replication, transcription, translation, codon, anticodon, peptide bond, gene, genetic engineering, recombinant DNA, plasmid, bacterial conjugation;*

2. identify the components of deoxyribonucleotides and ribonucleotides;

3. distinguish between DNA and RNA according to their structure and function;

4. describe DNA replication, transcription, and translation;

5. give the base sequence of DNA or RNA when presented with the complementary strand;

6. identify a codon and anticodon on RNA models and describe the location and function of each;

7. give the base sequence of an anticodon when presented with that of a codon, and vice versa;

8. describe what is meant by the *one-gene, one-polypeptide hypothesis;*

9. describe the process of DNA recombination by bacterial conjugation;

10. explain the difference between DNA recombination by bacterial conjugation and the technique by which eukaryotic gene products are produced by bacteria.

INTRODUCTION

By 1900 the patterns of inheritance had been demonstrated by Gregor Mendel, based solely on careful experimentation and observation. Mendel had no idea how the traits he observed were passed from generation to generation, although the seeds of that knowledge had been sown as early as 1869, when the physician-chemist Friedrich Miescher isolated the chemical substance of the nucleus. Miescher found the substance to be an acid with a large phosphorus content and named it "nuclein." Subsequently, nuclein was identified as **DNA,** short for **deoxyribonucleic acid.** Some seventy-five years would pass before the significance of DNA would be revealed.

Few would argue that the demonstration of DNA as the genetic material and subsequent determination of its molecular structure are among the most significant discoveries of the twentieth century. Since the early 1950s, when James Watson and Francis Crick built their first model of DNA, tremendous advances in molecular biology have occurred, many of them based upon the structure of DNA. Today we speak of gene therapy and genetic engineering in household conversations. In the minds of some, these topics raise hopes for curing or preventing many of the diseases plaguing humanity. For others, thoughts turn to "playing with nature," undoing the deeds of God, or creating monstrosities that will wipe humanity off the face of the earth.

This exercise will familiarize you with the basic structure of nucleic acids and their role in the cell. Understanding the function of nucleic acids—both DNA and **RNA (ribonucleic acid)**—is central to understanding life itself. We hope you will gain an understanding that will allow you to form educated opinions concerning what science should do with its newfound technology.

I. Isolation and Identification of Nucleic Acids

As we find ourselves on the doorstep of the twenty-first century, no scientific endeavor holds more potential impact for humanity than the "Human Genome Initiative," an effort by the National Institutes of Health, coordinated by James Watson, to identify and determine the location of every gene within our chromosomes. This fifteen-year, $3-billion project is being called biology's "moon shot," with the objective of unlocking the secrets of human life. Among the estimated fifty- to one hundred thousand genes are those that are associated with some four thousand human diseases.

The first step in this incredible undertaking is the isolation of DNA. In the following exercise we'll isolate and identify the nucleic acid components of *Halobacterium salinarum,* a bacterium that grows in habitats with extremely high salt (NaCl) concentrations.

Halobacterium is able to live in its specialized environment because of its cell wall, which differs from that of most other bacteria by maintaining its rodlike shape *only* at high salt concentrations. As NaCl levels drop, the cell shape first becomes irregular and finally spherical. At still lower concentrations, the cell ruptures because of osmotic effects (Exercise 5). We will take advantage of this response to allow us to release the cells' contents, including the nucleic acids, for isolation.

MATERIALS

Per student pair:
- culture tube of *Halobacterium salinarum*
- cotton applicator stick
- 10 mL 95% ethanol in a test tube
- glass rod
- inoculating needle
- 5- or 10-mL sterile pipette
- 10-mL graduated cylinder
- test tube
- test tube rack

Per student group (4):
- 400-mL beaker
- heat source
- 6 screw-cap test tubes
- china marker

Per lab bench:
- vortex mixer (optional)
- 4% NaCl solution*
- diphenylamine solution*
- orcinol solution*
- DNA standard solution*
- RNA standard solution*
- paper towels

Per lab room:
- source of distilled water (dH$_2$O)

PROCEDURE

Work in pairs for part A of this exercise.

A. Isolation of Nucleic Acids

1. With the 10-mL graduated cylinder, measure 1.5 mL of distilled water (dH$_2$O) into a clean test tube.

2. Remove the cap from the slant culture of *H. salinarum* and insert the cotton swab applicator stick into the culture tube. Gently swab the entire surface of the culture by carefully rotating the cotton swab over the pink bacterial colonies. Try to pick up as much of the bacterial colony as possible on the swab. Remove the cotton swab from the culture tube and replace the cap.

3. Transfer the cotton swab applicator stick to the test tube containing the distilled water. Release all the adhering cells by vigorously swirling the cotton swab in the distilled water and occasionally pressing the swab against the wall of the tube.

---

*These solutions should be in Wheaton bottles with Pi-pump–fitted, labeled 5- or 10-mL pipets inserted through a rubber stopper.

The bacterial cells rupture from osmotic shock, since the cell wall cannot withstand the change from conditions of extremely high salt concentration to those of salt absence.

4. Withdraw the swab from the test tube, pressing it against the tube wall to squeeze out as much fluid as possible. *The gelatinous fluid adhering to the swab's surface after wetting will contain a large concentration of nucleic acids. This fluid must be left in the test tube.* Discard the swab.

5. Wipe the surface of the glass rod with a piece of paper towel moistened with 95% ethanol. Insert the clean rod into the test tube containing the cell suspension and stir vigorously. This action will assure total cell lysis. Remove the glass rod.

6. Using the sterile pipet, add 3 mL of 95% ethanol one drop at a time down the *side* of the test tube containing the cell suspension. Any material adhering to the tube wall should be washed into the suspension. The alcohol should form a layer on top of the aqueous cell suspension; *be careful not to mix the water and alcohol layers.*

7. Clean the inoculating wire with a paper towel moistened in 95% ethanol. Insert the wire into the culture tube so that the hook is at the cell-suspension–alcohol interface and rotate the wire in a circular motion. The swirling should mix the contents only at the partition layer between the alcohol and cell suspension.

Nucleic acids precipitate and are extracted at this boundary between alcohol and water. Notice that strands of material adhere to the wire and trail off into the solution. The long linear-chain molecules of DNA appear stringy and form a cottony, viscous cloud around the wire.

The nucleic acids you have extracted are not pure; they contain cellular debris as well as adhering proteins. Nevertheless, we can test the extracted precipitate to identify its major components.

Work in groups of four for parts B and C of this exercise.

B. Identification: Test for DNA

1. Start a boiling water bath by placing 250 mL of tap water in your beaker and placing it on the heat source.

2. Pipet 3 mL of 4% NaCl into a clean screw-top test tube. With a china marker, label this test tube with the letters "P$_A$" to indicate that the precipitated material from one test tube will be placed in it.

3. Place the nucleic acid from the wire winder into the labeled test tube. Replace the cap and dissolve the viscous material in the NaCl solution, using a vortex mixer, if available.

4. Place 3 mL of the DNA standard into a second test tube (labeled "S$_A$" for Standard) and 3 mL of dH$_2$O into a third (labeled "C$_A$" for Control).

5. Pipet 3 mL of diphenylamine into each test tube.

Table 12-1 Identification of Nucleic Acids in *Halobacterium*

Tube Code	Tube Contents	Color After Boiling
P$_A$	Precipitate + diphenylamine	
S$_A$	DNA standard + diphenylamine	
C$_A$	dH$_2$O + diphenylamine	
P$_B$	Precipitate + orcinol	
S$_B$	RNA standard + orcinol	
C$_B$	dH$_2$O + orcinol	

CAUTION

Diphenylamine contains concentrated sulfuric and glacial acidic acids. Should you get any on your skin, flush the area with large amounts of running water and inform your instructor immediately.

6. *With the screw tops loose,* place the test tubes into a boiling water bath for 15 minutes.

7. Record the color of the contents of each test tube in table 12-1.

C. Identification: Test for RNA

1. Pipet 2 mL of 4% NaCl into a clean, screw-top test tube. With a china marker, label the tube with the letters "P$_B$".

2. Place the precipitated nucleic acid from a second culture into tube "P$_B$", attach the screw cap, and dissolve the viscous material in the NaCl solution, using a vortex mixer, if available.

3. Pipet 2 mL of an RNA standard into a second test tube and 2 mL of dH$_2$O into a third. Label these two tubes "S$_B$" and "C$_B$", respectively.

4. Pipet 2 mL of orcinol into each tube.

CAUTION

Orcinol contains concentrated hydrochloric acid. Should you get any on your skin, flush the affected area with large amounts of running tap water and inform your instructor immediately.

5. *With the screw caps loosened,* place the test tubes into a boiling water bath for 20 minutes.

6. Record the colors of the contents of each test tube in table 12-1.

What substances were present in the material you isolated?

II. Modeling the Structure and Function of Nucleic Acids and Their Products

MATERIALS

Per student pair:
• DNA puzzle kit

Per lab room:
• DNA model

PROCEDURE

In this exercise, we are concerned with three processes: *replication, transcription,* and *translation.* But before we study these three per se, let's formulate an idea of the structure of DNA itself.

A. Nucleic Acid Structure

Work in pairs.

NOTE

Clear your work surface of everything except your lab manual and the DNA puzzle kit.

1. Obtain a DNA puzzle kit. It should contain the following parts:

• 18 deoxyribose sugars
• 9 ribose sugars
• 18 phosphate groups
• 4 adenine bases
• 6 guanine bases
• 6 cytosine bases
• 4 thymine bases
• 2 uracil bases

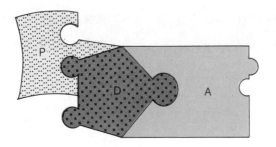

Figure 12-1 One deoxyribonucleotide.

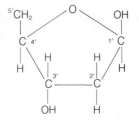

Figure 12-2 Deoxyribose.

- 3 transfer RNA (tRNA)
- 3 amino acids
- 3 activating units
- ribosome template sheet

2. Group the components into separate stacks. Select a single deoxyribose sugar, an adenine base (labeled "A"), and a phosphate, fitting them together as shown in figure 12-1. This is a single nucleotide (specifically a *deoxy*ribonucleotide), a unit consisting of a sugar (deoxyribose), a phosphate group, and a nitrogen-containing base (adenine).

Let's examine each component of the nucleotide.

Deoxyribose (fig. 12-2) is a sugar compound containing five carbon atoms. Four of the five are joined by covalent bonds into a ring. Each carbon is given a number, indicating its position in the ring. (These numbers are read "1-prime, 2-prime," and so on. "Prime" is used to distinguish the carbon atoms from the position of atoms that are sometimes numbered in the nitrogen-containing bases.)

This structure is usually illustrated in a simplified manner, without actually showing the carbon atoms within the ring (fig. 12-3).

Figure 12-3 Simplified representation of deoxyribose.

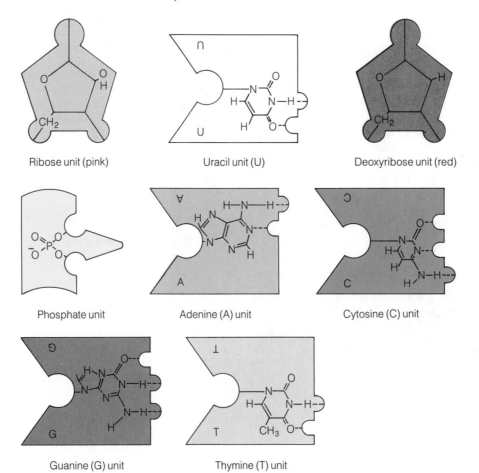

Ribose unit (pink)

Uracil unit (U)

Deoxyribose unit (red)

Phosphate unit

Adenine (A) unit

Cytosine (C) unit

Guanine (G) unit

Thymine (T) unit

There are four kinds of nitrogen-containing bases in DNA. Two are **purines** and are double-ring structures. Specifically, the two purines are *adenine* and *guanine* (abbreviated A and G, respectively; fig. 12-4).

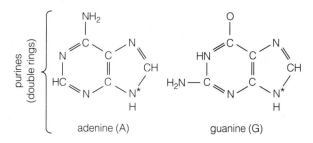

Figure 12-4 Double-ringed purines found in DNA.

The other two nitrogen-containing bases are **pyrimidines,** specifically *cytosine* and *thymine* (abbreviated C and T, respectively). Pyrimidines are single-ring compounds, as illustrated in figure 12-5.

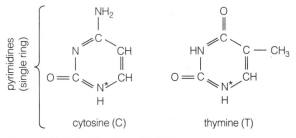

Figure 12-5 Pyrimidines found in DNA.

The symbol * indicates where a bond forms between each nitrogen-containing base and the 1' carbon atom of the sugar ring structure. Although deoxyribose and the nitrogen-containing bases are organic compounds (they contain carbon), the phosphate group is an inorganic compound, with the structural formula shown in figure 12-6:

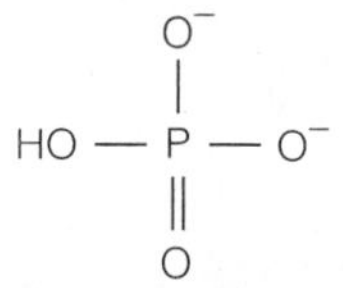

$$HO - \overset{\overset{\displaystyle O^-}{|}}{\underset{\underset{\displaystyle O}{\parallel}}{P}} - O^-$$

Figure 12-6 Phosphate group found in nucleic acids.

The phosphate end of the deoxyribonucleotide is referred to as the 5' end, because the phosphate group bonds to the 5' carbon atom.

There are four kinds of deoxyribonucleotides, each differing only in the type of base it possesses. Construct the other three kinds of deoxyribonucleotides. Draw each in figure 12-7b through d. Rather than drawing the somewhat complex shape of the model, in this and other drawings, just give the correct position and letters. Use *D* for deoxyribose, *P* for a phosphate group, and *A*, *C*, *G*, and *T* for the different bases (as shown in fig. 12-7a).

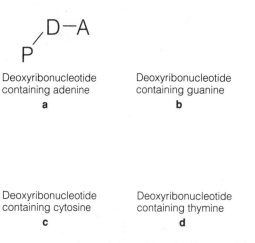

Deoxyribonucleotide containing adenine
a

Deoxyribonucleotide containing guanine
b

Deoxyribonucleotide containing cytosine
c

Deoxyribonucleotide containing thymine
d

Figure 12-7 Drawings of deoxyribonucleotides containing guanine, cytosine, and thymine.

Note the small notches and projections in the nitrogen-containing bases. Will the notches of adenine and thymine fit together?

_____

Of guanine and cytosine?

_____

Of adenine and cytosine?

_____

Of thymine and guanine?

_____

The notches and projections represent bonding sites. Make a conclusion about which bases will bond with one another.

_____

_____

_____

Will a purine base bond with another purine?

_____

Will a purine base bond with both types of pyrimidines?

_____

3. Assemble the three additional deoxyribonucleotides, linking them with the adenine-containing unit, to form a nucleotide strand of DNA. Note that the sugar backbone is bonded together by phosphate groups. Your strand should appear as shown in figure 12-8.

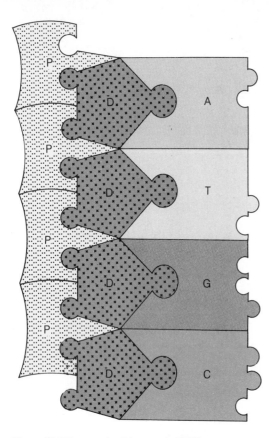

Figure 12-8 Four-nucleotide strand of DNA.

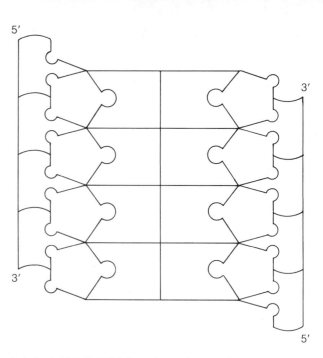

Labels: A, T, G, C, D, P (all used more than once)

Figure 12-9 Drawing of a double strand of DNA.

4. Now assemble a second four-nucleotide strand, similar to that of figure 12-8. However, this time make the base sequence T — A — C — G, from bottom to top.

DNA molecules consist of *two* strands of nucleotides, each strand the *complement* of the other.

5. Assemble the two strands by attaching (bonding) the nitrogen bases of complementary strands. Note that the adenine of one nucleotide always pairs with the thymine of its complement; similarly, guanine always pairs with cytosine. This phenomenon is called the **principle of base pairing.**

On figure 12-9 attach letters to the model pieces indicating the composition of your double-stranded DNA model.

What do you notice about the *direction* in which each strand is running? (That is, are both 5′ carbons at the same end of the strands?)

_____

_____

_____

(Does the second strand of your drawing show this? It should.)

In life, the purines and pyrimidines are joined together by hydrogen bonds. Note again that the sugar backbone is linked by phosphate groups. Your model illustrates only a very small portion of a DNA molecule. The entire molecule may be tens of thousands of nucleotides in length!

156

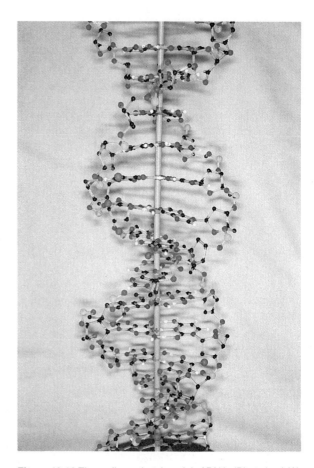

Figure 12-10 Three-dimensional model of DNA. (Photo by J. W. Perry.)

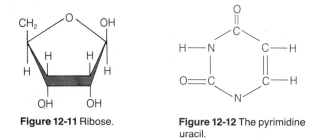

Figure 12-11 Ribose.

Figure 12-12 The pyrimidine uracil.

6. Slide your DNA segment aside for the moment.

7. Examine the three-dimensional model of DNA on display in the laboratory (fig. 12-10). Notice that the two strands of DNA are twisted into a spiral-staircase-like pattern.

This is why DNA is known as a *double helix*. Identify the deoxyribose sugar, nitrogen-containing bases, hydrogen bonds linking the bases, and the phosphate groups.

The second type of nucleic acid is RNA, short for ribonucleic acid. There are three important differences between DNA and RNA:

a. RNA is a *single strand* of nucleotides.

b. The sugar of RNA is **ribose.**

c. RNA lacks the nucleotide that contains thymine. Instead, it has one containing the pyrimidine uracil (U) (see fig. 12-12).

Compare the structural formulas of ribose (fig. 12-11) and deoxyribose (fig. 12-3). How do they differ?

_____

_____

_____

Why is the sugar of DNA called *deoxy*ribose?

_____

_____

_____

8. From the remaining pieces of your model kit, select four ribose sugars — an adenine, uracil, guanine, and cytosine — and four phosphate groups. Assemble the four **ribonucleotides** and draw each in figure 12-13. (Use the convention illustrated in fig. 12-7 rather than drawing the actual shapes.)

Disassemble the RNA models after completing your drawing.

B. Modeling DNA Replication

DNA **replication** takes place during the S stage of interphase of the cell cycle (see Exercise 9, page 97). Recall that the DNA is aggregated into chromosomes. Before mitosis, the chromosomes duplicate themselves so that the daughter nuclei formed by mitosis will have the same number of chromosomes (and hence the same amount of DNA) as did the parent cell.

Figure 12-13 Drawings of four possible ribonucleotides.

Replication begins when hydrogen bonds between nitrogen bases break and the two DNA strands "unzip." Free nucleotides within the nucleus bond to the exposed bases, thus creating *two* new strands of DNA (as described below). The process of replication is controlled by enzymes called **DNA polymerases.**

1. Construct eight more deoxyribonucleotides (two of each kind) but don't link them into strands.

2. Now return to the double-stranded DNA segment you constructed earlier. Separate the two strands, imagining the zipperlike fashion in which this occurs within the nucleus.

3. Link the free deoxyribonucleotides to each of the "old" strands. When you are finished, you should have two double-stranded segments.

Note that one strand of each is the parental ("old") strand and that the other is newly synthesized from free nucleotides. This illustrates the *semiconservative* nature of DNA replication. Each of the parent strands remains intact — it is *conserved* — and a new complementary strand is formed on it. Two "half-old, half-new" DNA molecules result.

4. Draw the two replicated DNA molecules in figure 12-14, labeling the old and new strands. (Once again, use the convention illustrated in fig. 12-7.)

C. Transcription: DNA to RNA

DNA is an "information molecule" residing *within* the nucleus. The information it provides is for assembling proteins *outside* the nucleus, within the cytoplasm. The information does not go directly from the DNA to the cytoplasm. Instead, RNA serves as an intermediary, carrying the information from DNA to the cytoplasm.

Synthesis of RNA takes place within the nucleus by **transcription.** During transcription, the DNA double helix unwinds and unzips, and a single strand of RNA, designated **messenger RNA (mRNA),** is assembled using the nucleotide sequence of *one* of the DNA strands as a pattern (template). Let's see how this happens.

To *transcribe* means to "make a copy of." Is transcription of RNA from DNA the formation of an *exact* copy?

_____

Explain.

_____

_____

_____

You will use this strand of mRNA in the next section. Keep it close at hand.

D. Translation: RNA to Polypeptides

Once in the cytoplasm, mRNA strands attach to *ribosomes*, on which translation occurs. To *translate* means to change from one language to another. In the biological sense, **translation** is the conversion of the linear message encoded on mRNA to a linear strand of amino acids to form a polypeptide. (A *peptide* is two or more amino acids linked by a peptide bond.)

Translation is accomplished by the interaction of mRNA, ribosomes, and **transfer RNA (tRNA)**, another type of RNA. The tRNA molecule is formed into a four-cornered loop. You can think of tRNA as a baggage-carrying molecule. Within the cytoplasm, tRNA attaches to specific free amino acids. This occurs with the aid of activating enzymes, represented in your model kit by the pieces labeled "glycine activating" or "alanine activating." The amino acid-carrying tRNA then positions itself on ribosomes where the amino acids become linked together to form polypeptides.

1. Obtain three tRNA pieces, three amino acid units, and three activating units.

2. Join the amino acids first to the activating units and then to the tRNA. Will a particular tRNA bond with *any* amino acid, or is each tRNA specific?

_____

3. Now let's do some translating. In the space below, list the sequence of bases on the *messenger* RNA strand, starting at the left.

(left, 3′ end) _____ (right, 5′ end)

Translation occurs when a *three*-base sequence on mRNA is "read" by tRNA. This three-base sequence on mRNA is called a **codon**. Think of a codon as a three-letter word, read right (5′) end to left (3′) end. What is the order of the rightmost (first) mRNA codon? (Remember to list the letters in the *reverse* order of that in the mRNA sequence.) The first codon on the mRNA model is

(5′ end) _____ (3′ end)

Figure 12-14 Drawing of two replicated DNA segments, illustrating their semiconservative nature.

1. Disassemble the replicated DNA strands into their component deoxyribonucleotides.

2. Construct a new DNA strand consisting of nine deoxyribonucleotides. With the purines and pyrimidines pointing away from you, lay the strand out horizontally in the following base sequence: T—G—C—A—C—C—T—G—C.

3. Now assemble RNA ribonucleotides on the exposed nitrogen bases of the DNA strand. Don't forget to substitute the pyrimidine uracil for thymine.

What is the sequence from left to right of nitrogen bases on the mRNA strand?

left _____ right

After the mRNA is synthesized within the nucleus, the hydrogen bonds between the nitrogen bases of the deoxyribonucleotides and ribonucleotides break.

4. Separate your mRNA strand from the DNA strand. (You can disassemble the deoxyribonucleotides now.) The mRNA now moves out of the nucleus and into the cytoplasm.

By what avenue do you suppose the mRNA exits the nucleus? (Hint: Reexamine the structure of the nuclear membrane, as described in Exercise 4.)

_____

_____

_____

4. Slide the mRNA strand onto the ribosome template sheet, with the first codon at the 5' end.

5. Find the tRNA-amino acid complex that complements (will fit with) the first codon. The complementary three-base sequence on the tRNA is the **anticodon.** Binding between codons and anticodons begins at the P site of the 40s subunit of the ribosome. The tRNA-amino acid complex with the correct anticodon positions itself on the P site.

6. Move the tRNA-amino acid complex onto the P site on the ribosome template sheet and fit the codon and anticodon together. In the boxes below, indicate the codon, anticodon, and the specific amino acid attached to the tRNA.

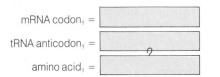

7. Now identify the second mRNA codon and fill in the boxes.

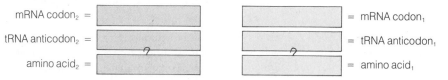

8. The second tRNA-amino acid complex moves onto the A site of the 40s subunit. Position this complex on the A site. An enzyme now catalyzes a condensation reaction, forming a **peptide bond** and linking the two amino acids into a dipeptide. (Water, HOH, is released by this condensation reaction.)

9. Separate amino acid$_1$ from its tRNA and link it to amino acid$_2$. (In reality, separation occurs somewhat later, but the puzzle does not allow this to be shown accurately; see below for correct timing.)

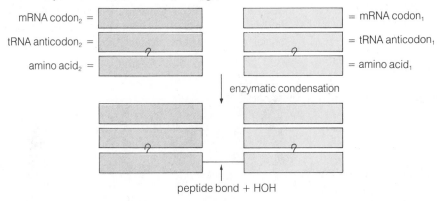

One tRNA-amino acid complex remains. It must occupy the A site of the ribosome in order to bind with its codon. Consequently, the dipeptide must move to the right.

10. Slide the mRNA to the right (so that tRNA$_2$ is on the P site) and fit the third mRNA codon and tRNA anticodon to form a peptide bond, creating a model of

a tripeptide. At about the same time that the second peptide bond is forming, the first tRNA is released from both the mRNA and the first amino acid. Eventually, it will pick up another specific amino acid.

tRNA$_1$ will pick up another (name the type of amino acid) _____ .

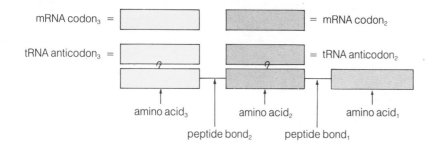

mRNA codon₃ = □   ▨ = mRNA codon₂

tRNA anticodon₃ = □   ▨ = tRNA anticodon₂

amino acid₃   amino acid₂   amino acid₁

peptide bond₂   peptide bond₁

Record above the tripeptide that you have just modeled.

You have created a short polypeptide. Polypeptides may be thousands of amino acids in length. As you see, the amino acid sequence is ultimately determined by DNA, because it was the original source of information.

Finally, let's turn our attention to the concept of a gene. A **gene** is a unit of inheritance. Our current understanding of a gene is that a gene codes for one polypeptide. This is appropriately called the **one-gene, one-polypeptide hypothesis.** Given this concept, do you think a gene consists of one, several, or many deoxyribonucleotides?

A gene probably consists of _____ deoxyribonucleotides.

NOTE

Please disassemble your models and return them to the proper location.

III. Principles of Genetic Engineering: Recombination of DNA

People suffering from diabetes are unable to produce enough insulin, a hormone that is synthesized by the pancreas and that is instrumental in regulating the amount of blood sugar. Therapy for severe diabetes includes daily injections of insulin. Until recently, that insulin was extracted from the pancreas of slaughtered pigs and cows. With the advent of techniques commonly referred to as genetic engineering, human insulin is now produced by bacteria. These organisms grow and reproduce rapidly, hence producing quantities of insulin en masse.

Genetic engineering is a convenient phrase to describe what is more properly called methods in recombinant DNA. **Recombinant DNA** is DNA into which a set of "foreign" nucleotides has been inserted. In the case of insulin production, researchers first located on human chromosomes the gene (set of nucleotides) that codes for insulin production. Once identified, the nucleotides were removed from the human DNA and inserted into the DNA of a bacterium. As this bacterial cell reproduced, each new generation contained the gene coding for insulin synthesis. The cells produced

the hormone, which was harvested. Thus, these recombinant bacteria are "insulin factories."

Bacteria have been exchanging genes with each other for millennia. In the process, new strains of bacteria may be produced. The following experiment will familiarize you with genetic recombination in bacteria, principles of which are the basis for genetic engineering.

Two strains of the bacterium *Escherichia coli* will be used in this experiment:

• J-53R carries a chromosomal gene that causes it to be resistant to the antibiotic rifampicin; it is susceptible to (killed by) another antibiotic, chloramphenicol.

• HT-99 is resistant to chloramphenicol but susceptible to rifampicin; the gene for resistance to chloramphenicol is located on a small extra–chromosomal (not on the chromosome) loop of DNA called a **plasmid.**

Like chromosomal DNA, plasmids can replicate. Insertion of a foreign DNA segment (set of nucleotides) results in the formation of a hybrid plasmid that can thereby replicate the foreign DNA as well.

Plasmids also code for the ability to transfer themselves from the host (donor) bacterium to a recipient cell by a process called **bacterial conjugation.** Thus, the plasmid acts both as a carrier of foreign DNA and as an agent (vector) for the introduction of that DNA into the recipient cell. Once in the recipient, the plasmid replicates, and the recipient bears the genes (and hence makes the gene products) formerly in the host.

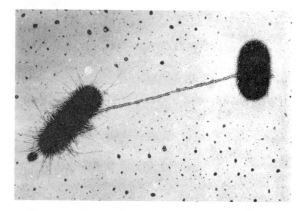

Figure 12-15 Conjugation between two bacteria. (Photo courtesy C. C. Brinton, Jr., and J. Carnahan.)

Plasmid transfer between host and recipient (in this case, two bacterial cells) occurs through a bridge formed by the host cell that connects it to the recipient. Figure 12-15 illustrates bacterial conjugation.

> **NOTE**
>
> This experiment may be set up as a demonstration. If so, skip to the questions at the end of section C.

MATERIALS

Per student pair:

- nutrient agar plate containing chloramphenicol
- nutrient agar plate containing rifampicin
- nutrient agar plate containing both chloramphenicol and rifampicin
- culture tube containing *E. coli* J-53R
- culture tube containing *E. coli* HT-99
- culture tube containing mating mixture of both J-53R and HT-99
- test tube rack
- bottle of 10% Clorox solution
- transfer loop
- burner (Bunsen or alcohol), striker or matches
- china marker

Per lab room:

- paper toweling
- 37°C incubator
- container for discarded cultures

Your instructor will demonstrate aseptic technique before you begin. Then proceed in the manner described below.

PROCEDURE

A. Chloramphenicol Plate

1. Pour 10% Clorox disinfectant onto the lab bench work area and wipe thoroughly with paper toweling.

2. Without removing the lid, turn the chloramphenicol-containing plate over and, using a china marker, draw a line on the bottom dividing the plate in half. Label one half "J-53R" and the other half "HT-99". Turn the plate right-side up.

3. Light the burner.

4. Sterilize the transfer loop by flaming, as illustrated in figure 12-16a on the following page.

5. Allow the loop to cool for fifteen seconds.

6. As shown in figure 12-16b, hold the sterile transfer loop and culture tube containing J-53R and remove the cap of the tube.

> **NOTE**
>
> Never set the cap on the bench surface.

7. Flame the mouth of the culture tube (fig. 12-16c).

8. Insert the loop into the culture tube, touch the bacterial colony, and withdraw the loop. Reflame the mouth of the culture tube and replace the cap (fig. 12-16d).

9. Lift one edge of the chloramphenicol-containing nutrient agar plate's cover (fig. 12-16e). Transfer the J-53R to the appropriately labeled side of the chloramphenicol plate. Pull the loop lightly across the agar surface in a close zigzag streaking pattern. Close the lid and set the plate aside.

> **NOTE**
>
> Do not dig into the agar.

10. Sterilize the transfer loop again by flaming. Allow it to cool.

11. Transfer HT-99 to the other half of the chloramphenicol plate, spreading it the same way as before.

12. Sterilize the loop.

B. Rifampicin Plate

1. With the china marker, draw a line down the center of the bottom of the rifampicin-containing culture plate, labeling one half "J-53R," the other "HT-99".

2. Using the same series of steps you used above, transfer J-53R and HT-99 to this plate.

> **NOTE**
>
> Don't forget to sterilize the loop before and after each transfer.

C. Chloramphenicol plus Rifampicin Plate

1. Proceed as above but do not divide this plate in half. Inoculate the culture plate containing both antibiotics with a drop of the culture solution that contains *both* J-53R and HT-99. This *mating solution* was prepared yesterday by mixing the two strains and then incubating it overnight.

2. Mark your name on the culture plates and place them in a 37°C incubator.

> **NOTE**
>
> Disinfect your work surface with Clorox.

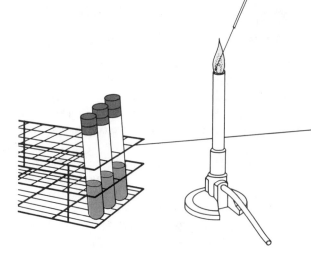

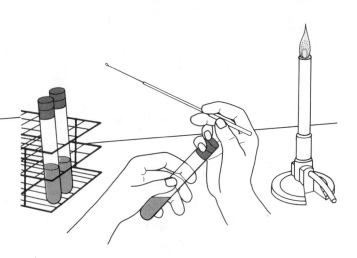

a Sterilize the loop by holding the wire in a flame until it is red hot. Allow it to cool before proceeding.

b While holding the sterile loop and the bacterial culture, remove the cap as shown.

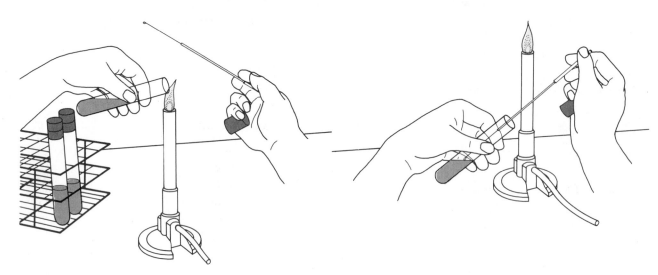

c Briefly heat the mouth of the tube in a burner flame before inserting the loop for an inoculum.

d Get a loopful of culture, withdraw the loop, heat the mouth of the tube, and replace the cap.

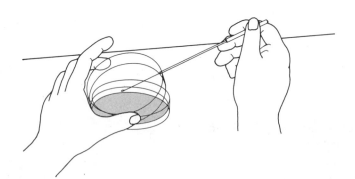

e To inoculate a solid medium in a Petri plate, place the plate on a table and lift one edge of the cover.

Figure 12-16 Procedure for inoculating a culture plate. (After Case and Johnson, 1984.)

Table 12-2 Bacterial Growth on Antibiotic-containing Plates			
	Growth of Bacteria on Nutrient Agar Containing:		
Strain of E. coli	**Chloram-phenicol**	**Rifampicin**	**Chloramphenicol plus Rifampicin**
HT-99 (Donor)			
J-53R (Recipient)			
"Mating mixture"			

After 24 hours remove the culture plates and examine them for growth. (Your instructor will provide demonstration plates for you to examine to recognize bacterial growth.)

Record your observations in table 12-2, using a "+" to indicate the growth of bacteria, a "−" to indicate absence of growth.

Discard your plates in the container provided.

Was HT-99 susceptible or resistant to chloramphenicol?

_____

To rifampicin?

_____

Was J-53R susceptible or resistant to chloramphenicol?

_____

To rifampicin?

_____

Make a conclusion about the presence of a gene in each of the two strains of *E. coli* for resistance to each of the antibiotics.

HT-99: _____

_____

J-53R: _____

Make a conclusion concerning what happened when the two strains were mixed together. Incorporate your observations concerning antibiotic resistance into your conclusion.

_____

_____

_____

_____

PRE-LAB QUESTIONS

____ 1. The individuals responsible for constructing the first model of DNA structure were (a) Wallace and Watson, (b) Lamarck and Darwin, (c) Aristotle and Socrates, (d) Crick and Watson.

____ 2. Deoxyribose is (a) a five-carbon sugar, (b) present in RNA, (c) a nitrogen-containing base, (d) one type of purine.

____ 3. A nucleotide may consist of (a) deoxyribose or ribose, (b) purines or pyrimidines, (c) phosphate groups, (d) all of the above.

____ 4. Which of the following is consistent with the principle of base pairing? (a) purine-purine, (b) pyrimidine-pyrimidine, (c) adenine-thymine, (d) guanine-thymine.

____ 5. Nitrogen-containing bases between two complementary DNA strands are joined by (a) polar covalent bonds, (b) hydrogen bonds, (c) phosphate groups, (d) deoxyribose sugars.

____ 6. The difference between deoxyribose and ribose is that ribose (a) is a six-carbon sugar, (b) bonds only to thymine, not uracil, (c) has one more oxygen atom than deoxyribose has, (d) all of the above.

____ 7. Replication of DNA (a) takes place during interphase, (b) results in two double helices from one, (c) is semiconservative, (d) all of the above.

____ 8. Transcription of DNA (a) results in formation of a complementary strand of RNA, (b) produces two new strands of DNA, (c) occurs on the surface of the ribosome, (d) is semiconservative.

____ 9. An anticodon (a) is a three-base sequence of nucleotides on tRNA, (b) is produced by translation of RNA, (c) has the same base sequence as does the codon, (d) is the same as a gene.

____ 10. Bacterial conjugation (a) may result in a new strain of bacteria, (b) occurs when DNA nucleotides are transferred from one bacterial strain to another, (c) may result in genetic hybridization, (d) all of the above.

Name _____ Section Number _____

E X E R C I S E   1 2

Nucleic Acids: Blueprints for Life

POST-LAB QUESTIONS

1. The illustration below represents some of the puzzle pieces used in this exercise.
   a. Assembled in this form, do they represent a(an) amino acid, base, portion of messenger RNA, or deoxyribonucleotide?

   b. Justify your answer.

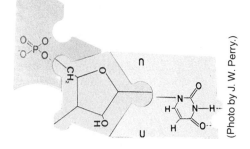

(Photo by J. W. Perry.)

2. Why is DNA often called a *double helix?*

3. a. What is the ratio of guanine to cytosine in a double-stranded DNA molecule?

   b. Of adenine to thymine?

4. Define:
   a. *replication*

   b. *transcription*

   c. *translation*

   d. *codon*

   e. *anticodon*

5. What does it mean to say that DNA replication is *semiconservative?*

6. a. If the base sequence on one DNA strand is ATGGCCTAG, what will the sequence be on the other strand of the helix?

   b. If the original strand serves as the template for transcription, what will the sequence be on the newly formed RNA strand?

7. The illustration at the right is an enlarged portion of the DNA model pictured in figure 12-10 (page 156). Identify parts a and b.

   a _____

   b _____

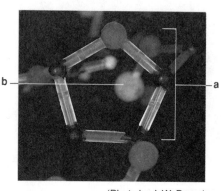

(Photo by J. W. Perry.)

8. a. What amino acid would be produced if *transcription* took place from a nucleotide with the three-base sequence ATA?

   b. Suppose a genetic mistake took place during *replication* and the new DNA strand had the sequence ATG. What would be the three-base sequence on an RNA strand transcribed from this series of nucleotides?

   c. Which amino acid would this codon result in?

   d. Explain.

9. a. What is a plasmid?

   b. How are plasmids used in genetic engineering?

10. How does bacterial conjugation differ from the process by which eukaryotic gene products are produced by bacteria?

EXERCISE 13
Evolutionary Agents

OBJECTIVES

After completing this exercise you will be able to:

1. define *evolutionary agent, natural selection, fitness, directional selection, stabilizing selection, disruptive selection, gene flow, divergence, speciation, mutation, genetic drift, bottleneck effect, founder effect;*

2. determine the allele frequencies for a gene in a model population;

3. calculate expected ratios of phenotypes based on Hardy-Weinberg proportions;

4. describe the effects of nonrandom mating, natural selection, migration, genetic drift, and mutation on a model population;

5. describe the effects of different selection pressures on identical model populations;

6. identify the level at which selection operates in a population;

7. describe the impact of the founder effect on the genetic structure of populations.

INTRODUCTION

The Hardy-Weinberg Principle says that heredity itself cannot cause changes in the frequencies of alternate forms of the same gene (alleles). If certain conditions are met, then the proportions of genotypes that make up a population of organisms should remain constant generation after generation according to Hardy-Weinberg equilibrium:

$$p^2 + 2pq + q^2 = 1.0 \text{ (for two alleles)}$$

If p is the frequency of one allele, A, and q is the frequency of the other allele, a, then

$$p + q = 1.0$$

If two alleles for coat color exist in a population of mice and the allele for white coats is present 70% of the time, then the alternate allele (black) must be present 30% of the time.

The Hardy-Weinberg Principle can be used to determine the proportions of phenotypes present in succeeding generations, as long as conditions do not change. In our example, since $p = 0.7$, we would expect 49% (p^2) of the mice in our population to be homozygous for white coats. Forty-two percent ($2pq$) would have one of each allele and would appear gray if the alleles are codominants (that is, both alleles have equal expression in the phenotype). What percentage of our population is homozygous for black coats?

_____ %

In nature, however, the frequencies of genes in populations are not static (that is, not unchanging). Natural populations never meet all of the assumptions for Hardy-Weinberg equilibrium. *Evolution is a process resulting in changes in the genetic makeup of populations through time;* therefore, factors that disrupt Hardy-Weinberg equilibrium are referred to as **evolutionary agents.** In random mating populations, natural selection, gene flow, genetic drift, and mutation can all result in a shift in gene frequencies predicted by the Hardy-Weinberg formula. Nonrandom mating can also result in such changes. This exercise will demonstrate the effect of these agents on the genetic structure of a simplified model population.

MATERIALS

Per student group (4):
- plastic dishpan (12″ × 7″ × 2″)
- 50 large (10-mm diameter) white beads
- 50 large red beads
- 50 large pink beads
- 1 large gray bead
- 4,000 small (8-mm diameter) white beads
- 4,000 small red beads (optional)
- pair of long forceps
- coarse sieve (9.5-mm)
- small bowl
- calculator

Per lab room:
- clock with a second hand

PROCEDURE

A. The General Model

The populations you will be working with are composed of colored beads. White beads in our model represent individuals that are homozygous for the white allele ($C^W C^W$). Red beads are homozygous for the red allele ($C^R C^R$), and pink beads are heterozygotes ($C^W C^R$). These beads exist in "ponds" that are represented by plastic dishpans filled with smaller beads. The smaller beads can be strained to retrieve all the "individuals" that make up the model population. When the individuals are recovered, the frequencies of the color alleles can be determined using the Hardy-Weinberg formula. The alleles in our population are codominant. Thus, each white bead contains two white alleles; each pink bead, one white and one

Table 13-1 Counts of Large Beads Before and After Four Rounds of Simulated Predation

	White	Pink	Red	Total
Initial population				
Before	10	20	10	40
After	_____	_____	_____	_____
Second population				
Before	_____	_____	_____	50
After	_____	_____	_____	_____
Third population				
Before	_____	_____	_____	50
After	_____	_____	_____	_____
Fourth population				
Before	_____	_____	_____	50
After	_____	_____	_____	_____

red; and each red bead, two red alleles. The total number of color alleles in a population of twenty individuals is forty. If such a population contains five white beads and ten pink beads, the frequency of the white allele is:

$$p = \frac{(2 \times 5) + 10}{40} = 0.5$$

Because $p + q = 1.0$, the frequency of the red allele (q) must also be 0.5 if there are only two color alleles in this population.

B. Natural Selection

Natural selection disturbs Hardy-Weinberg equilibrium by discriminating between individuals with respect to their ability to produce young. Those individuals that survive and reproduce will perpetuate more of their genes in the population. These individuals are said to exhibit greater **fitness** than those who leave no offspring or fewer offspring. We will model the effect of natural selection by simulating predation on our population.

1. Working in groups of four, establish the initial population. Place ten large white beads, ten large red beads, and twenty large pink beads into a dishpan filled with small white beads (to a depth of at least 5 cm).

2. One student is the predator. After the beads are mixed, the predator searches the pond and removes as many prey items (large beads) as possible in 30 seconds. In order to more closely model the handling time required by real predators, you must search for and remove beads with a pair of long forceps.

3. Because some of the large beads are cryptically colored (they blend into the environment), the propor-

tions of beads taken may not reflect the original proportions. Sift the pond with the sieve, count the number of large white, pink, and red beads, and record the totals in table 13-1. Use these counts to calculate the frequencies of the white (p) and red (q) alleles remaining in the population after selection and record them in table 13-2. For example, if five white, eight pink, and eight red beads remain, the frequency of the white allele is:

$$p = \frac{(2 \times 5) + 8}{42} = 0.43$$

4. Using the new values for allele frequencies, calculate genotype frequencies for homozygous white (p^2), heterozygous pink ($2pq$), and homozygous red (q^2) individuals, and record them in table 13-2. For example, if p now equals 0.43, the frequency of homozygous white individuals is

$$p^2 = (0.43)^2 = 0.18$$

Assuming that fifty individuals comprise the next generation, calculate the number of white, pink, and red individuals needed to create the population of a new pond and record these numbers in table 13-1. For example, if $p^2 = 0.18$, the number of white beads is:

$$p^2 \times 50 = 0.18 \times 50 = 9.0$$

Using these numbers, construct a new pond.

5. Repeat steps 2–4 for three more rounds. Stop when tables 13-1 and 13-2 are filled in completely. A different student should be the predator in each round. When you are finished, record the frequency of the red allele in table 13-3 and plot this data in figure 13-1.

Table 13-2 Allele and Genotype Frequencies due to Selection by Simulated Predation

Population	p	q	p^2	$2pq$	q^2
Initial population	0.5	0.5	0.25	0.5	0.25
First generation after selection	_____	_____	_____	_____	_____
Second generation after selection	_____	_____	_____	_____	_____
Third generation after selection	_____	_____	_____	_____	_____
Fourth generation after selection	_____	_____			

Table 13-3 Frequency of Red Allele due to Selection and Migration

Generation	Selection Alone (Section B)	Selection and Migration
1	$q = $ _____	$q = $ _____
2	$q = $ _____	$q = $ _____
3	$q = $ _____	$q = $ _____
4	$q = $ _____	$q = $ _____

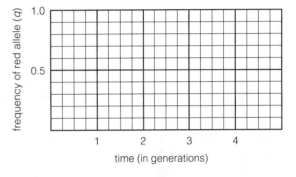

Figure 13-1 Effects of predation on allele frequencies.

If you had started with small red beads (which you may do if time permits) as a background, how would the gene frequencies change?

_____

_____

_____

Selection that favors one extreme phenotype over the other and causes allele frequencies to change in a predictable direction is known as **directional selection.** When selection favors an intermediate phenotype rather than one at the extremes, it is known as **stabilizing selection.** Selection that operates against the intermediate phenotype and favors the extreme ones is called **disruptive selection.**

It is important to realize that selection operates on the entire phenotype so that the overall fitness of an organism is based on the result of interactions of thousands of genes.

The model presented here is very simple. Occasionally simple genetic differences like the one you have modeled are critical to the survival of different phenotypes. For an example, read a description of natural selection in the peppered moth discovered by H. B. D. Kettlewell (for example, in *Biology: The Unity and Di-*

versity of Life, Starr and Taggart, 6th ed., p. 283 and *Biology: Concepts and Applications,* Starr, p. 184).

If two identical populations were in different environments (such as in our red and white ponds), how would the frequency of the color genes in each pond compare after a large number of generations?

_____

_____

_____

As two populations become genetically different through time (**divergence**), individuals from these populations may lose the ability to interbreed. If this happens, two species form from one ancestral species. This process is called **speciation.**

C. Gene Flow

The frequencies of alleles in a population also change if new organisms immigrate and interbreed, or when old breeding members emigrate. **Gene flow** due to *migration* may be a powerful force in evolution. To demonstrate its effect:

1. Establish an initial population as in section B.1.

2. Begin selection as before, except in this part of the exercise add five new red beads to each generation before the new allele frequencies are determined. These beads represent migrants from a population where the red allele confers greater fitness. For each generation, record in table 13-3 the frequencies of the red allele obtained with both selection and migration.

How does migration influence the effectiveness of selection in this example?

_____

_____

_____

How would migration have influenced the change in gene frequencies if white instead of red individuals had entered the population?

_____

_____

Some level of gene flow is necessary to keep local populations of the same species from becoming more and more different from each other. Things that serve as barriers to gene flow may accelerate the production of new species. Migration may also introduce new genes into a population and produce new genetic combinations. Imagine the result of a black allele being introduced into our model population and the new heterozygotes (perhaps gray and dark red) it would produce.

D. Mutation

> **NOTE**
>
> **If time is short this section may be done as a thought-experiment.**

Another way to introduce new genetic information into a population is through **mutation.** This usually represents an actual change in the information encoded by the DNA of an organism. As such, most mutations are harmful to the organism and will be eliminated by natural selection. Nevertheless, mutations do provide the raw material for evolution by introducing new genetic information.

1. Establish an initial population as in section B.1, but do not place these beads in a pond. Place them, instead, in a small bowl without small beads.

2. One member of the group should choose, without looking, twenty of these large beads at random.

3. Calculate the gene frequencies of the individuals selected:

$p =$ _____

$q =$ _____

4. Now replace one of the white beads with a gray bead. This represents a mutation in one of the parents of the next generation. The gray individual has new genetic information for color production. (Of course, the only mutations important in evolution are those that accumulate in gametes. That is the way changes in information can be passed on to future generations.)

5. Recalculate the new allele frequencies, with the frequency of the new color allele equal to r. (*Hint:* The gray bead must be a heterozygote unless both parents had the same mutation at the same time, a very unlikely event.)

$p =$ _____

$q =$ _____

$r =$ _____

If the next generation contains fifty individuals, how many of each phenotype would you expect? (*Hint:* Three alleles are present [$p + q + r = 1.0$], so the equilibrium formula must be expanded [$p^2 + 2pq + q^2 + 2pr + 2qr + r^2 = 1.0$] because in addition to white, pink, and red phenotypes we now have gray, dark red, and black.

white _____

pink _____

red _____

gray _____

dark red _____

black _____

Imagine a population made up of individuals in these proportions. What effect will natural selection have on these phenotypes in a white pond?

_____

_____

_____

How could conditions change to favor the selection of the rare black allele?

_____

_____

E. Genetic Drift

Chance is another factor that affects the kind of gametes in a population that are involved in fertilization. As a result, shifts in gene frequencies can occur between generations just because of the random aspects of fertilization. This phenomenon is known as **genetic drift.** In this portion of the experiment, we'll simulate genetic drift.

1. Establish an initial population as in section D.1.

Table 13-4 Allele Frequencies Produced by Genetic Drift

Expected Frequencies	Actual Frequencies (n = 10)	Actual Frequencies (n = 30)
p = 0.5	p = _____	p = _____
q = 0.5	q = _____	q = _____

Table 13-5 Allele Frequencies in a Founder Population

Initial Population	Founder Population
p = 0.5	p = _____
q = 0.5	q = _____

2. One student in the group should, without looking, place his or her hand in the bowl and choose ten beads at random.

3. Record the allele frequencies in table 13-4 that would result from the ten individuals you have chosen.

If the individuals you selected were the only individuals to reproduce in this generation, what would be the effect on allele frequencies compared to those initially present in the population?

_____

_____

_____

4. Now replace the ten beads you removed in step 2. Select beads at random again, but this time select thirty beads.

5. Calculate the allele frequencies from the thirty beads and record these frequencies in table 13-4. How do these frequencies compare to those generated in step 3?

_____

_____

_____

Generally, the larger the breeding population, the smaller the sampling effect that we call genetic drift. In small populations, genetic drift can cause fluctuations in gene frequencies that are great enough to eliminate an allele from a population, such that p becomes 0.0 and the other allele becomes fixed ($q = 1.0$). Genetic variation in such a population is reduced. Populations that become very small may lose much of their genetic variation. This is known as a **bottleneck effect.**

Another way in which chance affects allele frequencies in a population is when new populations are established by migrants from old populations.

6. To model this effect, choose at random six individuals from an initial population to represent the migrants.

7. Move these individuals to a new unoccupied pond. (It is not necessary to actually set up a new pond for this demonstration — use your imagination.)

8. Now calculate the allele frequencies in the new pond and record them in table 13-5. How do they compare with the frequencies that were characteristic of the pond from which these migrants came?

_____

_____

_____

The genetic makeup in future generations in the new population will more closely resemble the six migrants than the population from which the migrants came. This effect is known as the **founder effect.** The founder effect may not be an entirely random process because organisms that migrate from a population may be genetically different from the rest of the population to begin with. For example, if wing length in a population of insects is variable, one might expect insects with longer wings to be better at founding new populations because they may be carried farther by winds.

F. Nonrandom Mating

NOTE

If time is short this section may be done as a thought-experiment.

Hardy-Weinberg equilibrium is also disturbed if individuals in a population do not choose mates randomly. Some members of a population may show a strong preference for mates with similar genetic makeups. To model this effect, conduct the following exercise.

1. Establish an initial population as in section D.1.

2. Assume that all red individuals will mate with only other red individuals, and that white individuals select only other white members as mates. The pink heterozygotes will also mate only with each other. Arbitrarily assign sex to every bead so there are equal numbers of males and females in each color group.

3. If each pair of beads produces four offspring, how many of each phenotype will be present in the next generation? Remember that the pink pairs will produce one red, one white, and two pink individuals on the average:

red _____ white _____ pink _____

**Table 13-6 Genotype Frequency Changes
due to Nonrandom Mating**

Initial Generation	Next Generation
$p^2 = 0.25$	$p^2 =$ _____
$2pq = 0.5$	$2pq =$ _____
$q^2 = 0.25$	$q^2 =$ _____

4. Calculate the genotype frequencies in this generation, record them in table 13-6, and compare these with the frequencies in the initial generation.

What will happen to the frequency of the heterozygote genotype in subsequent generations?

_____

_____

Now that you have modeled the major factors that disrupt Hardy-Weinberg equilibrium in natural populations, attempt to answer the post-lab questions concerning the effect of these agents on the genetic makeup of such populations. Remember, the models you have used are very simple, while the phenotypes of real organisms are the result of interactions of thousands of genes. Also remember that most real populations are very complex mixtures of phenotypes, and the factors you have examined operate on these phenotypes as a whole — they do not affect some genes in an individual without affecting others.

PRE-LAB QUESTIONS

_____ 1. If all conditions of Hardy-Weinberg equilibrium are met, (a) allele frequencies move closer to 0.5 each generation, (b) allele frequencies change in the direction predicted by natural selection, (c) allele frequencies stay the same, (d) all allele frequencies increase.

_____ 2. If a population is in Hardy-Weinberg equilibrium and $p = 0.6$, (a) $q = 0.5$, (b) $q^2 = 0.4$, (c) $q = 0.16$, (d) $q^2 = 0.16$.

_____ 3. Natural selection operates directly on (a) the genotype, (b) individual alleles, (c) the phenotype, (d) color only.

_____ 4. The process that discriminates between phenotypes with respect to their ability to produce offspring is known as (a) natural selection, (b) gene flow, (c) genetic drift, (d) cytokinesis.

_____ 5. Two populations that have no gene flow between them are likely to (a) become more different with time, (b) become more alike with time, (c) become more alike if the directional selection pressures are different, (d) stay the same unless mutations occur.

_____ 6. A process that results in individuals of two populations losing the ability to interbreed is referred to as (a) stabilizing selection, (b) fusion, (c) speciation, (d) differential migration.

_____ 7. Two ways in which new alleles can become incorporated in a population are (a) mutation and genetic drift, (b) selection and genetic drift, (c) selection and mutation, (d) mutation and gene flow.

_____ 8. If a new allele appears in a population, the Hardy-Weinberg formula (a) cannot be used because no equilibrium exists, (b) can be used but only for two alleles at a time, (c) can be used by lumping all but two phenotypes in one class, (d) can be expanded by adding more terms.

_____ 9. A shift from expected allele frequencies, resulting from chance, is known as (a) natural selection, (b) genetic drift, (c) fission, (d) gene flow.

_____ 10. Genetic drift is a process that has a greater effect on populations that (a) are large, (b) are small, (c) are not affected by mutation, (d) do not go through bottlenecks.

E X E R C I S E   1 3

Evolutionary Agents

P O S T - L A B   Q U E S T I O N S

1. What two evolutionary agents are most responsible for decreases in genetic variation in a population?

2. How can selection cause two populations to become different with time?

3. What effect would increasing gene flow between two populations have on their genetic makeup?

4. Through what mechanisms can new genetic information be introduced into a population?

5. What kind of an effect can nonrandom mating exert on a population?

6. Describe how the effects of directional selection may be offset by gene flow.

7. What is the fate of most new mutations?

8. If a population has three color alleles, and the frequencies are $p = 0.5$, $q = 0.3$, and $r = 0.2$, how many phenotypes are possible?

9. In question 8, if the alleles are represented by yellow (p), red (q), and blue (r), what are the phenotypes and their proportions?

10. In humans, birth weight is an example of a character affected by stabilizing selection. What does this mean to the long-term average birth weight of human babies? How might the increasing number of Caesarean sections be affecting this character?

EXERCISE 14
Taxonomy: Classifying and Naming Organisms

OBJECTIVES

After completing this exercise you will be able to:

1. define *common name, scientific name, binomial, genus, specific epithet, species, taxonomy, phylogenetic system, dichotomous key, herbarium;*

2. distinguish common names from scientific names;

3. explain why scientific names are preferred over common names in biology;

4. identify the genus and specific epithet in a scientific binomial;

5. write out scientific binomials in the form appropriate to the Linnean system;

6. construct a dichotomous key;

7. explain the usefulness of an herbarium;

8. use a dichotomous key to identify plants, animals, or other organisms as provided by your instructor.

INTRODUCTION

We are all great classifiers. Every day, we consciously or unconsciously classify and categorize the objects around us. We recognize an organism as a cat or a dog, a pine tree or an oak tree. But there are numerous kinds of oaks, so we refine our classification, giving the trees distinguishing names such as "red oak," "white oak," or "bur oak." These are examples of **common names,** names with which you are probably most familiar.

Scientists are continually exchanging information about living organisms. But not all scientists speak the same language. The common name "white oak," familiar to an American, would probably be unfamiliar to a Spanish biologist, even though the tree we know as white oak may exist in Spain as well as in our own backyard. Moreover, even within our own language, the same organism may have several common names. For example, within North America a "gopher" may also be called a "ground squirrel," a "pocket mole," or a "groundhog." On the other hand, the same common name may actually describe many different organisms; there are more than 300 different trees called "mahogany"! To circumvent the problems associated with common names, biologists use **scientific names** that are unique to each kind of organism and that are used throughout the world.

An eighteenth-century Swedish naturalist, Carl von Linné (now most frequently known by the latinized form of his name, Linnaeus), is largely responsible for creating the system of scientific names that we use today. Linnaeus undertook the formidable task of naming and classifying all plants and animals, assigning each organism a two-part name called a **binomial.** The first word of the binomial designates the group to which the organism belongs; this is the **genus** name (the plural of genus is *genera*). All oak trees belong to the genus *Quercus,* a word derived from Latin, the universal scholarly language of Linnaeus's time. Each kind of organism within a genus is given a **specific epithet.** Thus, the scientific name in the Linnean system for white oak is *Quercus alba* (specific epithet is *alba*), while that of bur oak is *Quercus macrocarpa* (specific epithet is *macrocarpa*).

Notice that the genus name is always capitalized; the specific epithet usually is not capitalized (although it may be if it is the proper name of a person or place). The binomial is written in *italics* (since these are Latin names); if italics are not available, the genus name and specific epithet are underlined.

In this course you will hear discussion of species of organisms. For example, if you are on a field trip you may be asked "What species is this tree?" Assuming you are looking at a white oak, your reply would be *"Quercus alba."* Note that the scientific name of the **species** includes *both* the genus name and specific epithet.

If a species is named more than once within textual material, it is accepted convention to write out the full genus name and specific epithet the first time and to abbreviate the genus name every time thereafter. For example, if white oak is being described, the first use would be written *Quercus alba,* and each subsequent naming would appear as *Q. alba.*

Similarly, when a number of species, all of the same genus, are being listed, the accepted convention is to write both the genus name and specific epithet for the first species and to abbreviate the genus name for each species listed thereafter. Thus, it would be acceptable to list the scientific names for white oak and bur oak as *Quercus alba* and *Q. macrocarpa,* respectively.

Taxonomy is the science of classification (categorizing) and nomenclature (naming). Biologists prefer a system that indicates the evolutionary relationships among organisms. To this end, classification became a **phylogenetic system;** that is, one indicating the presumed evolutionary ancestry among organisms.

Most current taxonomic thought separates all living organisms into five kingdoms:

- Kingdom Monera (prokaryotic organisms)
- Kingdom Protista (euglenids, chrysophytes, diatoms, dinoflagellates, slime molds, and protozoans)
- Kingdom Fungi (fungi)
- Kingdom Plantae (plants)
- Kingdom Animalia (animals)

Let's consider the scientific system of classification, using ourselves as examples. All members of our species belong to:

- Kingdom Animalia (animals)
- Phylum Chordata (animals with a notochord)
- Class Mammalia (animals with mammary glands)
- Order Primates (mammals that walk upright on two legs)
- Family Hominidae (human forms)
- Genus *Homo* (mankind)
- Specific epithet *sapiens* (wise)
- Species: *Homo sapiens*

The more closely related evolutionarily two organisms are, the more categories they share. You and I are different individuals of the same species. We share the same genus and specific epithet, *Homo* and *sapiens*. A creature believed to be our closest extinct ancestor walked the earth 1.5 million years ago. That creature shared our genus name but had a different specific epithet, *erectus*. Thus, *Homo sapiens* and *Homo erectus* are *different* species.

Unfortunately, there is one bit of confusion in the classification system we currently use: While animal biologists recognize the category called a *phylum*, plant biologists use the term *division* instead.

Like all science, taxonomy is subject to change as new information becomes available. Modifications are made to reflect revised interpretations. This is particularly true in tropical biology, where our knowledge is exceptionally limited.

I. Constructing a Dichotomous Key

MATERIALS

Per lab room:
- several meter sticks or metric height charts taped to a wall

PROCEDURE

To classify organisms, you must first identify them. A *taxonomic key* is a device for identifying an object unknown to you but that someone else has described. The user makes choices between a set of alternative characteristics of the unknown object, and by making the correct choices he or she arrives at the name of the object.

Keys that are based upon successive choices between two alternatives are known as **dichotomous keys** (*dichotomous* means "to fork into two equal parts"). When using a key, always read both choices, even though the first appears to be the logical one. Don't guess at measurements; use a scale. Since living organisms vary in their characteristics, don't base your conclusion on a single specimen if more are available.

1. Suppose the geometric shapes below have unfamiliar names. Look at the dichotomous key following the figures. Notice there is a 1a and a 1b. Start with 1a. If the description in 1a fits the figure you are observing, then proceed to the choices listed under number 2, as shown at the end of line 1a. If 1a does not describe the figure in question, 1b does. Looking at the end of line 1b, you see that the figure would be called an Elcric.

2. Using the key provided, determine the hypothetical name for each object. Write the name beneath the object and then check with your instructor to see if you have made the correct choices.

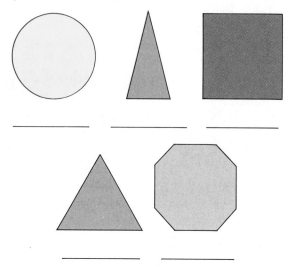

Key

1a. Figure with distinct corners 2

1b. Figure without distinct corners .... Elcric

2a. Figure with three sides 3

2b. Figure with four or more sides 4

3a. All sides of equal length .... Legnairt

3b. Only two sides equal .... Legnairtosi

4a. Figure with only right angles .... Eraqus

4b. Figure with other than right angles .... Nogatco

3. Now you will construct a dichotomous key, using your classmates as objects. The class should divide up into groups of eight (or as evenly as the class size will allow). Working with the individuals in your group, fill in table 14-1, measuring height with a metric ruler or the scale attached to the wall.

4. Let's use a branch diagram to see how we might plan a dichotomous key. If there are both men and women in a group, the most obvious first split is male/female (although other possibilities for the split could be chosen as well). Follow the course of splits for two of the men in the group (branch diagram at right).

Note that each choice has *only* two alternatives. Thus we split into "under 1.75 m" and "1.75 m or taller." Likewise, our next split is into "blue eyes" and "nonblue eyes" rather than all the possibilities.

5. On a separate sheet of paper, construct a branch diagram for your group using the characteristics in table 14-1 and then condense it into the dichotomous key below. When you have finished, exchange your key with that of an individual in another group. Key out the individuals in the other group without speaking until you believe you know the name of the individual you are examining. Ask that individual if you are correct. If not, go back to find out where you made a mistake, or possibly where the key was misleading. (Depending on how you construct your key, you may need more or fewer lines than have been provided below.)

Key to Students in Group _____

1a. _____

1b. _____

2a. _____

2b. _____

3a. _____

3b. _____

4a. _____

4b. _____

5a. _____

5b. _____

6a. _____

6b. _____

7a. _____

7b. _____

8a. _____

8b. _____

Table 14-1 Characteristics of Students

Student (name)	Sex (m/f)	Height (m)	Eye color	Hair color	Shoe size
1.					
2.					
3.					
4.					
5.					
6.					
7.					
8.					

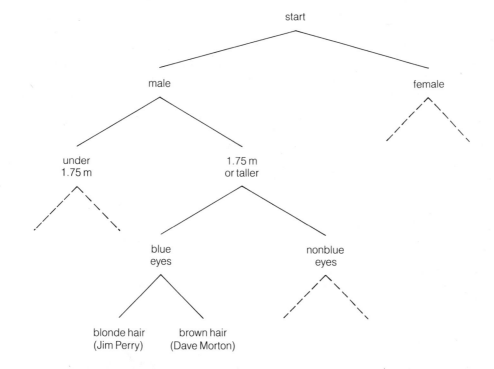

II. Use of a Taxonomic Key

A. Common Trees and Shrubs

MATERIALS

Per student group (table):

• set of eight tree twigs with leaves (fresh or herbarium specimens) *or*

• trees and shrubs in leafy condition (for an outdoor lab)

PROCEDURE

Suppose you were interested in identifying the trees growing on your campus or in your yard at home. Without having an expert present, you will now be able to do that, knowing how to use a taxonomic key. But how can you be certain that you have keyed your specimen correctly?

Typically, scientists compare the results of their keying against *reference specimens*, that is, preserved organisms that have been identified by an expert *taxonomist* (a person who names and classifies organisms). If you were doing fishes or birds, the reference specimen might be a bottled or mounted specimen with the name on it. In the case of plants, reference specimens most frequently take the form of *herbarium mounts* (fig. 14-1) of the plants. An **herbarium** (the plural is *herbaria*) is a repository, a museum of sorts, of preserved plants. The taxonomist flattens freshly collected specimens in a plant press. They are then dried and mounted on sheets of paper. Herbarium labels are affixed to the sheets, indicating the scientific name of the plant, the person who collected it, the location and date of collection, and oftentimes pertinent information about the habitat in which the plant was found.

It is likely that your school has an herbarium. If so, your instructor may show you the collection. To some this endeavor may seem boring, but herbaria serve an extremely critical function because the appearance or disappearance of plants from the landscape often gives a very good indication of environmental changes taking place. An herbarium records the diversity of plants in the area, at any point in history since the start of the collection.

Use the key that follows to identify the tree and shrub specimens that have been provided in the lab or that you find on your campus. Refer to the *Glossary to Accompany Tree Key* (pages 181–182) and figs. 14-2 through 14-9 (pages 179–180) when you encounter an unfamiliar term. When you have finished keying a specimen, confirm your identification by checking the herbarium mounts or asking your instructor.

> **NOTE**
>
> Some descriptions within the key have more characteristics than your specimen will exhibit. For example, the key may describe a fruit type when the specimen does not have a fruit on it. However, other specimen characteristics are described, and these should allow you to identify the specimen.

> **NOTE**
>
> The keys provided are for *selected* trees of your area. In nature, you will find many more genera than can be identified by use of these keys.

Common names within parentheses follow the scientific name. A metric ruler is provided on page 183 for use where measurements are required.

Label indicates name of specimen, site and date of collection, associated species at same site, name(s) of collector(s)

Figure 14-1 A typical herbarium mount. (Photo by J. W. Perry.)

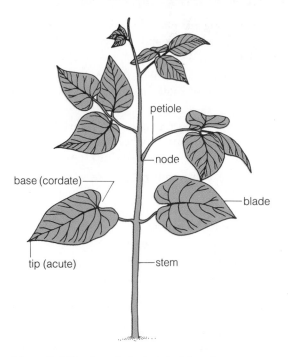

Figure 14-2 Structure of a typical plant (bean).

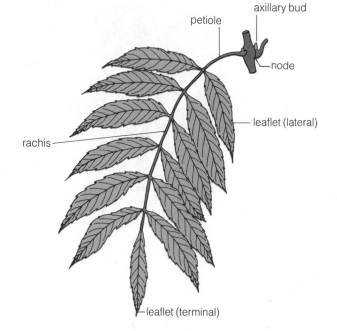

Figure 14-4 Pinnately compound leaf.

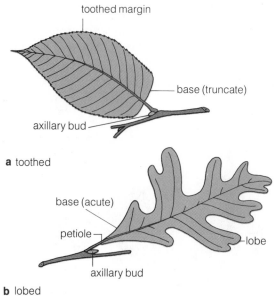

a toothed

b lobed

Figure 14-3 Simple leaves.

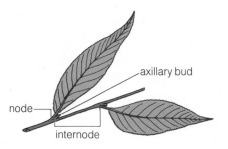

Figure 14-5 Simple leaves — alternate.

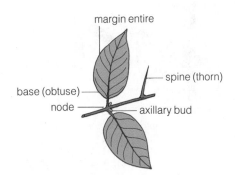

Figure 14-6 Simple leaves — opposite.

Key to Some Common Genera of Trees of the Midwestern and Eastern United States and Canada*

1a. Leaves broad and flat; plants producing flowers and fruits (angiosperms) 2

1b. Leaves needlelike or scalelike; plants producing cones, but no flowers or fruits (gymnosperms) 22

2a. Leaves compound 3

2b. Leaves simple 9

3a. Leaves alternate 4

3b. Leaves opposite 7

4a. Leaflets short and stubby, less than twice as long as broad; branches armed with spines or thorns; fruit a beanlike pod 5

*If you live in the Pacific region of the U.S. or Canada, your instructor will provide you with a key appropriate to your environment.

179

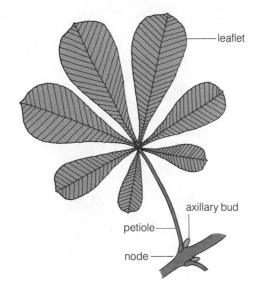

Figure 14-7 Palmately compound leaf.

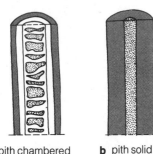

a pith chambered   **b** pith solid

Figure 14-8 Pith types.

4b. Leaflets long and narrow, more than twice as long as broad; trunk and branches unarmed; fruit a nut 6

5a. Leaflet margin without teeth; terminal leaflet present; small deciduous spines at leaf base .... *Robinia* (black locust)

5b. Leaflet margin with fine teeth; terminal leaflet absent; large permanent thorns on trunk and branches.................... *Gleditsia* (honey locust)

6a. Leaflets usually numbering less than eleven; pith of twigs solid.................. *Carya* (hickory)

6b. Leaflets numbering one or more, pith of twigs divided into chambers................................... .... *Juglans* (walnut, butternut)

7a. Leaflets pinnately arranged; fruit a light-winged samara 8

7b. Leaflets palmately arranged; fruit a heavy leathery spherical capsule .... *Aesculus* (buckeye)

8a. Leaflets numbering mostly three to five; fruit a schizocarp with curved wings .. *Acer* (box elder)

8b. Leaflets numbering mostly more than five; samaras borne singly, with straight wings .... *Fraxinus* (ash)

9a. Leaves alternate 10

9b. Leaves opposite 21

10a. Leaves very narrow, at least three times as long as broad; axillary buds flattened against stem .... *Salix* (willow)

a pome (apple, hawthorne)   **b** schizocarps (maple)

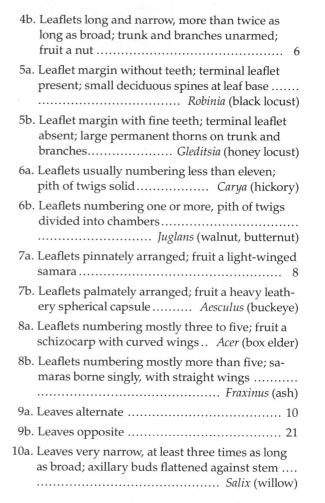

c samaras (ash)   **d** nuts (oak)

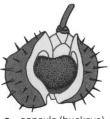

e capsule (buckeye)   **f** pods (pea)   **g** drupe (cherry)

Figure 14-9 Fruit types.

10b. Leaves broader, less than three times as long as broad... 11

11a. Leaf margin without small, regular teeth 12

11b. Leaf margin with small, regular teeth 13

12a. Fruit a pod with downy seeds; leaf blade obtuse at base; petioles flattened, or if rounded, bark smooth............ *Populus* (poplar, popple, aspen)

12b. Fruit an acorn; leaf blade acute at the base; petioles rounded; bark rough.......... *Quercus* (oaks)

13a. Leaves (at least some of them) with lobes or other indentations in addition to small, regular teeth.. 14

13b. Leaves without lobes or other indentations except for small, regular teeth 16

14a. Lobes asymmetrical, leaves often mitten-shaped .... *Morus* (mulberry)

14b. Lobes or other indentations fairly symmetrical 15

15a. Branches thorny (armed); fruit a small applelike pome .... *Crataegus* (hawthorne)

15b. Branches unarmed.................................... 17

16a. Bark smooth and waxy, often separating into thin layers; leaf base symmetrical .... *Betula* (birch)

16b. Bark rough and furrowed, leaf base asymmetrical.. *Ulmus* (elm)

17a. Leaf base asymmetrical, strongly heart-shaped, at least on one side .... *Tilia* (basswood or linden)

17b. Leaf base acute, truncate, or slightly cordate 18

18a. Leaf base asymmetrical; bark on older stems (trunk) often warty .... *Celtis* (hackberry)

18b. Leaf base symmetrical 19

19a. Leaf blade usually about twice as long as broad, generally acute at the base; fruit fleshy.......... 20

19b. Leaf not much longer than broad, generally truncate at base; fruit a dry pod .... *Populus* (poplar, popple, aspen)

20a. Leaf tapering to a pointed tip, glandular at base .... *Prunus* (cherry)

20b. Leaf spoon-shaped with a rounded tip, no glands at base .... *Crataegus* (hawthorne)

21a. Leaf margins with lobes and points, fruit a schizocarp .... *Acer* (maple)

21b. Leaf margins without lobes or points; fruit a long capsule .... *Catalpa* (catalpa)

22a. Leaves needlelike, with two or more needles in a cluster.. 23

22b. Leaves needlelike or scalelike, occurring singly 24

23a. Leaves more than five in a cluster, soft, deciduous, borne at the ends of conspicuous stubby branches................... *Larix* (larch, tamarack)

23b. Leaves two to five in a cluster .... *Pinus* (pines)

24a. Leaves soft, not sharp to the touch.............. 25

24b. Leaves stiff, sharp, often unpleasant to the touch 27

25a. Leaves rounded at tip, whitened beneath .... *Tsuga* (hemlock)

25b. Leaves pointed at tip 26

26a. Tree; without distinct petioles, with two white lines on undersurface; bases circular .... *Abies* (firs)

26b. Low shrub; leaves without petioles that follow down twigs, lighter green beneath but without distinct white lines .... *Taxus* (yew)

27a. Leaves more or less four-sided, neither in opposing pairs nor in whorls of three .... *Picea* (spruces)

27b. Leaves three-sided, either in opposing pairs or in whorls of three 28

28a. Twigs very strongly flattened; cones consisting of a few brown, dry scales .... *Thuja* (white cedar, arbor vitae)

28b. Twigs not flattened, easy to roll between the fingers; cones blue, spherical, berrylike............ 29

29a. Leaves about 1 cm long, all needlelike.............. .... *Juniperus* (juniper, Eastern red cedar)

29b. Leaves mostly less than 0.5 cm long, often needlelike and scalelike on the same individual .... *Thuja* (Western red cedar)

Glossary to Accompany Tree Key

• *Acorn* — The fruit of an oak, consisting of a nut and its basally attached cup (fig. 14-9d).

• *Acute* — Sharp-pointed (fig. 14-2).

• *Alternate* — Describing the arrangement of leaves or other structures that occur singly at successive nodes or levels; not opposite or whorled (fig. 14-5).

• *Angiosperm* — A flowering seed plant (for example, bean plant, maple tree, grass).

• *Armed* — Possessing thorns or spines.

• *Asymmetrical* — Not symmetrical.

• *Axil* — The upper angle between a branch or leaf and the stem from which it grows.

• *Axillary bud* — A bud occurring in the axil of a leaf (figs. 14-3–14-7).

• *Basal* — At the base.

• *Blade* — The expanded, more or less flat portion of a leaf (fig. 14-2).

• *Bract* — A much reduced leaf.

• *Capsule* — A dry fruit that splits open at maturity (for example, buckeye; fig. 14-9e).

• *Compound leaf* — Blade composed of two or more separate parts (leaflets) (figs. 14-4, 14-7).

• *Cordate* — Heart-shaped (fig. 14-2).

- *Deciduous* — Falling off at the end of a functional period (such as a growing season).
- *Drupe* — Fleshy fruit containing a single hard stone that encloses the seed (for example, cherry, peach, or dogwood; fig. 14-9g).
- *Fruit* — A ripened ovary, in some cases with associated floral parts (figs. 14-9a–g).
- *Glandular* — Bearing secretory structures (glands).
- *Gymnosperm* — Seed plant lacking flowers and fruits (for example, pine tree).
- *Lateral* — On or at the side (fig. 14-4).
- *Leaflet* — One of the divisions of the blade of a compound leaf (figs. 14-4, 14-7).
- *Lobed* — Separated by indentations (sinuses) into segments (lobes) larger than teeth (fig. 14-3b).
- *Node* — Region on a stem where leaves or branches arise (figs. 14-2–14-7).
- *Nut* — A hard, one-seeded fruit that does not split open at maturity (for example, acorn; fig. 14-9d).
- *Obtuse* — Blunt (fig. 14-6).
- *Opposite* — Describing the arrangement of leaves of other structures that occur two at a node, each separated from the other by half the circumference of the axis (fig. 14-6).
- *Palmately compound* — With leaflets all arising at apex of petiole (fig. 14-7).
- *Petiole* — Stalk of a leaf (figs. 14-2, 14-3, 14-4, 14-7).
- *Pinnately compound* — A leaf constructed somewhat like a feather, with the leaflets arranged on both sides of the rachis (fig. 14-4).
- *Pith* — Internally, the centermost region of a stem (figs. 14-8a,b).
- *Pod* — A dehiscent, dry fruit; a rather general term sometimes used when no other more specific term is applicable (fig. 14-9f).
- *Pome* — Fleshy fruit containing several seeds (for example, apple or pear; fig. 14-9a).
- *Rachis* — Central axis of a pinnately compound leaf (fig. 14-4).
- *Samara* — Winged, one-seeded, dry fruit (for example, ash fruits; fig. 14-9c).
- *Schizocarp* — Dry fruit that splits at maturity into two one-seeded halves (fig. 14-9b).
- *Simple leaf* — One with a single blade, not divided into leaflets (figs. 14-3, 14-5, 14-6).
- *Spine* — Strong, stiff, sharp-pointed outgrowth on a stem or other organ (fig. 14-6).
- *Symmetrical* — Capable of being divided longitudinally into similar halves.
- *Terminal* — Last in a series (fig. 14-4).
- *Thorn* — Sharp, woody, spinelike outgrowth from the wood of a stem; usually a reduced, modified branch.

- *Tooth* — Small, sharp-pointed marginal lobe of a leaf (fig. 14-3a).
- *Truncate* — Cut off squarely at end (fig. 14-3a).
- *Unarmed* — Without thorns or spines.
- *Whorl* — A group of three or more leaves or other structures at a node.

B. Some Microscopic Members of the Freshwater Environment

MATERIALS

Per student:
- compound microscope
- microscope slide
- coverslip
- dissecting needle

Per student group (table):
- cultures of freshwater organisms
- 1 disposable plastic pipet per culture
- methylcellulose in dropping bottle

PROCEDURE

Suppose you wished to identify the specimens in some pond water. The easiest way would be to key them out with a dichotomous key, now that you know how to use one. Let's do just that.

1. Obtain a clean glass microscope slide and clean coverslip.

2. Using a disposable plastic pipet or dissecting needle, withdraw a small amount of the culture provided.

3. Place *one* drop of the culture on the center of the slide.

4. Gently lower the coverslip onto the liquid.

5. Using your compound light microscope, observe your wet mount. Focus first with the low-power objective and then with the medium or high-dry objective, depending on the size of the organism in the field of view.

6. Concentrate your observation on a single specimen, keying out the specimen using the "Key to Selected Freshwater Inhabitants" below.

7. In the space provided, write the scientific name of each organism you identify. After each identification, have your instructor verify your conclusion.

8. Clean and reuse your slide and coverslip after each identification.

Key to Selected Freshwater Inhabitants

1a. Filamentous organism consisting of green, chloroplast-bearing threads 2

1b. Organism consisting of a single cell or nonfilamentous colony 4

2a. Filament branched, each cell mostly filled with green chloroplast .... *Cladophora*

2b. Filament unbranched................................ 3

3a. Each cell of filament containing one or two spiral-shaped green chloroplasts................. *Spirogyra*

3b. Each cell of filament containing two star-shaped green chloroplasts.......................... *Zygnema*

4a. Organism consisting of a single cell 5

4b. Organism composed of many cells aggregated into a colony 6

5a. Motile, teardrop-shaped or spherical organism .... *Chlamydomonas*

5b. Nonmotile, elongate cell on either end; clear, granule-containing regions at ends .... *Closterium*

6a. Colony a hollow round ball of more than 500 cells; new colonies may be present inside larger colony .... *Volvox*

6b. Colony consisting of less than fifty cells 7

7a. Organism composed of a number of tooth-shaped cells .... *Pediastrum*

7b. Colony a loose square or rectangle of four to thirty-two spherical cells...................... *Gonium*

Organism 1 is _____

Organism 2 is _____

Organism 3 is _____

Organism 4 is _____

Organism 5 is _____

Organism 6 is _____

Organism 7 is _____

Organism 8 is _____

OPTIONAL

What Species Is Your Christmas Tree?

A "Yuletide Evergreen Key" is contained within the Instructor's Manual. Ask your instructor to copy the key and give it to you if you wish to key out your Christmas tree (even if it's not "real").

PRE-LAB QUESTIONS

_____ 1. The name "human" is an example of a (a) common name, (b) scientific name, (c) binomial, (d) polynomial.

_____ 2. The person primarily responsible for the scientific nomenclature used today is (a) Darwin, (b) Linnaeus, (c) Watson, (d) Hooke.

_____ 3. The scientific name for the ruffed grouse is *Bonasa umbellus*. *Bonasa* is (a) the family name, (b) the genus, (c) the specific epithet, (d) all of the above.

_____ 4. A binomial is always a (a) genus, (b) specific epithet, (c) scientific name, (d) two-part name.

_____ 5. The science of classifying and naming organisms is known as (a) taxonomy, (b) phylogeny, (c) morphology, (d) physiology.

_____ 6. Which scientific name for the wolf is presented correctly? (a) Canis lupus, (b) canis lupus, (c) *Canis lupus,* (d) Canis Lupus.

_____ 7. A road that dichotomizes is a (an) (a) intersection of two crossroads, (b) road that forks into two roads, (c) road that has numerous entrances and exits, (d) road that leads nowhere.

_____ 8. Most scientific names are derived from (a) English, (b) Latin, (c) Italian, (d) French.

_____ 9. One objection to common names is that (a) many organisms may have the same common name; (b) many common names may exist for the same organism; (c) the common name may not be familiar to an individual not speaking the language of the common name; (d) all of the above.

_____ 10. Phylogeny is the apparent (a) name of an organism, (b) ancestry of an organism, (c) nomenclature, (d) dichotomy of a system of classification.

|....|....|....|....|....|....|....|....|....|....|....|....|....|....|....|
0   1   2   3   4   5   6   7   8   9   10   11   12   13   14   15

centimeters

EXERCISE 14

Taxonomy: Classifying and Naming Organisms

POST-LAB QUESTIONS

1. If you were to use a binomial system to identify the members of your family (mother, father, sisters, brothers), how would you write their names so that your system would most closely approximate that used to designate species?

2. If you owned a large, varied music collection, how might you keep track of all your different kinds of music?

3. Describe several advantages of the use of scientific names over common names.

4. Based upon the following classification scheme, which two organisms are most closely phylogenetically related? Why?

	Organism 1	Organism 2	Organism 3	Organism 4
Kingdom	Animalia	Animalia	Animalia	Animalia
Phylum	Arthropoda	Arthropoda	Arthropoda	Arthropoda
Class	Insecta	Insecta	Insecta	Insecta
Order	Coleoptera	Coleoptera	Coleoptera	Coleoptera
Genus	*Caulophilus*	*Sitophilus*	*Latheticus*	*Sitophilus*
Specific epithet	*oryzae*	*oryzae*	*oryzae*	*zeamaize*
Common name	broadnosed grain weevil	rice weevil	longheaded flour beetle	maize weevil

Consider the drawing of Plants A and B in answering questions 5 to 7.

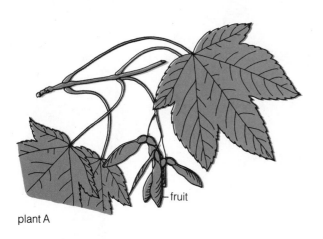

plant A

cone

plant B

5. Using the taxonomic key, identify the two plants as either angiosperms or gymnosperms.

Plant A is a (an) ———————————————— .

Plant B is a (an) ———————————————— .

6. To what genus does Plant A belong? What is its common name?

genus ————————————————

common name ————————————————

7. To what genus does Plant B belong? What is its common name?

genus ————————————————

common name ————————————————

Consider the drawing of Plants C and D in answering questions 8 to 10.

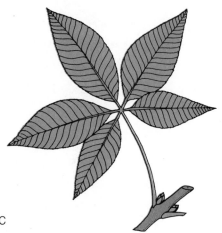

plant C

plant D

8. As completely as possible, describe the leaf of Plant C.

9. To what genus does Plant C belong? What is its common name?

genus _____

common name _____

10. Is the leaf of Plant D simple or compound?

_____

What is the genus of Plant D?

genus _____

EXERCISE 15
Monerans and Protistans

OBJECTIVES

After completing this exercise you will be able to:

1. define *pathogen, decomposer, producer, consumer, Gram stain, antibiotic, symbiont, symbiosis, parasitism, commensalism, mutualism, nitrogen fixation, pellicle, diatomaceous earth, red tide, obligate mutualism, phagocytosis, vector, plasmodium;*

2. describe characteristics distinguishing monerans from protistans;

3. identify and classify the organisms studied in this exercise;

4. identify structures (those in **boldface** within the procedure sections) in the organisms studied;

5. distinguish Gram-positive and Gram-negative bacteria, indicating their susceptibility to certain antibiotics;

6. suggest measures that might be used to control malaria.

INTRODUCTION

The monerans (kingdom Monera) and protistans (kingdom Protista) are among the simplest of living organisms. Both kingdoms consist of unicellular organisms, but that's where the similarity ends. The members of the kingdom Monera are *prokaryotic* organisms, meaning that their DNA is free in the cytoplasm, unbounded by a membrane. They lack organelles. By contrast, the kingdom Protista consists of unicellular *eukaryotic* organisms: The genetic material contained within the nucleus and many of their cellular components are compartmentalized into membrane-bound organelles. Prokaryotic and eukaryotic organization was introduced in Exercise 4.

Monerans are such organisms as bacteria and cyanobacteria. Most bacteria are heterotrophic, dependent upon an outside source for nutrition, while cyanobacteria are autotrophic (photosynthetic), able to produce their own carbohydrates. The protistans are also a diverse assemblage of organisms, both green (photosynthetic) and nongreen (heterotrophic).

Some heterotrophic bacteria are **pathogens,** causing plant and animal diseases, but most are **decomposers,** breaking down and recycling the waste products of life. Others are nitrogen-fixers, capturing the gaseous nitrogen in the atmosphere and making it available to plants via a symbiotic association with their roots.

Both monerans and protistans are at the base of the food chain. Many are autotrophic **producers,** capturing the energy of the sun. They are eaten by the heterotrophic **primary consumers;** these in turn are eaten by heterotrophic **secondary consumers** and so on. From an ecological standpoint, these simple organisms are among the most important organisms on our planet. Ecologically, they're much more important than we are.

I. Kingdom Monera

MATERIALS

Per student:
- nutrient agar culture plate
- sterile cotton swab
- china marker
- bacteria type slide
- microscope slide
- coverslip
- dissecting needle
- compound microscope

Per group:
- distilled water (dH_2O) in dropping bottle
- transparent adhesive tape

Per lab room:
- Gram-stained bacteria (three demonstration slides)
- *Oscillatoria* — living culture; disposable pipet
- *Azolla* — living plants

PROCEDURE

A. Bacteria (Heterotrophic Monerans)

1. Obtain a petri dish containing sterile nutrient agar. Open the dish and expose it to an environment of the classroom by first running a sterile cotton swab over the surface of the object you wish to sample and then over the surface of the agar. Some examples of things you might wish to sample include the surface of your lab bench, the floor, and the sink. Be creative!

> **NOTE**
> Be careful that you do not break the agar surface.

2. Replace the cover of the dish. Using a china marker, label your culture with your name and item sampled.

3. Tape the lid securely to the bottom half of the dish and place the culture in a desk drawer to incubate until the next class period. At that time, examine your culture for bacterial colonies, noting the color and texture of the bacterial growth.

4. Describe what you see.

Source of sample: _____

Description: _____

_____

_____

Bacteria come in three shapes: *coccus* (the plural is *cocci;* spherical), *bacillus* (the plural is *bacilli;* rods), and *spirillum* (the plural is *spirilla;* spirals).

5. Study a bacteria type slide illustrating these three shapes. You'll need to use the highest magnification available on your compound microscope. In figure 15-1, draw the bacteria you are observing.

In addition to being differentiated on the basis of their shape, bacteria can be separated according to how they react to a staining procedure called **Gram stain,** in honor of a nineteenth-century microbiologist, Hans Gram. **Gram-positive** bacteria are purple after being stained by the Gram stain procedure, while **Gram-negative** bacteria appear pink. The Gram stain reaction is important to bacteriologists because it is one of the first steps in identifying an unknown bacterium. Furthermore, the Gram stain reaction indicates a bacterium's susceptibility or resistance to certain **antibiotics,** substances that inhibit the growth of bacteria.

6. Examine the *demonstration slides* illustrating Gram-stained bacteria. Gram-positive bacteria are susceptible to penicillin, while Gram-negative bacteria are not. In table 15-1, list the species of bacteria that you have examined and their staining characteristics.

Table 15-1 Gram Stain Reaction of Various Bacteria	
Bacterial Species	**Gram Reaction (+ or −)**

B. Cyanobacteria (Blue-green Algae)

The cyanobacteria (sometimes called blue-green al-gae) are distinguished from the heterotrophic bacteria by being photosynthetic.

1. From the culture provided, obtain filaments of *Oscillatoria.*

2. Make a wet mount slide and examine it with your compound microscope, starting with the medium-power objective and finally with the highest magni-fication available (oil-immersion, if possible). Note that the individual cells are joined and so form the filament.

Do all the *Oscillatoria* cells look alike, or is there differentiation of certain cells within the filament?

_____

Oscillatoria is widespread, often forming a black ooze on the surface of flower pots or other surfaces that are usually wet. The color is a consequence of the photosynthetic accessory pigments that for the most part mask the chlorophyll.

3. Draw a portion of the filament in figure 15-2.

Some cyanobacteria live as **symbionts** within other organisms. Literally, *symbiosis* means "living to-gether." There are three types of symbiosis. In a para-sitic symbiosis **(parasitism),** one organism lives at the expense of the other; that is, the parasite benefits while the host is harmed. A commensalistic sym-biosis **(commensalism)** occurs when effects are posi-tive for one species and neutral for the other. In a mutualistic symbiosis **(mutualism)** both organisms

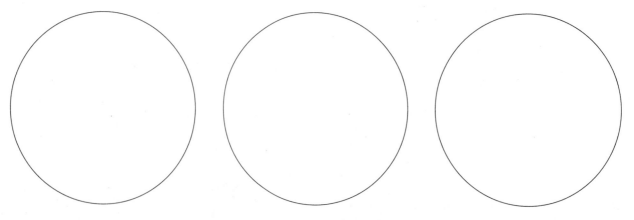

Figure 15-1 Drawings of the three bacterial shapes (_____ ×).

EXERCISE 15

benefit from living together. Let's examine one such symbiotic relationship.

4. Place a leaf of the tiny water fern *Azolla* on a clean glass slide. Use a dissecting needle to crush the leaf into very small pieces. Now add a drop of water and a coverslip.

5. Scan your preparation with the medium-power objective of your compound microscope, looking for long chains (composed of numerous beadlike cells) of the filamentous cyanobacterium *Anabaena*. Switch to higher magnification when you find *Anabaena*.

6. Within the filament, locate the **heterocysts,** cells that are a bit larger than the other cells. (Good features to look for in attempting to identify heterocysts are the *polar nodules* at either end of the cell. They appear dark in figure 15-3, but in living cells they are bright spots.)

Heterocysts convert nitrogen in the air (or water) to a form that the cyanobacterium can use for cellular metabolism. This process is called **nitrogen fixation.** Presumably, the nitrogen fixed by *Anabaena* is harvested by the water fern, which in turn uses it for its own metabolic needs.

Which type of symbiosis is the association between *Anabaena* and the water fern?

_____

7. Examine figure 15-3, an electron micrograph of *Anabaena*. The single large cell with the electron-dense regions at either end is the heterocyst. In the other cells, note the numerous wavy **thylakoids,** membranes on and in which the photosynthetic pigments are found. Identify the large electron-dense **storage granules** within the cytoplasm and the cell wall.

Is a nucleus present within the cells of *Anabaena?* Explain.

_____

Based upon this electron micrograph, would you hypothesize that the heterocyst is photosynthetic?

_____

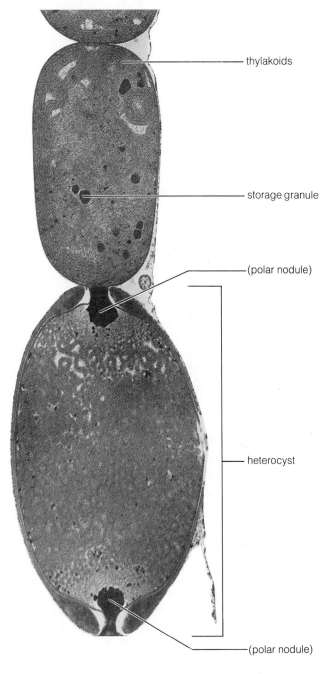

— thylakoids

— storage granule

— (polar nodule)

— heterocyst

— (polar nodule)

Figure 15-3 Electron micrograph of *Anabaena* (17,000 ×).
(Photo courtesy R. D. Warmbrodt.)

II. Kingdom Protista

MATERIALS

Per student:
- plate culture of slime mold (*Physarum*)
- prepared slide of a dinoflagellate (for example, *Gymnodinium, Ceratium,* or *Peridinium*)
- prepared slide of *Trypanosoma* in blood smear
- prepared slide of *Trichonympha*
- depression slide

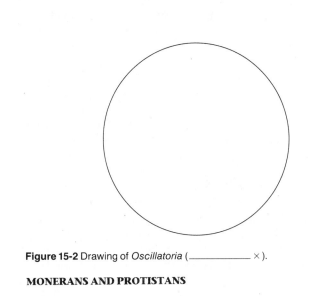

Figure 15-2 Drawing of *Oscillatoria* (_____ ×).

MONERANS AND PROTISTANS

191

- microscope slide
- coverslip
- dissecting needle
- compound microscope

Per group (table):
- methylcellulose in dropping bottles (2)
- diatomaceous earth
- distilled water (dH$_2$O) in dropping bottles (2)
- carmine, in screw-cap bottle (2)
- tissue paper
- acetocarmine stain in dropping bottles (2)
- box of toothpicks

Per lab room:
- demonstration of slime mold (*Physarum*) sporangia
- *Euglena* — living culture; disposable pipet
- diatom — living culture; disposable pipet
- dinoflagellate — living culture; disposable pipet (optional)
- *Amoeba* — living culture on demonstration at dissecting microscope; disposable pipet
- demonstration slide of *Plasmodium vivax,* sporozoites
- demonstration slide of *P. vivax,* merozoites
- demonstration slide of *P. vivax,* immature gametocytes
- *Paramecium caudatum* — living culture; disposable pipet
- Congo red — yeast mixture; disposable pipet

A. Phylum Gymnomycota: Slime Molds

The slime molds have both plantlike and animal-like characteristics. Because they engulf their food and lack a cell wall in their vegetative (nonreproductive) state, they are placed in the kingdom Protista. However, when they reproduce, they produce spores with a rigid cell wall.

PROCEDURE

Physarum: A Plasmodial Slime Mold

The vegetative (nonreproductive) body of the plasmodial slime molds consists of a naked multinucleate mass of protoplasm known as a **plasmodium.**

1. Obtain a petri dish culture of *Physarum* (fig. 15-4) and remove the cover. After examining it with your unaided eye, place the culture dish on the stage of your compound microscope and examine it with the low-power objective.

2. Watch the cytoplasm. The motion that you see within the plasmodium is cytoplasmic streaming (Exercise 4). Is the cytoplasmic streaming unidirectional, or does the flow reverse?

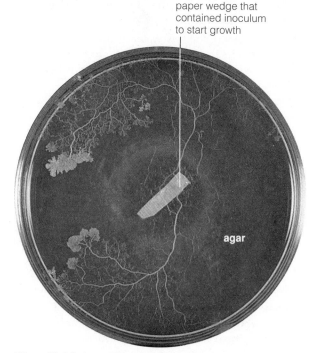

Figure 15-4 Culture dish containing the plasmodial slime mold, *Physarum* (0.8×). (Photo by J. W. Perry.)

As you noticed, the *Physarum* culture was stored in the dark. That's because light (along with other factors) stimulates the plasmodium to switch to the reproductive phase.

3. Examine the spore-containing sporangia of *Physarum* that are on demonstration. Spores released from the sporangia germinate, producing a new plasmodium.

Because the plasmodium is multinucleate (whereas the spores are uninucleate), what event must occur following spore germination?

PROCEDURE

B. Phylum Euglenophyta: Euglenids

Euglenids are motile, unicellular, photosynthetic protistans.

1. From the culture provided, prepare a wet mount slide of *Euglena* and observe with the medium-power objective of your compound microscope. Notice the motion of these green cells as they swim through the medium. If they're swimming too rapidly, prepare another slide, but add a drop of methylcellulose to the cell suspension before adding a coverslip. Switch to the high-dry objective for more detailed observation. Figure 15-5a will serve as a guide in your study. This specimen was photographed by a special technique to give it a three-dimensional appearance.

2. Within the **cytoplasm,** identify the green **chloroplasts** and, if possible, the centrally located **nucleus.** By closing the microscope's diaphragm to increase the contrast, you may be able to locate the **flagellum** at one

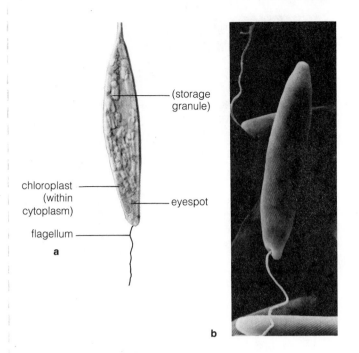

chloroplast (within cytoplasm)

(storage granule)

eyespot

flagellum

a

b

Figure 15-5 *Euglena.* (**a**) Light micrograph (550 ×). (Photo by J. W. Perry.) (**b**) Scanning electron micrograph (1,700 ×). (Photo from Shih/Kessel: *Living Images*, © 1982 Science Books International. Reprinted with permission of present publisher, Jones and Bartlett Publishers.)

end of the cell. Search for the orange **eyespot,** a photoreceptive organelle located within the cytoplasm at the base of the flagellum.

In the world of microscopic swimmers, there are two types of flagella. One type, the *whiplash flagellum,* pushes the organism through the medium. An example of this is a human sperm cell. The other type, with which you are less likely to be familiar, is the *tinsel flagellum,* which pulls the organism through its watery environment. Tinsel flagella have tiny hairlike projections, visible only with the electron microscope.

Notice the direction of motion of the *Euglena* cells. Which type of flagellum does *Euglena* have?

---

3. Besides seeing the swimming motion caused by the flagellum, you may observe a contractionlike motion of the entire cell (*euglenoid movement*). *Euglena* is able to perform this contortion because it lacks a rigid cell wall. Instead, flexible helical interlocking proteinaceous strips within the cell membrane delimit the cytoplasm. These strips plus the cell membrane form the **pellicle.** Euglenoid movement provides a means of locomotion for mud-dwelling organisms.

4. Examine figure 15-5b, a scanning electron micrograph (SEM) showing the *flagellum* and the helical strips of the pellicle.

5. In figure 15-6, make a series of sketches illustrating the different shapes *Euglena* takes on during euglenoid movement.

Figure 15-6 Different shapes possible in living *Euglena* exhibiting euglenoid movement.

C. Phylum Chrysophyta: Diatoms

Often considered as algae (Exercise 16), the diatoms are called the organisms that live in glass houses because their cell walls are composed largely of opaline *silica* ($SiO_2 \cdot nH_2O$). Diatoms are important as primary producers in the food chain of aquatic environments, and their cell walls are used for a wide variety of industrial purposes, ranging from the polishing agent in toothpaste to a reflective roadway paint additive. Massive deposits of cell walls of long-dead diatoms make up **diatomaceous earth** (fig. 15-7).

1. Examine microscopically a bit of diatomaceous earth by preparing a wet mount slide. Then prepare a wet mount of living diatoms. Use the high-dry objective to note the pigmentation within the cytoplasm and the numerous perforations in the cell walls (shells).

2. In figure 15-8 make a sketch of several of the diatoms you are observing.

3. Now obtain a prepared slide of a freshwater diatom. These cells have been "cleaned," making the perforations in the cell wall especially obvious if you close

Figure 15-7 Diatomaceous earth quarry near Quincy, Washington. (Photo courtesy Dan Williams.)

Figure 15-8 Drawing of diatoms (_____ ×).

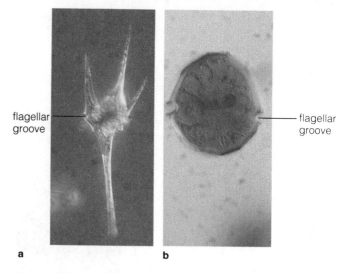

a   b

Figure 15-9 Two representative dinoflagellates. (**a**) *Ceratium* (240 ×). (**b**) *Peridinium* (640 ×). (Photos by J. W. Perry.)

the iris diaphragm on your microscope's condenser to increase the contrast. Study with the high-dry objective.

The diameter of the holes in the walls are characteristic of a given species. Before the advent of electronic techniques, microscopists observed diatom walls to assess the quality of microscope lenses. The resolving power (see discussion of resolving power in Exercise 3) could be determined if one knew the diameter of the holes under observation.

D. Phylum Pyrrophyta: Dinoflagellates

Dinoflagellates are commonly called *whirling whips* because of the spinning motion they exhibit. On occasion, populations of certain dinoflagellates may increase dramatically, causing the seas to turn red or brown. These are the **red tides,** which may devastate fish populations because neurotoxins produced by the dinoflagellates poison fish that feed on them.

With the high-dry objective of your compound microscope, examine a prepared slide or living representative (fig. 15-9a and b). Dinoflagellates are encased in stiff cellulosic plates. The junction of these plates forms two grooves in which flagella are located. Find the plates and grooves. If you are examining living specimens, chloroplasts may be visible beneath the cellulose plates.

E. Phylum Mastigophora: Flagellated Protozoans

Protozoans are heterotrophic, generally motile, single-celled organisms. Some cause human disease. Amoebic dysentery and giardiasis are examples of illnesses caused by drinking water contaminated with the causal protozoans.

What type of symbiosis is exemplified by these disease-causing organisms? (See definitions, p. 190.)

Figure 15-10 *Trypanosoma*. (From Stanier, Ingraham, Wheelis, Painter, *The Microbial World*, 5th ed., © 1986. Reprinted by permission of Prentice-Hall, Inc.)

1. Examine a prepared slide of human blood that contains the parasitic flagellate *Trypanosoma* (fig. 15-10), the cause of African sleeping sickness. This flagellate is transmitted from host to host by the bloodsucking tsetse fly. Note the **flagellum** arising from one end of the cell.

2. Another example of a flagellated protozoan is the termite-inhabiting *Trichonympha* (fig. 15-11). Study a prepared slide of these organisms. Examine the gut of the termite with the high-dry objective to find *Trichonympha*. Note the large number of **flagella** covering the upper portion of the cell, the more or less centrally located **nucleus** and wood fragments in the cytoplasm.

The association of the termite and *Trichonympha* is an example of **obligate mutualism,** neither organism being capable of surviving without the other. Termites lack the enzymes to metabolize cellulose, a major component of wood. Wood particles ingested by termites are engulfed by *Trichonympha,* whose enzymes break the cellulose into soluble carbohydrates that are released for use by the termite.

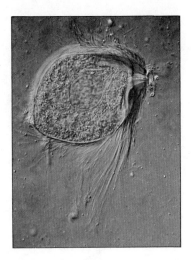

Figure 15-11 *Trichonympha* (1,000 ×). (Photo by M. Abbey.)

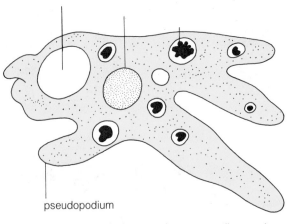

pseudopodium

Labels: ectoplasm, endoplasm, nucleus, contractile vacuole, food vacuole

Figure 15-12 *Amoeba*.

F. Phylum Sarcodina: Amoeboid Protozoans

The best-known amoeboid protozoans are the amoebas, organisms that are continually changing shape through formation of projections called *pseudopodia* (the singular is *pseudopodium,* "false foot").

1. Observe the *Amoeba*-containing culture on the stage of a demonstration dissecting microscope. (One species of amoeba has the scientific name *Amoeba.*) The microscope has been focused on the bottom of the culture dish, where the amoebas are located. Look for gray, irregularly shaped masses moving among the food particles in the culture.

2. Using a clean pipet, remove an amoeba and place it, along with some of the culture medium, in a depression slide.

3. Examine it with your compound microscope using the medium-power objective. You will need to adjust the diaphragm to increase the contrast (see Exercise 3) because *Amoeba* is nearly transparent.

4. Locate the **pseudopodia.** At the periphery of the cell, identify the **ectoplasm,** a thin, clear layer that surrounds the inner, granular **endoplasm.** Watch the organism as it changes shape. Which region of the endoplasm appears to stream, the outer or the inner?

This region, called the *plasmasol,* consists of a fluid matrix that can undergo phase changes with the semi-solid *plasmagel,* the outer layer of the endoplasm. Pseudopodium formation occurs as the plasmasol flows into new environmental frontiers and then changes to plasmagel. (This phenomenon was discussed in Exercise 4.)

Numerous granules will be found within the endoplasm. Some of these are organelles; others are food particles.

5. Within the endoplasm, try to locate the **nucleus,** a densely granular, spherical structure around which the cytoplasm is streaming. You should be able to see clear, spherical **contractile vacuoles,** which regulate water balance within the cell. Watch for a minute or two to observe the action of contractile vacuoles. Label figure 15-12.

Amoeba feeds by a process called **phagocytosis,** engulfing its food. Pseudopodia form around food particles, and then the pseudopodia fuse, creating a **food vacuole** within the cytoplasm. Enzymes are then emptied into the food vacuole, where the food particle is digested into a soluble form that can pass through the vacuolar membrane.

6. You can stimulate feeding behavior by drawing carmine under the coverslips: Place a drop of distilled water (dH₂O) against one edge of the coverslip; pick up some carmine crystals by dipping a dissecting needle into the bottle; and deposit them into the water droplet. Draw the suspension beneath the coverslip by holding a piece of absorbent tissue against the coverslip on the side *opposite* the carmine suspension. Observe the *Amoeba* again—you may catch it in the act.

G. Phylum Apicomplexa: Sporozoans

All sporozoans are parasites, infecting a wide range of animals, including humans. *Plasmodium vivax* causes one type of malaria in humans. We will study its life cycle (fig. 15-13) with demonstration slides.

P. vivax is transmitted to humans through the bite of an infected female *Anopheles* mosquito. The mosquito serves as a **vector,** a means of transmitting the organism from one host to another. Male mosquitos cannot serve as vectors, because they lack the mouth

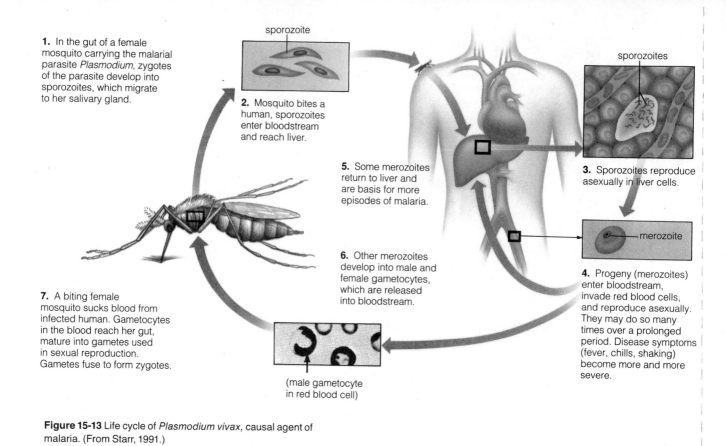

1. In the gut of a female mosquito carrying the malarial parasite *Plasmodium*, zygotes of the parasite develop into sporozoites, which migrate to her salivary gland.

sporozoite

2. Mosquito bites a human, sporozoites enter bloodstream and reach liver.

sporozoites

3. Sporozoites reproduce asexually in liver cells.

merozoite

5. Some merozoites return to liver and are basis for more episodes of malaria.

4. Progeny (merozoites) enter bloodstream, invade red blood cells, and reproduce asexually. They may do so many times over a prolonged period. Disease symptoms (fever, chills, shaking) become more and more severe.

6. Other merozoites develop into male and female gametocytes, which are released into bloodstream.

7. A biting female mosquito sucks blood from infected human. Gametocytes in the blood reach her gut, mature into gametes used in sexual reproduction. Gametes fuse to form zygotes.

(male gametocyte in red blood cell)

Figure 15-13 Life cycle of *Plasmodium vivax*, causal agent of malaria. (From Starr, 1991.)

parts for piercing skin and sucking blood. If the mosquito is carrying the pathogen, **sporozoites** enter the host's bloodstream with the saliva of the mosquito.

1. Examine the demonstration slide of sporozoites.

The sporozoites travel through the bloodstream to the liver, where they penetrate certain cells, grow, and multiply. When released from the liver cells, the parasite is in the form of a **merozoite** and infects the red blood cells.

(There are two intervening stages between sporozoites and merozoites. These are the *trophozoites* and *schizonts,* both developmental stages in red blood cells.)

2. Examine the demonstration slide illustrating merozoites in red blood cells.

Within the red blood cells, merozoites divide, increasing the merozoite population. At intervals of forty-eight or seventy-two hours, the infected red blood cells break down, releasing the merozoites. At this time, the infected individual exhibits disease symptoms, including fever, chills, and shaking caused by the release of merozoites and metabolic wastes from the red blood cells. Some of these merozoites return to the liver cells, where they repeat the cycle and are responsible for recurrent episodes of malaria.

Merozoites within the bloodstream may develop into **gametocytes.** For development of a gametocyte to be completed, the gametocyte must enter the gut of the mosquito. This occurs when a mosquito feeds upon an infected (diseased) human.

3. Observe the demonstration slide of an **immature gametocyte** in a red blood cell.

Within the gut of the mosquito, the gametocyte matures into gametes. When gametes fuse, they form a zygote that matures into an **oocyst.** Within each oocyst, sporozoites form, completing the life cycle of *Plasmodium vivax.* These sporozoites migrate to the mosquito's salivary glands to be injected into a new host.

H. Phylum Ciliophora: Ciliated Protozoans

Most members of the phylum Ciliophora are covered with numerous short locomotory structures called *cilia.* One of the largest ciliates is the predatory *Paramecium caudatum.*

1. From the culture provided, prepare a wet mount of *Paramecium* on a clean microscope slide. Observe their rapid motion with the medium-power objective of your microscope.

Paramecium possesses two nuclei. One is a larger, centrally located **macronucleus,** which is associated with nonreproductive functions. The other is a smaller **micronucleus** adjacent to the macronucleus. The micronucleus regulates reproductive functions.

2. Add a drop of acetocarmine stain, drawing it beneath the coverslip by touching a piece of absorbent tissue to the opposite side of the coverslip. The acetocarmine should make the nuclei more visible.

3. On another slide, add a drop of methylcellulose and a drop of Congo red yeast mixture. Stir the mixture with a toothpick, then add a drop of *Paramecium* culture. Place a coverslip on your wet mount and observe with the medium-power objective.

4. Use the drawing of *Paramecium* (fig. 15-14) to identify the structures described below.

Paramecium is covered with numerous **cilia,** which serve two functions: locomotion and directing food into the opening of the digestive tract. This opening is at the base of a depression called the **oral groove.** Locate the oral groove. The actual entrance (cytostome)

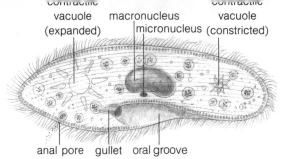

anal pore   gullet   oral groove

Figure 15-14 *Paramecium caudatum*. (From Starr and Taggart, 1987.)

is probably not obvious, but the tubule leading from the oral groove, the **gullet,** should be identified.

5. Observe the yeast cells as they are eaten by *Paramecium*. The yeast will be taken into **food vacuoles,** where they are digested. Congo red is a pH indicator. As digestion occurs within the food vacuoles, the indicator will turn blue because of the increased acidity.

6. At either end of the organism, find the **contractile vacuoles,** which regulate water content within the organism. Watch them in operation.

PRE-LAB QUESTIONS

_____ 1. Members of the kingdom Monera *lack* (a) a nucleus, (b) organelles, (c) chloroplasts, (d) all of the above.

_____ 2. Unicellular eukaryotic organisms are placed in the kingdom (a) Monera, (b) Protista, (c) Animalia, (d) Plantae.

_____ 3. Which of the following organisms are autotrophic? (a) all monerans, (b) all protistans, (c) *Oscillatoria*, (d) *Trypanosoma*.

_____ 4. Which organisms are characterized as decomposers? (a) bacteria, (b) diatoms, (c) amoebas, (d) dinoflagellates.

_____ 5. Organisms capable of nitrogen fixation (a) include some bacteria, (b) include some cyanobacteria, (c) may live as symbionts with other organisms, (d) all of the above.

_____ 6. A spherical bacterium would be called a (a) bacillus, (b) coccus, (c) spirillum, (d) none of the above.

_____ 7. Gram stain would be used to distinguish between different (a) bacteria, (b) protistans, (c) dinoflagellates, (d) all of the above.

_____ 8. Which of the organisms (or parts of the organisms) listed below might you find as an ingredient in toothpaste? (a) cyanobacteria, (b) diatoms, (c) amoebas, (d) *Trypanosoma vivax*.

_____ 9. Red tides are caused by (a) dinoflagellates, (b) bacteria, (c) diatoms, (d) monerans.

_____ 10. Those organisms that are covered by numerous, tiny locomotory structures belong to the phylum (a) Gymnomycota, (b) Sarcomastigophora, (c) Apicomplexa, (d) Ciliophora.

EXERCISE 15

Monerans and Protistans

COMPARE/CONTRAST
How 3 organisms
alike; how different?

P O S T - L A B   Q U E S T I O N S

1. Both monerans and protistans are unicellular organisms. What major characteristic distinguishes the organisms within the two kingdoms?

2. Examine the photomicrograph below, taken using an oil-immersion objective, the highest practical magnification of a light microscope. (The final magnification is $770\times$.)

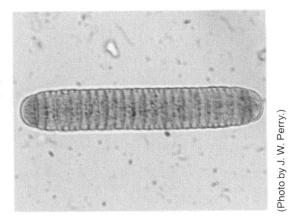

(Photo by J. W. Perry.)

   a. Based upon observation, identify the kingdom to which the organism belongs.

   b. Justify your answer.

3. On a field trip to a stream, you collect a leaf that has fallen into the water and scrape some of the material from its surface, and prepare a wet mount. You examine your preparation with the high-dry objective of your compound microscope, finding the organism pictured below at $750\times$. What is it?

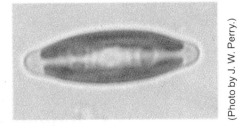

(Photo by J. W. Perry.)

4. Of what significance are thylakoids in photosynthetic organisms?

5. The protistan *Euglena* is often studied in plant-related courses because it is photosynthetic. What characteristic of the pellicle makes *Euglena* different from true plants?

6. Observe the photomicrograph below (290×) of an organism that was found growing symbiotically within the leaves of the water fern *Azolla*.

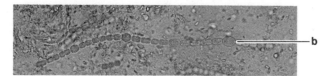

(Photo by J. W. Perry.)

a. What is the common name given an organism of this type?

b. Give the name *and* function of the cell depicted by the line.

7. What characteristic separates the euglenids, diatoms, and dinoflagellates from protozoans?

8. How does phagocytosis differ from endocytosis?

9. Based upon your knowledge of the life history of *Plasmodium vivax,* suggest two methods for controlling malaria. Explain why each method would work.

   a.

   b.

10. Why would an organism such as *Paramecium* need contractile vacuoles?

EXERCISE 16

Fungi

OBJECTIVES

After completing this exercise you will be able to:

1. define *parasite, saprophyte, mycologist, pathogen, gametangium, oogonium, antheridium, hypha, mycelium, multinucleate, monoecious, dioecious, sporangium, rhizoid, zygospore, ascus, conidium, ascospore, ascocarp, basidium, basidiospore, basidiocarp;*

2. recognize representatives of the major divisions of fungi;

3. distinguish structures that are used to place various representatives of the fungi in their proper divisions;

4. list reasons why fungi are important;

5. distinguish between the structures associated with asexual and sexual reproduction described in this exercise;

6. identify the structures (those in **boldface**) of the fungi examined.

INTRODUCTION

As you walk into the woods following a warm rain you are likely to be met by a vast assemblage of colorful fungi. Some of them are growing on dead or diseased trees, some on the surface of the soil, some in pools of water. Some are edible, some deadly poisonous.

Fungi (kingdom Fungi) are *heterotrophic* organisms; that is, they are incapable of producing their own food material. They secrete enzymes that digest their food source, which is then taken into the body, primarily by diffusion. Some are **parasites,** organisms that obtain their nutrients from the organic material of another living organism. Others are **saprophytes,** growing on nonliving organic matter.

Fungi, along with the bacteria, are essential components of the ecosystem as decomposers. These organisms recycle the products of life, making the products of death available so that life may continue. Without them we would be hopelessly lost in our own refuse. Fungi and fungal metabolism are responsible for some of the food products that enrich our lives — the mushrooms of the field, the blue cheese of the dairy case, even the citric acid used in making soft drinks.

Individuals who specialize in the study of fungi are called **mycologists** (*myco-* is a prefix coming from the Greek word meaning "fungus"). Most divide the kingdom Fungi into three separate divisions. Further taxonomic separation of the divisions results in several classes, as described in the following table.

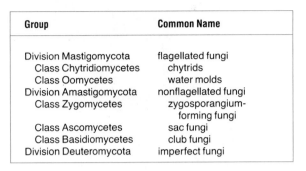

Group	Common Name
Division Mastigomycota	flagellated fungi
Class Chytridiomycetes	chytrids
Class Oomycetes	water molds
Division Amastigomycota	nonflagellated fungi
Class Zygomycetes	zygosporangium-forming fungi
Class Ascomycetes	sac fungi
Class Basidiomycetes	club fungi
Division Deuteromycota	imperfect fungi

I. Division Mastigomycota: The Flagellated Fungi

All members of this division produce structures that bear flagella, which serve to propel the structure through a watery medium. The division gets its scientific name from the Greek root *mastig,* which means "a whip." Indeed, many flagella resemble tiny whips, which thrash back and forth, pushing the organism along. Others, however, actually pull the organism through the water; there are different types of flagella.

A. Class Chytridiomycetes

Commonly known as chytrids, these fungi are among the most simple in body form. While most are saprophytes, some are parasitic on economically important plants, resulting in damage and/or death of the host plant. Parasitic organisms that cause disease are called **pathogens.**

MATERIALS

Per lab room:
- preserved specimen of potato tuber with black wart disease

PROCEDURE

Examine the preserved specimen (if available) and figure 16-1 of the potato tuber that exhibits the disease known as black wart. Notice the warty eruptions on the surface of the tuber. The warts are caused by the presence of numerous cells of a chytrid infecting the tuber.

B. Class Oomycetes: The Water Molds

Members of the Oomycetes are commonly referred to as water molds because they are primarily aquatic organisms. If you have ever kept goldfish, you may have seen a white, cottony mass growing on the sides of

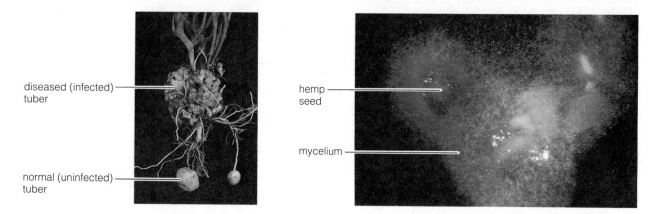

Figure 16-1 Potato tubers infected by black wart pathogen (0.25 ×). (Photo courtesy Earl Wade.)

Figure 16-2 Mycelium of water mold growing on hemp seed (1 ×). (Photo by J. W. Perry.)

some of the fish. This is a parasitic water mold that is easily controlled by the addition of chemicals to the aquarium.

While not all members of this division grow in free-standing water, they all do rely on the presence of water for spread of their asexual spores.

During sexual reproduction all members produce large nonmotile female gametes, the eggs. These eggs are contained in sex organs (**gametangia;** the singular is *gametangium*) called **oogonia.** By contrast the male gametes are nothing more than *sperm nuclei* contained in **antheridia,** the male gametangia.

MATERIALS

Per student:

• culture of *Saprolegnia* or *Achlya*

• culture of *Phytophthora cactorum*

• glass microscope slide

• dissecting needle

• compound microscope

Per lab room:

• refrigerator

or

Per student group (4):

• ice bath

PROCEDURE

1. *Saprolegnia* or *Achlya:* A Water Mold

a. Examine the water cultures of either *Saprolegnia* or *Achlya.* These fungi are chiefly saprophytes and are growing on a sterilized hemp seed that provides a carbohydrate source. Examine its life cycle (fig. 16-6) as you study this organism.

Notice the numerous filamentous hyphae that radiate from the hemp seed. A **hypha** (the plural is

hyphae) is the basic unit of the fungal body. Collectively, all the hyphae constitute the **mycelium** (the plural is *mycelia;* figs. 16-2, 16-6a).

The mycelium of the water mold is **multinucleate,** meaning that there are many nuclei in each cell, not just one. Each nucleus is diploid (2n). Moreover, cross walls separating the mycelium into distinct cells are infrequent, forming only when reproductive organs are formed.

b. Place a glass slide on the stage of your compound microscope. This slide will serve as a platform for the culture dish that contains the fungal culture, allowing you to use the mechanical stage of the microscope to move the culture (if the microscope is so equipped).

c. Remove the lid from the culture dish, carefully place the culture dish on the platform, and examine with the low-power objective. Look first at the edge of the mycelium, that is, at the tips of the youngest hyphae. Find the tips that appear more dense (darker) than the rest of the hyphae. These dense tips are cells specialized for asexual reproduction. They are the **zoosporangia** (the singular is *zoosporangium;* figs. 16-3, 16-6b), which produce biflagellate **zoospores** (fig. 16-6c). Note the cross wall separating the zoosporangium from the rest of the hypha.

Zoospores are released from the zoosporangia, swim about for a period of time, lose their flagella, and encyst (fig. 16-6d); that is, they form a thick wall around the cytoplasm. When the encysted zoospore germinates, it will produce a second type of zoospore (fig. 16-6e) that also will eventually encyst (fig. 16-6f). When the second cyst stage germinates, it produces a hypha (fig. 16-6g) that will proliferate into a new mycelium (fig. 16-6a). As you see, this is asexual reproduction; no sex organs were involved in the formation of zoospores.

d. Sexual reproduction in the water molds occurs in the older portion of the mycelium. Scan the colony to find the spherical female gametangia, the oogonia (the singular is **oogonium;** figs. 16-4, 16-6h). Meiosis takes place within the oogonium to form haploid eggs. Switch to the medium-power objective and study a single oogonium in greater detail. Depending upon the stage of development, you will find either **eggs** (fig. 16-6i) or **zygotes** (fertilized eggs) (figs. 16-5, 16-6j) within the oogonium.

e. The male gametangium, the **antheridium,** is a short fingerlike hypha that attaches itself to the wall of the oogonium (figs. 16-4, 16-6i), much as you would wrap your finger around a baseball. Find an antheridium. Because the nuclei of the antheridium have undergone meiosis, each nucleus is haploid.

Fertilization takes place when tiny fertilization tubes penetrate the wall of the oogonium (fig. 16-6i). Rather than forming special male gametes, the haploid nuclei within the antheridium flow through the fertilization tubes to fuse with the egg nuclei.

f. Search your culture to see if you can locate any thick-walled zygotes within oogonia.

After fertilization, each zygote forms a thick wall (figs. 16-5, 16-6j). Following a maturation period that may last several months, the zygote germinates by forming a germ tube (fig. 16-6k) that grows into a new mycelium (fig. 16-6a).

Notice that both male and female gametangia were produced on the same mycelium. The term describing this condition is **monoecious** (from the Greek words for "one house"). Organisms producing one type of sex organ on one body and the other type of sex organ on another body are **dioecious** (from the Greek words for "two" and "house"). Are humans monoecious or dioecious? _____

2. *Phytophthora*

Some of the most notorious and historically important plant pathogens known to humans are water molds. Included in this group is *Phytophthora infestans,* the fungus that causes the disease known as late blight of potato. This disease spread through the potato fields of Ireland between 1845 and 1847. Most of the potato plants died. One million Irish working-class citizens who had come to depend on potatoes as their primary food source starved to death. Another two million emigrated, many to the United States.

Perhaps no other plant pathogen so beautifully illustrates the importance of environmental factors in causing disease. While *Phytophthora infestans* had been present previous to 1845 in the potato-growing fields of Ireland, it was not until the region experienced several consecutive years of wet and especially cool growing seasons that late blight became a major problem.

We can easily observe the importance of these climate-associated factors by studying another *Phytophthora* species, *P. cactorum.*

a. Obtain a culture of *P. cactorum* that has been flooded with distilled water. Note that the agar has been removed from the edges of the petri dish and

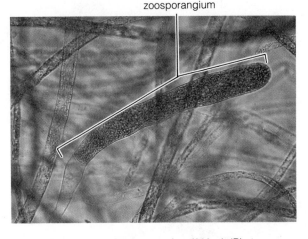

Figure 16-3 Water mold zoosporangium (230 ×). (Photo courtesy C. A. Taylor III.)

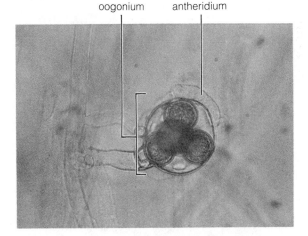

Figure 16-4 Gametangia of water mold (230 ×). (Photo courtesy C.A. Taylor III.)

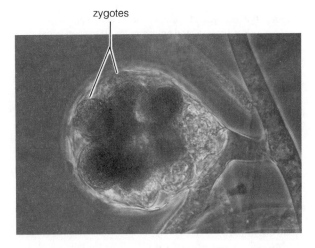

Figure 16-5 Oogonium with zygotes (230 ×). (Photo courtesy C. A. Taylor III.)

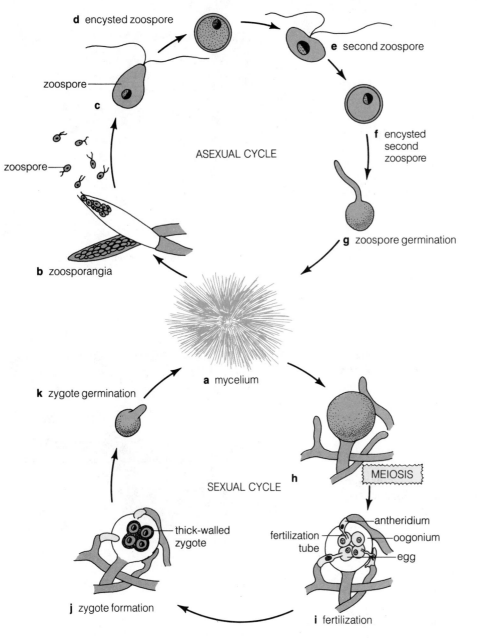

Figure 16-6 Life cycle of a water mold such as *Saprolegnia* or *Achlya*. (Structures colored green are 2n.) (Modified from artwork by Carolina Biological Supply Company.)

that the mycelium has grown from the agar edge into the water.

b. Place a glass slide on the stage of your compound microscope. This slide will serve as a platform for the culture, allowing you to use the mechanical stage of the microscope to move the culture (if the microscope is so equipped).

c. Remove the lid from the culture, carefully place the culture dish on the platform, and examine the culture with the low-power objective.

d. Search the surface of the mycelium, especially at the edges, until you find the rather *pear-shaped* **zoosporangia.** Switch to the medium-power objective for closer observation and then draw in figure 16-7 a single zoosporangium.

Labels: zoosporangium, zoospores

Figure 16-7 Drawing of zoosporangium and zoospores of *Phytophthora cactorum* (_____ ×).

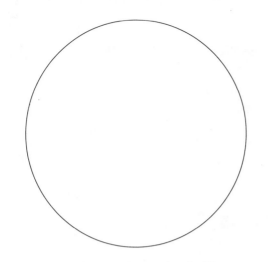

Labels: oogonium, eggs (zygotes), antheridium

Figure 16-8 Drawing of the gametangia of *Phytophthora cactorum* (_____ ×).

e. Return your microscope to the low-power objective, remove the culture, replace the cover, and place it in a refrigerator or on ice for 15–30 minutes.

f. After the incubation time, again observe microscopically the zoosporangia. Find one in which the **zoospores** are escaping from the zoosporangium. Each zoospore has the potential to grow into an entirely new mycelium! Draw the zoospores in figure 16-7.

Hypothesize what probably took place in the potato fields of Ireland between 1845 and 1847 that led to the destructive explosion in disease.

_____

_____

Like the other water molds, *Phytophthora* reproduces sexually. Your cultures contain the sexual structures as well as asexual zoosporangia.

g. Using a dissecting needle, cut a section about 1 cm square from the agar colony and *invert* it on a glass slide (that is, so that the bottom side of the agar is now uppermost). Place a coverslip on the agar block and observe with your compound microscope, first with the low-power objective, then with the medium-power, and finally with the high-dry objective. Identify the **spherical oogonia** that contain **eggs** or thick-walled **zygotes** (depending upon the stage of development). If present, the **antheridia** are club-shaped and plastered to the wall of the oogonium. In figure 16-8, draw an oogonium, eggs (zygotes), and antheridia.

II. Division Amastigomycota: The Nonflagellated Fungi

None of the fungi in the division Amastigomycota has flagellum-bearing structures. The prefix *a-* means "without." If you recall that the Greek word *mastig* means "a whip," you can combine the root words, deriving "without-a-whip fungi."

A. Class Zygomycetes: Zygosporangium-forming Fungi

All members of the Zygomycetes produce a thick-walled zygote, a **zygosporangium.** Most fungi in this division are saprophytes, including the common black bread mold, *Rhizopus*. Before the introduction of chemical preservatives into bread, *Rhizopus* was an almost certain invader, especially if the humidity was high.

Growth of two different mycelia in close proximity is necessary before sexual reproduction will occur. (The difference in the mycelia is genetic rather than structural. Because they are impossible to distinguish, the mycelia are simply referred to as + and − strains, as indicated in figure 16-13.)

MATERIALS

Per student:
- culture of *Rhizopus*
- prepared slide of *Rhizopus*
- dissecting needle
- glass microscope slide
- coverslip
- compound microscope

Per student pair:
- distilled water (dH$_2$O) in dropping bottle

Per lab room:
- demonstration culture of *Rhizopus* zygosporangia, on dissecting microscope

PROCEDURE

Rhizopus: A Bread Mold

Examine figure 16-13, a diagram of the life cycle of *Rhizopus*, as you study this organism.

1. Obtain a petri dish culture containing the mycelium, which consists of many hyphae (fig. 16-9). The numerous black "dots" are the **sporangia** (the singular

Figure 16-9 Bread mold culture. (Photo by J. W. Perry.)

205

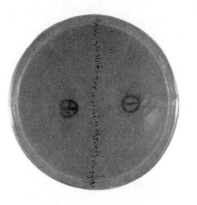

Figure 16-10 Sexual reproduction in the bread mold. Black line consists of zygosporangia (0.5 ×). (Photo by J. W. Perry.)

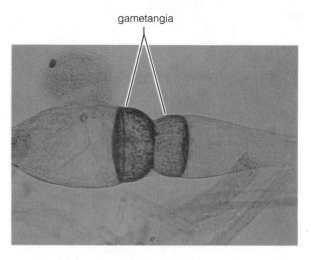

Figure 16-11 Gametangia of a bread mold (230 ×). (Photo by J. W. Perry.)

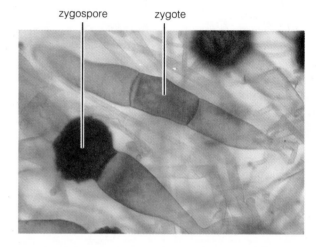

Figure 16-12 Zygote and zygosporangium of the bread mold (230 ×). (Photo by J. W. Perry.)

is *sporangium;* figs. 16-13a, 13b). Sporangia are containers of **spores** (figs. 16-13b, 13c, 13d) by which *Rhizopus* reproduces asexually.

2. Using a dissecting needle, remove a small portion of the culture to prepare a wet mount. Examine your preparation with the high-dry objective of your compound microscope. It's likely that when you added the coverslip, you crushed the sporangia, liberating the spores. Are there many or few spores within a single sporangium?

Identify the **rhizoids** (fig. 16-13a) at the base of the sporangium-bearing hypha. Rhizoids serve to anchor the mycelium to the substrate.

3. Now observe the demonstration culture illustrating sexual reproduction in *Rhizopus* (fig. 16-10). The black line running down the center of the culture plate consists of numerous **zygosporangia,** the products of sexual reproduction.

Sexual reproduction in *Rhizopus* (figs. 16-13f through 16-13j) occurs when two sexually compatible mycelia are in close proximity.

As the hyphae from each mating type grow close, chemical messengers signal them to produce protuberances (fig. 16-13f). When the protuberances make contact, gametangia (figs. 16-11, 16-13g) are produced at their tips. Each gametangium contains many haploid nuclei of a single mating type. The wall between the two gametangia then dissolves, and the cytoplasms of the gametangia mix.

Eventually the many haploid nuclei from each gametangium fuse (fig. 16-13h). The resulting cell contains many diploid nuclei resulting from the fusion of gamete nuclei of opposite mating types. Each diploid nucleus is considered a *zygote*. This multinucleate cell is called a zygosporangium. Eventually, a thick, bumpy wall forms about this diploid cell (figs. 16-12, 16-13i).

4. On prepared slides, find the stages of sexual reproduction in *Rhizopus,* including gametangia, zygotes, and zygospores. Use the medium-power and high-dry objectives of your compound microscope in making your observations.

Meiosis occurs within the thick-walled zygosporangium, which then germinates to produce a sporangiophore and a sporangium (fig. 16-13j). Some of the spores give rise to mycelia of one mating type, others to the other mating type (fig. 16-13e).

B. Class Ascomycetes: Sac Fungi

Members of the Ascomycetes produce spores in a sac, the **ascus** (the plural is *asci*), which develops as a result of sexual reproduction. Asexual reproduction takes place by means of production of asexual spores called **conidia** (the singular is *conidium*). The division includes organisms of considerable importance, such as the yeasts responsible for the baking and brewing industries, as well as numerous plant pathogens. A few are highly prized for food, including morels and truffles.

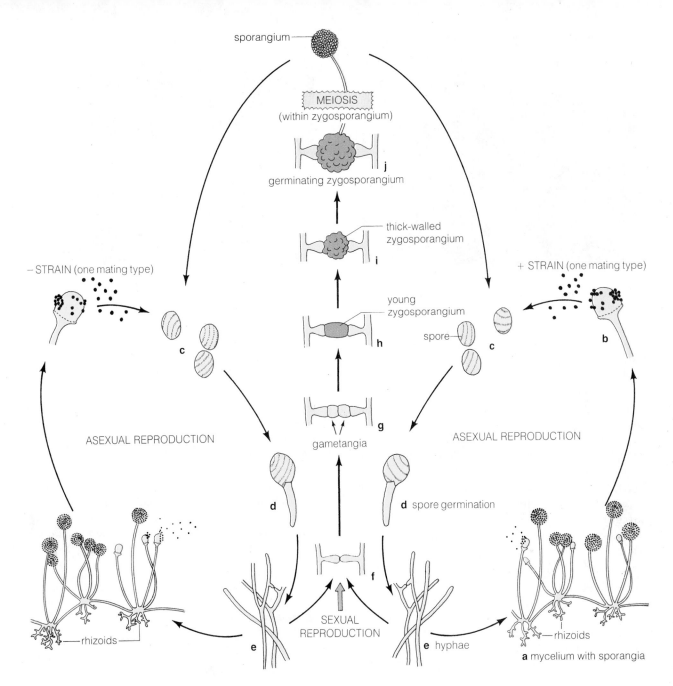

Figure 16-13 Life cycle of *Rhizopus*, black bread mold. (Green structures are 2n.) (Modified from artwork by Carolina Biological Supply Company.)

MATERIALS

Per student:

- glass microscope slide
- coverslip
- dissecting needle
- prepared slide of *Peziza*
- compound microscope

Per student pair:

- culture of *Eurotium*
- distilled water (dH₂O) in dropping bottle
- large preserved specimen of *Peziza* or another cup fungus

PROCEDURE

1. *Eurotium:* A Blue Mold

a. The blue mold, *Eurotium*, gets its common name from the production of blue-walled asexual **conidia**. From the culture provided, using a dissecting needle, scrape some conidia from the agar surface and prepare a wet mount. (Try to avoid the yellow bodies; more about them in a bit.)

b. Observe with the high-dry objective of your compound microscope (fig. 17-14). Note that the conidia are produced at the end of a specialized hypha that has a swollen tip. This arrangement has been named

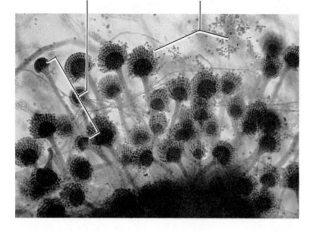

conidiophore without conidia   conidia

Figure 16-14 Conidiophore and conidia of the blue mold, *Eurotium* (150 ×). (Photo by Ripon Microslides, Inc.)

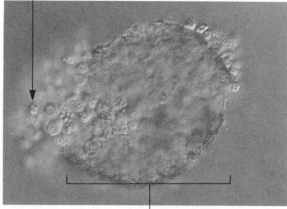

ascus containing ascospores

ascocarp

Figure 16-15 Crushed ascocarp of *Eurotium* (287 ×). (Photo by J. W. Perry.)

Aspergillus. This structure, the **conidiophore,** resembles somewhat an aspergillum used in the Roman Catholic Church to sprinkle holy water, from which its name is derived.

(You may be confused about why *Aspergillus* is in italics. The reason is that this is a scientific name. There are fungi other than *Eurotium* that produce the same type of asexual structure. Hence the asexual structure *itself* is given a scientific name.)

These tiny conidia are carried by air currents to new environments, where they germinate to form new mycelia.

c. Note the yellow bodies on the culture medium. These are the *"fruiting bodies,"* known to mycologists as **ascocarps,** which are the products of sexual reproduction.

d. With your dissecting needle, remove an ascocarp from the culture and prepare a wet mount. Using your thumb, carefully press down on the coverslip to rupture the ascocarp. Observe your preparation with the medium-power and high-dry objectives of your compound microscope. Identify the **asci,** which contain dark-colored, spherical **ascospores** (fig. 16-15).

The sexual cycle of the sac fungi is somewhat complex and is summarized by figure 16-16. The female gametangium, the **ascogonium** (fig. 16-16a; the plural is *ascogonia*) is fertilized by male nuclei from antheridia (fig. 16-16a). The male nuclei (darkened circles in fig. 16-16a) pair with but do not fuse immediately with the female nuclei (open circles). Papillae grow from the ascogonium, and the paired nuclei flow into these papillae (fig. 16-16b). Now cell walls form between each pair of sexually compatible nuclei (fig. 16-16c). Subsequently, the two nuclei fuse; the resultant cell is the diploid ascus (fig. 16-16d). (Consequently, some mycologists consider the ascus to be a zygote.) The nucleus of the ascus undergoes meiosis to form four nuclei (fig. 16-16e). Mitosis then produces eight nuclei from these four (fig. 16-16f). (Notice that cytokinesis—

cytoplasmic division—did not follow meiosis.) Next, a cell wall forms about each nucleus, some of the cytoplasm of the ascus being included within the cell wall (fig. 16-16g). Thus, eight uninucleate ascospores (fig. 16-16i) have been formed. While asci and ascospores were being formed, surrounding hyphae proliferated to form the ascocarp (fig. 16-16h) around the asci.

As you see, two different types of ascospores are produced during meiosis. Each ascospore gives rise to a mycelium having only one mating type (fig. 16-16i). Unlike many species, a *single* mycelium is capable of producing *both* ascogonia and antheridia. Based on this information, is *Eurotium* monoecious or dioecious?

___

2. *Peziza:* A Cup Fungus

a. The cup fungi are found commonly on soil during cool early spring and fall weather. Observe the preserved specimen of a cup fungus (fig. 16-17).

Actually, the structure we identify as a cup fungus is the "fruiting body," produced as a result of sexual reproduction by the fungus. Most of the organism is present within the soil as an extensive mycelium. Specifically, the fruiting body is called an **ascocarp.**

b. Obtain a prepared slide of the ascocarp of *Peziza* or a related cup fungus. Examine the slide with the medium-power and high-dry objectives of your compound microscope. Identify the elongate fingerlike **asci,** which contain dark-colored, spherical **ascospores** (fig. 16-18).

C. Class Basidiomycetes: Club Fungi

Members of this group of fungi are probably what the average person thinks of as fungi, because the division contains those organisms called mushrooms. Actually the mushroom is only a portion of the fungus—

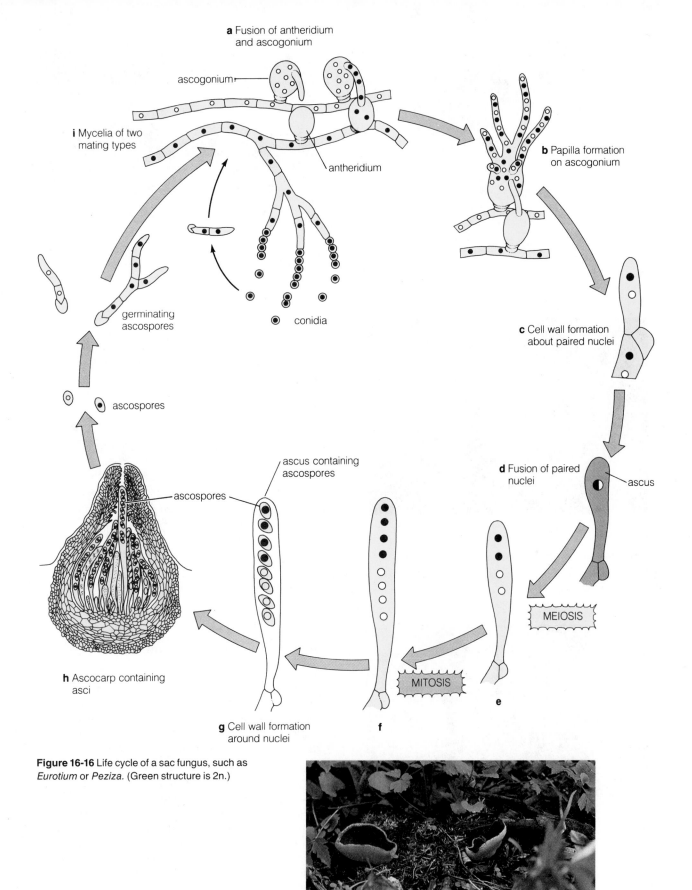

a Fusion of antheridium and ascogonium

ascogonium

i Mycelia of two mating types

antheridium

b Papilla formation on ascogonium

germinating ascospores

conidia

c Cell wall formation about paired nuclei

ascospores

d Fusion of paired nuclei

ascus

ascus containing ascospores

ascospores

MEIOSIS

MITOSIS

h Ascocarp containing asci

g Cell wall formation around nuclei

f

e

Figure 16-16 Life cycle of a sac fungus, such as *Eurotium* or *Peziza.* (Green structure is 2n.)

Figure 16-17 Cup fungi (0.25 ×). (Photo by J. W. Perry.)

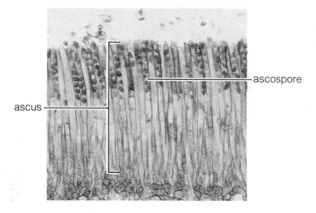

Figure 16-18 Cross section of an ascocarp from a cup fungus (186×). (Photo by J. W. Perry.)

Figure 16-19 Mushroom basidiocarps. The one on the left is younger than that on the right (0.25×). (Photo by J. W. Perry.)

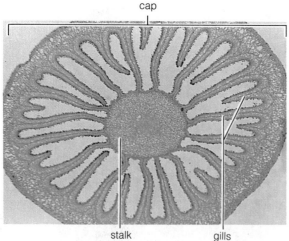

Figure 16-20 Cross section of mushroom cap (23×). (Photo courtesy Triarch, Inc.)

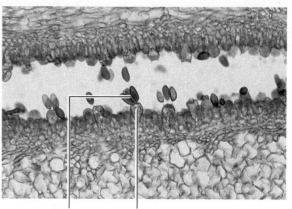

Figure 16-21 High magnification of a mushroom gill (287×). (Photo by J. W. Perry.)

it's the "fruiting body," specifically a **basidiocarp,** containing the sexually produced haploid **basidiospores.** These basidiospores are produced by a club-shaped **basidium** for which the group is named. Much (if not most) of the fungal mycelium grows out of sight, within the substrate upon which the basidiocarp is found.

MATERIALS

Per student:

• commercial mushroom

• prepared slide of mushroom pileus (cap), cross section (*Coprinus*)

• compound microscope

Per lab room:

• demonstration specimens of various club fungi

PROCEDURE

1. Gill Fungi: The Mushrooms

a. Obtain a fresh fruiting body, more properly called a **basidiocarp** (fig. 16-19). Identify the **stalk** and **cap.** Look at the bottom surface of the cap, noting the numerous gills. It is on the surface of these gills that the haploid basidiospores are borne. Remember that all the structures you are looking at are composed of aggregations of fungal hyphae.

b. Study a prepared slide of a cross section of the cap of a mushroom (fig. 16-20) as you proceed. Observe the slide first with the low-power objective of your compound microscope. In the center of the cap identify the stalk. The gills radiate from the stalk to the edge of the cap, much as spokes of a bicycle wheel radiate from the hub to the rim.

c. Switch to the high-dry objective to study a single gill (fig. 16-21). Note that the component hyphae produce club-shaped structures at the edge. These are the

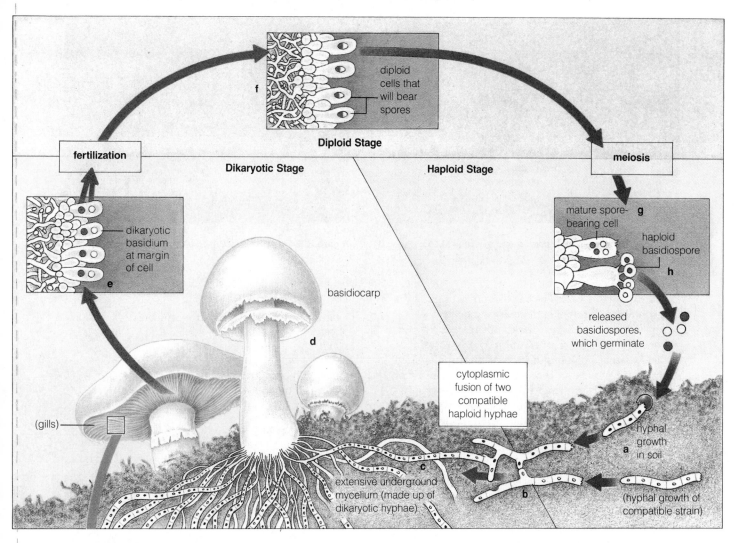

Figure 16-22 Life cycle of a mushroom.

basidia (the singular is *basidium*). Each basidium produces four haploid **basidiospores**. Find them. (All four may not be in the same plane of section.)

Each basidiospore is attached to the basidium by a tiny hornlike projection. As the basidiospore matures, it is shot off the projection due to buildup of turgor pressure within the basidium.

The life cycle of a typical mushroom is illustrated in figure 16-22. There are two different mating strains of basidiospores, because of genetic differences. These two different types of nuclei are represented in figure 16-22 as open and closed (darkened) circles.

When a haploid basidiospore (fig. 16-22a) germinates, it produces a haploid *primary mycelium* (fig. 16-22b). The primary mycelium is incapable of producing a fruiting body. Fusion between two sexually compatible mycelia (fig. 16-22b) must occur to continue the life cycle.

Surprisingly, the nuclei of the two mycelia do not fuse immediately; thus, each cell of this so-called *secondary mycelium* (fig. 16-22c) contains two genetically

different nuclei. Such a cell is said to be *dikaryotic*. (The prefix *di-* is Greek for "two," while *karyon* refers to the nucleus.)

The secondary mycelium forms an extensive network within the substrate. An environmental or genetic trigger eventually stimulates the formation of the aerial *basidiocarp* (fig. 16-22d). Each cell of the basidiocarp is dikaryotic, including the *basidia* (fig. 16-22e) on the gills.

Within the basidia, the two nuclei fuse; the basidia are now diploid (fig. 16-22f). These diploid nuclei undergo meiosis, forming genetically distinct nuclei (fig. 16-22g). Each nucleus flows with a small amount of cytoplasm through the hornlike projections at the tip of the basidium to form a basidiospore (fig. 16-22a).

Note that the gilled mushrooms do not reproduce by means of asexual conidia.

2. Other Club Fungi

Examine the representatives of fruiting bodies of other members of the club fungi that have been put on demonstration. These include puffballs (fig. 16-23) and

Figure 16-23 Puffballs. Note pore for spore escape (0.25 ×). (Photo by J. W. Perry.)

Figure 16-24 Shelf fungus (0.1 ×). (Photo by J. W. Perry.)

shelf fungi (fig. 16-24). The basidiospores of puffballs are contained within a spherical basidiocarp that develops a pore at the apex. Basidiospores are released when the puffball is crushed or hit by driving rain.

The presence of a basidiocarp of the familiar shelf fungi is an indication that an extensive network of fungal hyphae is growing within a tree, digesting the cells of the wood. As a forester assesses a woodlot to determine the amount of usable wood it might yield, one of the things noted is the presence of shelf fungi, indicating low-value (diseased) trees.

The most common shelf fungi are called polypores because of the numerous holes or pores on the lower surface of the fruiting body. These pores are lined with basidia bearing basidiospores. Observe the undersurface of the basidiocarp that is on demonstration at a dissecting microscope and note the pores.

III. Division Deuteromycota: Imperfect Fungi

The Deuteromycota consists of fungi for which no sexual stage is known and which are hence called imperfect. Reproduction takes place primarily by production of asexual conidia.

These fungi are among the most economically important, producing antibiotics (for example, one species of *Penicillium* produces penicillin). Others produce citric acid used in the soft-drink industry and are used in the manufacture of cheese. Some are important as pathogens of both plants and animals.

MATERIALS

Per student:
- dissecting needle
- glass microscope slide
- coverslip
- prepared slide of *Penicillium* conidia (optional)
- compound microscope

Per student pair:
- distilled water (dH₂O) in dropping bottle
- culture of *Alternaria*

Per lab room:
- demonstration of *Penicillium*-covered foodstuff *and/or* plate cultures

PROCEDURE

A. *Penicillium**

1. Examine demonstration specimens of moldy oranges or other foodstuffs. The blue color is attributable to a pigment in the numerous conidia produced by this fungus, *Penicillium*.

2. With a dissecting needle, scrape some of the conidia from the surface of the moldy specimen (or from a culture plate containing *Penicillium*) and prepare a wet mount. Observe your preparation using the high-dry objective. (Prepared slides may also be available.)

3. Identify the **conidiophore** and the numerous tiny, spherical **conidia** (fig. 16-25). *Penicillium* comes from the Latin word *penicillus*, meaning "a brush." (Appropriate, isn't it?)

B. *Alternaria*

Perhaps no other fungus causes more widespread human irritation than does *Alternaria*, an allergy-causing organism. During the summer, many weather programs announce the daily pollen (from flowering plants, Exercise 21) and *Alternaria* counts as an index of air quality for allergy sufferers.

---

*Some species of *Penicillium* reproduce sexually, forming ascocarps. These species are classified in the class Ascomycetes. However, not *all* fungi producing the conidiophore form called *Penicillium* reproduce sexually. Those that reproduce only by asexual means are classified in the Deuteromycota.

Figure 16-25 *Penicillium* (300 ×). (Photo courtesy Ripon Microslides, Inc.)

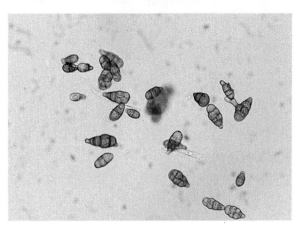

Figure 16-26 Conidia of *Alternaria* (287 ×). (Photo by J. W. Perry.)

From the culture plate provided, remove a small portion of the mycelium to prepare a wet mount. Examine with the high-dry objective of your compound microscope to find the **conidia** (fig. 16-26). Produced in chains, *Alternaria* conidia are multicellular, unlike those of *Penicillium* or *Aspergillus*. This makes identification very easy, since the conidia are quite distinct from all others produced by the fungi.

O P T I O N A L

Experiment: Bread Mold and Food Preservatives

Your instructor may provide you with an experiment allowing you to determine the effect of food preservatives on fungal growth.

PRE-LAB QUESTIONS

_____ 1. An organism that grows specifically on non-living organic material is called a (an) (a) autotroph, (b) heterotroph, (c) parasite, (d) saprophyte.

_____ 2. Taxonomic separation into fungal divisions is based upon (a) sexual reproduction, or lack thereof; (b) whether or not the fungus is a parasite or saprophyte; (c) the production of certain metabolites, like citric acid; (d) the edibility of the fungus.

_____ 3. Fungi is to fungus as _____ is to _____. (a) mycelium, mycelia; (b) hypha, hyphae; (c) mycelia, mycelium; (d) zoospore, zoospores.

_____ 4. A fungus that is dioecious (a) requires two different, sexually compatible mycelia for sexual reproduction; (b) produces both sex organs on the same mycelium; (c) reproduces only by asexual means; (d) all of the above.

_____ 5. Antheridia and oogonia would be found in the (a) Zygomycetes, (b) Oomycetes, (c) Deuteromycota, (d) Basidiomycetes.

_____ 6. Asexual reproduction in the water molds takes place by the production of (a) eggs, (b) zoospores, (c) oogonia, (d) conidia.

_____ 7. Which of the following is *not* true of the zygospore-forming fungi? (a) they are in the class Zygomycetes; (b) ascospores would be found in an ascus; (c) *Rhizopus* is a representative genus; (d) a zygospore is formed after fertilization.

_____ 8. Which structures would you find in a sac fungus? (a) ascogonium, antheridium, zygospores, (b) ascospores, oogonia, asci, ascocarps, (c) basidia, basidiospores, basidiocarps, (d) ascogonia, asci, ascocarps, ascospores.

_____ 9. The club fungi are placed in the class Basidiomycetes (a) because of their social nature, (b) because they form basidia, (c) because of the presence of an ascocarp, (d) because they are dikaryotic.

_____ 10. Which of the following is *not* true of the Deuteromycota? (a) they reproduce sexually by means of conidia; (b) they form an ascocarp; (c) sex organs are present in the form of oogonia and antheridia; (d) all of the above.

EXERCISE 16

Fungi

POST-LAB QUESTIONS

1. a. Define *parasite*.

   b. Give an example of a fungal parasite.

2. Distinguish between a *hypha* and a *mycelium*.

3. Examine the following photomicrograph of a fungal structure that you studied.

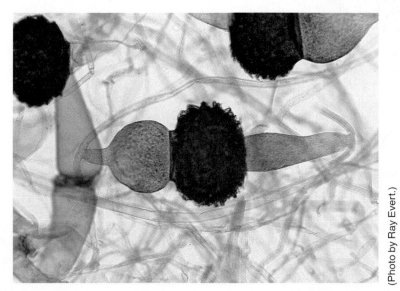

(Photo by Ray Evert.)

(287×).

   a. Is this structure the product of sexual or asexual reproduction?

   b. What is the black structure in the middle of the field?

4. a. Distinguish between sexual and asexual reproduction.

b. Sexual reproduction is "cost-intensive," requiring a large expenditure of energy for the production of one to a few offspring. On the other hand, asexual reproduction as occurs by conidia results in numerous offspring with minimal energy expenditure. What advantage is there to sexual reproduction that warrants this large expenditure of energy?

5. In the blanks provided, give the correct singular or plural form of the word provided.

Singular	Plural
a. hypha	_____
b. _____	mycelia
c. zygospore	_____
d. _____	asci
e. basidium	_____
f. _____	conidia

6. You've never seen the fungus whose sexual structures appear below, but you have seen one very similar to it. To which class of fungi does it belong?

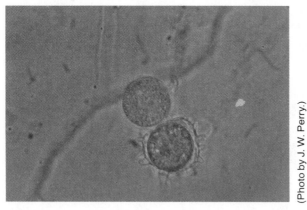

(Photo by J. W. Perry.)

(287×).

7. Distinguish among an *ascus*, an *ascospore*, and an *ascocarp*.

8. What type of spores are produced by the fungus pictured below?

(Photo by J. W. Perry.)

9. Walking in the woods, you find a cup-shaped fungus that you bring back to the lab. You remove a small portion from what appears to be its fertile surface and crush it on a microscope slide, preparing the wet mount that appears below.

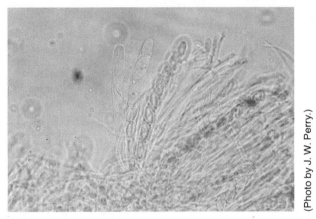

(Photo by J. W. Perry.)

 (287×).

   a. To which fungal class does this organism belong?

   b. Identify the fingerlike structures present on the slide.

10. Several decades ago, the organism *Eurotium chevelieri* was known as *Aspergillus chevelieri*. Why was the genus name changed?

Algae

OBJECTIVES

After completing this exercise you will be able to:

1. define *monoecious, dioecious, phytoplankton, phycologist, phycobilin, agar, fucoxanthin, algin, kelp, gametangium, isogamous, oogamous, oogonia, antheridia, colony, fragmentation, thallus;*

2. recognize selected members of the red, brown, and green algae, stoneworts, and brittleworts;

3. identify the pigments that are characteristic of each algal division;

4. distinguish between the structures associated with asexual and sexual reproduction described in this exercise;

5. identify the structures of the algae that are presented in **boldface** within the procedure section of this exercise.

INTRODUCTION

Algae — pond scum, frog spittle, seaweed, the stuff that clogs your aquarium if it's not cleaned routinely, the debris on an ocean beach after a storm at sea; the nuisance organisms of a lake. These are the images that probably pop into your mind when you first think about the organisms called algae. But let's consider the algae from another point of view. **Phytoplankton,** the weakly swimming or floating algae, are at the base of the aquatic food chain. They are among the smallest of the photosynthetic organisms producing their own food, themselves serving as food for animal life. As photosynthesizers, algae return vast amounts of oxygen to the water and in turn, to the atmosphere.

The term *algae* was once a bona fide taxonomic category. More recently, however, the organisms once grouped under this term have been placed in separate kingdoms. The organisms we will examine in this exercise are placed in the kingdom Plantae. (Other algae were considered in the exercise concerning the Monera and Protista kingdoms.) Today, the term *algae* (the singular is *alga*) is still used commonly by **phycologists** (*phyco-* is the Green prefix meaning "seaweed"), those who study the algae, without taxonomic implication to include all photosynthetic aquatic microbes as well as larger nonvascular aquatic plants.

The classification system we will use is as follows:

Division	Common Name
Rhodophyta	red algae
Phaeophyta	brown algae
Chlorophyta	green algae
Charophyta	stoneworts/brittleworts

This exercise will acquaint you with these four divisions. These divisions are composed of mostly unrelated organisms that share few features. All possess chlorophyll a as the primary photosynthetic pigment (but so do most other photosynthetic organisms). Perhaps the one feature that sets the algae apart from other plants is their lack of multicellular sex organs (and there are even exceptions to this rule).

Three characteristics are typically used to separate the divisions: (1) the photosynthetic pigments other than chlorophyll a; (2) the type of stored food materials; and (3) the characteristics of locomotory structures, when present.

I. Division Rhodophyta: Red Algae

Characteristics:

- Photosynthetic pigments: chlorophylls a and d, phycobilins
- Stored food: floridean starch
- Motile cells: none

MATERIALS

Per lab room:

- demonstration slide of *Porphyridium*
- demonstration specimen and slide of *Porphyra* (nori)

PROCEDURE

Although commonly called red algae, members of the Rhodophyta vary in color from red to green to purple to greenish-black. The color depends upon the quantity of their accessory pigments, the **phycobilins,** which are blue and red. These accessory pigments allow capture of light energy across the entire visible spectrum. This energy is passed on to chlorophyll for photosynthesis. One phycobilin, the red phycoerythrin, allows some red algae to live at great depths where red wavelengths, those of primary importance for green and brown algae, fail to penetrate.

Which wavelengths (colors) would be absorbed by a red pigment?

_____

Most abundant in warm marine waters, red algae are the source of **agar,** a substance extracted from their cell walls. Agar is the solidifying agent in media on which some microorganisms are cultured.

Representatives of the red algae are on demonstration.

Figure 17-1 Drawing of *Porphyridium* (Rhodophyta; _____ ×).

1. *Porphyridium:* a unicellular red alga. Unicellular red algae are exemplified by *Porphyridium*. Examine these cells at the demonstration microscope, noting the reddish chloroplast. In figure 17-1, sketch a cell of *Porphyridium*.

2. *Porphyra:* a multicellular membranous form. At the other extreme of the morphological spectrum is *Porphyra*, a membranous form. Examine a portion of this organism and the wet mount specimen on demonstration. *Porphyra* is used extensively as a food substance in Asia, where it is commonly sold under the name *nori*. In Japan, nori production is valued at $20 million annually. In figure 17-2a, make a sketch of the appearance of nori. Then study the demonstration slide of its microscopic appearance, illustrated in fig. 17-2b. Note that the clear areas are actually the cell walls.

II. Division Phaeophyta: Brown Algae

Characteristics:

• Photosynthetic pigments: chlorophylls a and c, xanthophylls, including fucoxanthin

• Stored food: laminarin

• Motile cells: zoospores and gametes

MATERIALS

Per lab room:

• demonstration specimen of *Laminaria*

• demonstration specimen of *Macrocystis*

• demonstration specimen of *Fucus*

PROCEDURE

The vast majority of the brown algae are found in cold, marine environments. All members are multicellular, and most are macroscopic. Their color is due to the accessory pigment **fucoxanthin,** which is so abundant

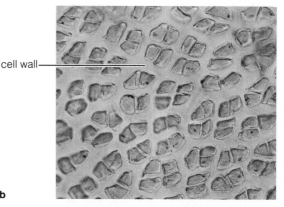

a

cell wall

b

Figure 17-2 (a) Drawing of the macroscopic appearance of *Porphyra* (nori) (Rhodophyta; _____ ×). **(b)** Microscopic appearance of *Porphyra* (nori) (Rhodophyta; 278 ×). (Photo by J. W. Perry.)

blade

stipe

holdfast

Figure 17-3 *Laminaria.* (Photo by J. W. Perry.)

that it masks the green chlorophylls. Some species are used as food, while others are harvested for fertilizers. Of primary economic importance is **algin,** a cell wall component of brown algae that is used to make ice cream smooth, cosmetics soft, and paint uniform in consistency, among other uses.

A. Kelps: *Laminaria* and *Macrocystis*

Kelps are large (up to 100 meters long), complex brown algae. Examine specimens of *Laminaria* (fig. 17-3) and *Macrocystis* (fig. 17-4). On each, identify the rootlike **holdfast** that anchors the alga to the substrate; the **stipe,** a stemlike structure; and the leaflike **blades.**

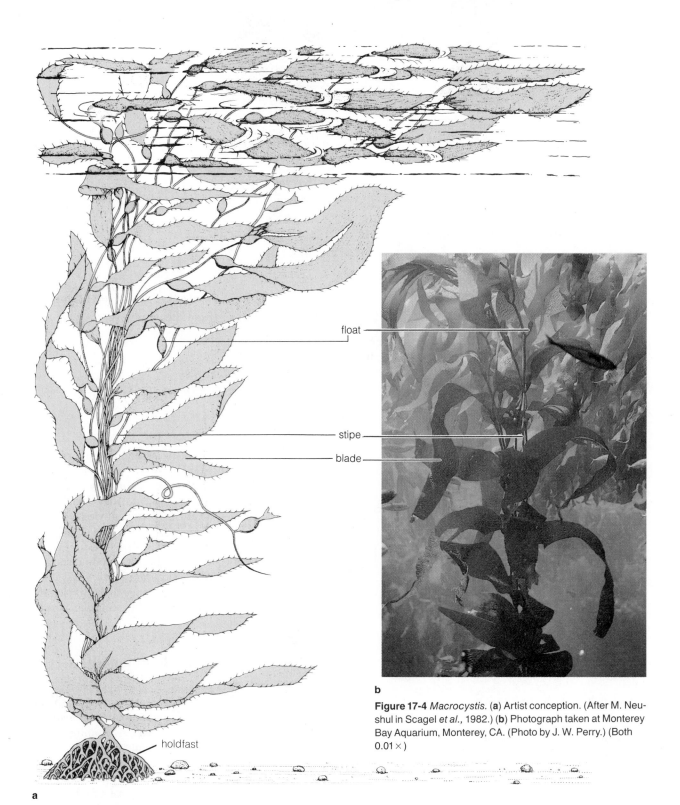

float

stipe

blade

holdfast

a

b

Figure 17-4 *Macrocystis.* (**a**) Artist conception. (After M. Neushul in Scagel *et al.*, 1982.) (**b**) Photograph taken at Monterey Bay Aquarium, Monterey, CA. (Photo by J. W. Perry.) (Both 0.01×)

(a red alga)

(other brown algae)

Figure 17-5 The rockweed, *Fucus* (Phaeophyta; 0.25 ×). (Photo by Susan Carpenter.)

B. Rockweed: *Fucus*

Fucus is a common brown alga of the coastal shore, especially abundant attached to rocks where the plants are periodically wetted by splashing waves and the tides.

Examine demonstration specimens of *Fucus* (fig. 17-5), noting the branching nature of the plant body. Locate the **holdfast,** short **stipe,** and **blade.** The tips of the blades are swollen and inflated, housing the sex organs of the plant and apparently also serving to keep the plant buoyant at high tide. Notice the numerous tiny dots on the surface of these inflated ends. These are the openings through which motile sperm cells swim to fertilize the enclosed, nonmotile egg cells.

III. Division Chlorophyta: Green Algae

Characteristics:

• Photosynthetic pigments: chlorophylls a and b, carotenoids: xanthophyll, carotene
• Stored food: starch
• Motile cells: zoospores and gametes

MATERIALS

Per student:

• clean microscope slides
• coverslips
• depression slide
• small culture dish
• dissecting needle
• prepared slide of *Oedogonium*
• prepared slide of *Spirogyra*
• compound microscope

Per student pair:

• diluted India ink in dropping bottle
• tissue paper
• I₂KI in dropping bottle
• dissecting microscope

Per lab room:

• living culture of *Chlamydomonas*, disposable pipet
• living culture of *Volvox*, disposable pipet
• demonstration slide of *Volvox* zygotes
• living culture of *Oedogonium*
• living cuture of *Spirogyra*
• demonstration specimen of *Ulva*

PROCEDURE

The Chlorophyta is a diverse assemblage of green organisms, ranging from motile and nonmotile unicellular forms to colonial, filamentous, membranous, and multinucleate forms. Not only do they include the most species, they are also important phylogenetically because ancestral green algae are believed to have given rise to the land plants. We will examine the green algae from a morphological standpoint, starting with unicellular forms. For some, we'll study their modes of reproduction and identify their characteristic sex organs, the **gametangia** (the singular is *gametangium*).

A. *Chlamydomonas:* A Motile Unicell

1. Prepare a wet mount of *Chlamydomonas* cells from the culture provided. Examine first with the low-power objective of your compound microscope. Notice the numerous small cells swimming across the field of view. It will be difficult to study the fast-swimming cells, so kill the cells by adding a drop of I₂KI to the edge of the coverslip; draw the I₂KI under the coverslip by touching a folded tissue to the opposite edge. To observe the cells, switch to the highest power objective available.

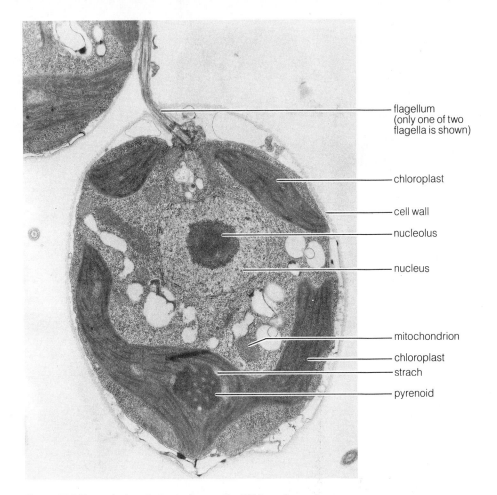

Figure 17-6 Transmission electron micrograph of *Chlamydomonas* (9,750×). (Photo courtesy H. Hoops.)

Most of the *Chlamydomonas* cytoplasm is filled with a large, green cup-shaped **chloroplast**. Identify it.

2. *Chlamydomonas* stores its excess photosynthate as *starch grains*, which will appear dark blue or black when stained with I₂KI. Locate the starch grains. The green algae have a specialized center for starch synthesis located within their chloroplast, the **pyrenoid**. The starch grains you are observing are clustered around the pyrenoid. If the orientation of the cell is just right, you may be able to detect an orange **stigma** (eyespot) that serves as a light receptor. Finally, find the two **flagella** at the anterior end of the cell.

3. Examine figure 17-6, a transmission electron micrograph of *Chlamydomonas*. Notice the magnification. The electron microscope makes much more obvious the structures you could barely see with your light microscope.

The type of sexual reproduction in *Chlamydomonas* is species-dependent. Some are **isogamous,** meaning that there is no visible differentiation between male and female gametangia. Others are **oogamous,** in which relatively large, non-motile egg cells are produced within gametangia called **oogonia,** while numerous small motile sperm are produced in gametangia called **antheridia.**

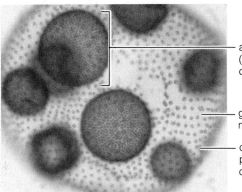

Figure 17-7 *Volvox*, with autocolonies (56×). (Photo by J. W. Perry.)

B. *Volvox:* A Motile Colony

1. From the culture provided, place a drop of *Volvox*-containing culture solution on a depression slide. (A prepared slide may be substituted if living specimens are not available.) Observe the large motile **colonies** (fig. 17-7), first with the microscope's low- and medium-power objectives. The colony is a hollow cluster of mostly identical, *Chlamydomonas*-like cells that are

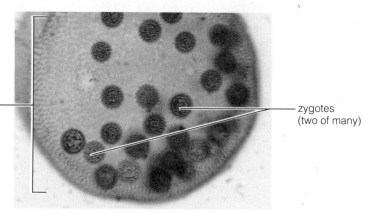

Figure 17-8 Zygotes in *Volvox* (185×). (Photo by J. W. Perry.)

held together by a **gelatinous matrix.** Identify the gelatinous matrix, which appears as the transparent region between individual cells.

Each cell possesses flagella. As the flagella beat, the entire colony rolls through the water. (The scientific name, *Volvox*, comes from the Latin word *volvere*, which means to roll.)

2. Asexual reproduction takes place by **autocolony (daughter colony) formation.** Certain cells within the colony divide and then round up into a sphere, the **autocolony.** Find the autocolonies within your specimen.

Sexual reproduction in *Volvox* is oogamous. When sperm fertilize the eggs, **zygotes** are produced.

3. Observe the demonstration slide of a spiny-walled zygote of *Volvox* (fig. 17-8).

C. *Oedogonium:* A Nonmotile Filament

Oedogonium (fig. 17-9) is a filamentous, zoospore-producing green alga that is found commonly attached to sides of aquaria and slow-moving freshwater streams.

1. Prepare a wet mount slide from the living culture provided and observe with the medium-power and high-dry objectives of your compound microscope. Within each cell, locate the single, netlike **chloroplast** with its many **pyrenoids.** The cell that attaches the filament to the substrate is specialized as a **holdfast.** Scan your specimen to determine if any holdfasts are present at the ends of the filaments. Each cell in the filament contains a single nucleus and a large central vacuole (both of which are difficult to distinguish).

Asexual reproduction takes place by **zoospores.** If present, zoosporangia (the cells producing zoospores) will be found within the filament.

Sexual reproduction is oogamous. (See definition on p. 223.)

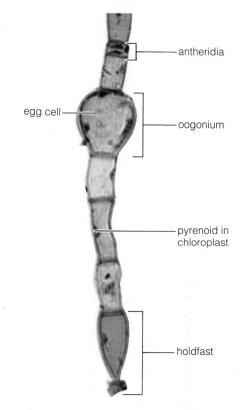

Figure 17-9 *Oedogonium*, with gametangia (370×). (Photo by Ripon Microslides, Inc.)

2. Obtain a prepared slide of a monoecious species of *Oedogonium*, which bears both male and female gametangia on the same filament. Find the large, spherical **oogonia.**

Oogonia are

_____ (male or female) sex organs. Within each oogonium locate a single, large *egg cell*, which virtually fills the oogonium.

3. Now find the male gametangia, the **antheridia,** which appear as short, boxlike cells.

Antheridia are

_____ (male or female) gametangia.

Figure 17-10 *Spirogyra* (570 ×). (Photos courtesy C. A. Taylor, III.)

scum because it forms a bright green, frothy mass on and just below the surface of the water.

1. Observe the *Spirogyra* in the large culture dish. Pick up some of the mass, noting the slimy sensation. This is due to the watery sheath surrounding each filament.

2. Using a dissecting needle, place a few filaments on a slide and add a drop of diluted India ink before adding a coverslip. (Prepared slides may be used if living filaments are not available.) Observe the filaments with the medium-power and high-dry objectives of your compound microscope (fig. 17-10). In living filaments the **sheath** will appear as a bright area off the edge of the cell wall. Note the spiral-shaped chloroplast with the numerous **pyrenoids.** Each cell contains a large central **vacuole** and a single **nucleus.** Remember, the chloroplast is located within the cytoplasm, as is the nucleus. The nucleus, however, is suspended in strands of cytoplasm, much as a spider might be found in the center of a web. Locate the nucleus.

Asexual reproduction occurs by means of **fragmentation;** that is, a small portion of the filament simply breaks off and continues to grow. Zoospores are not formed.

Sexual reproduction is isogamous. (See definition on p. 223.)

3. Obtain a prepared slide illustrating sexual reproduction in *Spirogyra* (fig. 17-11).

Find two filaments that are joined by cytoplasmic bridges known as **conjugation tubes.** The entire cytoplasmic contents serve as **isogametes** (that is, gametes of similar size) in *Spirogyra*, with one isogamete moving through the conjugation tube into the other cell, where it fuses with the other gamete. Find stages illustrating conjugation as in figure 17-11.

Each antheridium produces two *sperm*.

With fertilization, a *zygote* is formed. The zygote develops a thick, heavy wall. Some oogonia on your slide may contain zygotes. When the zygote germinates, the diploid nucleus undergoes meiosis, producing motile spores. When the spore settles down, mitosis and cell division occur, producing the haploid filament you have examined.

D. *Spirogyra:* A Nonmotile Filament

Another filamentous green alga common to freshwater ponds is *Spirogyra*. This alga is often called pond

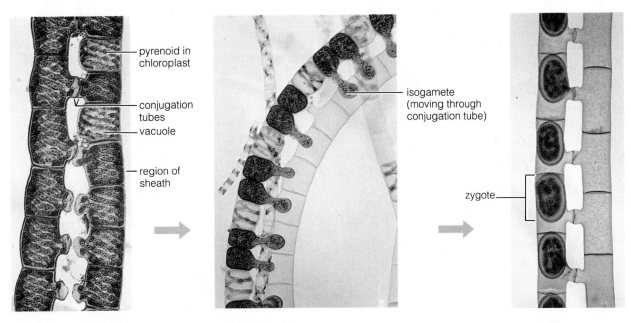

Figure 17-11 *Spirogyra*, stages in sexual reproduction (160 ×). (Photos courtesy Ripon Microslides, Inc.)

Figure 17-12 The sea lettuce, *Ulva* (Chlorophyta; 0.5×). (Photo by J. W. Perry.)

Eventually, the two nuclei of each gamete fuse to form a **zygote,** which develops a thick wall. This thick-walled zygote serves as an overwintering structure. In the spring, the zygote nucleus undergoes meiosis. Three of the four nuclei die, leaving one functional, haploid nucleus. Germination of this haploid cell results in the formation of a haploid filament.

E. *Ulva:* A Membranous Form

A final representative of the green algae illustrates the fourth morphological form in the group, those having a membranous (tissuelike) body. Examine living or preserved specimens of *Ulva,* commonly known as sea lettuce (fig. 17-12). The broad, leaflike body is called a **thallus,** a general term describing a vegetative body with relatively little cell differentiation. The thallus originates from a single cell that undergoes cell division in three planes, but only one division occurs in each of the planes, giving rise to a two-cell-thick body.

IV. Division Charophyta: Stoneworts and Brittleworts

Characteristics:
• Photosynthetic pigments: chlorophylls a and b, carotenoids: xanthophyll, carotene
• Stored food: starch
• Motile cells: male gametes (sperm)

MATERIALS

Per student:
• small culture dish
• dissecting microscope

Per lab room:
• living culture (or preserved) *Chara* or *Nitella*
• demonstration slide of *Chara* with sex organs

sex organs at tip of plant

Figure 17-13 The stonewort, *Chara,* with sex organs (2×). (Photo by J. W. Perry.)

PROCEDURE

The division Charophyta contains plants that are sometimes included in the division Chlorophyta because of the similarities they share with the green algae, as you can see by comparing their characteristics listed above. Some phycologists believe they are sufficiently dissimilar to warrant classification in a separate division, an approach we adopt here. The principal reason for separating the charophytes is their distinctive body form and sex organs, which are unlike those found in the green algae. Some botanists consider them to be a link between the green algae and the land plants.

Regardless of how they are classified, the stoneworts and brittleworts are interesting organisms found "rooted" in brackish and fresh waters, particularly those high in calcium. They get their common names by virtue of being able to precipitate calcium carbonate over their surfaces, encrusting them and rendering them somewhat brittle. (The suffix -*wort* is from a Greek word meaning "herb.")

1. From the classroom culture provided, obtain some of the specimen and place it in a small culture dish partially filled with water. Observe the specimen with a dissecting microscope. Note that the stoneworts resemble what we would think of as a plant. They are divided into "stems" and "branches" (fig. 17-13). Search for flask-shaped and spherical structures along the stem. These are the sex organs. If none is present on the specimen, observe the demonstration slide that has been selected to show these structures.

2. The flask-shaped structures are **oogonia,** each of which contains a single, large egg (fig. 17-14). Notice that the oogonium is covered with cells that twist over the surface of the gametangium. Because of the presence of these cells, the oogonium is considered to be a *multicellular* gametangium.

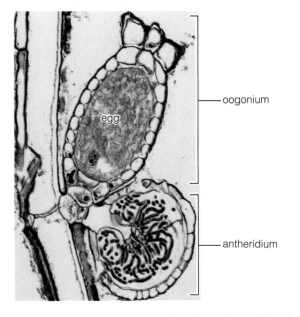

Figure 17-14 Antheridia and oogonia of *Chara* (80×). (Photo by J. W. Perry.)

Based upon your study of previously examined specimens, would you say the egg is motile or nonmotile?

_____

3. Now find a spherical **antheridium** (fig. 17-14). Like the oogonium, the antheridium is covered by cells and is also considered to be multicellular. Cells within the interior of the antheridium produce numerous flagellated sperm cells.

Fertilization of eggs by sperm produces a *zygote* within the oogonium. The zygote-containing oogonium eventually falls off from the parent plant. The zygote may remain dormant for some time before the nucleus undergoes meiosis in preparation for germination. Apparently, three of the four nuclei produced during meiosis disintegrate. Thus stoneworts and brittleworts are _____ (diploid or haploid). Upon germination, a small plant is produced that develops *rhizoids,* anchoring it to the substrate.

PRE-LAB QUESTIONS

____ 1. Small, green, floating or weakly swimming organisms are specifically called (a) algae, (b) plankton, (c) phytoplankton, (d) zooplankton.

____ 2. Specifically, a person who specializes in the study of the algae is called a (a) botanist, (b) phycologist, (c) biologist, (d) mycologist.

____ 3. Agar is derived from (a) red algae, (b) brown algae, (c) green algae, (d) all of the above.

____ 4. Which set of characteristics is found in the green algae? (a) chlorophyll a and d, phycobilins, floridean starch; (b) chlorophyll a and b, fucoxanthin, laminarin; (c) chlorophyll a and c, fucoxanthin, laminarin; (d) chlorophyll a and b, carotenoids, starch.

____ 5. Which is the correct plural form of the word for the organisms studied in this exercise? (a) alga, (b) algae, (c) algas, (d) algaes.

____ 6. Phycobilins are (a) photosynthetic pigments; (b) found in the red algae; (c) blue and red pigments; (d) all of the above.

____ 7. The cell wall component algin is (a) found in the brown algae; (b) used in the production of ice cream; (c) used as a medium on which microorganisms are grown; (d) a and b above.

____ 8. Specifically, female sex organs are known as (a) oogonia; (b) gametangia; (c) antheridia; (d) a zygote.

____ 9. A reagent that would stain the stored food of a green alga black is (a) India ink; (b) I_2KI; (c) methylene blue; (d) a and b above.

____ 10. Which of the following distinguishes the division Charophyta from the division Chlorophyta? (a) the Charophyta contains organisms that might be considered to have multicellular gametangia; (b) their photosynthetic pigments are different; (c) their stored food material is different; (d) the Charophyta is in a different kingdom from the Chlorophyta.

EXERCISE 17

Algae

POST-LAB QUESTIONS

1. While wading in the warm salt water off the beaches of the Florida Keys on spring break, you stoop down to look at the feathery alga shown here. Since you've just completed the exercise on the algae in your biology class, you know that

   this alga is a member of the division _____ .

(Photo by J. W. Perry.)

(0.25×)

2. Your class takes a field trip to a freshwater stream, where you collect the organism shown microscopically below.

(160×). (Photo courtesy C. A. Taylor III.)

   a. To which division of algae does it belong?

   b. Identify and give the function of the structure within the chloroplast at the end of the leader (line).

3. a. Which group of algae studied in this lab possesses phycobilins as accessory pigments? b. Which other group of organisms studied in a previous lab had phycobilins as accessory pigments?

a. This lab _____

b. Other lab _____

The presence of the same accessory pigments in two separate groups of organisms is believed to have phylogenetic significance.

c. Speculate on what the relationship between these two otherwise dissimilar groups might be.

4. While walking along the beach at Point Lobos, California, the fellow below walks up to you with alga in hand. Figuring you to be a college student who has probably had a good introductory biology course, he asks if you know what it is.

While you don't know the scientific name, you can tell him that it's a _____

(color) alga belonging to the division _____ .

(Photo by J. W. Perry.)

5. As was mentioned in the exercise, some botanists consider the division Charophyta to be a link between the higher plants and the algae. As you will learn in future exercises, higher plants, such as the mosses, have both haploid and diploid stages that are *multicellular.*

a. Describe the multicellular organism in the charophytes.

b. Is this organism haploid or diploid?

c. Is the zygote haploid or diploid?

d. Is the zygote unicellular or multicellular?

6. a. What color are the marker lights at the edge of an airport taxiway?

_____

   b. Are the wavelengths of this color long or short, relative to the other visible wavelengths?

_____

   c. Which wavelengths penetrate deepest into water, long or short?

_____

   d. Make a statement regarding why phycobilin pigments are present in deep-growing red algae.

   e. What benefit is there to the color of airport taxiway lights for a pilot attempting to taxi during foggy weather?

7. a. What are the principal photosynthetic pigments in the green algae?

   b. The most highly evolved land plants, the flowering plants, also contain the same principal photosynthetic pigments as do the green algae. Phylogenetically, of what importance might this be?

8. a. How is an algal holdfast similar to a root?

   b. How is it different?

9. Why do you suppose the Swedish automobile manufacturer, Volvo, chose this company name?

10. List three reasons why algae are important to life.
   a.

   b.

   c.

EXERCISE 18

Bryophytes:

Liverworts and Mosses

OBJECTIVES

After completing this exercise you will be able to:

1. define *alternation of generations, isomorphic, heteromorphic, dioecious, antheridium, archegonium, hygroscopic, protonema;*

2. produce a cycle diagram illustrating alternation of generations;

3. distinguish between alternation of isomorphic and heteromorphic generations;

4. list evidence supporting the evolution of land plants from green algae;

5. recognize mosses and liverworts;

6. identify the sporophytes, gametophytes, and associated structures of liverworts and mosses (those in **boldface** in the procedure sections);

7. describe the function of the sporophyte and the gametophyte;

8. postulate why bryophytes are restricted to environments where free water is often available.

INTRODUCTION

It's generally agreed that land plants arose from the green algae. Evidence for this includes identical food reserves (starch), the same photosynthetic pigments (chlorophylls a and b, carotenes, and xanthophylls), and similarities in structure of their flagella. Some biologists have gone so far as to suggest that the land plants are nothing more than highly evolved green algae.

One major mystery is the origin of a feature common to *all* land plants, **alternation of generations.** In alternation of generations, two distinct phases exist: A diploid **sporophyte** alternates with a haploid **gametophyte,** as summarized in figure 18-1.

At first, alternation of generations is difficult to envision. As animals, we find this concept foreign. But think of it as the existence of two body forms of the same organism. The primary reproductive function of one body form, the gametophyte, is to produce gametes (eggs and/or sperm) by *mitosis.* The primary reproductive function of the other, the sporophyte, is to produce spores by *meiosis.*

A fundamental distinction exists between the green algae and land plants in respect to alternation of generations. Although some green algae have alternation of generations, the sporophyte and gametophyte look *identical.* To the naked eye they are indistinguishable. This is called alternation of iso-

morphic generations. (*Iso-* comes from a Greek word meaning "equal"; *morph-* is Greek for "form.")

By contrast, the alternation of generations in land plants is **heteromorphic** (*hetero-* is Greek for "different"). The sporophytes and gametophytes of land plants, including the bryophytes, are distinctly different from one another. The contrast between isomorphic and heteromorphic alternation of generations is modeled in figure 18-2.

During the course of evolution, two major lines of divergence took place in the plant kingdom. The plants in one line had as the dominant phase the gametophytic generation, meaning that the sporophyte never was free-living but was permanently attached to and dependent upon the gametophyte for nutrition. Today these plants are represented by the bryophytes, mosses and their relatives. It seems that this line is an example of *dead-end evolution,* with no other group of plants present today arising from it.

In the other line, the sporophyte led an independent existence, the gametophyte being quite small and inconspicuous. Figure 18-3 summarizes these two evolutionary lines.

As you will see in this exercise, liverworts and mosses are often fairly dissimilar in appearance. They do share similarities that result in placing these plants (kingdom Plantae) in the division Bryophyta. These similarities include the following:

1. Both exhibit alternation of heteromorphic generations in which the gametophyte is the dominant organism. The sporophyte remains attached to the gametophyte, deriving most of its nutrition from the gametophyte.

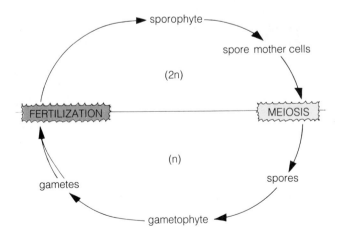

Figure 18-1 Summary of alternation of generations.

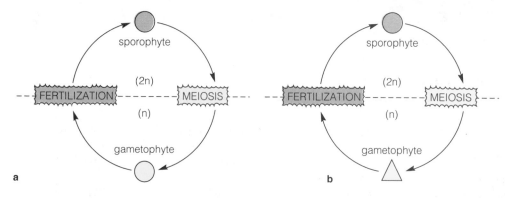

Figure 18-2 Types of alternation of generations. (**a**) Isomorphic. (**b**) Heteromorphic.

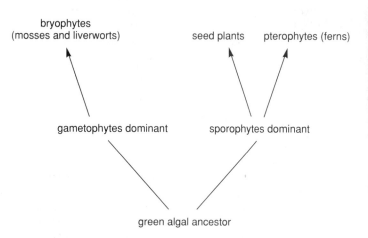

Figure 18-3 Evolution of land plants.

Figure 18-4 Thallus of a liverwort with gemma cups (1 ×). (Photo by J. W. Perry.)

2. Both are dependent upon water for fertilization, since their sperm must swim to a nonmotile egg.

3. Both lack true vascular tissues, xylem and phloem, and hence are relatively small organisms.

Step into almost any moist forest, look down, and what do you see? More than likely, covering the bases of tree trunks, on decaying logs, and on rocks you'll find mosses. Growing on rocks along streams you'll see flattened liverworts. This exercise will acquaint you with the liverworts and mosses that are part of today's flora.

I. Liverworts

MATERIALS

Per student:

- living or preserved *Marchantia* thalli, with gemma cups, antheridiophores, archegoniophores, and mature sporophytes
- dissecting microscope

PROCEDURE

1. Examine the living or preserved plants of the liverwort, *Marchantia,* that are on demonstration. The body of this plant is called a **thallus** (the plural is *thalli*), because it is flattened and has little internal tissue differentiation. Is the thallus haploid or diploid?

_____

2. Obtain and examine one thallus of the liverwort. Notice that the thallus is lobed. Centuries ago, herbalists believed that plants that looked like portions of the human anatomy could be used to treat ailments of that portion of the body. This plant reminded them of the liver, and hence the plant was called a liverwort. (The suffix *-wort* is derived from a Greek word meaning "herb," which is a small, nonwoody plant.)

3. Now place the thallus on the stage of a dissecting microscope. Looking first at the top surface, find the *pores* that lead to the interior of the plant and serve as avenues of exchange for atmospheric gases (CO_2 and O_2; fig. 18-4). Identify **gemma cups** (fig. 18-4) on the upper surface as well. Look closely within a gemma cup to find the **gemmae** inside (the singular is *gemma*).

antheridiophore

a

archegoniophore

thallus

b

Figure 18-5 Liverwort thalli with gametangiophores. (**a**) Male thalli with antheridiophores. (**b**) Female thalli with archegonio-phores (0.5×). (Photos by J. W. Perry.)

Gemmae are produced by mitotic divisions of the thallus. They are dislodged by splashing water. If they land on a suitable substrate, they grow into new thalli; hence they are a means of asexual reproduction. Are gemmae haploid or diploid?

---

4. Turn over the plant to find the **rhizoids** that anchor the organism to the substrate.

The thalli are gametophytes, so you may see elevated *gametangiophores* attached to the upper surface. (The suffix *-phore* is derived from a Greek word meaning "branch.") Those that look like flattened umbrellas are **antheridiophores** (fig. 18-5). The antheridia (the singular is *antheridium*), male sex organs, are within the flattened splash platform.

5. On female plants, find the **archegoniophores** (fig. 18-5); these look somewhat like an umbrella that has lost its fabric and only the ribs remain. Underneath these ribs are borne the female gametangia, archegonia (the singular is *archegonium*).

OPTIONAL
Experiment: Effect of Photoperiod

Your instructor may provide you with an experiment allowing you to determine the effect of photoperiod on the production of gametangiophores.

II. Mosses

MATERIALS

Per student:

- *Polytrichum*, male and female gametophytes, the latter with attached sporophytes
- prepared slide of moss antheridial head, l.s.
- prepared slide of moss archegonial head, l.s.
- prepared slide of moss sporangium (capsule), l.s.
- glass microscope slide
- coverslip
- compound microscope
- dissecting microscope
- dissecting needle

Per student pair:

- distilled water (dH₂O) in dropping bottle

Per student group (table):

- moss protenemata growing on culture medium

PROCEDURE

As you study the stages of the life history of the mosses, refer to figure 18-12, a diagrammatic representation of their life cycle.

1. Obtain a living or preserved specimen of a moss that grows in your area. The hairy-cap moss, *Polytrichum*, is a good choice because one of the ten species in this genus is certain to be found wherever you live in North America.

Lacking vascular tissues, the mosses do not have true roots, stems, or leaves, although they do have structures that are rootlike, stemlike, and leaflike, and which function for the same purposes as do the true organs.

2. Identify the rootlike **rhizoids** at the base of the plant. What function do you think rhizoids serve?

Notice that the leaflike organs are arranged more or less radially about the stemlike *axis*. You are examining the **gametophyte generation** of the moss. In terms of the life cycle of the organism, what function does the gametophyte serve?

Polytrichum is **dioecious,** meaning that there are separate male and female plants. The male gametophyte can usually be distinguished by the flattened rosette of leaflike structures at its tip.

3. Examine a male gametophyte (figs. 18-6, 18-12a), noting this feature. Embedded within the rosette are the male sex organs, **antheridia** (the singular is *antheridium*).

Figure 18-6 Colony of male gametophytes (0.25×). (Photo by J. W. Perry.)

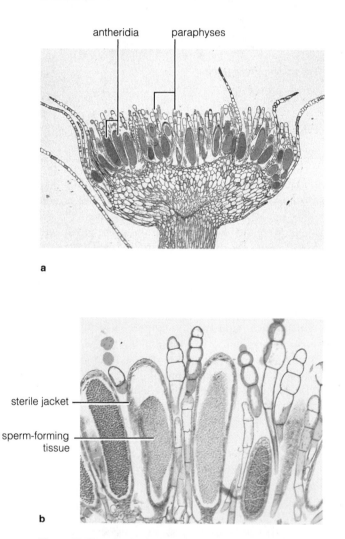

Figure 18-7 Longitudinal section of antheridial head of a moss (**a**) 8×. (**b**) 93×. (Photos by J. W. Perry.)

4. With the low- and medium-power objectives of your compound microscope, examine a prepared slide with the antheridia of a moss (fig. 18-12c). Identify the *antheridia* (fig. 18-7). Study a single antheridium (fig. 18-7), using the high-dry objective. Locate the jacket

layer surrounding the sperm-forming tissue that, with maturity, gives rise to numerous biflagellate *sperm*. Scattered among the antheridia, find the numerous sterile *paraphyses*. These do not have a reproductive role (and hence are called sterile) but instead function to hold capillary water, preventing the sex organs from drying out.

5. Examine a female gametophyte of *Polytrichum* (fig. 18-14b). Before the development of the sporophyte, the female gametophyte can usually be distinguished by the absence of the rosette at its tip. Nonetheless, the apex of the female gametophyte contains the female sex organs, **archegonia** (the singular is *archegonium*).

6. Obtain a prepared slide of the archegonial tip of a moss (figs. 18-8, 18-12d). Start with the low-power objective of your compound microscope to gain an impression of the overall organization. Find the sterile *paraphyses* and the *archegonia*. Switch to the medium-power objective to study a single archegonium, identifying the long **neck** and the slightly swollen base. Within the base, locate the **egg cell** (fig. 18-12d).

Remember, you are looking at a *section* of a three-dimensional object. The archegonium is very much like a long-necked vase, except that it's solid. The venter is analogous to the base of the vase, while the egg cell is like a marble suspended in the middle of the base.

The central core of the archegonial neck contains cells that break down when the egg is mature, liberating a fluid that is rich in sucrose and that attracts sperm that are swimming in dew or rainwater. (Sperm are capable of swimming only short distances and so must be present close by.) Fertilization of the egg produces the diploid zygote (fig. 18-12e), the first cell of the **sporophyte generation.** Numerous mitotic divisions produce an **embryo** (embryo sporophyte, fig. 18-12f), which differentiates into the **mature sporophyte** (fig. 18-12g) that protrudes from the tip of the gametophyte. What is the function of the sporophyte?

_____

7. Now examine a female gametophyte that has an attached *mature sporophyte* (figs. 18-9, 18-12g). Is the sporophyte green?

_____

What would you conclude about its ability to produce at least a portion of its own food?

_____

8. Grasp the stalk of the sporophyte and detach it from the gametophyte. The base of the stalk absorbs water and nutrients from the gametophyte. At the tip of the sporophyte locate the **sporangium** covered by a papery hood. The cover is a remnant of the tissue that surrounded the archegonium and is covered with tiny hairs: hence the common name, hairy-cap moss, for *Polytrichum*.

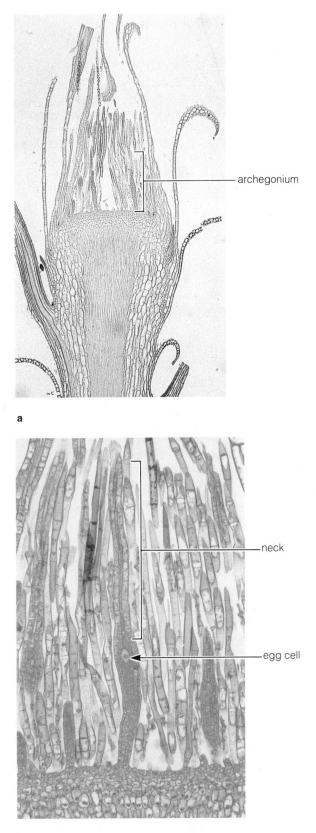

a

Figure 18-9 Colony of female gametophytes with attached sporophyte (0.25×). (Photo by J. W. Perry.)

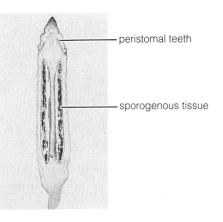

peristomal teeth

sporogenous tissue

Figure 18-10 Longitudinal section of sporangium of a moss (12×). (Photo by J. W. Perry.)

archegonium

neck

egg cell

b

Figure 18-8 Longitudinal section of archegonial head of a moss (**a**) 24×. (**b**) Detail of archegonium (96×). (Photos by J. W. Perry.)

9. Remove the hood to expose the sporangium (fig. 18-12h). Notice the small cap at the top of the sporangium. Remove the cap and observe the interior of the sporangium with a dissecting microscope. Find the *peristomal teeth* that point inward from the margin of the opening. The peristomal teeth are **hygroscopic,** meaning that they change shape as they absorb water. As the teeth dry, they arch upward, loosening the cap over the spore mass. The teeth may subsequently shrink or swell, thus regulating how readily the spores inside may escape.

10. Study a prepared slide of a longitudinal section of a sporangium (fig. 18-10). At the top you will find sections of the peristomal teeth. Internally, locate the **sporogenous tissue,** which when mature differentiates into *spores*. The sporogenous tissue of the sporangium is diploid, but the spores are haploid and are the first cells of the gametophyte generation. What type of nuclear division must take place for the sporogenous tissue to become spores?

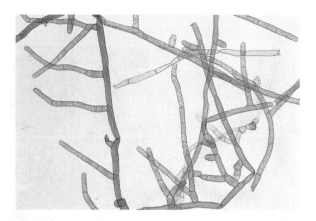

Figure 18-11 Moss protonema, whole mount (70 ×). (Photo by Ripon Microslides, Inc.)

When the spores are shed from the sporangium, they are carried by wind and water currents to new sites. If conditions are favorable, the spore germinates to produce a filamentous **protonema** (figs. 18-11, 18-12j; the plural is *protonemata*) that looks much like a filamentous green alga.

Is the protonema part of the gametophyte or sporophyte generation?

_____

11. Use a dissecting needle to remove a protonema from the culture provided and make a wet mount. Examine it with the medium-power and high-dry objectives of your compound microscope. Note the cellular composition and numerous green *chloroplasts*. If the protonema is of sufficient age, you should find *buds* (fig. 18-12k) that grow into the leafy gametophyte (figs. 18-12l, 12a, 12b).

O P T I O N A L

Experiment: Effect of Light Quality

Your instructor may provide you with an experiment allowing you to determine which wavelengths of light are effective for the germination of moss spores.

PRE-LAB QUESTIONS

_____ 1. Land plants are believed to have evolved from (a) mosses, (b) ferns, (c) green algae, (d) fungi.

_____ 2. In the bryophytes, the sporophyte is (a) the dominant generation, (b) dependent upon the gametophyte generation, (c) able to produce all of its own nutritional requirements, (d) a and c above.

_____ 3. Liverworts and mosses utilize which of the following pigments for photosynthesis? (a) chlorophylls a and b, (b) carotenes, (c) xanthophylls, (d) all of the above.

_____ 4. Which of the following *best* describes the concept of alternation of generations? (a) one generation of plants is skipped every other year; (b) there are two phases, a sporophyte and a gametophyte; (c) the parental generation alternates with a juvenile generation; (d) a green sporophyte phase produces food for a nongreen gametophyte.

_____ 5. Alternation of heteromorphic generations (a) is found only in the bryophytes, (b) is common to all land plants, (c) is typical of most green algae, (d) occurs in the liverworts but not mosses.

_____ 6. An organ that is hygroscopic is (a) sensitive to changes in moisture, (b) exemplified by the peristomal teeth in the sporophyte of mosses, (c) may aid in spore dispersal in mosses, (d) all of the above.

_____ 7. Gemmae function for (a) sexual reproduction, (b) water retention, (c) anchorage of a liverwort thallus to the substrate, (d) asexual reproduction.

_____ 8. Sperm find their way to the archegonium (a) by swimming, (b) due to a chemical gradient diffusing from the archegonium, (c) as a result of sucrose being released during the breakdown of the neck canal cells of the archegonium, (d) all of the above.

_____ 9. A protonema (a) is part of the sporophyte generation of a moss, (b) is the product of spore germination of a moss, (c) looks very much like a filamentous brown alga, (d) produces the sporophyte when a bud grows from it.

_____ 10. The suffix -*phore* is derived from a Greek word meaning (a) branch, (b) moss, (c) liverwort, (d) male.

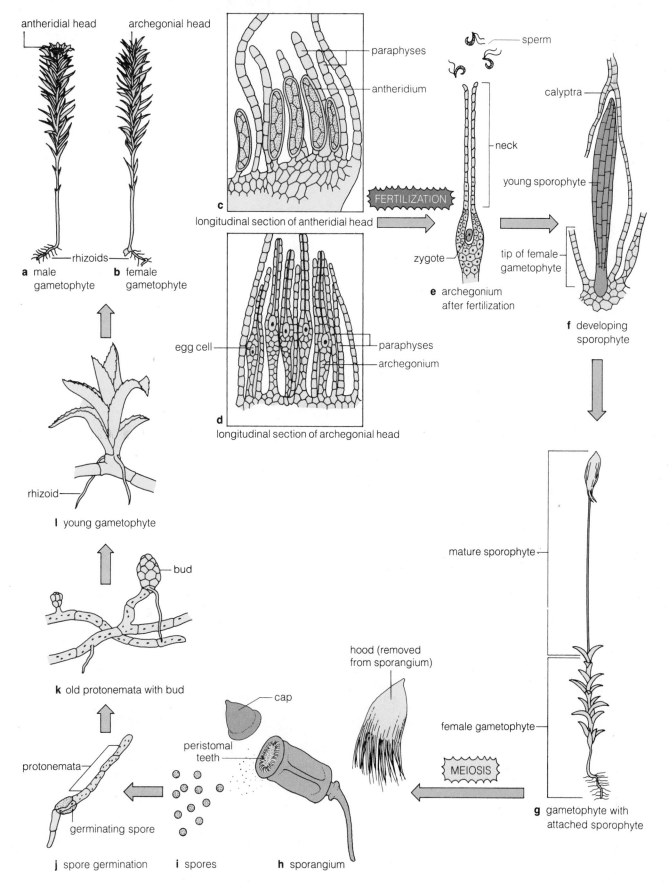

anteridial head   **archegonial head**

a male gametophyte
b female gametophyte

rhizoids

paraphyses

antheridium

c longitudinal section of antheridial head

sperm

neck

zygote

e archegonium after fertilization

FERTILIZATION

calyptra

young sporophyte

tip of female gametophyte

f developing sporophyte

egg cell

paraphyses

archegonium

d longitudinal section of archegonial head

rhizoid

l young gametophyte

bud

k old protonemata with bud

protonemata

peristomal teeth

cap

hood (removed from sporangium)

mature sporophyte

female gametophyte

MEIOSIS

g gametophyte with attached sporophyte

germinating spore

j spore germination   **i** spores   **h** sporangium

Figure 18-12 Life cycle of a representative moss. (Green structures are 2n.) (Modified from Carolina Biological Supply Company diagrams.)

BRYOPHYTES: LIVERWORTS AND MOSSES

239

EXERCISE 18

Bryophytes: Liverworts and Mosses

POST-LAB QUESTIONS

1. Explain why water must be present for the bryophytes to complete the sexual portion of their life cycle.

2. a. While the plant illustrated below is one that you did not study specifically in lab, you should be able to identify it as a moss or a liverwort. Do so.

(Photo by J. W. Perry.)

(1×).

   b. Why did you make the choice you did for answer a?

3. In what structure does meiosis occur in the bryophytes? How does this compare with the location of meiosis in the majority of the algae?

4. The ancestors of the bryophytes are believed to have been green algae. Cite four distinct lines of evidence to support this belief.

   a.

   b.

   c.

   d.

5. Identify the type of gametangium illustrated below.

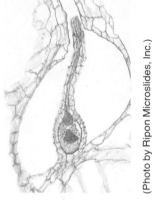

(Photo by Ripon Microslides, Inc.)

(150×).

6. Walking along a stream in a damp forest, you see the plants illustrated below.

(Photo by J. W. Perry.)

(0.25×).

   a. Are these mosses or liverworts?

   b. Why did you make the choice you did for answer a?

7. Complete this diagram of a "generic" alternation of generations.

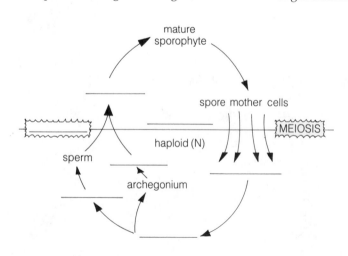

mature
sporophyte

spore mother cells

MEIOSIS

haploid (N)

sperm

archegonium

242

8. a. What are the golden stalks seen in the figure below?

(0.25×).

(Photo by J. W. Perry.)

   b. Are they the products of meiosis or fertilization?

9. Identify the plants in the figure below as male or female, gametophyte or sporophyte, moss or liverwort.

(0.25×).

(Photo by J. W. Perry.)

10. Describe in your own words the difference between a sporophyte
    and a gametophyte.

EXERCISE 19

Seedless Vascular Plants:

Fern Allies and Ferns

OBJECTIVES

After completing this exercise you will be able to:

1. define *tracheophyte, rhizome, sporophyte, sporangium, gametophyte, gametangium, antheridium, archegonium, epiphyte, strobilus, node, internode, frond, sorus, annulus;*

2. recognize whisk ferns, club mosses, horsetails, and ferns when you see them, placing them in the proper taxonomic division;

3. identify the structures of the fern allies and ferns that are presented in **boldface** in the procedure section of the exercise;

4. describe the life cycle of ferns;

5. explain the mechanism by which spore dispersal occurs from a fern sporangium;

6. describe the significant differences between the life cycles of the bryophytes and the ferns and their allies.

INTRODUCTION

The previous exercise, on liverworts and mosses, introduced the concept of alternation of heteromorphic generations. Indeed, this theme is present in all plants morphologically and evolutionarily "above" the algae. However, a major distinction exists between the bryophytes (liverworts and mosses) and the fern allies, ferns, gymnosperms, and flowering plants: Whereas the dominant and conspicuous portion of the life cycle in the bryophytes was the **gametophyte** (the gamete-producing part of the life cycle), in all other plants to be examined in this and subsequent exercises it is the **sporophyte** (that portion of the life cycle producing spores).

A second, and perhaps more important distinction also exists between those plants previously examined and the fern allies, ferns, gymnosperms, and flowering plants: The latter group contains *vascular tissue.* Vascular tissues include *phloem,* the tissue that conducts the products of photosynthesis, and *xylem,* the tissue conducting water and minerals. As a result of the presence of xylem and phloem, fern allies, ferns, gymnosperms, and flowering plants are sometimes called **tracheophytes.**

Push gently on the region of your throat at the base of your larynx (voicebox or Adam's apple). If you move your fingers up and down you should be able to feel the cartilage rings in your *trachea* (windpipe). To visualize its structure, imagine that your trachea is a pipe

with donuts inside of it (fig. 19-1). This same arrangement exists within some cell types within the xylem of plants. Thus, early botanical microscopists called plants having such an arrangement *tracheophytes.*

In this exercise we will study the tracheophytes that lack seeds. Four divisions of plants make up the seedless vascular plants:

Division	Common Names
Psilophyta	whisk ferns
Lycophyta	club mosses
Sphenophyta	horsetails
Pterophyta	ferns

The whisk ferns, club mosses, and horsetails are frequently called *fern allies,* perhaps because they are often found in the same habitat as the more conspicuous ferns. In this exercise, we will examine representatives of all four divisions; but we will study most closely the life cycle of the ferns, since they are common in our environment and have gametophytes and sporophytes that illustrate most beautifully the concept of alternation of heteromorphic generations.

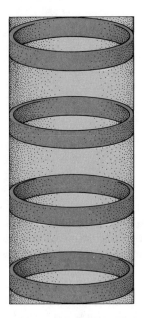

Figure 19-1 Three-dimensional representation of the cartilage in your trachea and some xylem cells of vascular plants.

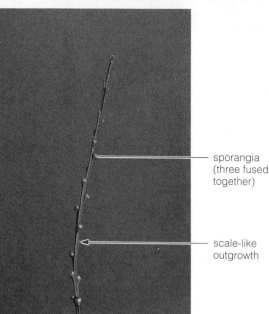

sporangia
(three fused
together)

scale-like
outgrowth

a   b

Figure 19-2 Sporophytes of *Psilotum*. (**a**) Single stem without
sporangia. (**b**) Portion of stem with sporangia (0.3×). (Photo by
J. W. Perry.)

I. Division Psilophyta: Whisk Ferns

MATERIALS

Per lab room:

- *Psilotum*, living plant
- *Psilotum*, herbarium specimen showing sporangia
  and rhizome
- *Psilotum* gametophyte
- dissecting microscope

PROCEDURE

The whisk ferns consist of only two genera of plants,
Psilotum (the *P* in the name is silent) and *Tmesipteris*
(the *T* is silent). Neither has any economic importance,
but they have intrigued botanists for a long time, especially because *Psilotum* resembles the first vascular
plants that colonized the earth. We'll examine only
Psilotum.

1. Observe first the sporophyte in a potted *Psilotum*
(fig. 19-2). Within the natural landscapes of the United
States, this plant grows abundantly in parts of Florida
and Hawaii.

   If the pot contains a number of stems, you can see
how it got its common name, the whisk fern, because
it looks a bit like a whisk broom. Examine more closely
a single aerial stem. What color is it?

From this observation, make a conclusion regarding
one function of this stem.

2. Observe the herbarium specimen (mounted plant)
of *Psilotum*. Identify the nongreen underground stem,
called a **rhizome.**

   Psilotum is unique among vascular plants in that it
lacks roots. Absorption of water and minerals takes
place through small rhizoids attached to the rhizome.
Additionally, a fungus surrounds and penetrates into
the outer cell layers of the rhizome. The fungus absorbs water and minerals from the soil and transfers
them to the rhizome. This is a beneficial association,
unlike that of the parasitic plant pathogens described
in the fungi exercise. A beneficial relationship such as
this one between the fungus and *Psilotum* is called a
mutualistic symbiosis.

3. On either the herbarium specimen or the living
plant, identify the tiny scalelike outgrowths that are
found on the aerial stems. Because these lack vascular
tissue, they are not considered true leaves. In any
case, their size would preclude any major role in
photosynthesis.

4. The plant you are observing is a sporophyte. Thus,
it must produce spores. Find the three-lobed structures on the stem (fig. 19-2b). Each lobe is a single
sporangium containing spores.

   If these spores germinate after being shed from the
sporangium, they produce a small and infrequently
found gametophyte that grows beneath the soil surface. The gametophyte survives beneath the soil

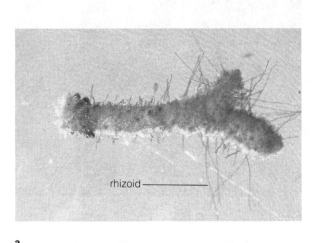

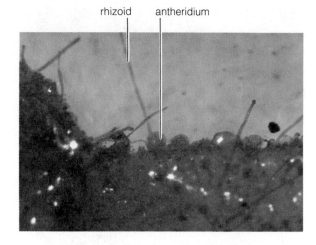
rhizoid   antheridium

a

b

Figure 19-3 *Psilotum* gametophyte. (**a**) 7.5×. (**b**) Higher magnification showing antheridia (19×). (Photo by J. W. Perry.)

thanks to a symbiotic relationship similar to that described for the sporophyte's rhizome.

5. If a gametophyte is available, observe it with the aid of a dissecting microscope (fig. 19-3). Note the numerous **rhizoids**. If you look very carefully, you may be able to distinguish **gametangia** (sex organs; the singular is *gametangium*). The male sex organs are **antheridia** (fig. 19-3b; the singular is *antheridium*); the female sex organs are **archegonia** (the singular, *archegonium*). Both antheridia and archegonia are on the same gametophyte. Thus, is the gametophyte dioecious or monoecious?

Fertilization of an egg within an archegonium results in the production of a new sporophyte.

II. Division Lycophyta: Club Mosses

Lycopodium and *Selaginella* are the most commonly found genera in the division Lycophyta, the club mosses. The common name of this division comes from the presence of the so-called **strobilus** (the plural is *strobili*), a region of the stem specialized for the production of spores. The strobilus looks like a very small club. Strobili are sometimes also called cones.

MATERIALS

Per lab room:

• *Lycopodium*, living, preserved, or herbarium specimens with strobili

• *Lycopodium* gametophyte, preserved (optional)

• *Selaginella*, living, preserved, or herbarium specimens

• *Selaginella lepidophylla,* resurrection plant, 2 dried specimens

• culture bowl containing water

PROCEDURE

A. *Lycopodium*

1. The club moss *Lycopodium* is a small, forest-dwelling plant. Some call it ground pine or trailing evergreen. Observe the living, preserved, or herbarium specimens of *Lycopodium* (fig. 19-4).

2. On your specimen, identify the true **roots, stems,** and **leaves.** The adjective *true* is used to indicate that the organs contain vascular tissue—xylem and phloem. Identify the **rhizome** to which the upright stems are connected. In the case of *Lycopodium*, the rhizome may be either beneath or on the soil surface, depending upon the species. Notice that the rhizome is covered by leaves, as are the upright stems.

strobilus

Figure 19-4 Sporophyte of *Lycopodium* (0.6×). (Photo by J. W. Perry.)

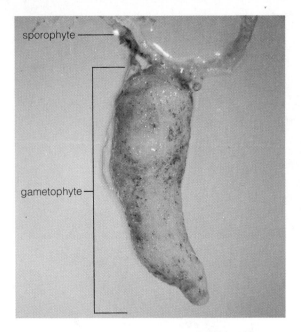

Figure 19-5 Gametophyte with young sporophyte of *Lycopodium* (2×). (Photo courtesy Dean P. Whittier.)

3. At the tip of an upright stem, find the **strobilus** (fig. 19-4). Look closely at the strobilus. As you see, it too is made up of leaves, but these leaves are much more tightly aggregated than the sterile (nonreproductive) leaves on the rest of the stem. The leaves of the strobilus produce *spores* within a *sporangium,* the spore container.

As was pointed out in the introduction, the plant you are looking at is a sporophyte. Is the sporophyte haploid (n) or diploid (2n)?

___

Since the spores are haploid, what process must have taken place within the sporangium?

___

As maturation occurs, the internodes between the sporangium-bearing leaves elongate slightly, the sporangium opens, and the spores are carried by wind away from the parent plant. If they land in a suitable habitat, the spores germinate to produce small and inconspicuous subterranean gametophytes, which bear sex organs—antheridia and archegonia. In order for fertilization and the development of a new sporophyte to take place, free water must be available, since the sperm are flagellated structures that must swim to the archegonium.

4. If available, examine a preserved *Lycopodium* gametophyte (fig. 19-5).

B. *Selaginella*

1. Examine living representatives of *Selaginella* (fig. 19-6). Some species are grown ornamentally for use in terraria. Although the strobili are not as obvious as

a

strobili

b

Figure 19-6 Sporophyte of *Selaginella* (**a**) 0.4×. (**b**) Higher magnification showing strobili at branch tips (0.6×). (Photos by J. W. Perry.)

they were in *Lycopodium,* they are present at the tips of most branches. Identify the **strobili.**

2. Examine the specimens of the resurrection plant, a species of *Selaginella* sold as a novelty, often in grocery stores. A native of the southwestern United States, this plant grows in environments that are subjected to long periods without moisture. It becomes dormant during these periods. Describe the color and appearance of the dried specimen.

___

3. Now place a dried specimen in a culture bowl containing water. Observe what happens in the next hour or so, describing the change in appearance of the plant.

___

III. Division Sphenophyta: Horsetails

A single genus, *Equisetum,* is the only living representative of this division. During the age of the dinosaurs, tree-sized representatives of this division flourished. But like the dinosaurs, they have become extinct. Different species of *Equisetum* are common throughout North America. Many are highly branched, giving

strobilus

leaves at node

branches

Figure 19-7 Sporophyte of the horsetail, *Equisetum* (0.2 ×). (Photo by J. W. Perry.)

the appearance of a horse's tail, and hence the common name. (*Equus* is Latin for "horse;" *saeta* means "bristle.")

MATERIALS

Per lab room:

• *Equisetum*, living, preserved, or herbarium specimens with strobili

• *Equisetum* gametophytes, living or preserved

PROCEDURE

1. Examine the available specimens of *Equisetum* (fig. 19-7). Depending upon the species, it will be more, or less, branched. Note that the plant is divided into **nodes** (places on the stem where the leaves arise) and **internodes** (regions on the stem between nodes). If yours is a highly branched species, don't confuse the branches with leaves. The leaves are small, scalelike structures, oftentimes brown. (They *do* have vascular tissue, so they are true leaves.) Distinguish the leaves.

2. On the herbarium mount, identify the underground **rhizome** bearing **roots.** Examine both the aerial stem and rhizome closely. Do both have nodes?

_____ Do both have leaves? _____

Which portion of *Equisetum* is primarily concerned

with photosynthesis? _____

3. Find the **strobilus** (fig. 19-7). Where on the plant is it located?

_____

Based upon the knowledge you gained from the study of *Lycopodium,* what would you expect to find within the strobilus?

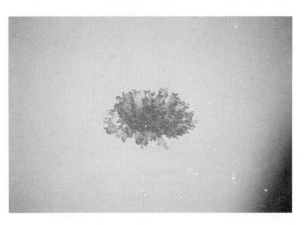

Figure 19-8 Horsetail gametophyte (4 ×). (Photo courtesy Dean Whittier.)

When spores fall to the ground, what would you expect them to grow into after germination?

_____

4. Now observe the horsetail gametophytes (fig. 19-8).

What color are they? _____

Would you expect them to be found on or below the soil surface? Why?

_____

_____

IV. Division Pterophyta: Ferns

MATERIALS

Per student:

• fern sporophytes, fresh, preserved, or herbarium specimens

• prepared slide of fern rhizome, cross section

• fern gametophytes, living, preserved, or whole mount prepared slides

• fern gametophyte with young sporophyte, living or preserved

• microscope slide

• compound microscope

• dissecting microscope

Per lab room:

• squares of fern sori, in moist chamber (*Polypodium aureum* recommended)

• demonstration slide of fern archegonium, median l.s.

• other fern sporophytes, as available

Figure 19-9 Morphology of a typical fern (0.25×). (Photo by J. W. Perry.)

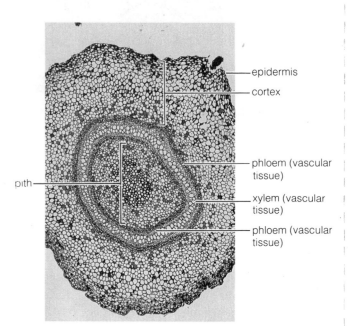

Figure 19-10 Cross section of a fern rhizome (10×). (Photo by J. W. Perry.)

PROCEDURE

1. Obtain a fresh or herbarium specimen of a typical fern **sporophyte**. As you examine the structures described below, refer to figure 19-16, a diagram representing the life cycle of a fern.

The sporophyte of many ferns (figs. 19-9, 19-16a) consists of *true* roots, stems, and leaves; that is, these possess vascular tissue. Identify the horizontal stem, the **rhizome** (which produces true **roots**), and upright leaves. The leaves of ferns are called **fronds** and are often compound.

2. Ferns, unlike bryophytes, are vascular plants, their sporophytes containing xylem and phloem. With the low-power objective of your compound microscope, examine a prepared slide of a cross section of a fern stem (rhizome). Using figure 19-10 as a reference, find the **epidermis, cortex,** and **vascular tissue**. Within the vascular tissue, distinguish between the **phloem** (of which there are outer and inner layers) and the thick-walled **xylem** sandwiched between the phloem layers.

3. Now examine the undersurface of the frond. Locate the dotlike **sori** (the singular is *sorus;* figs. 19-9, 19-16a). Each sorus is a cluster of **sporangia**. Using a dissecting microscope, study an individual sorus. Identify the sporangia (figs. 19-11, 19-16b).

Each sporangium contains *spores* (fig. 19-16b). Although the sporangium is part of the diploid (sporophytic) generation, spores are the first cells of the

haploid (gametophytic) generation. What process occurred within the sporangium to produce the haploid spores?

_____

4. Obtain a single sorus-containing square of the hare's foot fern *(Polypodium aureum);* place it sorus-side up on a glass slide (DON'T ADD A COVERSLIP); and examine it with the low-power objective of your compound microscope. Note the row of brown, thick-walled cells running over the top of the sporangium, the **annulus** (fig. 19-11). The annulus is hygroscopic. Changes in moisture content within the cells of the annulus cause the sporangium to crack open. Watch what happens as the sporangium dries out.

As the water evaporates from the cells of the annulus, a tension develops that pulls the sporangium apart. Separation of the halves of the sporangium begins at the thin-walled *lip cells* (fig. 19-16b). As the water continues to evaporate, the annulus pulls back the top half of the sporangium, exposing the spores. The sporangium continues to be pulled back until the tension on the water molecules within the annulus exceeds the strength of the hydrogen bonds holding the water molecules together. When this happens, the

Figure 19-11 Sorus containing sporangia (16 ×). (Photo by J. W. Perry.)

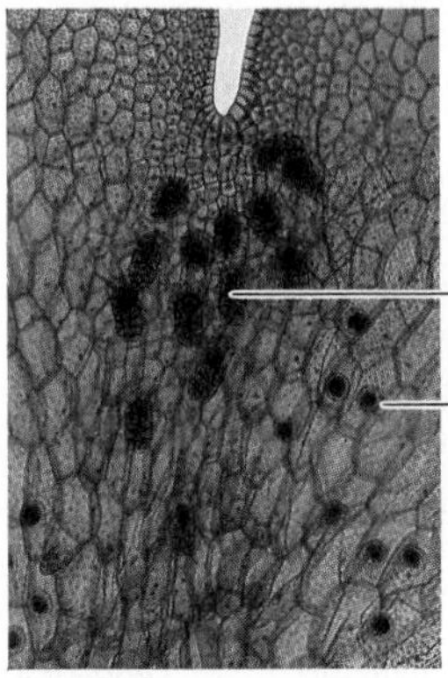

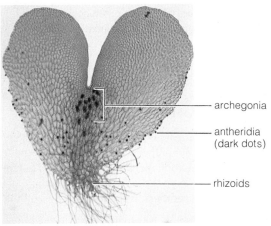

Figure 19-12 Whole mount of fern gametophyte undersurface (10 ×). (Photo by J. W. Perry.)

archegonia

antheridia (dark dots)

rhizoids

top half of the sporangium flies forward, throwing the spores out. The fern sporangium is a biological catapult!

5. If spores land in a suitable environment, one that is generally moist and shaded, they germinate (figs. 19-16d, 19-16f), eventually growing into the heart-shaped adult **gametophyte**. Using your dissecting microscope, examine a living, preserved, or prepared slide whole mount of the gametophyte (figs. 19-12, 19-16g). What color is the gametophyte?

_____

What does the color indicate relative to the ability of the gametophyte to produce its own carbohydrates?

_____

Figure 19-13 Gametangia (antheridia and archegonia) on undersurface of fern gametophyte (38 ×). (Photo by J. W. Perry.)

archegonium

antheridium

6. Examine the undersurface of the gametophyte. Find the **rhizoids,** which serve to anchor the gametophyte and perhaps absorb water.

7. Locate the _gametangia_ (sex organs) clustered among the rhizoids (fig. 19-16g). There are two types of gametangia: **antheridia,** which produce the flagellated _sperm;_ and **archegonia** (figs. 19-13, 19-16h), which produce _egg cells._

8. Study the demonstration slide of an archegonium (figs. 19-14, 19-16). Identify the **egg** within the swollen basal portion of the archegonium. Note that the _neck_ of the archegonium protrudes from the surface of the gametophyte.

The archegonia secrete chemicals that attract the flagellated sperm, which swim in a water film down a canal within the neck of the archegonium. One sperm fuses with the egg to produce the first cell of the sporophyte generation, the _zygote._ With subsequent cell divisions, the zygote develops into an embryo (embryo sporophyte; fig. 19-16j). As the embryo grows, it pushes out of the gametophyte and develops into a young sporophyte (fig. 19-16k).

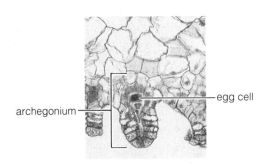

archegonium

egg cell

Figure 19-14 Longitudinal section of a fern archegonium (120 ×). (Photo by J. W. Perry.)

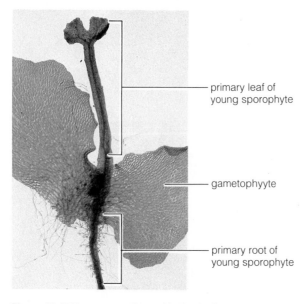

primary leaf of
young sporophyte

gametophyyte

primary root of
young sporophyte

Figure 19-15 Fern gametophyte with attached sporophyte
(10 ×). (Photo by Ripon Microslides, Inc.)

9. Obtain a specimen of a young sporophyte that is attached to the gametophyte (figs. 19-15, 19-16l). Identify the **gametophyte,** *primary leaf,* and *primary root* of the young sporophyte. As the sporophyte continues to develop, the gametophyte withers away.

10. Examine any other specimens of ferns that may be on demonstration, noting the incredible diversity in form. Look for sori on each specimen.

OPTIONAL

Experiment: Hormones and Gametophyte Growth

Your instructor may provide you with an experiment allowing you to determine the effect of certain hormones on growth of fern gametangia.

PRE-LAB QUESTIONS

_____ 1. The sporophyte is the dominant and conspicuous generation in the (a) fern allies and ferns; (b) gymnosperms; (c) flowering plants; (d) all of the above.

_____ 2. A tracheophyte is a plant that has (a) xylem and phloem; (b) a windpipe; (c) a trachea; (d) the gametophyte as its dominant generation.

_____ 3. *Psilotum* lacks (a) roots; (b) a mechanism to take up water and minerals; (c) vascular tissue; (d) alternation of heteromorphic generations.

_____ 4. Spore germination followed by cell divisions results in the production of a (an) (a) sporophyte; (b) antheridium; (c) zygote; (d) gametophyte.

_____ 5. Which phrase *best* describes a plant that is an epiphyte? (a) a plant that grows upon another plant; (b) a parasite; (c) a plant with the gametophyte generation dominant and conspicuous; (d) a plant that has a mutually beneficial relationship with another plant.

_____ 6. Club mosses (a) are placed in the division Sphenophyta; (b) are so called because of the social

nature of the plants; (c) do not produce gametophytes; (d) are so called because most produce strobili.

_____ 7. The resurrection plant (a) is a species of *Selaginella;* (b) grows in the desert Southwest of the United States; (c) is a member of the division Lycophyta; (d) all of the above.

_____ 8. Which of the following statements is *not* true? (a) nodes are present on horsetails; (b) the rhizome on horsetails bears roots; (c) the internode of a horsetail is the region where the leaves are attached; (d) horsetails are members of the division Sphenophyta.

_____ 9. In ferns (a) xylem and phloem are present in the sporophyte, (b) the sporophyte is the dominant generation, (c) the leaf is called a frond, (d) all of the above.

_____ 10. The spores of a fern are (a) produced by mitosis within the sporangium, (b) diploid cells, (c) the first cells of the gametophyte generation, (d) a and b above.

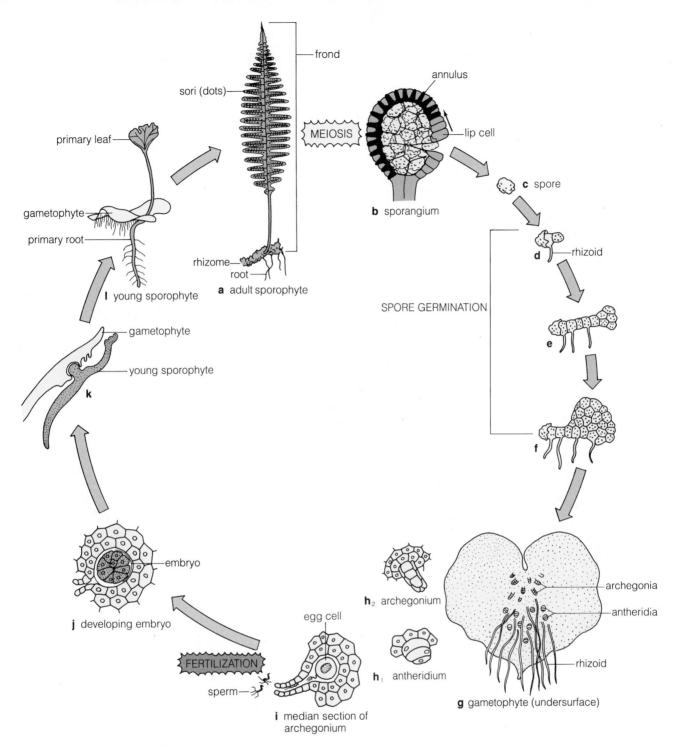

frond

sori (dots)

primary leaf

MEIOSIS

annulus

lip cell

gametophyte

primary root

b sporangium

c spore

l young sporophyte

rhizome
root

a adult sporophyte

d rhizoid

SPORE GERMINATION

gametophyte

young sporophyte

k

e

f

embryo

j developing embryo

h₂ archegonium

archegonia

antheridia

egg cell

FERTILIZATION

h₁ antheridium

sperm

i median section of archegonium

rhizoid

g gametophyte (undersurface)

Figure 19-16 Life cycle of a typical fern. (Green structures are 2n.)

E X E R C I S E   1 9

Seedless Vascular Plants: Fern Allies and Ferns

POST-LAB QUESTIONS

1. List two features distinguishing the seedless vascular plants from the bryophytes.

   a.

   b.

2. Using a biological or scientific dictionary, or a reference in your textbook, determine the meaning of the root word *psilo,* relating it to the appearance of *Psilotum.*

3. Both *Lycopodium* and *Equisetum* have strobili, roots, and rhizomes. What did you learn in this exercise that would allow you to distinguish these two plants?

4. Some species of *Lycopodium* produce gametophytes that grow beneath the surface of the soil, while others grow on the soil surface. Basing your answer upon what you have learned from other plants in this exercise, make a prediction concerning how each respective type of *Lycopodium* gametophyte would obtain its nutritional needs.

5. The environments in which ferns grow range from standing water to very dry areas. Nonetheless, all ferns are dependent upon free water in order to complete their life cycles. Explain why this is the case.

6. Examine the figure to the right. You've studied this genus in lab, but the illustrations were of a different species. Nonetheless, based upon the characteristics obvious in this figure, identify the plant, giving its common and scientific names.

(0.15×).

(Photo by J. W. Perry.)

7. Below is an illustration of a clone of plants growing in a moist woods. To which taxonomic division studied in this exercise do the plants belong?

(Photo by J. W. Perry.)

(0.15×).

8. Explain the distinction between a *node* and an *internode*.

9. After consulting a biological or scientific dictionary, explain the derivation from the Greek of the word *symbiosis*.

10. a. Give the taxonomic division of the plant illustrated here.

b. What is the structure indicated by the leader (line)?

b

(Photo by J. W. Perry.)

(0.15×).

EXERCISE 20
Seed Plants I:
Gymnosperms

OBJECTIVES

After completing this exercise you will be able to:

1. define *gymnosperm, pulp, heterosporous, homosporous, pollination, fertilization, dioecious, monoecious;*

2. describe the characteristics that distinguish seed plants from other vascular plants;

3. produce a cycle diagram illustrating heterosporous alternation of generations;

4. list uses for conifers;

5. recognize the structures in **boldface** print and describe the life cycle of a pine;

6. distinguish between a male and a female pine cone;

7. describe the method by which pollination occurs in pines;

8. describe the process of fertilization in pines;

9. recognize members of the divisions Cycadophyta, Ginkgophyta, and Gnetophyta.

INTRODUCTION

During the evolution of vascular plants, the development of the seed was one of the most striking events to occur. Seeds have remarkable survival value and seem to be one of the reasons for the dominance of seed plants today.

Let's examine the characteristics of seeds and seed plants.

1. All seed plants produce **pollen grains.** Pollen grains serve as carriers for sperm. This characteristic is one factor accounting for the widespread distribution of seed plants. As a consequence of pollen production, the sperm of seed plants do *not* need free water to swim to the egg. Thus, seed plants are capable of reproducing in harsh climates where nonseed plants are much less successful.

2. Virtually all seeds have some type of **stored food** that the embryo uses as it emerges from the seed during germination. (The sole exception is orchid seeds, which rely on symbiotic association with a fungus to obtain nutrients.)

3. All seeds have a **seed coat,** a protective covering enclosing the embryo and its stored food.

A seed coat and stored food are particularly important for survival. An embryo within a seed is protected from an inhospitable environment. Consider, for example, that a seed may be produced during a severe drought. Water is necessary for growth of the embryo. If none is available, the seed may remain dormant until growing conditions are favorable. When germination occurs, a ready food source is present to get things underway, providing nutrients until the developing plant can produce its own carbohydrates by photosynthesis.

4. As was the case with the bryophytes, fern allies, and ferns, seed plants exhibit **alternation of generations.**

5. All seed plants are **heterosporous;** that is, they produce *two* types of spores. Bryophytes and most fern allies and ferns are **homosporous,** producing only *one* spore type.

Examine figure 20-1, a diagram representing heterosporous alternation of generations. Contrast it with the diagram you completed in Exercise 18, post-lab question 7.

Gymnosperms are one of two groups of seed plants. *Gymnosperm* translates literally as "naked seed," referring to the production of seeds on the *surface* of reproductive structures. This contrasts with the situation in the angiosperms (see Exercise 21), whose seeds are contained within a fruit.

The general assemblage of plants known as gymnosperms contains plants in four separate divisions:

Division	Common Name
Coniferophyta	conifers
Cycadophyta	cycads
Ginkgophyta	ginkgos
Gnetophyta	gnetophytes

By far the most commonly recognized gymnosperms are the conifers. Among the conifers, perhaps the most common is the pine (*Pinus*). Many people believe conifers and pines to be one and the same. However, while all members of the genus *Pinus* are conifers, not all conifers are pines. Give the *scientific name* (genus) of a conifer that is not a pine. (If you cannot think of one, turn to Exercise 14, which will assist you.)

I. Division Coniferophyta: Conifers

The conifers are among the most important plants economically, because their wood is used in building construction. Millions of hectares (1 hectare = 2.47

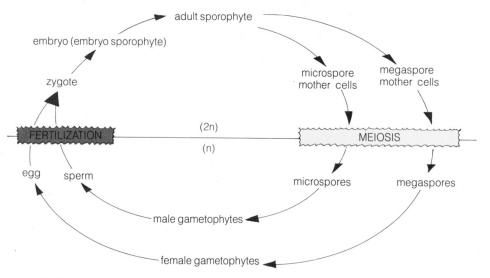

Figure 20-1 Heterosporous alternation of generations.

acres) are devoted to growing conifers for this purpose, not to mention the numerous plantations that grow Christmas trees. In many areas conifers are used for **pulp,** the moistened cell-wall-derived cellulose used to manufacture paper.

The structures and events associated with reproduction in pine (*Pinus*) will be studied here as a representative conifer.

MATERIALS

Per student:

- cluster of male cones
- prepared slide of male strobilus, l.s., with microspores
- young female cone
- prepared slide of female strobilus, l.s., with megaspore mother cell
- prepared slide of pine seed, l.s.
- compound microscope
- dissecting microscope
- single-edged razor blade

Per student group (table):

- young sporophyte, living or herbarium specimen

Per lab room:

- demonstration slide of female strobilus with archegonium
- demonstration slide of fertilization
- pine seeds, soaking in water
- pine seedlings, 12 weeks old
- pine seedlings, 36 weeks old

PROCEDURE

As you do the exercise, examine figure 20-12, representing the life cycle of a pine tree. To refresh your memory, look at a specimen of a small pine tree. This is the adult **sporophyte** (figure 20-12a). Identify the stem and leaves. You probably know the main stem of a woody plant as the trunk. The leaves of conifers are often called needles because most are shaped like needles.

A. Male Reproductive Structures and Events

1. Obtain a cluster of **male cones** (figures 20-2, 20-12b). The function of male cones is to produce **pollen;** consequently, they are typically produced at the ends of branches, where the wind currents can catch the pollen as it is being shed. Note all the tiny scalelike structures that make up the male cones. These are *microsporophylls*.

Figure 20-2 Cluster of male cones shedding pollen (0.66 ×). (Photo by J. W. Perry.)

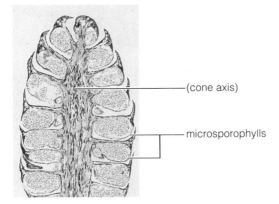

Figure 20-3 Male pine cone, l.s. (8 ×). (Photo by J. W. Perry.)

Labels on figure 20-3: (cone axis), microsporophylls

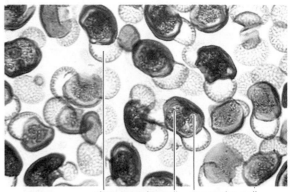

wing · tube cell · generative cell

Figure 20-4 Pine pollen grains (400 ×). (Photo by Ripon.)

Figure 20-5 Female cone of pine (0.5 ×). (Photo by J. W. Perry.)

Translated literally, a sporophyll would be a "spore-bearing leaf." The prefix *micro-* refers to "small." But rather than having the literal interpretation "small, spore-bearing leaf," a microsporophyll is one that will produce *male* spores, called *microspores*. These develop into winged, immature *male gametophytes* called **pollen grains.** (Why they are immature is a logical question. The male gametophyte is not mature until it produces sperm.)

2. Remove a single microsporophyll and examine its lower surface with a dissecting microscope (fig. 20-12c). Identify the two **microsporangia,** also called **pollen sacs.**

3. Study a prepared slide of a longitudinal section of a male cone (also called a male *strobilus*), first with a dissecting microscope to gain an impression of the cone's overall organization and then with the low-power objective of your compound microscope (figs. 20-3, 20-12d). Identify the *cone axis* bearing numerous **microsporophylls.**

4. Switch to the medium-power objective to observe more closely a single microsporophyll. Note that it contains a cavity; this is the **microsporangium** (also called a *pollen sac*), which contains numerous *pollen grains.*

As the male cone grows, several events occur in the microsporophylls that lead to the production of pollen grains. Microspore mother cells within the microsporangia undergo meiosis to form *microspores.* Cell division within the microspore wall and subsequent differentiation result in the formation of the pollen grain.

5. Examine a single **pollen grain** with the high-dry objective (figs. 20-4, 20-12e). Identify the earlike *wings* on either side of the body. The body consists of four cells, the two most obvious of which are the **tube cell** and smaller **generative cell.** (The nucleus of the tube cell is almost as large as the entire generative cell.) **Pollination,** the transfer of pollen from the male cone to the female cone, is accomplished by wind, occurring in the spring of the year. Pollen grains are caught in a sticky *pollination droplet* produced by the female cone.

B. Female Reproductive Structures and Events

Development and maturation of the female cone take two to three years, the exact time depending upon the species. Female cones are typically produced on higher branches of the tree. Because the individual tree's pollen is generally shed downward, this arrangement favors crossing between *different* individuals.

1. Obtain a young **female cone** (figs. 20-5, 20-12f), noting the arrangement of the cone scales. Unlike the male cone, the female cone is a complex structure, each scale consisting of an **ovuliferous scale** fused atop a *sterile bract.*

2. Remove a single scale-bract complex (fig. 20-12g). On the top surface of the complex find the two **ovules,** the structures that eventually will develop into the **seeds.**

3. Examine a prepared slide of a longitudinal section of a female cone (figs. 20-6, 20-12h) first with the dissecting microscope. Note the spiral arrangement of the scales on the cone axis. Distinguish the smaller *sterile bract* from the *ovuliferous scale.*

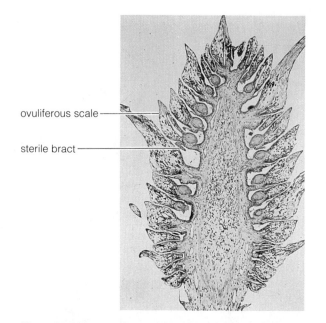

Figure 20-6 Female pine cone, l.s. (5×). (Photo by J. W. Perry.)

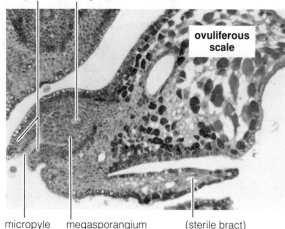

Figure 20-7 Portion of female cone showing megaspore mother cell (75×). (Photo by J. W. Perry.)

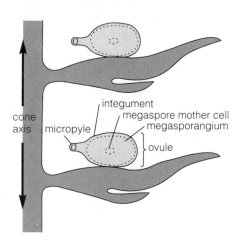

Figure 20-8 Ovule and ovuliferous scale/sterile bract complex.

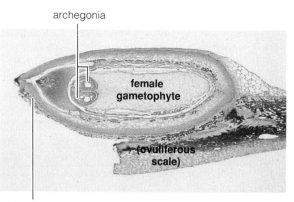

Figure 20-9 Pine ovule with female gametophyte and archegonia, l.s. (12×). (Photo by J. W. Perry.)

4. Now examine the slide with the low-power objective of your compound microscope. Look for a section through an ovuliferous scale containing a very large cell; this is the **megaspore mother cell** (figs. 20-7, 20-12i). The tissue surrounding the megaspore mother cell is the **megasporangium**. Protruding inward toward the cone axis are "flaps" of tissue surrounding the megasporangium, the **integument.** Find the integuments and the opening between them, the **micropyle.**

Think for a moment about the three-dimensional nature of the ovule: it's much like a short vase lying on its side on the ovuliferous scale. The neck of the vase is the integument, the opening the micropyle. The integument extends around the base of the vase. If

you poured liquid rubber inside the base of a vase, suspended a marble in the middle, and allowed the rubber to harden, you'd have a model of the megasporangium and the megaspore mother cell. Figure 20-8 gives you an idea of the three-dimensional structure.

The megaspore mother cell undergoes meiosis to produce four haploid *megaspores* (fig. 20-12j), but only one survives, the other three degenerating. The functional megaspore repeatedly divides mitotically to produce the multicellular **female gametophyte** (fig. 20-12k). At the same time, the female cone is continually increasing in size to accommodate the developing female gametophytes. (Remember, there are numerous ovuliferous scale/sterile bract complexes on each cone.)

The female gametophyte of pine is produced

_____ (within *or* outside of)

the megasporangium.

5. Archegonia eventually develop within the female gametophyte. Examine the demonstration slide showing archegonia (figs. 20-9, 20-12l). Identify the single

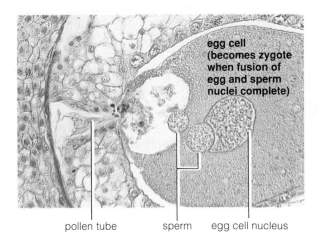

Figure 20-10 Fertilization in pine, l.s. (175×). (Photo by Ripon Microslides, Inc.)

large **egg cell** that fills the entirety of the archegonium. (The nucleus of the egg cell may be visible as well. The other generally spherical structures are protein bodies within the egg.)

Recall that the pollen grains produced within male cones were caught in a sticky pollination droplet produced by the female cone. As the pollination droplet dries, the pollen grain is drawn through the micropyle and into a cavity called the *pollen chamber* (fig. 20-12l).

Fertilization — the fusion of egg and sperm — occurs after the *pollen tube*, an outgrowth of the pollen grain's tube cell, penetrates the megasporangium and enters the archegonium. The generative cell of the pollen grain has divided to produce two sperm, one of which fuses with the egg. (The second sperm nucleus degenerates.)

6. Examine the demonstration slide illustrating fertilization in *Pinus*. Identify the **zygote,** the product of fusion of egg and sperm (figs. 20-10, 20-12l).

After fertilization, numerous mitotic divisions of the zygote take place, eventually producing an **embryo** (*embryo sporophyte*). Fertilization also triggers changes in the integument, causing it to harden and become the seed coat.

7. With the low-power objective of your compound microscope, study a prepared slide of a longitudinal section through a pine seed (figs. 20-11, 20-12m). Starting from the outside, identify the **seed coat, megasporangium** (a very thin, papery remnant), the female gametophyte, and **embryo** (embryo sporophyte).

8. Within the embryo, identify the **hypocotyl-root axis** and numerous **cotyledons,** in the center of which is the **epicotyl.** (*Hypo-* and *epi-* are derived from Greek, meaning "under" and "over," respectively. Thus, these terms refer to orientation with reference to the cotyledons.) The female gametophyte will serve as a food source for the embryo sporophyte when germination takes place.

9. Obtain a pine seed that has been soaked in water to soften the seed coat. Remove it and make a freehand longitudinal section with a sharp razor blade. Again, identify the papery remnant of the **megasporangium,** the white **female gametophyte,** and **embryo** (embryo sporophyte). How many cotyledons are present?

---

10. Examine the culture of pine seeds that were planted in sand twelve weeks ago. Note the germinating seeds (fig. 20-12n), finding the **hypocotyl-root axis, cotyledons, female gametophyte,** and **seed coat.** The cotyledons serve two functions. One is to absorb the nutrients stored in the female gametophyte during germination. As the cotyledons are exposed to light, they turn green. What then is the second function of the cotyledons?

---

11. Finally, examine the 36-week-old sporophyte seedlings. Notice that eventually the cotyledons wither away as the epicotyl produces new leaves.

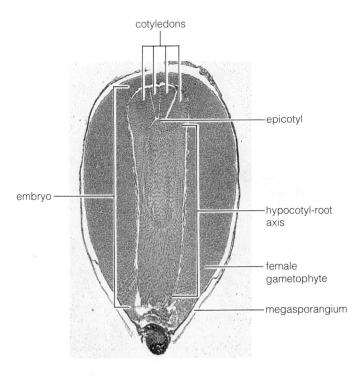

Figure 20-11 Pine seed, l.s. (24×). (Photo by Ripon Microslides, Inc.)

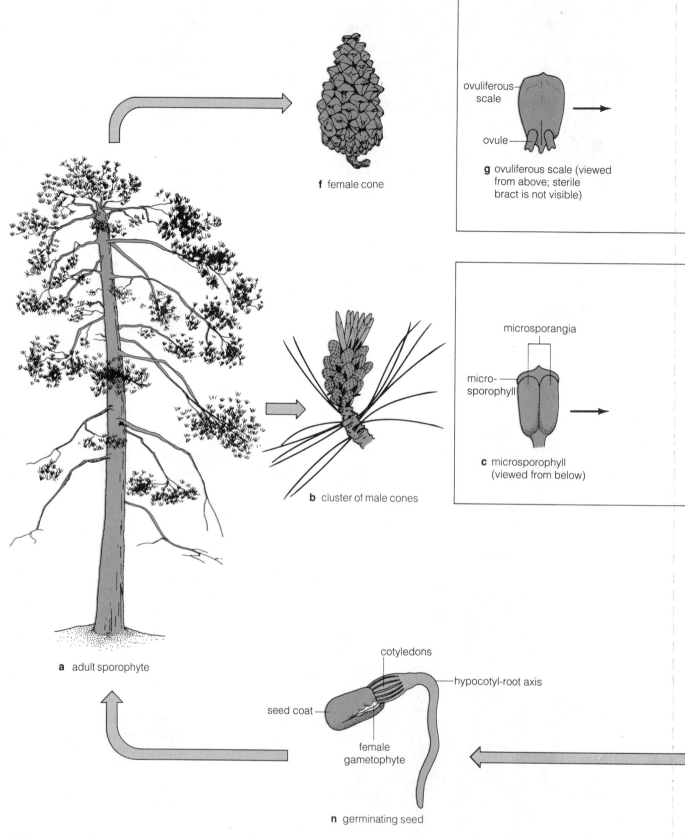

f female cone

ovuliferous scale

ovule

g ovuliferous scale (viewed from above; sterile bract is not visible)

b cluster of male cones

microsporangia

micro-sporophyll

c microsporophyll (viewed from below)

a adult sporophyte

cotyledons

hypocotyl-root axis

seed coat

female gametophyte

n germinating seed

Figure 20-12 Pine life cycle. Green and brown structures are 2n.

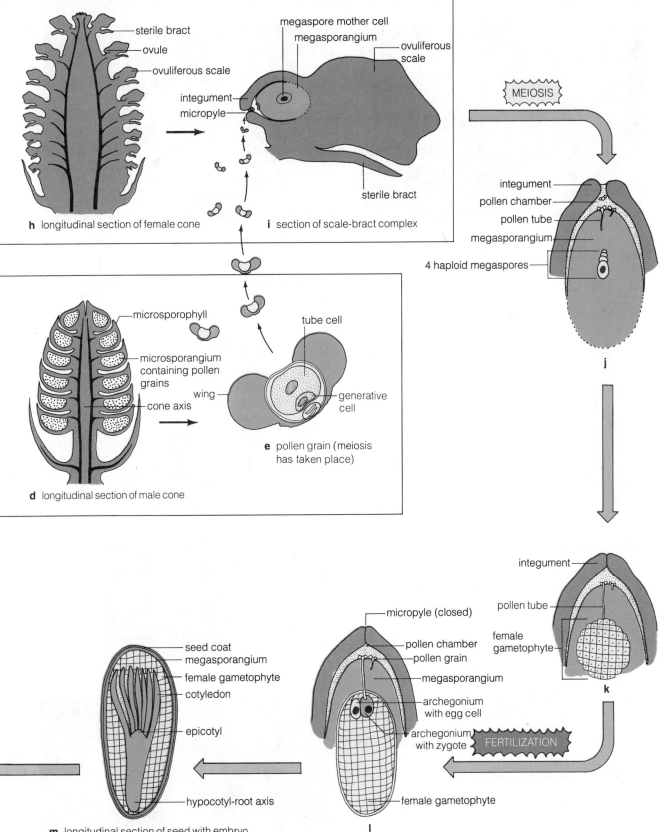

h longitudinal section of female cone

sterile bract
ovule
ovuliferous scale

megaspore mother cell
megasporangium
ovuliferous scale
integument
micropyle
sterile bract

i section of scale-bract complex

MEIOSIS

integument
pollen chamber
pollen tube
megasporangium
4 haploid megaspores

j

microsporophyll
microsporangium containing pollen grains
cone axis

tube cell
wing
generative cell

e pollen grain (meiosis has taken place)

d longitudinal section of male cone

integument
pollen tube
female gametophyte

k

FERTILIZATION

seed coat
megasporangium
female gametophyte
cotyledon

epicotyl

hypocotyl-root axis

m longitudinal section of seed with embryo

micropyle (closed)
pollen chamber
pollen grain
megasporangium
archegonium with egg cell
archegonium with zygote
female gametophyte

l

a

b

Figure 20-13 Cycads. (**a**) *Cycas* (0.01 ×). (**b**) *Zamia* (0.1 ×). (Photos by J. W. Perry.)

II. Other Gymnosperms

Although the conifers are the most important and widely distributed gymnosperms, there are three other divisions. Let's take a look at representative sporophytes of each of these.

MATERIALS

Per lab room:

• demonstration specimens of: *Zamia* and/or *Cycas, Ginkgo, Ephedra* and/or *Gnetum*

PROCEDURE

A. Division Cycadophyta: Cycads

1. Examine the demonstration specimen of *Zamia* and/or *Cycas* (fig. 20-13). Both have the common name cycad. Do these plants resemble any of the conifers you know?

Zamia and *Cycas* are limited to the subtropical regions of North America. They're often planted as ornamentals in Florida, Gulf Coast states, and California. During the age of the dinosaurs (200 million years ago), cycads were extremely numerous.

All of the cycads are **dioecious;** that is, there are distinct male and female plants. *Zamia* produces male and female cones, while *Cycas* has only male cones, the female structures being much more leaflike. (Pine is **monoecious,** meaning that both male *and* female structures are produced on the *same* plant.)

The female cones of some other genera become extremely large, weighing as much as 30 kilograms!

2. Notice the leaves of the cycads. They resemble more closely the leaves of the ferns than those of the conifers.

B. Division Ginkgophyta: Ginkgos

1. Examine the demonstration specimen of *Ginkgo biloba*, the maidenhair tree (fig. 20-14a). *Ginkgo* is the only living representative of the division. It is some-

a

b

Figure 20-14 *Ginkgo*. (**a**) Tree. (**b**) Leaves (0.5 ×). (Photos by J. W. Perry.)

a

b

Figure 20-15 *Ephedra*. (**a**) Several plants (0.1 ×). (**b**) Close-up of stems (0.5 ×). (Photos by J. W. Perry.)

times called a living fossil because it has changed little in the last 80 million years. At one time, in fact, it was believed to be extinct; the Western world knew it from the fossil record before trees were discovered in remote China.

2. Note the fan-shaped leaves (fig. 20-14b). *Ginkgo* is a highly prized ornamental tree that is now commonly planted in our urban areas. The tree has a reputation for being resistant to most insect pests and atmospheric pollution.

Like the cycads, *Ginkgo* is dioecious. Male trees are preferred as ornamentals, because the female trees produce seeds whose seed coat, when mature, has a disagreeable odor.

C. Division Gnetophyta: Gnetophytes

1. Examine the demonstration specimen of *Ephedra* (fig. 20-15). Most species are found in desert or arid regions of the world. In the desert Southwest of the United States, *Ephedra* is a common shrub known as Mormon tea, because its stems were once harvested and used to make a tea by Mormon settlers in Utah.

2. A second representative gnetophyte is the genus *Gnetum* (pronounced "neat-um"). A few North American college and university greenhouses — and some tropical gardens — keep these specimens. *Gnetum* is native to Brazil, tropical west Africa, India, and Southeast Asia. Different species vary in form from vines to trees (fig. 20-16a,b). Examine the living specimen if one is on display, noting particularly the broad, flat leaves (fig. 20-16c).

Many scientists now believe that the Gnetophyta is very closely related to the flowering plants. Look for the reproductive structures on the specimens before you. They look very much like flowers. Moreover, the structure of their water-conducting tissue (xylem) is more like that of the flowering plants than is true of the other gymnosperms.

a

b

c

Figure 20-16 *Gnetum*. (**a**) A species that is a vine (0.01 ×). (**b**) A species that is a tree (0.01 ×). (**c**) Leaves (0.25 ×). (Photos by J. W. Perry.)

_____ 1. Which of the following statements about conifers is *not* true? (a) conifers are gymnosperms, (b) all conifers belong to the genus *Pinus*, (c) all conifers have naked seeds, (d) conifers are heterosporous.

_____ 2. Seed plants (a) have alternation of generations, (b) are heterosporous, (c) develop a seed coat, (d) all of the above.

_____ 3. A pine tree is (a) a sporophyte, (b) a gametophyte, (c) diploid, (d) a and c above.

_____ 4. The male pine cone (a) produces pollen, (b) contains a female gametophyte, (c) bears a megasporangium containing a megaspore mother cell, (d) gives rise to a seed.

_____ 5. The male gametophyte of a pine tree (a) is produced within a pollen grain, (b) produces sperm, (c) is diploid, (d) a and b above.

_____ 6. Which of the following are produced *directly* by meiosis in pine? (a) sperm cells, (b) pollen grains, (c) microspores, (d) microspore mother cells.

_____ 7. An ovule (a) is the structure that develops into a seed, (b) contains the microsporophyll, (c) is produced on the surface of a male cone, (d) all of the above.

_____ 8. The process by which pollen is transferred to the ovule is called (a) transmigration, (b) fertilization, (c) pollination, (d) all of the above.

_____ 9. Which of the following is true of the female gametophyte of pine? (a) it's a product of repeated cell divisions of the functional megaspore; (b) it's haploid; (c) it serves as the stored food to be used by the embryo sporophyte upon germination; (d) all of the above.

_____ 10. The seed coat of a pine seed (a) is derived from the integuments, (b) was produced by the micropyle, (c) surrounds the male gametophyte, (d) is divided into the hypocotyl-root axis and epicotyl.

E X E R C I S E   2 0

Seed Plants I: Gymnosperms

POST-LAB QUESTIONS

1. Think about the structures you've seen in the seed plants you've examined in this exercise. What survival advantage does a seed have that has allowed the seed plants to be the most successful of all plants?

2. List four uses for conifers.

   a.

   b.

   c.

   d.

3. Below is a diagrammatic representation of a seed. Give the ploidy level (n or 2n) of each part listed.

   seed coat _____

   megasporangium _____

   female gametophyte _____

   embryo sporophyte _____

4. a. While snowshoeing through the winter woods, you stop to look at the tree branch pictured here. Are you looking at a conifer, cycad, ginkgo, or gnetophyte?

   b. Specifically, what are the brown structures hanging from the branch?

(Photo by J. W. Perry.)

5. Distinguish between *pollination* and *fertilization*.

6. How does the complexity of the gametophyte generation of the Coniferophyta compare with that of the Pterophyta? Assuming that the Coniferophyta represents greater evolutionary development than the Pterophyta, what can be said about the changes that have taken place in the gametophytic generation over the course of evolution?

7. Are antheridia present in conifers? Archegonia?

8. What environmental factor necessary for fertilization in bryophytes, fern allies, and ferns is not required in conifers?

9. Distinguish between a homosporous and a heterosporous type of life cycle.

10. a. Suppose you saw the seedling in the figure below while walking in the woods. To which gymnosperm division does the plant belong?

    b. Identify structures A and B.

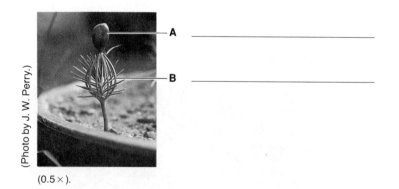

(Photo by J. W. Perry.)

A _____

B _____

(0.5×).

Seed Plants II:

Angiosperms

OBJECTIVES

After completing this exercise you will be able to:

1. define *angiosperm, fruit, pollination, double fertilization, endosperm, seed, germination, annual, biennial, perennial;*

2. describe the significance of the flower, fruit, and seed for the success of the angiosperms;

3. identify the structures of the flower;

4. recognize the structures and events (those in **boldface**) that take place in angiosperm reproduction;

5. describe the origin and function of fruit and seed;

6. identify the characteristics distinguishing angiosperms from gymnosperms.

INTRODUCTION

The **angiosperms,** seed plants that produce flowers, are placed in the division Anthophyta. The word *angiosperm* literally means "vessel seed," referring to the seeds being borne in a fruit.

There are more flowering plants in the world today than any other group of plants. Assuming that numbers indicate success, it must be said that flowering plants are the most successful plants to have evolved.

The most important characteristic that distinguishes the Anthophyta from other seed plants is the presence of flower parts that mature into a **fruit,** a container that protects the seeds, allowing them to be dispersed without coming into contact with the rigors of the external environment for some time. In many instances, the fruit also contributes to the dispersal of the seed. For example, some fruits stick to fur (or clothing) of animals and are brushed off some distance from the plant that produced them. Others are eaten by animals. The undigested seeds may pass out of the digestive tract, falling into an environment often far removed from the seeds' source.

Our lives and diets revolve around flowering plants. Fruits enrich our lives and include such things as apples, oranges, tomatoes, beans, peas, corn, wheat, walnuts, pecans . . . the list goes on and on. Moreover, even when we are not eating fruits, we're eating flowering plant parts. Cauliflower, broccoli, potatoes, celery, and carrots all are parts of flowering plants.

The number of different kinds of flowers is so large it's difficult to pick a single example to be representative of the entire division. Nonetheless, there is enough similarity among flowers that, once you've learned the structure of one representative, you'll be able to recognize the parts of most.

Flower parts are believed to have originated as leaves modified during the course of evolution to increase the probability for fertilization. For instance, some flower parts are colorful, attracting animals that serve to transfer the sperm-producing pollen to the receptive female parts.

Figure 21-15 represents the life cycle of a typical flowering plant. Refer to it as you study the specimens in this exercise.

I. External Structure of the Flower

MATERIALS

Per student:

• flower for dissection (gladiolus or hybrid lily, for example)

• single-edged razor blade

• dissecting microscope

PROCEDURE

1. Obtain a flower provided for dissection. The flower parts are arranged in whorls atop a swollen stem tip, the **receptacle.** The outermost whorl is frequently green (although not always) and is the **calyx.** Individual components of the calyx are called **sepals.** The calyx surrounds the rest of the flower in the bud stage. Identify these parts. See figure 21-15a.

2. Moving inward, locate the next whorl of the flower, the usually colorful **corolla** made up of **petals.** It is usually the petals that we appreciate for their color. Remember, however, that the evolution of colorful flower parts was associated with the presence of color-visioned *pollinators,* animals that carry pollen from one flower to another. The colorful flowers attract those animals; thus, the plants stand a good chance of being pollinated, producing seeds, and perpetuating their species.

Both the calyx and corolla are sterile; that is, they do not produce gametes.

3. The next whorl of flower parts consists of the male, pollen-producing parts, the **stamens** (also called *microsporophylls;* "microspore-bearing leaves"; figure 21-15b). Examine a single stamen in greater detail. Each stamen consists of a stalklike **filament** and an **anther.** The anther consists of four **microsporangia** (also called *pollen sacs*).

4. Next locate the female portion of the flower, the **pistil** (figs. 21-15a, 15c). A pistil consists of one or more

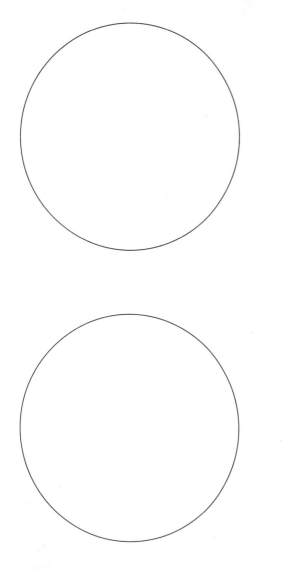

Label: ovules

Figure 21-1 Drawings of cross section and longitudinal section of an ovary.

carpels, also called *megasporophylls,* "megaspore-bearing leaves." If the pistil consists of more than one carpel, they are usually fused together, making it difficult to distinguish the individual components.

5. Identify the different parts of the pistil (fig. 21-15c): at the top, the **stigma,** which serves as a receptive region on which pollen is deposited; a necklike **style;** and a swollen **ovary.** Note that the only members of the plant kingdom to have ovaries are angiosperms.

6. With a sharp razor blade, make a section of the ovary. (Some students should cut the pistil longitudinally; others should cut the ovary crosswise. Then compare the different sections.)

7. Examine the sections with a dissecting microscope, finding the numerous small **ovules** within the ovary. The illustration in figure 21-15c is oversimpli-

fied; many flowers have more than one ovule per ovary. In figure 21-1, make and label two sketches: one of the cross section and the other of a longitudinal section of the ovary. After fertilization, the ovules will develop into *seeds,* and the ovary will enlarge and mature into the *fruit.* Notice that the ovules are completely enclosed within the ovary.

There are two groups of flowering plants, monocotyledons and dicotyledons. The number of flower parts indicates to which group a plant belongs. Generally, monocots have the flower parts in threes or multiples of three. Dicots have their parts in fours or fives or multiples thereof. Count the number of petals or sepals in the flower you have been examining. Are you studying a monocot or dicot?

II. Microsporangia and the Male Gametophyte

MATERIALS

Per student:

• prepared slide of young lily anther, cross section
• prepared slide of mature lily anther (pollen grains), c.s.
• *Impatiens* flowers, with mature pollen
• glass microscope slide
• coverslip
• compound microscope

Per student pair:

• 0.5% sucrose, in dropping bottle

PROCEDURE

1. With the low-power objective of your compound microscope, examine a prepared slide of a cross section of an immature anther (figs. 21-2, 21-15d). Find sections of the four **microsporangia** (also called *pollen*

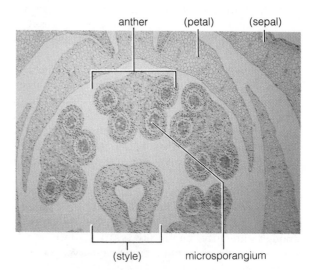

Figure 21-2 Immature anthers within flower bud, c.s. (12×). (Photo by J. W. Perry.)

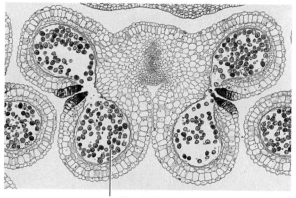

pollen grains within
a microsporangium

Figure 21-3 Mature anther, c.s. (21 ×). (Photo by Ripon Micro-slides, Inc.)

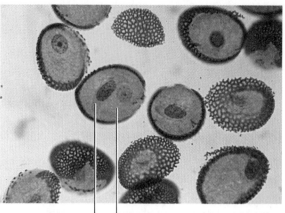

generative cell   tube cell

Figure 21-4 Pollen grain, c.s. (145 ×). (Photo by J. W. Perry.)

sacs), which appear as four clusters of densely stained cells within the anther. Study the contents of a single microsporangium. Depending upon the stage of development, you will find either diploid *microspore mother cells* (fig. 21-15e) or haploid *microspores* (figs. 21-15f, 15g).

2. Obtain a prepared slide of a cross section of a mature anther (fig. 21-3). Observe it first with the low-power objective, noting that the walls have split open to allow the pollen grains to be released, as illustrated in figure 21-15i.

3. Pollen grains are immature *male gametophytes,* and very small ones at that. Switch to the high-dry objective to study an individual pollen grain more closely (fig. 21-4). The pollen grain consists of only two cells. Identify the large **tube cell** and a smaller, crescent-shaped **generative cell** that floats freely in the cytoplasm of the tube cell (fig. 21-15h).

Note the ridged appearance of the outer wall layer of a pollen grain. Within the ridges and valleys of the wall, glycoproteins are present that are believed to play a role in recognition between the pollen grain and the stigma.

Transfer of pollen from the microsporangia to the stigma, called **pollination,** is effected by various means — wind, insects, and birds being the most common carriers of pollen. When a pollen grain lands on the stigma of a compatible flower, it germinates, producing a **pollen tube** that grows down the style (fig. 21-15j). The generative cell flows into the pollen tube, where it divides to form two *sperm* (fig. 21-15k). Because it bears two gametes, the pollen grain is now considered to be a *mature* male gametophyte.

4. Obtain an *Impatiens* flower and tap some pollen onto a clean microslide. Add a drop of 0.5% sucrose, cover with a coverslip, and observe with the medium-power objective of your compound microscope. Look for the *pollen tube* as it grows from the pollen grain. (You may wish to set the slide aside for a bit and re-examine it 15 minutes later to see what's happened to the pollen tubes.)

In figure 21-5, draw a sequence showing the germination of an *Impatiens* pollen grain.

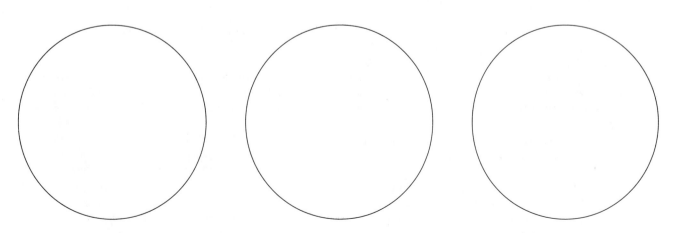

Figure 21-5 Germination of *Impatiens* pollen grain.

271

III. Megasporangia and the Female Gametophyte

MATERIALS

Per student:

• prepared slide of lily ovary, c.s., megaspore mother cell

• compound microscope

Per lab room:

• demonstration slide of lily ovary, c.s., seven-celled, eight-nucleate gametophyte

• demonstration slide of lily ovary, c.s., double fertilization

PROCEDURE

1. With the medium-power objective of your compound microscope, examine a prepared slide of a cross section of an ovary (figs. 21-6, 21-15l). Find the several **ovules** that have been sectioned. One ovule probably will be sectioned in a plane so that the very large, diploid **megaspore mother cell** is obvious (figs. 21-7, 21-15m). The megaspore mother cell is contained within the **megasporangium,** the outer cell layers of which form two flaps of tissue called **integuments.** Identify the structures in boldface print. After fertilization, the integuments develop into the *seed coat.* The ovule is attached to the ovary wall in a region known as the **placenta.**

2. As development of the megasporangium proceeds, the integuments grow, enveloping the megasporangium. However, a tiny circular opening remains. This opening is the **micropyle.** (The micropyle is obvious in fig. 21-15n.) Remember, the micropyle is an opening in the globose ovule. After pollination, the pollen grain germinates on the surface of the stigma, and the pollen tube grows down the style, through the space surrounding the ovule, through the micropyle, and penetrates the megasporangium (fig. 21-15p). Identify the micropyle present in the ovule you are examining.

3. Considerable variation exists in the next sequence of events, but the pattern found in lily is that described.

The diploid megaspore mother cell undergoes meiosis, producing four haploid *nuclei* (fig. 21-15n). (Note that cytokinesis does *not* follow meiosis, and thus only nuclei—not cells—are formed.) The cell containing the four nuclei (the old megaspore mother cell) is now called the **female gametophyte** (or *embryo sac*).

Three of these four nuclei fuse. Thus, the female gametophyte contains one triploid (3n) nucleus and one haploid (n) nucleus (fig. 21-15o). Subsequently, the nuclei undergo two *mitotic* divisions, forming eight nuclei in the female gametophyte. Now cell walls form around six of the eight nuclei; the large cell remaining—the **central cell**—contains two nuclei, one of

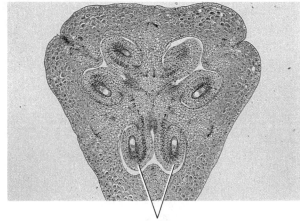

Figure 21-6 Lily ovary, c.s. (19×). (Photo by J. W. Perry.)

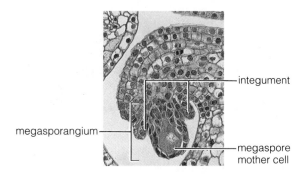

Figure 21-7 Megaspore mother cell within megasporangium (94×). (Photo by J. W. Perry.)

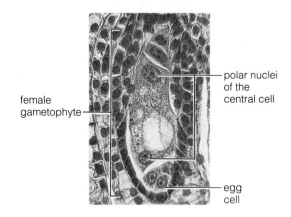

Figure 21-8 Seven-celled, eight-nucleate female gametophyte (170×). (Photo by Ripon Microslides, Inc.)

which is triploid, the other haploid. These two nuclei are called **polar nuclei.** At the micropylar end of the ovule there are three cells, all haploid. One of these is the **egg cell.** Opposite the micropylar end are three 3n cells. This stage of development is often called the *seven-celled, eight-nucleate female gametophyte* (figs. 21-8, 21-15p). Fertilization takes place at this stage.

4. Study the demonstration slide of the seven-celled, eight-nucleate female gametophyte. Identify the placenta, integuments, micropyle, egg cell, central cell, and polar nuclei (figs. 21-8, 21-15p).

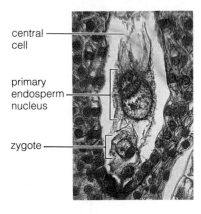

central cell

primary endosperm nucleus

zygote

Figure 21-9 Double fertilization (170 ×). (Photo by Ripon Microslides, Inc.)

As the pollen tube penetrates the female gametophyte, it discharges the sperm; one of the sperm nuclei fuses with the haploid egg nucleus, forming the *zygote*. Figure 21-15q represents the female gametophyte after fertilization has occurred.

The zygote is a

_____ (haploid, diploid) cell.

The other sperm nucleus enters the central cell and fuses with the two polar nuclei, forming the **primary endosperm nucleus** (fig. 21-15q).

Thus, the primary endosperm nucleus is

_____ (haploid, diploid, triploid,

tetraploid, pentaploid).

5. The cell containing the primary endosperm nucleus (the old central cell) is now called the **endosperm mother cell.** Traditionally, the process of fusion of one sperm nucleus with the egg nucleus and the fusion of the other sperm nucleus and the two polar nuclei has been called **double fertilization.** Observe the demonstration slide illustrating double fertilization (fig. 21-9), identifying the zygote, primary endosperm nucleus, and central cell of the female gametophyte.

Numerous mitotic and cytoplasmic divisions of the endosperm mother cell form the **endosperm,** a tissue used for nutrition of the embryo sporophyte as it develops within the seed.

IV. Embryogeny

The zygote undergoes mitosis and cytokinesis to produce a two-celled **embryo** (also called the *embryo sporophyte*). Numerous subsequent divisions produce an increasingly large and complex embryo.

MATERIALS

Per lab room:

• demonstration slides of *Capsella* embryogeny: globular embryo, emerging cotyledons, torpedo-shaped embryo, mature embryo

PROCEDURE

1. Observe the series of four demonstration slides illustrating the development of the embryo in the female gametophyte of *Capsella*. Figure 21-10 illustrates the stages. The first slide shows the so-called globular stage (fig. 21-10a), in which the *embryo*, a spherical mass of cells, is attached to the wall of the female gametophyte (embryo sac) by a chain of cells (the *suspensor*). The very enlarged cell at the base of the suspensor is the *basal cell* and is active in uptake of nutrients to be used by the developing embryo. Note the endosperm within the female gametophyte.

2. The second slide illustrates the heart-shaped stage (fig. 21-10b). Now you can distinguish the emerging **cotyledons** (seed leaves), the "lobes" of the heart. In many plants the cotyledons absorb nutrients from the endosperm, serving as a food reserve to be used during seed germination.

3. Further development of the embryo has occurred in the third slide, the torpedo stage (fig. 21-10c). Notice that the entire embryo has elongated. Find the *cotyledons*. Between the cotyledons, locate the **epicotyl,** which is the *apical meristem of the shoot*. Beneath the epicotyl and cotyledons find the **hypocotyl-root axis.** At the tip of the hypocotyl-root axis, locate the *apical meristem of the root* and the **root cap** covering it.

4. The final slide (fig. 21-10d) shows a mature embryo, neatly packaged inside the **seed coat.** Identify the seed coat and other regions previously identified in the torpedo stage.

How many cotyledons were there in the slides you examined?

_____

Thus, *Capsella* is a

_____ (monocot or dicot).

V. Fruit and Seed

Simply stated, a **fruit** is a matured ovary, while a **seed** is a matured ovule. Bear in mind that for each seed produced, a pollen grain had to fertilize the egg cells in the ovules. Fertilization not only causes the integuments of the ovule to develop into a seed coat, it also causes the ovary wall to expand into the fruit.

MATERIALS

Per student:

• bean fruits
• soaked bean seeds
• iodine solution (I₂KI), in dropping bottle

Per lab room:

• herbarium specimen of *Capsella*, with fruits
• demonstration slide of *Capsella* fruit, c.s.

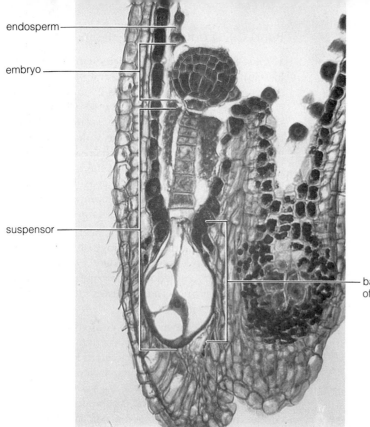

endosperm

embryo

suspensor

basal cell
of suspensor

a Globular embryo stage (250 ×).

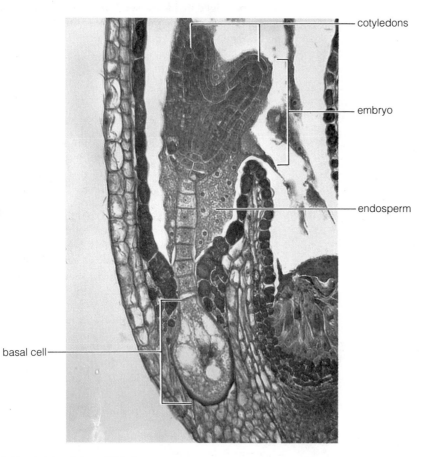

cotyledons

embryo

endosperm

basal cell

b Heart-shaped stage (250 ×).

Figure 21-10 Embryogeny in *Capsella* (shepherd's purse).
(Photos courtesy Ripon Microslides, Inc.)

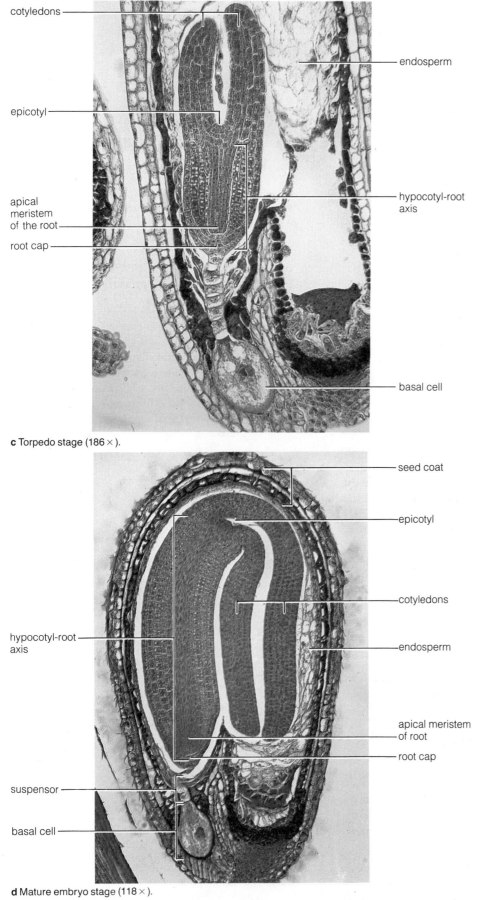

cotyledons

endosperm

epicotyl

apical
meristem
of the root

root cap

hypocotyl-root
axis

basal cell

c Torpedo stage (186 ×).

seed coat

epicotyl

cotyledons

hypocotyl-root
axis

endosperm

apical meristem
of root

root cap

suspensor

basal cell

d Mature embryo stage (118 ×).

Figure 21-10 *continued*

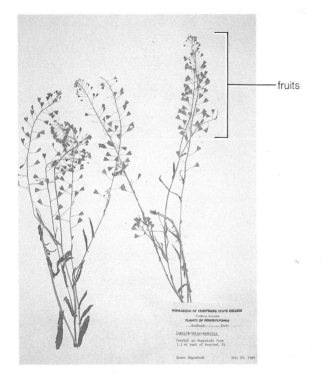

Figure 21-11 Herbarium specimen of *Capsella*, shepherd's purse (0.25×). (Photo by J. W. Perry.)

PROCEDURE

1. Examine the herbarium specimen and demonstration slide of the fruits of *Capsella*. On the herbarium specimen, identify the **fruits,** which are shaped like the bag that shepherds carried at one time (fig. 21-11; the common name of this plant is "shepherd's purse.")

2. Now study the demonstration slide of a cross section through a single fruit (fig. 21-12). Note the numerous **seeds** in various stages of embryo development.

3. Obtain a bean pod and carefully split it open along one seam.

The pod is a matured ovary and thus is a

_____ .

Find the *sepals* at one end of the pod and the shriveled *style* at the opposite end. The "beans" inside are

_____ .

Note the point of attachment of the bean to the pod. What is the point of attachment of the ovule to the ovary wall called?

_____ .

In figure 21-13, draw the split-open bean pod, labeling it with the correct scientific terms. Figure 21-15r represents a section of a typical fruit.

4. Study more closely one of the beans from within the pod or a bean that has been soaked overnight to soften it. Find the scar left where the seed was attached to the fruit wall. Near the scar, look for the tiny

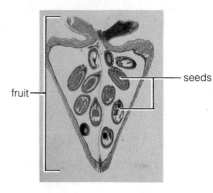

Figure 21-12 Cross section of *Capsella* fruit (8×). (Photo by J. W. Perry.)

opening left in the seed coat. What is this tiny opening? (*Hint:* The pollen tube grew through it.)

_____

5. Remove the seed coat to expose the two large **cotyledons**. Split the cotyledons apart to find the **epicotyl** and **hypocotyl-root axis**. During maturation of the bean embryo, the cotyledons absorb the endosperm. Thus, bean cotyledons are very fleshy, storing carbohydrates that will be used during seed germination. Add a drop of iodine solution (I_2KI) to the cotyledon. What substance is located in the cotyledon?

_____

(*Hint:* Return to Exercise 7 if you've forgotten what is stained by I_2KI.)

Labels: fruit, seeds, placenta

Figure 21-13 Drawing of an open bean pod.

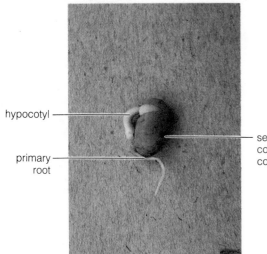

hypocotyl

primary root

seed coat covering cotyledons

Figure 21-14 Germinating bean seed (0.5 ×). (Photo by J. W. Perry.)

VI. Seedling

When environmental conditions are favorable for growth (adequate moisture, oxygen, and proper temperatures), the seed **germinates;** that is, growth of the seedling (young sporophyte) begins.

MATERIALS

Per student:
• germinating bean seeds
• bean seedlings

Per table:
• a dishpan of water

PROCEDURE

1. Obtain a germinating bean seed from the culture provided (fig. 21-14). Wash the root system in the dishpan provided, *not* in the sink.

Identify the **primary root** with the smaller **secondary roots** attached to it. Emerging in the other direction will be the **hypocotyl,** the **cotyledons,** the **epicotyl** (above the cotyledons), and the first **true leaves** above the epicotyl. Identify these. When some seeds (like the pea) germinate, the cotyledons remain *below* ground. Others, like the bean, emerge from the ground because of elongation of the hypocotyl-root axis.

2. Now, obtain a bean seedling from the growing medium. Be careful as you pull it up so as not to damage the root system. Wash the root system in the dishpan provided.

Your seedling should be in a stage of development similar to that shown in figure 21-15s. Much of the growth that has taken place is the result of cellular elongation of the parts present in the seed.

3. On the root system, identify the **primary** and **secondary roots.** As the root system merges into the shoot (above-ground) system, find the **hypocotyl.** (The prefix *hypo-* is derived from Greek, meaning "below" or "underneath.")

4. Next, identify the cotyledons. Notice their shriveled appearance. Knowing the function of the cotyledons from your previous study, why do you suppose the cotyledons are shriveled?

_____

5. Above the cotyledons find that portion of the stem called the **epicotyl.** Knowing what you do about the prefix *hypo-*, speculate on what the prefix *epi-* means.

_____

As noted above, the seedling has *true leaves*. Contrast the function of the cotyledons ("seed leaves") with the true leaves.

_____

Depending upon the stage of development, your seedling may have even more leaves that have been produced by the shoot apex.

The amount of time between seed germination and flowering depends largely upon the particular plant. Some plants produce flowers and seeds during their first growing season, completing their life cycle in that growing season. These plants are called **annuals.** Marigolds are an example of an annual. Others, known as **biennials,** grow vegetatively during the first growing season and do not produce flowers and seeds until the second growing season (carrots, for example). Both annuals and biennials die after seeds are produced.

Perennials are plants that live several to many years. The time between seed germination and flowering (seed production) varies, some requiring many years. Moreover, perennials do not usually die after producing seed, but flower and produce seeds many times during their lifetime.

OPTIONAL

An Investigative Approach to the Life Cycle of Angiosperms

Your instructor may provide you with an alternative means for studying the angiosperm life cycle. If so, the objectives and introduction at the beginning of the exercise here apply to the alternative also.

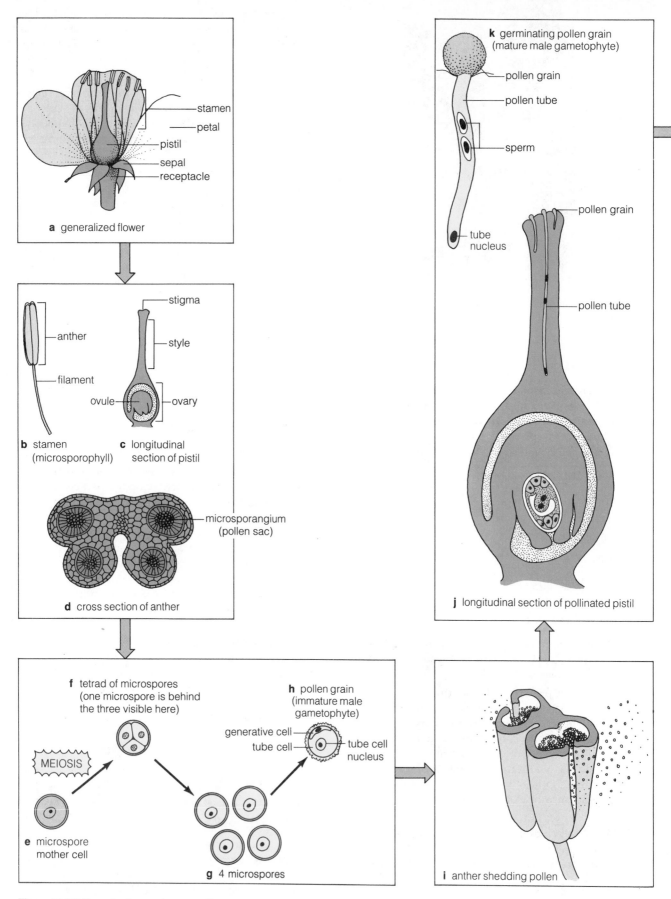

Figure 21-15 Life cycle of an angiosperm. (Color scheme: Haploid (n) structures are yellow; diploid (2n) are green, gold, or red; triploid (3n) are purple.)

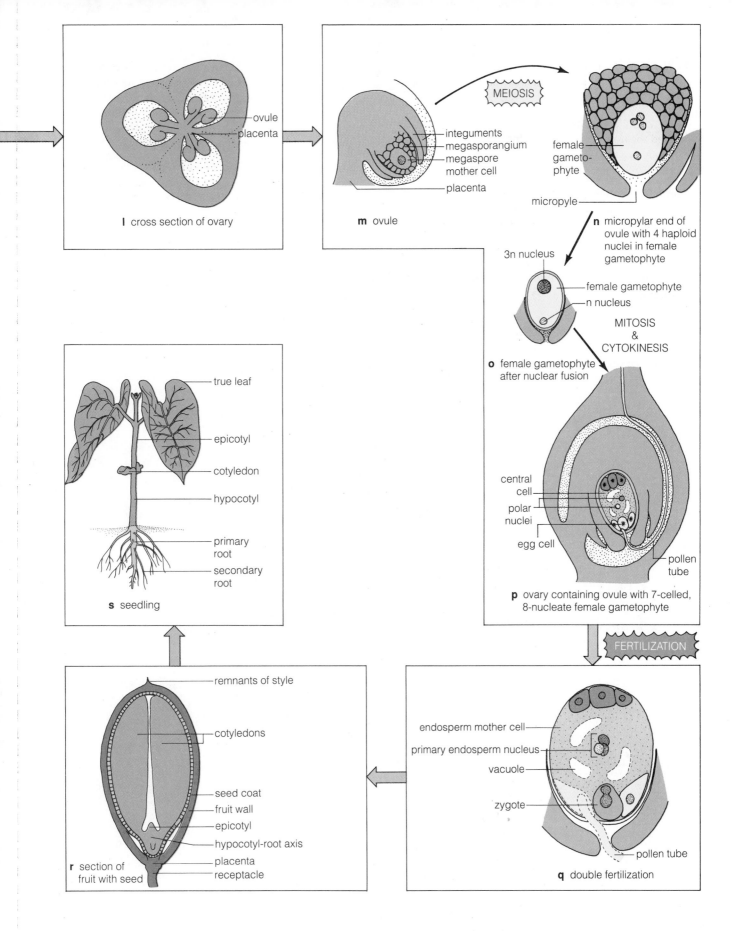

l cross section of ovary

m ovule

integuments
megasporangium
megaspore mother cell
placenta

MEIOSIS

female gameto-phyte

micropyle

n micropylar end of ovule with 4 haploid nuclei in female gametophyte

3n nucleus

female gametophyte

n nucleus

MITOSIS & CYTOKINESIS

o female gametophyte after nuclear fusion

central cell
polar nuclei
egg cell
pollen tube

p ovary containing ovule with 7-celled, 8-nucleate female gametophyte

FERTILIZATION

true leaf
epicotyl
cotyledon
hypocotyl
primary root
secondary root

s seedling

remnants of style
cotyledons
seed coat
fruit wall
epicotyl
hypocotyl-root axis
placenta
receptacle

r section of fruit with seed

endosperm mother cell
primary endosperm nucleus
vacuole
zygote
pollen tube

q double fertilization

_____ 1. Plants that produce flowers are (a) members of the Anthophyta, (b) angiosperms, (c) seed producers, (d) all of the above.

_____ 2. Collectively, all of the petals of a flower are called the (a) corolla, (b) stamens, (c) receptacles, (d) calyx.

_____ 3. Which of the following refer to the microsporophyll, the male portion of a flower? (a) ovary, stamens, pistil; (b) stigma, style, ovary; (c) anther, stamen, filament; (d) megasporangium, microsporangium, ovule.

_____ 4. A carpel is the (a) same as a megasporophyll, (b) structure producing pollen grains, (c) component making up the anther, (d) synonym for microsporophyll.

_____ 5. The portion of the flower containing pollen grains is the (a) pollen sac, (b) microsporangium, (c) anther, (d) all of the above.

_____ 6. Which of the following is in the correct developmental sequence? (a) microspore mother cell, meiosis, megaspore, female gametophyte; (b) microspore mother cell, meiosis, microspore, pollen grain; (c) megaspore, mitosis, female gametophyte, meiosis, endosperm mother cell; (d) all of the above.

_____ 7. Where would germination of a pollen grain occur in a flowering plant? (a) in the anther, (b) in the micropyle, (c) on the surface of the corolla, (d) on the stigma.

_____ 8. Double fertilization refers to (a) fusion of two sperm nuclei and two egg cells, (b) fusion of one sperm nucleus with two polar nuclei and fusion of another with the egg cell nucleus, (c) maturation of the ovary into a fruit, (d) none of the above.

_____ 9. Ovules mature into _____, while ovaries mature into _____. (a) seeds, fruits; (b) stamens, seeds; (c) seeds, carpels; (d) fruits, seeds.

_____ 10. A bean pod is (a) a seed container, (b) a fruit, (c) a part of the stamen, (d) a and b above.

E X E R C I S E   2 1

Seed Plants II: Angiosperms

POST-LAB QUESTIONS

1. There are two major groups of seed plants, gymnosperms and angiosperms. Compare these two groups of seed plants with respect to:

Feature	Gymnosperms	Angiosperms
a. type of reproductive structure		
b. source of nutrition for developing embryo		
c. enclosure of mature seed		

2. Some biologists contend that the term *double fertilization* is a misnomer and that the process should be called *fertilization* and *triple fusion*. Why do they argue that the fusion of the one sperm nucleus and the two polar nuclei is *not* fertilization?

3. After doing this lab, suppose you and your roommate go to the grocery store. Your roommate says that you need vegetables and asks you to pick up tomatoes. To your roommate's surprise, you say a tomato is not a vegetable, but a fruit. Explain.

4. What *event*, critical to the production of seeds, is shown below?

(Photo by J. W. Perry.)

5. Distinguish between *pollination* and *fertilization*.

6. Identify the parts of the trumpet creeper flower shown here.

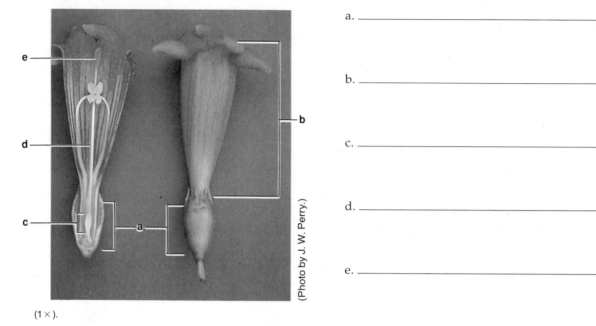

(1×).

(Photo by J. W. Perry.)

a. _____

b. _____

c. _____

d. _____

e. _____

7. The figure below illustrates the flower of the pomegranate some time after fertilization. Identify the parts shown.

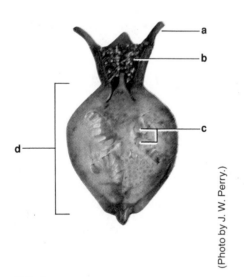

(0.3×).

(Photo by J. W. Perry.)

a. _____

b. _____

c. _____

d. _____

8. The illustration below is a cross section of a (an) _____ . The numerous circles within the four cavities are _____ .

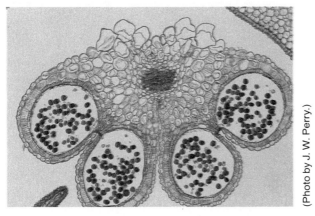

(Photo by J. W. Perry.)

(74×).

9. Plants of a particular bamboo species, the preferred food of giant pandas, grow wild in the mountains of southern China. Throughout the country stands of this bamboo, which had been growing vegetatively for nearly 200 years, recently produced flowers and then died, much to the dismay of botanists (and presumably giant pandas).

a. Is this bamboo an annual, biennial, or perennial?

_____

b. Why?

10. Based upon observation of the stigma of the daylily flower in the figure below, how many carpels would you expect to comprise the ovary?

(Photo by J. W. Perry.)

EXERCISE 22

Sponges and Cnidarians

OBJECTIVES

After completing this exercise you will be able to:

1. define *larva, spongocoel, osculum, spicules, collar cell, monoecious, budding, spongin, radial symmetry, polyp, tentacles, medusa, nematocyst, dioecious, mesoglea;*

2. list the characteristics of animals;

3. describe how the phyla Porifera and Cnidaria show the cell-specialization and tissue levels of organization, respectively;

4. explain the basic body plan of members of the phyla Porifera and Cnidaria;

5. describe the natural history of members of the phyla Porifera and Cnidaria;

6. identify representatives of the classes Calcarea and Demospongiae of the phylum Porifera and classes Hydrozoa, Scyphozoa, and Anthozoa of the phylum Cnidaria;

7. identify structures (and indicate associated functions) of representatives of these classes.

INTRODUCTION

Early in our lives, as occurred early in the history of the systematic study of organisms, we learned to identify animals as organisms that exhibit considerable movement. In his *Scala Naturae,* Aristotle distinguished plants from animals on the basis of the extent of movement and response to stimulation. Until relatively recently, every living organism was assigned to either the plant or animal kingdom, partly on the basis of whether animal-like movement was exhibited. Even today, when organisms are divided among five kingdoms in Whittaker's classification scheme (see Exercise 14), organisms outside the animal kingdom (for example, *Euglena;* fig. 15-4) are sometimes described as animal-like because of the movement they exhibit.

Evidence accumulated from centuries of study supports the notion that animals evolved from a group of protistanlike ancestors distinct from those that gave rise to the plants and fungi. Current classification schemes rely less heavily on movement as a diagnostic feature of animals because we have learned that some organisms that show considerable motility are unrelated to animals. *Animals are* now recognized as *multicellular, heterotrophic organisms that usually have several levels of organization, including specialized cells, tissues, organs, and systems.* Most animals are *diploid* and *reproduce sexually,* although asexual reproduction is also common. The animal life cycle includes a *period of embryonic development, often with a* **larva** — a sexually immature, free-living form that grows and transforms into an adult — or equivalent stage. During the embryonic development of most animals, three *primary germ layers* — *ectoderm, mesoderm,* and *endoderm* — form. The primary germ layers give rise to the adult tissues. See table 37-1 on page 492.

CAUTION

Preserved specimens are kept in a formalin-based or other preservative solution. Wash any part of your body exposed to this solution with copious amounts of water. If preservative solution is splashed into your eyes, wash them with a safety eyewash bottle for fifteen minutes.

I. Phylum Porifera: Sponges

Habitat:	aquatic, mostly marine
Body Arrangement:	asymmetrical or crude radial symmetry
Level of Organization:	cell specialization
Body Support:	spicules
Life Cycle:	sexual reproduction (mostly monoecious), larva (amphiblastula), asexual reproduction (budding)
Special Characteristics:	spongocoel, osculum, gemmules

Aristotle said of sponges, "In the sea there are things which it is hard to label as either animal or vegetable." This was a reference to their sensitivity to stimulation combined with a relative lack of motility.

Sponges evolved directly from protistanlike ancestors and diverged early from the main line of animal evolution (fig. 22-9). Most sponges are marine, and all forms are *aquatic.* As adults, sponges are *sessile* — which means they are immobile and attached to a surface such as the sea bottom — but many disperse as free-swimming larvae. Although sponges possess a variety of different types of cells, they are atypical animals in their lack of definite tissues. They exhibit the *cell-specialization level of organization,* as there is a division of labor among the different cell types. Cells are organized into layers, but these associations of cells do not show all the characteristics of tissues. Sponges have a *crude radial symmetry* (see below) *or are asymmetrical* — cannot be cut into two like halves.

Three body plans exist in sponges (fig. 22-1). They are referred to as *asconoid, syconoid,* and *leuconoid,* increasing in complexity in this order.

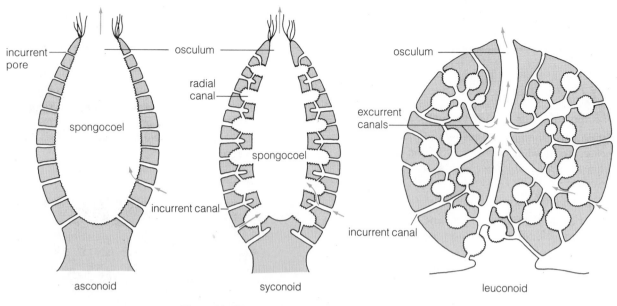

Figure 22-1 Three body plans of sponges. The arrows indicate the direction of water flow. (After Villee, Walker, and Barnes, 1973.)

Water enters a vase-shaped asconoid sponge through incurrent *pores;* flows directly to a large central internal cavity, the **spongocoel;** and exits through a larger opening, the **osculum.** A syconoid sponge is similar, except that the body wall is everted to form *incurrent canals* between (and *radial canals* within) the resulting pockets. In a leuconoid sponge, the pores open into numerous chambers that empty into *excurrent canals,* which in turn lead to the osculum.

MATERIALS

Per student:

• dissection microscope

• hand lens

• compound microscope, lens paper, a bottle of lens cleaning solution (optional), a lint-free cloth (optional)

• safety goggles

• clean microscope slide and coverslip

Per student pair:

• preserved specimen of *Scypha*

• prepared slide of a longitudinal section of *Scypha*

• prepared slide of a whole mount of *Leucosolenia*

• prepared slide of sponge gemmules

Per student group (4):

• 100-mL beaker of heatproof glass

• bottle of 5% sodium hydroxide (NaOH) or potassium hydroxide (KOH)

• hotplate

• squeeze bottle of distilled water

Per lab room:

• demonstration collection of commercial sponges

• squeeze bottle of 50% vinegar and water

• demonstration of living freshwater sponges

PROCEDURE

A. Class Calcarea: Calcareous Sponges

Members of this class contain **spicules** (skeletal elements) composed of calcium carbonate ($CaCO_3$). All three body plans are represented. You will study members of the genus *Scypha,* which all have the syconoid body plan.

1. With a hand lens and the dissection scope, examine the general morphology of *Scypha* (fig. 22-2a). Identify the osculum, the excurrent opening of the spongocoel. Note the pores in the body wall. Observe the long spicules surrounding the osculum and the shorter spicules protruding from the surface of the sponge.

> ### CAUTION
> Hot 5% hydroxide solutions are corrosive. Wear safety goggles. If splashed, wash exposed skin for fifteen minutes in running tap water; then flood the area with a 50% vinegar-and-water solution.

2. *Calcium carbonate spicules of Scypha.* Place a small piece of a specimen of *Scypha* in a 100-mL heatproof beaker. Put on a pair of safety goggles. Add enough 5% sodium hydroxide or potassium hydroxide solution to the beaker to cover the piece and bring this to a boil on the hotplate. Allow the beaker to cool. **Your**

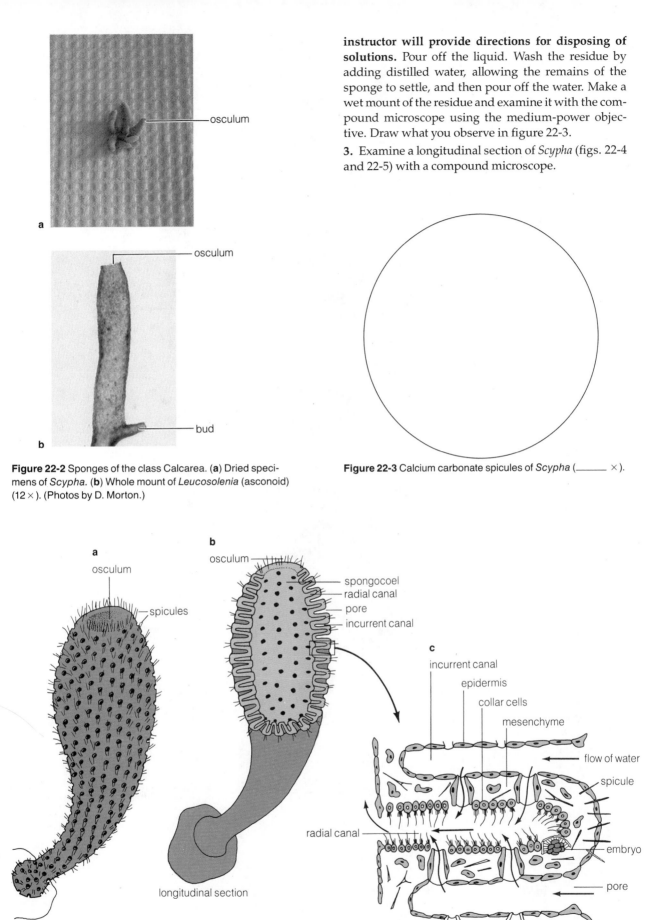

osculum

osculum

bud

a

b

Figure 22-2 Sponges of the class Calcarea. (**a**) Dried specimens of *Scypha*. (**b**) Whole mount of *Leucosolenia* (asconoid) (12×). (Photos by D. Morton.)

instructor will provide directions for disposing of solutions. Pour off the liquid. Wash the residue by adding distilled water, allowing the remains of the sponge to settle, and then pour off the water. Make a wet mount of the residue and examine it with the compound microscope using the medium-power objective. Draw what you observe in figure 22-3.

3. Examine a longitudinal section of *Scypha* (figs. 22-4 and 22-5) with a compound microscope.

Figure 22-3 Calcium carbonate spicules of *Scypha* (_____ ×).

a

osculum

spicules

external view

b

osculum

spongocoel
radial canal
pore
incurrent canal

longitudinal section

c

incurrent canal

epidermis

collar cells

mesenchyme

flow of water

spicule

radial canal

embryo

pore

Figure 22-4 *Scypha*. (**a**) External view. (**b**) Longitudinal section. (**c**) Magnified to show detail. (After Lytle and Wodsedalek, 1984.)

287

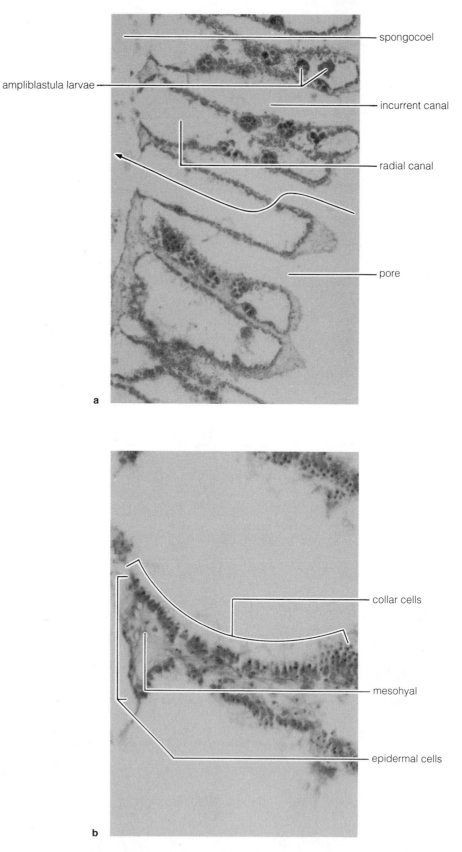

spongocoel

ampliblastula larvae

incurrent canal

radial canal

pore

collar cells

mesohyal

epidermal cells

Figure 22-5 Longitudinal sections of *Scypha*. (**a**) The arrow indicates the path of water flow (96 ×). (**b**) 384 × . (Photos by D. Morton.)

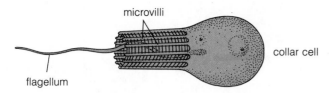

microvilli

flagellum

collar cell

Figure 22-6 Collar cell. (After Starr, 1991.)

In your section, trace the path of water as it would have flowed into and through the sponge. Water enters an incurrent canal through a pore. From the incurrent canal it flows through many smaller pores into several radial canals. From here it flows past the collar cells, which line the radial canals, into the large central spongocoel. Once inside the spongocoel, the water flows out of the animal through the osculum.

The **collar cells** will appear as numerous small, dark bodies lining the radial canal. Each bears a flagellum that projects into the radial canal, but it is unlikely that you will be able to see the flagella, even with the high-dry objective. Examine the structure of a collar cell in figure 22-6.

The beating of the flagella of collar cells is thought to create currents that move water through the sponge. The microvilli—fingerlike projections that comprise their collars—play a major role in filtering fine food particles out of the water. Collar cells and perhaps other cells engulf and digest food fragments carried into the sponge by the flow of water.

Flat protective cells cover both the outer surface and incurrent canals to form the *epidermis* and line the spongocoel. Between the epidermis and the cells lining the internal spaces lies a region of gelatinous material sometimes called the *mesohyal,* embedded in which are amoeboid cells and spicules.

Sexual reproduction occurs in sponges. Most sponges are **monoecious** or *hermaphroditic*—an individual has both male and female reproductive structures, producing eggs and sperm from amoebocytes at different times in the life cycle. In addition, sperm may be produced from collar cells. Although it is unlikely that you will observe gametes, you may be able to detect the early developmental stages of embryos in the body wall, protruding into the radial canals of *Scypha.* These appear as relatively large, dark structures (fig. 22-4) that soon would have developed into *amphiblastula larva,* broken loose, and exited the sponge through the osculum. Larvae may be seen in the radial canals (fig. 22-5).

4. Asexual reproduction occurs by **budding.** Budding is a process whereby parental cells differentiate and grow outward from the parent to form a bud, finally breaking off to become a new individual. *Scypha* often buds from its base. If the buds do not separate, a cluster of individuals may result. Look for buds on the specimen. You also may find buds on the prepared slide of a whole mount of the asconoid sponge *Leucosolenia,* another member of this class (fig. 22-2b).

a

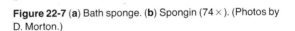

b

Figure 22-7 (**a**) Bath sponge. (**b**) Spongin (74 ×). (Photos by D. Morton.)

B. Class Demospongiae: Commercial and Freshwater Sponges

This class contains the majority of sponge species. Most sponges in this class are supported by **spongin** (fig. 22-7b, a flexible substance similar chemically to human hair), by spicules of silica, or both. Silica is silicon dioxide, a major component of sand and of many rocks, such as flint and quartz. All sponges in this class are leuconoid and are able to hold and pass large volumes of water in and through their bodies because of their complex canal systems. Some of those that have a skeleton composed only of spongin fibers are commercially valuable as bath sponges. Bath sponges have been largely replaced by synthetic sponges in today's marketplace.

1. Examine demonstration specimens of commercial sponges (fig. 22-7a).

2. Freshwater and some marine sponges in this class reproduce asexually by forming *gemmules,* resistant internal buds. In the freshwater varieties, groups of cells are surrounded by a shell of silica and spicules. During unfavorable conditions, as in winter, disintegration of the sponge releases the gemmules. Under favorable conditions, as in spring, cells emerge from the gemmules and develop into new individuals. Find

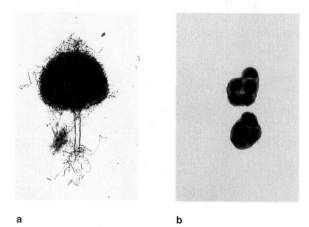

a   b

Figure 22-8 Gemmules. (**a**) 30×. (Photo courtesy Ripon Microslides, Inc.) (**b**) 14×. (Photo by D. Morton.)

a gemmule on the prepared slide of sponge gemmules (fig. 22-8).

3. Examine the demonstration of living freshwater sponges. They look like yellowish-brown, wrinkled scum with perforations and sometimes have a greenish tint due to the presence of symbiotic algae. They are fairly easy to find in nature. Search for them in clear, well-oxygenated water, encrusting submerged plant stems and debris such as a water-logged twig. Slow streams and the edges of ponds where there is some wave action are good places to find freshwater sponges.

II. Phylum Cnidaria: Cnidarians

Habitat:	aquatic, mostly marine
Body Arrangement:	radial symmetry, gut (incomplete)
Level of Organization:	tissue (derived from ectoderm and endoderm)
Body Support:	hydrostatic (gastrovascular cavity and epitheliomuscular cells)
Life Cycle:	sexual reproduction (monoecious or dioecious), larva (planula), asexual reproduction (budding)
Special Characteristics:	tissues, two adult body forms—polyp (sessile) and medusa (free-swimming)—nematocysts, nerve net

This phylum contains some of the most beautiful organisms in the seas, many of which are brightly colored and superficially plantlike or flowerlike, with tentacles that move in response to water currents.

Cnidarians are derived from the protistanlike ancestors that gave rise to the main line of evolution from which other animals sprang. Cnidarians diverged early from this evolutionary lineage (fig. 22-9).

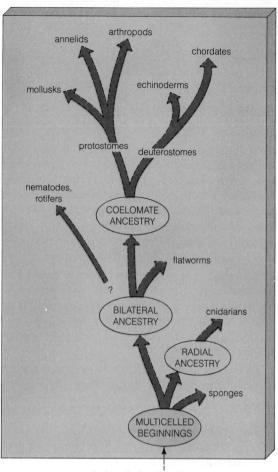

single-celled, protistanlike ancestors

Figure 22-9 Evolutionary lineage of animals. (After Starr, 1991.)

These animals are *aquatic* and found mostly in shallow marine environments, with the notable exception of the freshwater hydras. There are definite tissues; therefore, Cnidarians show a *tissue level of organization.* Cnidarians are mostly **radially symmetrical** (fig. 22-10); that is, any cut from the oral (mouth) to the other end through the center of the animal will yield two like halves. Members of this phylum have a *gut* and a definite *nerve net.*

Two main body forms exist in Cnidarians (fig. 22-11). The usually sessile **polyp** is cylindrical in shape, with the oral end and **tentacles** directed upward and the other end attached to a surface. The free-swimming **medusa** (jellyfish) is bell- or umbrella-shaped, with the oral end and tentacles directed downward. Either or both body forms may be present in the Cnidarian life cycle. Also found throughout the group is a larval form called a *planula.*

Cnidarians are efficient predators. All are carnivorous and possess tentacles that capture unwary invertebrate and vertebrate prey. Food is captured with the aid of stinging elements called **nematocysts** (fig. 22-12). The nematocysts are discharged by a combination of mechanical and chemical stimuli. The prey is either pierced by the nematocyst or entangled in its

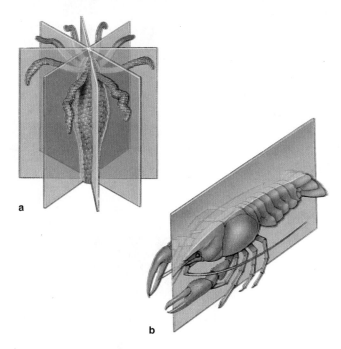

Figure 22-10 Planes of symmetry. (**a**) Radial symmetry in a polyp. (**b**) Bilateral symmetry in a crayfish. (Starr and Taggart, 1989.)

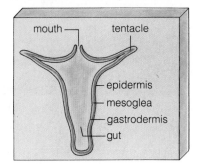

polyp

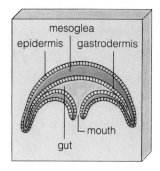

medusa

Figure 22-11 Two body forms of Cnidarians. (After Starr and Taggart, 1989.)

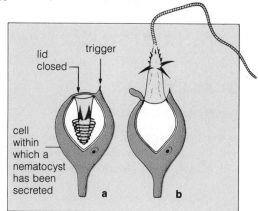

Figure 22-12 One type of nematocyst, a capsule with an inverted thread inside. This one has a bristlelike trigger (**a**). When prey (or predators) touch the trigger, the capsule becomes more "leaky" to water. As water diffuses inward, pressure inside the capsule increases and the thread is forced to turn inside out (**b**). The thread's tip may penetrate the prey, releasing a toxin as it does this. (After Starr, 1991.)

filament and pulled toward the mouth by the tentacles. In some species, nematocysts discharge a toxin that paralyzes the prey. The mouth opens to receive the food, which is deposited in the gut. Digestion is started in the gut cavity by enzymes secreted by its lining. It is completed inside the lining cells after they engulf the partially digested food.

MATERIALS

Per student:

• dissection microscope

• compound microscope

• clean microscope slide and coverslip

Per student pair:

• prepared slides of whole mounts of *Hydra* (showing testes, ovaries, buds, and embryos)

• prepared slide of a cross section of *Hydra*

• prepared slides of whole mounts of *Obelia* (showing colonial polyps and medusa)

Per student group (4):

• small finger bowl

• scalpel or razor blade

• preserved specimen of *Gonionemus*

• prepared slide of a whole mount of the medusa (*ephyra*) of *Aurelia*

• prepared slide of a whole mount of planula larvae of *Aurelia*

• prepared slide of a whole mount of scyphistoma of *Aurelia*

• prepared slide of a whole mount of the strobila of *Aurelia*

Per lab room:

• demonstration specimen of *Physalia*

• demonstration collection of corals

• demonstration specimen of *Metridium*

Per lab section:

• culture of live *Hydra*

• container of fresh stream water

• dropper bottle of glutathione

• culture of live copepods or cladocerans

• dropper bottle of vinegar

291

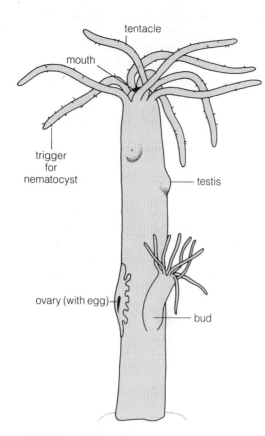

Figure 22-13 External view of *Hydra*. (After Hickman and Hickman, 1978.)

PROCEDURE

A. Class Hydrozoa: Hydrozoans

This is mostly a marine group, except for the hydras. A polyp, medusa, or both may be present in the life cycle of a hydrozoan. Marine polyps tend to be colonial.

1. *Hydra*

Hydra is a solitary polyp, and it does not have a medusae stage in its life cycle. It is a freshwater hydrozoan.

a. Place a live specimen of *Hydra* in a finger bowl containing fresh stream water. Examine the live *Hydra* with a dissection microscope, along with prepared slides of whole mounts with the compound microscope, noting its polyp body form and other structural details (fig. 22-13).

b. Observe the tentacles at the oral end. Find the elevation of the body at the base of the tentacles, in the center of which lies a mouth. Note the swellings on the tentacles; each is a stinging cell that contains a nematocyst. The clear area in the center of the body represents the gut or *gastrovascular cavity.*

c. Look for evidence of sexual reproduction: one to several *testes* or an *ovary* on the body, both of which may occur on the same individual. Testes (fig. 22-14a) are conical and located more orally than the broad, flattened ovary (fig. 22-14b). You may be able to see a

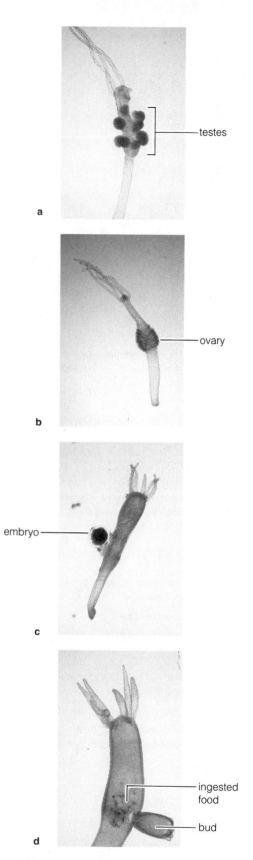

Figure 22-14 Whole mounts of *Hydra* (12 ×). (**a**) Testes. (**b**) Ovary. (**c**) Embryo. (**d**) Bud. (Photos by D. Morton.)

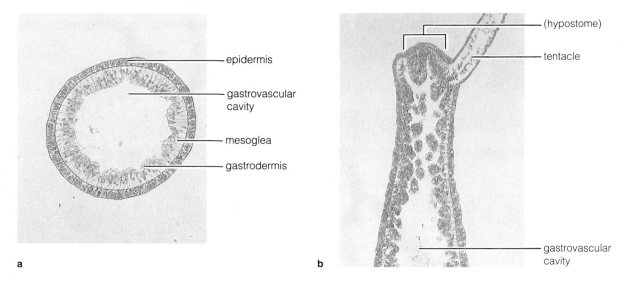

Figure 22-15 Sections of *Hydra* (70 ×). (Photos courtesy Ripon Microslides, Inc.)

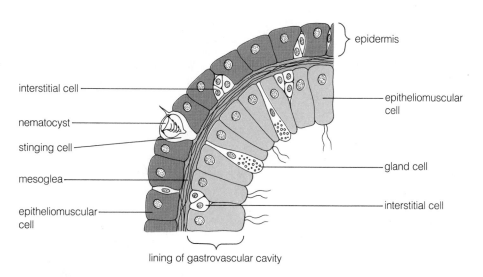

Figure 22-16 Cross section of the body wall of *Hydra*. (After Hickman, 1961.)

zygote or embryo marginally attached to or detached from the body of the *Hydra* (fig. 22-14c). Is *Hydra* monoecious or **dioecious**—male and female reproductive structures in separate individuals?

Monoecious

d. *Hydra* also reproduces asexually by budding (fig. 22-14d). Look for buds on your specimens.

e. Examine a cross section of *Hydra* (figs. 22-15 and 22-16) and locate the *epitheliomuscular cells* and *interstitial cells* comprising the epidermis. The epidermis is derived from ectoderm, one of the primary germ layers. An occasional stinging cell may be seen in this layer. The lining of the gastrovascular cavity, or *gastrodermis*, is derived from the endoderm and composed of epitheliomuscular cells (sometimes called

nutritive muscular cells), gland cells, and interstitial cells. Between the two is a thin gelatinous membrane called the **mesoglea**. In *Hydra*, the mesoglea is without cells.

The two layers of epitheliomuscular cells contract in opposite directions, although in *Hydra* the longitudinally arranged contractile filaments in the epidermis are more developed than the circular filaments of the gastrodermis. How might two layers of epitheliomuscular cells function to maintain and change the body shape of a Cnidarian?

The two layers can contract and expand in opposite directions allowing for the change in shape of the organism.

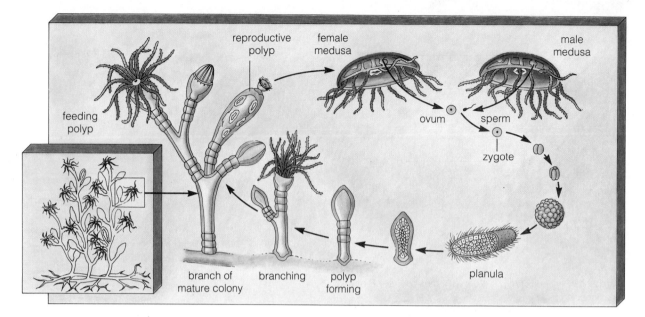

Figure 22-17 Life cycle of the marine hydrozoan *Obelia*. The inset shows a colony, actual size. On its branches are feeding polyps and reproductive polyps. The medusa stage, produced asexually by a reproductive polyp, is free-swimming. Fusion of gametes from male and female medusae leads to a zygote, which develops into a planula. The swimming or crawling planula develops into a polyp, which starts a new colony. As growth continues, new polyps are formed. (After Starr and Taggart, 1989.)

What additional function could you suggest for the epitheliomuscular cells of the gastrodermis? (Hint: What happens to partially digested food?)

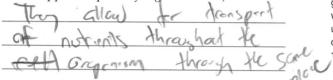

Interstitial cells can give rise to sperm, eggs, and any other type of cell in the body.

f. If it is not already there, return the live *Hydra* in a small finger bowl to the dissection microscope. Add a drop of glutathione and a number of live copepodan or cladoceran crustaceans to the water. The glutathione stimulates the feeding response of *Hydra*. Look for feeding behavior by the *Hydra*. Note the action of the tentacles and the nematocysts. If you do not observe the discharge of nematocysts, place a drop of vinegar in the finger bowl and observe carefully. Describe what you observe.

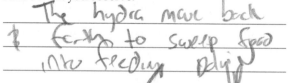

g. Are any of your *Hydra* green? If so, use a scalpel or razor blade to remove a tentacle. Make a wet mount of the tentacle, being certain to crush it with the cover slip. What do you observe? Can you explain the green color of your *Hydra*?

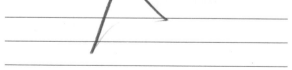

2. *Obelia* and *Gonionemus*

These are common examples of hydrozoans with both a polyp and a medusae stage in their life cycles (figs. 22-17 and 22-18).

a. Examine prepared slides of whole mounts of *Obelia*. The life cycle of this hydrozoan alternates between a stage with colonial polyps and one with solitary medusae. Identify the structures indicated in figure 22-18.

b. Examine the preserved specimen of the hydrozoan medusa *Gonionemus* (fig. 22-19).

Note the tentacles (which contain nematocysts and adhesive pads) around the margin of the umbrellalike *bell*. Extending downward from the center of the bell is a saclike structure containing the gastrovascular cavity, with the mouth at its tip surrounded by four *oral lobes*. Extending inward from the lower margin of the bell is a thin flap, the muscular *velum*. Four *radial canals* extend from the gastrovascular cavity to the *ring canal* at the margin of the bell. Between the bases of the tentacles are the *statocysts,* balancing organs. Gonads, either ovaries or testes, are attached to the radial canals. The medusa moves by "jet propulsion." Contractions of the body bring about a pulsing motion that alternately fills and empties the cavity of the bell.

3. *Physalia: Portuguese Man-of-War*

Look at the preserved specimen of the Portuguese man-of-war (fig. 22-20).

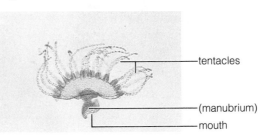

Figure 22-18 labels:
- (feeding polyp)
- (reproductive polyp)
- tentacle
- (gonad)
- mouth
- bell
- tentacles
- (manubrium)
- mouth

a

b

c

Figure 22-18 Whole mounts of *Obelia*. (**a**) Colonial polyps (14 ×). (**b**) A side view of a medusa (60 ×). (**c**) A top view of a medusa (60 ×). (Photos courtesy Ripon Microslides, Inc.)

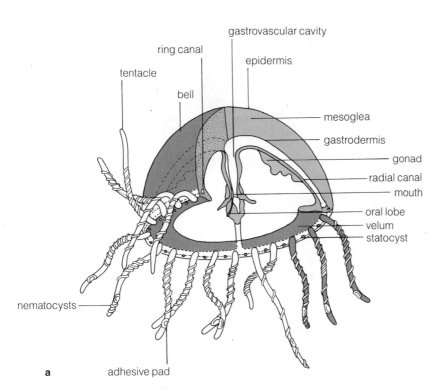

Figure 22-19a labels:
- gastrovascular cavity
- ring canal
- epidermis
- tentacle
- bell
- mesoglea
- gastrodermis
- gonad
- radial canal
- mouth
- oral lobe
- velum
- statocyst
- nematocysts
- adhesive pad

a

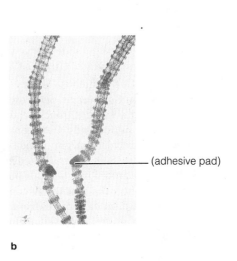

(adhesive pad)

b

Figure 22-19 (**a**) Hydrozoan medusa, *Gonionemus*. (After Boolootian and Stiles, *College Zoology*. Copyright © 1981, Macmillan Publishing Company. Reprinted by permission.) (**b**) Details of the tentacles (12 ×). (Photo courtesy Ripon Microslides, Inc.)

This organism is really a colonial hydrozoan that is composed of several types of modified polyps and medusae. One conspicuous feature is a gas-filled bladder. Its sting is very painful, and a neurotoxin associated with the nematocysts of the tentacles can be fatal to humans.

B. Class Scyphozoa: True Jellyfish

This marine class contains the true jellyfish. The medusa is the predominant stage of the life cycle and is generally larger than hydrozoan medusae. The nematocysts in the tentacles and oral arms of jellyfish such as the sea nettle can produce painful stings. A

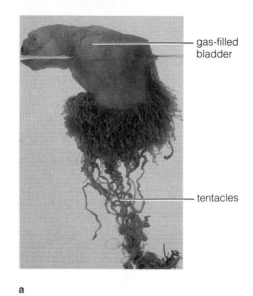

a

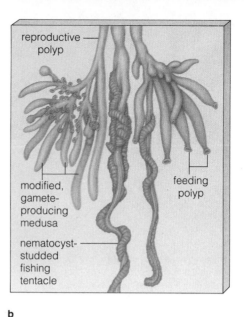

b

Figure 22-20 The Portuguese man-of-war. (**a**) Preserved specimen. (Photo by D. Morton.) (**b**) The tentacles contain several types of modified polyps and medusae. (After Starr and Taggart, 1989.)

tropical group called the sea wasps are more dangerous to humans than the hydrozoan Portuguese man-of-war. Some forms are bioluminescent, producing flashes of light as a result of chemical reactions. Bioluminescence may serve to lure prey toward the jellyfish or to scare potential predators.

1. *Aurelia*

a. Examine a prepared slide of a whole mount of the medusa (ephyra) of *Aurelia* (fig. 22-21).

The structure of this medusa is very similar to that of the hydrozoan, *Gonionemus*. *Sense organs* — consisting of cells sensitive to light, touch, chemicals, and balance — occur periodically between short tentacles around the muscular margin of the bell. Note the absence of a velum. Four *oral arms* used in capturing prey arise from the square mouth. Four horseshoe-shaped gonads surround the mouth. A complex system of branching radial canals radiates from the center of the animal to the ring canal in the margin of the bell. Can you suggest functions for the radial canals?

_____

_____

_____

b. Sexes are separate, and fertilization is internal. The zygote develops into a planula larva, which swims for a while before becoming a sessile polyp called a *scyphistoma*. This is an active, feeding stage, equipped with tentacles and a mouth. At certain times of the

Figure 22-21 Scyphozoan jellyfish, *Aurelia* (5×). (Photo courtesy Ripon Microslides, Inc.)

year young medusae are produced asexually by transverse budding. This stage is called a *strobila*, and the process is called strobilization. Examine prepared slides of planula larvae, a scyphistoma, and a strobila (fig. 22-22).

C. Class Anthozoa: Corals and Sea Anemones

Class Anthozoa is a marine group in which all members are heavy-bodied polyps that feed on small fish and invertebrates. The largest organisms in this class are the anemones. Both corals and sea anemones attach to hard surfaces or burrow in the sand or mud.

1. Corals

The epidermis of each colonial coral polyp secretes an exoskeleton of calcium carbonate (fig. 22–23a). After a polyp dies, another polyp uses the remaining skeleton of the dead polyp as a foundation and secretes its own skeleton. In this fashion, corals have built up various types of reefs, atolls, and islands. Examine the demonstration exoskeletons of a variety of corals.

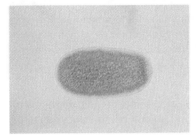

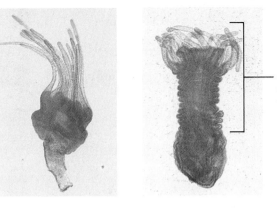

developing
medusae
(ephyrae)

a

Figure 22-22 Stages in the life cycle of *Aurelia*. (**a**) Planular larvae (143 ×). (**b**) Scyphistoma (18 ×). (**c**) Strobila (18 ×). (Photos courtesy Ripon Microslides, Inc.)

b   **c**

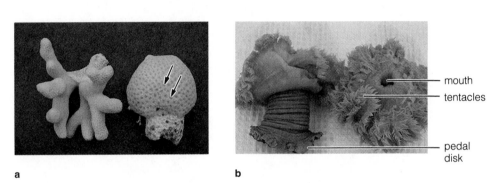

mouth

tentacles

pedal
disk

a   **b**

Figure 22-23 (**a**) Exoskeleton of hard coral. The arrows indicate sites of individual polyps. (**b**) Top and side views of preserved specimens of the common sea anemone, *Metridium*. (Photos by D. Morton.)

2. Sea Anemones

Some species of anemones form symbiotic relationships with certain small fishes that live among their tentacles. The fish receive protection from the tentacles, which for some reason do not discharge their nematocysts into these fish. In return, when the fish dart from and back to the protective cover of the tentacles to capture prey, the anemones benefit by ingesting food particles not eaten by the fish.

Examine the demonstration of a preserved or living specimen of *Metridium*, the common sea anemone (fig. 22–23b). Note its stout, stumplike body, with the mouth and surrounding tentacles at the oral end and the *pedal disk* at the other end. What is the function of the pedal disk?

_____

_____

PRE-LAB QUESTIONS

___A___ 1. Which of the following is characteristic of animals? (a) heterotrophy, (b) autotrophy, (c) photosynthesis, (d) b and c.

___C___ 2. Sponges are atypical animals because they (a) reproduce sexually, (b) are sessile as adults, (c) lack definite tissues, (d) lack a "head."

___C___ 3. The "skeleton" of a typical sponge is composed of (a) bones, (b) cartilages, (c) spicules, (d) scales.

___C___ 4. Many sponges produce both eggs and sperm and thus are (a) monoecious, (b) dioecious, (c) hermaphroditic, (d) a and c.

___D___ 5. Freshwater sponges reproduce asexually by forming resistant internal buds called (a) spicules, (b) larvae, (c) ovaries, (d) gemmules.

___B___ 6. Cnidarians evolved from an ancestral group of (a) sponges, (b) protists, (c) fungi, (d) plants.

SPONGES AND CNIDARIANS

__C__ 7. A free-swimming jellyfish has a body form known as a (a) polyp, (b) strobila, (c) medusa, (d) hydra.

__B__ 8. Cnidarians capture their prey with the use of (a) poison claws, (b) nematocysts, (c) oral teeth, (d) pedal disks.

__A__ 9. Which of the following is true for scyphozoan medusae? (a) they are the dominant body form; (b) they are generally smaller than hydrozoan medusae; (c) most live in fresh water; (d) they are scavengers.

__C__ 10. Coral reefs, atolls, and islands are the result of the buildup of coral (a) polyps, (b) medusae, (c) exoskeletons, (d) wastes.

E X E R C I S E 2 2

Sponges and Cnidarians

P O S T - L A B   Q U E S T I O N S

1. List six characteristics of the animal kingdom.

2. Define *cell specialization* and indicate how sponges exhibit this phenomenon.

3. How do the three body types of sponges differ from each other?

4. Describe three ways that sponges reproduce.

5. Identify the illustration below. It is magnified 186 × . _____

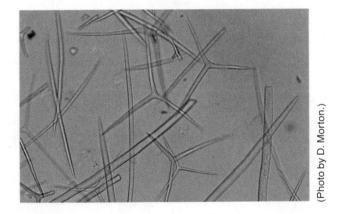

(Photo by D. Morton.)

6. Define the *tissue level of organization* and indicate how Cnidarians exhibit this phenomenon.

7. Name and illustrate the two body forms of Cnidarians.

8. Describe the means by which the Cnidarians seize and eat organisms that are faster than they are.

9. Define *radial symmetry.*

10. Compare the mechanisms of food digestion in sponges and Cnidarians.

Flatworms, Roundworms, and Rotifers

OBJECTIVES

After completing this exercise you will be able to:

1. define: *bilateral symmetry, acoelomate, cephalization, phototaxis, regeneration, host, intermediate host, cuticle, scolex, proglottid, bladder worm, pseudocoelomate, cloaca, wheel organ, parthenogenesis;*

2. describe the organ-system level of organization, with tissues derived from all three primary germ layers;

3. explain the basic body plan of members of the phyla Platyhelminthes, Nematoda, and Rotifera;

4. describe the natural history of members of the phyla Platyhelminthes, Nematoda, and Rotifera;

5. identify representatives of the classes Turbellaria, Trematoda, and Cestoda of the phylum Platyhelminthes; and representatives of the phyla Nematoda and Rotifera;

6. identify structures (and indicate associate functions) of representatives of these groups;

7. outline the life cycles of a human liver fluke and the pork tapeworm;

8. outline the life cycle of *Ascaris;*

9. compare and contrast the anatomy and life cycles of free-living and parasitic flatworms and roundworms.

INTRODUCTION

If you have ever spent any time turning over rocks and leaves in a clear, cool, shallow stream or roadside ditch, you may have observed representatives of phylum Platyhelminthes without realizing it. Do you recall any dark, smooth, soft-bodied, flattened or ovoid masses approximately 1 cm or less in length attached to the bottom of the rocks or leaves? This description fits that of a free-living flatworm or planarian. Parasitic flukes and tapeworms also belong to this phylum.

The roundworms of the phylum Nematoda are widespread and abundant, occurring in large numbers in marine and freshwater bottom sediments and in water films around soil particles. Most are microscopic, and some are parasitic. Besides being especially troublesome for our pet dogs and cats, members of this phylum parasitize a wide variety of animals and plants. Nonetheless, most roundworms are free-living and beneficial; great numbers are characteristic of rich soils, where they are important in nutrient and mineral cycling.

Rotifers are microscopic, the smallest animals. They are mostly bottom-dwelling, freshwater forms, and they are an important prey of larger animals in aquatic food chains.

CAUTION

Hot 5% hydroxide solutions are corrosive. Wear safety goggles. If splashed, wash exposed skin for fifteen minutes in running tap water; then flood the area with a 50% vinegar-and-water solution.

I. Phylum Platyhelminthes: Flatworms

Habitat:	aquatic, moist terrestrial, or parasitic
Body Arrangement:	bilateral symmetry, acoelomate, head (cephalization), gut (incomplete)
Level of Organization:	organ-system (tissues derived from all three primary germ layers); nervous system with brain; excretory system with flame cells; no circulatory, respiratory, or skeletal systems
Body Support:	muscles
Life Cycle:	sexual reproduction (mostly monoecious with internal fertilization); parasites have very complicated life cycles; some asexual reproduction (mostly transverse fission)
Special Characteristics:	bilateral symmetry, head, organ systems

Flatworms have soft, flat, wormlike bodies with **bilateral symmetry** (fig. 22-10). This is the type of symmetry exhibited by most advanced animals, in which the body can be cut through the center in only one plane, parallel to the main axis, to yield two equal halves. Bilateral symmetry is associated with greater motility. The flatworms are **acoelomates;** they lack a body cavity between the gut and body wall (fig. 23-1).

The *tissues* of flatworms are *derived from all three primary germ layers:* ectoderm, mesoderm, and endoderm. Organizationally, these animals represent an important evolutionary transition because they are the simplest forms to exhibit the organ-system level of organization (fig. 23-2) and **cephalization** (a definite head with sense organs). In an evolutionary sense, cephalization goes hand in hand with bilateral symmetry. When an animal with bilateral symmetry moves, the head generally leads, its sensors quickly detecting environmental stimuli that indicate the presence of food or danger ahead.

Although there are definite organs, the *digestive system* is still of the *gastrovascular type, incomplete with a mouth and no anus.*

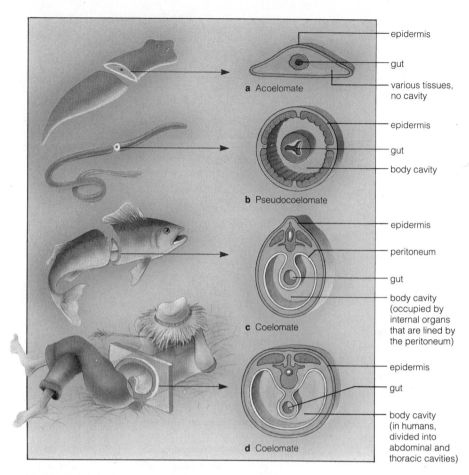

Figure 23-1 Body plans of animals with bilateral symmetry.
(**a**) *Acoelomate* refers to the lack of a body cavity (*coelom*).
(**b**) Pseudocoelomates have an incomplete lining of tissues derived from mesoderm. (**c**) This lining is complete in coelomates, and a peritoneum, a thin membrane derived from mesoderm, is present. (**d**) Mammals have a transverse muscular diaphragm, which divides the body cavity. (After Starr, 1991.)

Labels in figure:
- **a** Acoelomate — epidermis, gut, various tissues, no cavity
- **b** Pseudocoelomate — epidermis, gut, body cavity
- **c** Coelomate — epidermis, peritoneum, gut, body cavity (occupied by internal organs that are lined by the peritoneum)
- **d** Coelomate — epidermis, gut, body cavity (in humans, divided into abdominal and thoracic cavities)

MATERIALS

Per student:

- dissection microscope
- compound microscope, lens paper, a bottle of lens-cleaning solution (optional), a lint-free cloth (optional)
- scalpel
- blunt probe

Per student pair:

- prepared slide of a whole mount of *Dugesia* (planaria)
- prepared slide of a whole mount of planaria with the digestive system stained
- prepared slide of a cross section of planaria (pharyngeal region)
- prepared slide of a whole mount of *Fasciola hepatica*
- prepared slide of a whole mount of *Clonorchis sinensis*
- prepared slide of a whole mount of *Taenia pisiformis*
- prepared slide of a bladder worm of *Taenia*

Per student group (4):

- penlight or microscope light source
- glass petri dish

Per lab room:

- chunk of beef liver
- container of fresh stream water
- container of live planarians
- collection of preserved tapeworms

PROCEDURE

A. Class Turbellaria: Planarians

Most of the species in this class are free-living flatworms, although a few species are parasitic or symbiotic with other organisms. Free-living flatworms are found in marine and fresh waters on the underside of submerged rocks, leaves, and sticks. They are especially abundant in cool freshwater streams and along the ocean shoreline. Some live in wet or moist terrestrial habitats.

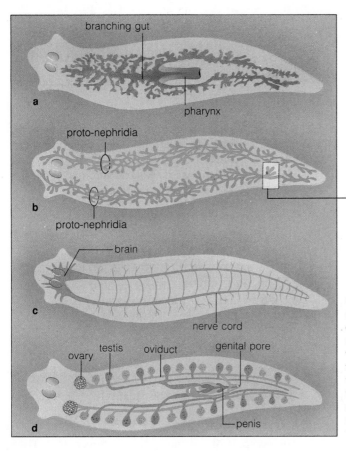

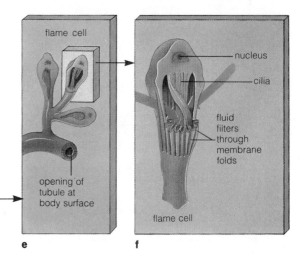

Figure 23-2 Organ systems in a planarian, a type of flatworm. (**a**) Its branching, saclike gut has a pharynx that opens to the outside. Between feedings, the pharynx is retracted into a narrow chamber in the body; it extends out past the body surface while the worm feeds. (**b**) The flatworm system of controlling the volume and composition of body fluids. There are two networks of branching tubules (protonephridia), which use ciliated flame cells to drive excess water from the body. (**c**) The nervous system, with two nerve cords and a rudimentary brain. (**d**) The reproductive system, with male and female parts. (After Starr, 1991.)

The freshwater planarians are the representatives of this class typically studied in biology courses. They are relatively small flatworms, usually two centimeters or less in length. *Dugesia* (planaria) is the genus most commonly studied.

1. Examine a whole mount of planaria and refer to figure 23-3. Locate the *head,* the *eyespots,* and the *auricles,* lateral projections of the head that are organs sensitive to touch and to molecules dissolved in the water.

2. Obtain a slide with the digestive system specially stained so that it is clearly visible. You should be able to see the protractile *pharynx,* with the mouth at its terminus. The pharynx protrudes from the ventral (belly) side of the animal to suck up the organic morsels (insect larvae, small crustaceans, and other small living and dead animals) that comprise its food. The *pharyngeal cavity* within the pharynx is continuous with the cavity of the *intestine.* The intestine has a front and two rear divisions; all are elaborately branched and together occupy much of the animal's body. For what is the extensive branching of the intestine an adaptation?

Allows for greater digestion

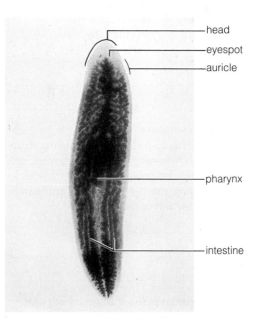

Figure 23-3 Whole mount of planaria, with the digestive system highlighted (12 x). (Photo courtesy Ripon Microslides, Inc.)

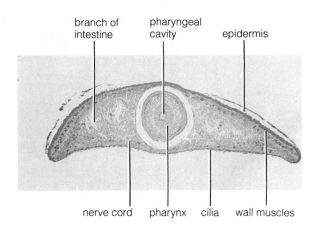

branch of intestine | pharyngeal cavity | epidermis

nerve cord | pharynx | cilia | wall muscles

Figure 23-4 Cross section of planaria (46 x). (Photo courtesy Ripon Microslides, Inc.)

3. Place a small piece of beef liver in a glass petri dish containing clean, fresh stream water. Place several live planaria in the dish and observe their feeding behavior with the dissection microscope. Describe what you see.

You may observe the elimination of waste from the gastrovascular cavity. If you do, describe what you see; if not, state the route by which this must occur.

4. Examine a cross section of planaria with the compound microscope (fig. 23-4). Find the following structures: *pharynx, nerve cords, transverse muscles, epidermis, intestine,* and the several layers of muscles in the body wall. Along the bottom of the animal are *cilia.* What is the function of these cilia?

Locomotion

The body wall contains three layers of muscle: an outer circular layer, a middle diagonal layer, and an inner longitudinal layer. You may not be able to distinguish all three layers.

5. With the dissection microscope, observe a living specimen in a glass petri dish with clean water and note its general size and shape. Touch the animal with a probe. What is its reaction? How does its shape

change? Which muscles contract to produce this shape change?

Rolls into ball

6. Phototaxis is movement of an individual in response to light. Use a light source (penlight or light source for a microscope) to determine phototaxis in planaria. In otherwise dark surroundings, shine the light at its head. What is its response?

The organism moves away from light

7. From the side of the animal, point the light toward its head. What is the animal's response?

moves away

Is the planaria positively or negatively phototactic?

The eyespots of a planaria are photoreceptors. They are sensitive to light but do not form images.

Planarians are monoecious—an individual has both male and female reproductive structures—but apparently do not self-fertilize. Individuals possess testes, ovaries, a penis, and a vagina; and copulation occurs. Planaria exhibit a high degree of **regeneration,** in which lost body parts are gradually replaced.

B. Class Trematoda: Flukes

Flukes are small, leaf-shaped, parasitic forms with complex life cycles. They are generally internal parasites of vertebrates, including humans. They are covered by a protoplasmic but noncellular **cuticle** instead of epidermis. For what is the cuticle an adaptation?

1. Examine a whole mount of a sheep liver fluke, *Fasciola hepatica,* noting its general morphology. Identify the structures indicated in figure 23-5.

2. Examine a whole mount of an adult *Clonorchis sinensis,* a human liver fluke (fig. 23-6). This animal is a common parasite in Asia, and humans are the final **host.** *Clonorchis* has two suckers for attachment to its host: an *oral sucker* surrounding the mouth and a *ven-*

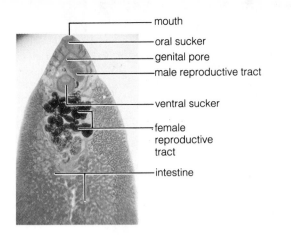

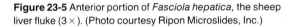

Figure 23-5 Anterior portion of *Fasciola hepatica*, the sheep liver fluke (3 ×). (Photo courtesy Ripon Microslides, Inc.)

tral sucker. The digestive tract begins with the *mouth.* This is followed by a muscular *pharynx,* a short *esophagus,* and the blind two-part *intestine.* At the tip of the back end, locate the *excretory pore* through which nitrogenous wastes are excreted.

Clonorchis is monoecious. Find the following female reproductive structures: *ovary, yolk glands, seminal receptacle,* and *uterus.* Find the following male reproductive structures: *testes* and *seminal vesicle.* During copulation, sperm is conveyed from the testes to the *genital pore* by the paired vasa efferentia, which fuse to form the vas deferens. These ducts cannot be seen in figure 23-6 but may be visible in your specimen. Copulating individuals exchange sperm at the genital pore, and the sperm is stored temporarily in the seminal receptacle. The yolk glands provide yolk for nourishment of the developing eggs. *Eggs* can be seen as black dots in the uterus, where they mature.

The life cycle of *Clonorchis sinensis* (fig. 23-7) involves two **intermediate hosts,** a snail and the golden carp. Intermediate hosts harbor sexually immature stages of a parasite. Fertilized eggs are expelled from the adult fluke into the human host's bile duct and are eventually voided with the feces of the host. If the eggs are shed in the appropriate moist or aquatic environment, they may be ingested by snails (the first intermediate host). A larva emerges from the egg inside the snail. The larva passes through several developmental stages and gives rise to a number of tadpolelike larvae. These free-swimming larvae leave the snail, encounter a golden carp (the second intermediate host), and burrow into this fish to encyst in muscle tissue. If raw or improperly cooked fish is eaten by humans, cats, dogs, or some other mammal, the larvae emerge from the cysts and make their way up the host's bile duct into the liver, where they mature into adult flukes. The cycle is then repeated. What is the infectious stage (for humans) of this parasite?

_____

What two precautions would be effective in reducing the frequency of contact of humans and other domesticated mammals with this parasite?

_____

_____

_____

_____

_____

_____

_____

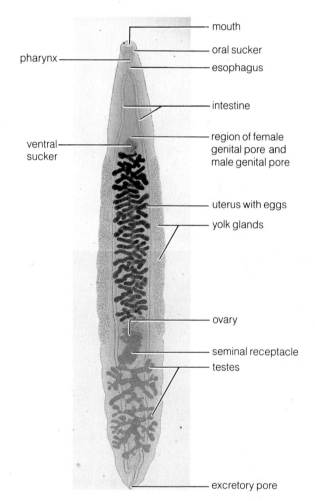

Figure 23-6 Photomicrograph of a whole mount of *Clonorchis sinensis*, a human liver fluke (9 ×). (Photo courtesy Ripon Microslides, Inc.)

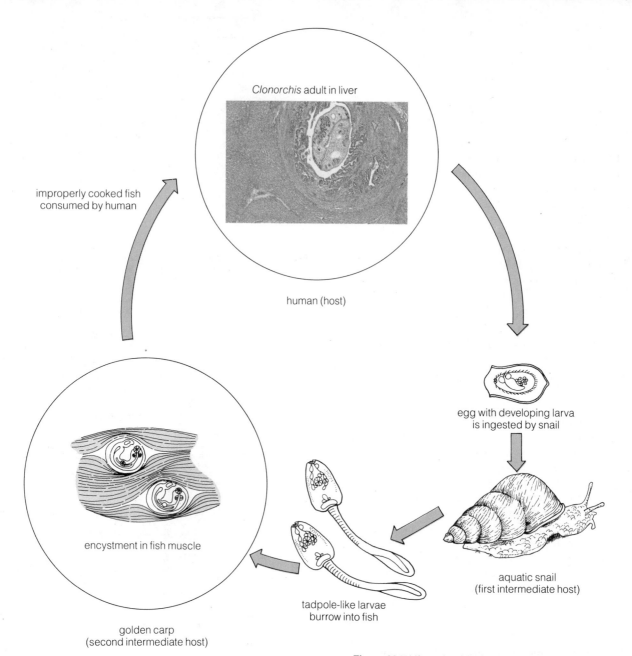

Clonorchis adult in liver

improperly cooked fish
consumed by human

human (host)

egg with developing larva
is ingested by snail

encystment in fish muscle

golden carp
(second intermediate host)

tadpole-like larvae
burrow into fish

aquatic snail
(first intermediate host)

Figure 23-7 Life cycle of the human liver fluke, *Clonorchis sinensis.* (Photo courtesy Ripon Microslides, Inc.)

C. Class Cestoda: Tapeworms

Tapeworms are flattened internal parasites that inhabit the intestine of vertebrates. Their general appearance resembles that of a segmented egg noodle. Tapeworms have no digestive system; instead, they absorb across their body walls predigested nutrients provided by the host. Their bodies are essentially reproductive machines, with extensive, well-developed reproductive organs occupying much of each mature body segment. The body is covered by a thick **cuticle,** similar in construction to that found in trematodes.

1. Examine preserved specimens of tapeworms taken from a variety of vertebrate hosts.

2. Examine the preserved specimens and whole mounts of *Taenia pisiformis,* a tapeworm of dogs and cats (fig. 23-8). The flattened body is divided into three general regions: the **scolex,** *neck,* and body. The scolex is an anterior holdfast organ with hooks and suckers. For what are the hooks and suckers used by the tapeworm?

_____

_____

_____

Behind the scolex is a short neck. The body is composed of successive units, the **proglottids.** Because

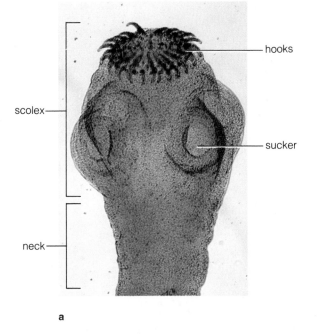

hooks

scolex

sucker

neck

a

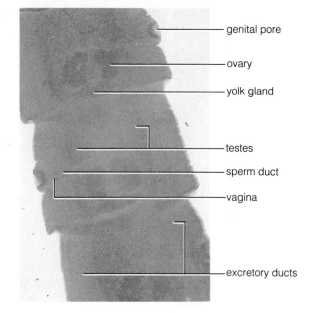

genital pore

ovary

yolk gland

testes

sperm duct

vagina

excretory ducts

b

Figure 23-8 The tapeworm, *Taenia pisiformis*. (**a**) Scolex (35 ×). (**b**) Immature proglottid (16 ×). (**c**) Diagram of an immature proglottid. (**d**) Mature proglottid (9 ×). (Photos courtesy Ripon Microslides, Inc.)

proglottids are added to the animal at the neck, the hindmost proglottids are the oldest. Mature proglottids are almost completely filled with reproductive organs and eggs ready for release.

Tapeworms are monoecious. They are one of the few animals that may self-fertilize, perhaps a result of their isolation from each other.

3. Examine the mature proglottids with a dissection microscope. The lateral *genital pore* opens into the *vagina*. The *uterus* should be visible as a fine line running top to bottom in the middle of the proglottid. The *ovaries* are represented by a dark, diffuse mass of tissue. The *vas deferens* leads from the testes to the genital pore. The *testes* are a paler, more diffuse mass of tissue than the ovaries. Flanking the reproductive organs are the *excretory canals*.

4. The life cycle of the pork tapeworm, *Taenia solium,* involves two host organisms, the pig and a human (fig. 23-9). Mature proglottids containing embryos are voided with the feces of a human. If a proglottid is ingested by a pig, the embryos escape from the proglottid and bore through the intestinal mucosa into the circulatory or lymphatic system. From there they make their way to muscle, where they encyst as larvae known as **bladder worms.** If improperly cooked pork is ingested by a human, a bladder worm is released from its encasement (bladder) and attaches with its scolex to the intestinal mucosa of the human host. Here it matures and adds proglottids to complete the life cycle.

Examine the bladder worm stage of *Taenia* with the dissection microscope.

excretory canal

uterus

testes

vas deferens

ovary

yolk gland

genital pore

vagina

c

zygotes in expanded uterus

d

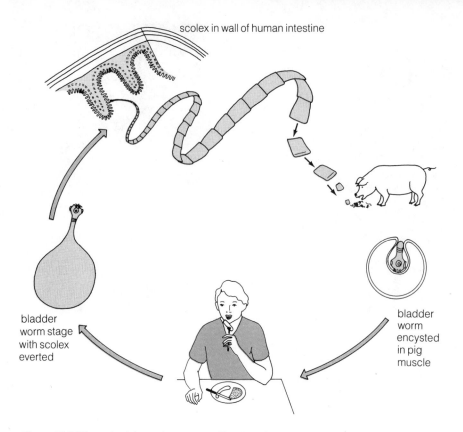

Figure 23-9 Life cycle of the pork tapeworm, *Taenia solium.*

Which mammal is the host?

_____

Which mammal is the intermediate host?

_____

II. Phylum Nematoda: Roundworms

Habitat:	aquatic, moist terrestrial, or parasitic
Body Arrangement:	bilateral symmetry, pseudo-coelomate, head, gut (complete)
Level of Organization:	organ-system (tissues derived from all three primary germ layers); nervous system with brain; no circulatory, respiratory, or skeletal systems
Body Support:	hydrostatic (pseudocoelom)
Life Cycle:	sexual reproduction (mostly dioecious with internal fertilization)
Special Characteristics:	pseudocoelomate, complete gut

Nematodes or *roundworms* are **pseudocoelomates,** which means they have a false body cavity or *pseudo-coel* (fig. 23-1). This term is somewhat unfortunate, because there *is* a body cavity. It is not considered a true body cavity or *coelom* because it is incompletely lined by tissues derived from mesoderm. Unlike flat-worms, roundworms have a complete gut with both a mouth and an anus. Together, the pseudocoel and the complete digestive system comprise a tube-within-a-tube arrangement, a body plan that is seen in most of the phyla yet to be discussed. Roundworms have slender, cylindrical bodies that taper at both ends. They are covered by a complex cuticle of protein.

Parasitic roundworms include *Ascaris* (see below), the hookworms, *Trichinella* worms of mammals (fig. 23-10), pinworms of humans, and the filarial worms of humans that cause elephantiasis by obstructing lymphatic vessels.

MATERIALS

Per student:
- dissection microscope
- dissecting needle

Per student pair:
- finger bowl

Per lab room:
- collection of free-living roundworms
- tray of moist soil
- container of fresh stream water
- demonstration dissection of a female *Ascaris*
- demonstration dissection of a male *Ascaris*

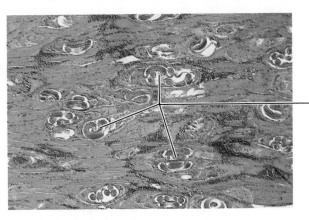

—encysted
larvae

Figure 23-10 *Trichinella spiralis* larvae encysted in muscle
(46×). This organism causes the disease trichinosis, which is
usually contracted by eating undercooked pork or wild game.
(Photo courtesy Ripon Microslides, Inc.)

PROCEDURE

A. Free-living Roundworms

1. Examine the demonstration collection of free-living
roundworms.

2. Place some of the moist soil your instructor has
provided in a finger bowl with some fresh stream
water. Using a dissection microscope, look for translucent roundworms in the soil sample. Describe the
behavior of any roundworms you find and make a
note of any visible anatomical features so that you can
compare these to the features of the internal parasite,
Ascaris, described below.

_____

_____

_____

_____

_____

B. *Ascaris*

Ascaris is a common parasite of humans, pigs, horses,
and other mammals. Adults range from 15 to 40 cm in
length, with males being smaller and having a sharp
bend in the posterior part of the body. In this section
you will study the anatomy of both sexes. Your lab
instructor has provided demonstration dissections of
both female and male *Ascaris*.

> **CAUTION**
>
> **Wash your hands and instruments thoroughly after ex-
> amining the female *Ascaris*; some eggs may still be
> alive even though the specimen has been preserved.**

1. Examine a female *Ascaris* (figs. 23-11a, 23-12a and
b). At the front end, the triangular *mouth* is surrounded by three *lips*. The opening on the ventral
surface near the other end of the body is the *anus*. A
genital pore is located about one-third the distance
from the mouth.

The *pseudocoel* is the fluid-filled space containing
the internal organs. The lips surround a *mouth* that
opens into a muscular *pharynx (esophagus)*. The esophagus leads into a flat *intestine*, running nearly the
length of the body. The intestine ends in a *rectum* and
opens into the *anus*.

Find the paired *excretory tubes* on each side of the
body. These tubes open to the outside through a single
excretory pore located near the mouth on the ventral
surface. The *ovaries* are the threadlike ends of a Y-
shaped reproductive tract. The ovary leads to a larger
oviduct and a still larger *uterus*. The two uteri converge
to form a short, muscular *vagina* that opens to the
outside through the *genital pore*.

2. Examine a male *Ascaris* (figs. 23-11b, 23-12c and d).
The lips and mouth are as in the female. However, the
male has a *cloacal opening* instead of an anus. A **cloaca**
is an exit for both reproductive and digestive products. A pair of *copulatory spicules* protrudes from this
opening.

Excretion is as in the female. The reproductive system superficially resembles that of the female. It consists of a *single* tubular structure that includes a
threadlike *testis* joined to a larger *vas deferens (sperm
duct)*, an even larger *seminal vesicle*, and terminally a
short, muscular *ejaculatory duct* that empties into the
cloaca.

After copulation, the eggs are fertilized in the female oviduct. The eggs, surrounded by a thick shell,
are expelled through the genital pore to the outside;
as many as two hundred thousand may be shed into
the host's intestinal tract per day.

The eggs are eliminated in the host's feces. They
are very resistant and may live for years under adverse
conditions. Infection with *Ascaris* results from ingesting eggs in food or water contaminated by feces or

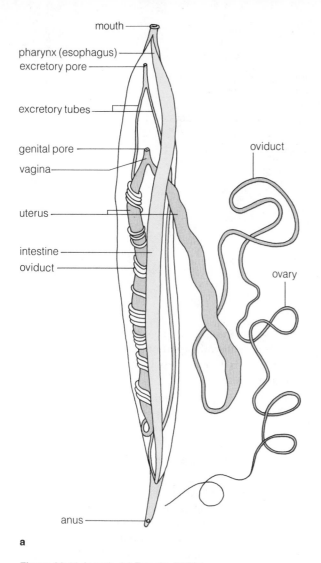

mouth
pharynx (esophagus)
excretory pore
excretory tubes
genital pore
vagina
uterus
intestine
oviduct
anus
oviduct
ovary

a

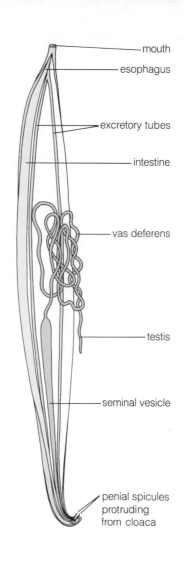

mouth
esophagus
excretory tubes
intestine
vas deferens
testis
seminal vesicle
penial spicules
protruding
from cloaca

b

Figure 23-11 *Ascaris.* (**a**) Female. (**b**) Male.

soil. The eggs pass through the stomach and hatch in the small intestine. The small larvae migrate through the bloodstream, exit the circulatory system into the respiratory system, crawl up the respiratory tubes and out the trachea into the throat. There they are swallowed to reinfect the intestinal tract and mature. *Sometimes larvae exit the nasal passages rather than being swallowed!* The entire journey of the larva from intestine back to intestine takes about ten days.

III. Phylum Rotifera: Rotifers

Habitat:	aquatic or moist terrestrial, mostly freshwater
Body Arrangement:	bilateral symmetry, pseudo-coelomate, head, gut (complete)
Level of Organization:	organ-system (tissues derived from all three primary germ layers); nervous system with brain; excretory system with flame cells; no circulatory, respiratory, or skeletal systems
Body Support:	hydrostatic (pseudocoelom)
Life Cycle:	sexual (dioecious with internal fertilization), some parthenogenesis, males much smaller than females or absent
Special Characteristics:	microscopic animals, wheel organ, mastax

The body wall of rotifers is very thin, usually transparent, and is covered by a cuticle composed of protein. Like the nematodes, the rotifers are pseudocoelomates.

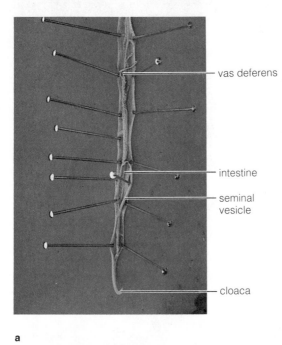

a

vas deferens

intestine

seminal
vesicle

cloaca

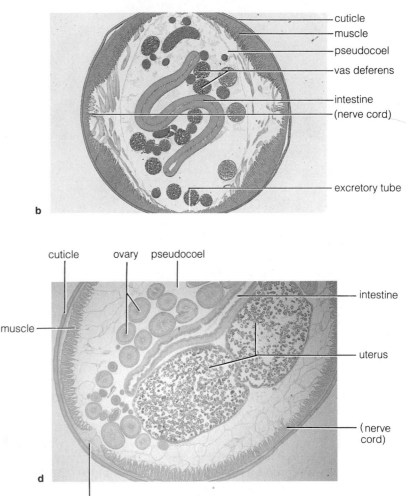

b

cuticle
muscle
pseudocoel
vas deferens
intestine
(nerve cord)

excretory tube

cuticle   ovary   pseudocoel

muscle

intestine

uterus

(nerve
cord)

d

excretory tube

Figure 23-12 (**a**) Dissection of male *Ascaris*. (Photo by D. Morton.) (**b**) Cross section of male *Ascaris* (12 ×). (Photo courtesy Ripon Microslides, Inc.) (**c**) Dissection of female *Ascaris*. (Photo by D. Morton.) (**d**) Cross section of female *Ascaris* (19 ×). (Photo by D. Morton.)

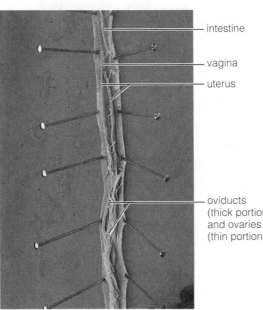

c

intestine

vagina

uterus

oviducts
(thick portions)
and ovaries
(thin portions)

Rotifers are dioecious. However, many species have no males, and the eggs develop parthenogenetically. **Parthenogenesis** is the development of unfertilized eggs, generally in response to seasonal chemical or temperature changes in the aquatic environment.

MATERIALS

Per student:

• clean microscope slide and coverslip

• compound microscope

Per lab room:

• live culture of rotifers

• disposable transfer pipet

PROCEDURE

You will examine rotifers from a living culture. Rotifers can be easily found in rain gutters and spouts, and in the slimy material around the bases of buildings. Many of the details of the digestive, excretory, and reproductive systems of these animals will not be distinguishable, and prepared slides are usually no better (and frequently worse) in revealing these details. However, the general body outline and structures described below should be apparent.

1. Because rotifers are microscopic, use a disposable transfer pipet to place a drop from the rotifer culture on a slide, cover with a coverslip, and observe with the compound microscope. The elongate, cylindrical body is divisible into the *trunk* and the *foot* (fig. 23-13).

The foot bears two to several *toes* for adhering to objects. The superior end of the trunk bears the **wheel**

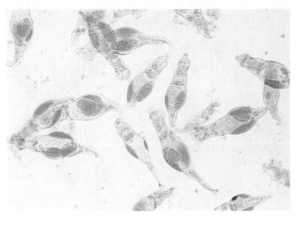

a

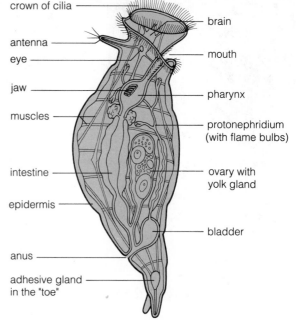

crown of cilia
brain
antenna
eye
mouth
jaw
pharynx
muscles
protonephridium
(with flame bulbs)
intestine
ovary with
yolk gland
epidermis
bladder
anus
adhesive gland
in the "toe"

b

Figure 23-13 (a) Prepared whole mount of rotifers (81 ×).
(Photo courtesy Ripon Microslides, Inc.) **(b)** Rotifer structure.
(After Starr, 1991.)

organ, which is encircled by cilia. These cilia beat in such a fashion as to give the appearance of a wheel (or two) turning; hence the common reference to these animals as wheel animals. The cilia are responsible for creating a current that sweeps organic food particles into the mouth at the center of the wheel organ. The cilia also enable the organism to move.

2. Inside the trunk, below the wheel organ, you may be able to see the movement of a grinding organ, the *mastax*. This structure is equipped with jaws made hard by a substance called chitin. For what is the mastax an adaptation?

_____

_____

_____

_____

OPTIONAL

IV. Experiment: Regeneration

Your instructor may provide you with an experiment about regeneration.

PRE-LAB QUESTIONS

_B_ 1. Cephalization is (a) the division of the trunk into a scolex, neck, and body, (b) the presence of a head with sense organs, (c) sexual reproduction involving self-fertilization, (d) the ability to replace lost body parts.

_D_ 2. Flatworms (a) are radially symmetrical, (b) possess a body cavity, (c) reproduce by parthenogenesis, (d) have tissues derived from all three primary germ layers.

_A_ 3. Free-living flatworms (planaria) move by using (a) cilia, (b) flagella, (c) pseudopodia, (d) a muscular foot.

_A_ 4. The flukes and tapeworms are covered by a protective (a) cuticle, (b) shell, (c) scales, (d) slime layer.

_D_ 5. Tapeworms do not have a (a) reproductive system, (b) head, (c) excretory system, (d) digestive system.

_B_ 6. The encysted larva of the pork tapeworm is known as a (a) glochidium, (b) bladder worm, (c) gemmule, (d) fluke.

_B_ 7. Nematodes and rotifers have a body cavity that is (a) called a coelom, (b) called a pseudocoelom, (c) completely lined by tissues derived from mesoderm, (d) a and c.

_C_ 8. Filarial roundworms cause elephantiasis by (a) encysting in muscle tissue, (b) promoting the growth of fatty tumors, (c) obstructing lymphatic vessels, (d) laying numerous eggs in the joints of their hosts.

_D_ 9. Digestive wastes of the male *Ascaris* roundworm exit the body through the (a) anus, (b) cloaca, (c) mouth, (d) excretory pore.

_B_ 10. Rotifers obtain their food by the action of (a) tentacles, (b) cilia on the wheel organ, (c) a muscular pharynx, (d) pseudopodia.

EXERCISE 23

Flatworms, Roundworms, and Rotifers
POST-LAB QUESTIONS

1. Define the following terms:

   a. *acoelomate*

   b. *cephalization*

   c. *cloaca*

   d. *parthenogenesis*

   e. *intermediate host*

2. Define the *organ-system level of organization* and indicate how flatworms exhibit this phenomenon.

3. In your own words, describe the characteristics of flatworms.

4. Compare the digestive systems of free-living flatworms and roundworms.

5. How is bilateral symmetry different from radial symmetry?

6. Describe several adaptations of parasitic flatworms to their external environment.

7. The nematodes and rotifers are called *pseudocoelomates,* which means "false body cavity." Why is their body cavity not considered to be a true body cavity?

8. Identify and label this illustration. It is magnified 9 ×.

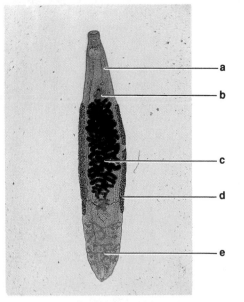

a _____

b _____

c _____

d _____

e _____

(Photo courtesy Ripon Microslides, Inc.)

9. Explain the relationship between the tremendous numbers of eggs produced by flukes, tapeworms, and parasitic roundworms and the complexity of their life cycles.

10. The relatively simple animals *Hydra* and *Dugesia* can regenerate lost body parts, but humans generally cannot. Discuss this in terms of tissue differentiation and the comparative levels of structural complexity of these organisms.

EXERCISE 24

Mollusks, Segmented Worms, and Joint-legged Animals

OBJECTIVES

After completing this exercise you will be able to:

1. define *protostome, deuterostome, coelomate, coelom, foot, mantle, segmentation, peritoneum, setae, clitellum, cocoon, exoskeleton, hemolymph, chitin, carapace, sexual dimorphism, gills, tracheae;*

2. differentiate between protostomes and deuterostomes;

3. describe the natural history of members of the phyla Mollusca, Annelida, and Arthropoda;

4. identify representatives of the classes Pelecypoda, Gastropoda, and Cephalopoda of the phylum Mollusca;

5. outline the life cycle of a freshwater clam or mussel;

6. identify representatives of the classes Oligochaeta, Polychaeta, and Hirudinea of the phylum Annelida;

7. outline the life cycle of an earthworm;

8. identify representatives of the classes Crustacea, Insecta, Chilopoda, Diplopoda, and Arachnida of the phylum Arthropoda;

9. identify structures (and indicate their associated functions) of the representatives of these phyla and classes.

INTRODUCTION

Some time after the origin of the first bilaterally symmetrical animals, two divergent paths were taken by the animals that subsequently evolved (fig. 22-9). The mollusks, annelids, and arthropods followed one of these paths, and the echinoderms and chordates the other. The mollusks, annelids, and arthropods comprise a collection of animals known as **protostomes,** animals whose blastopore (the first opening) in the gastrula stage of development ultimately becomes a mouth. The anus develops later. The echinoderms and chordates comprise the **deuterostomes,** whose blastopore develops into an anus. A second opening becomes the mouth.

Animals in both groups are **coelomates,** possessing a true body cavity or **coelom** that is lined by tissue derived from the mesoderm (fig. 23-1). The appearance of the coelom has great evolutionary significance, for it provides room for organs, as well as allowing an increase in overall body size and flexibility, and contributing to the development of systems primar-

ily derived from mesoderm. This exercise describes the protostomes. The next two exercises cover the deuterostomes.

> **CAUTION**
>
> Preserved specimens are kept in a formalin-based or other preservative solution. Wash any part of your body exposed to this solution with copious amounts of water. If preservative solution is splashed into your eyes, wash them with the safety eyewash bottle for fifteen minutes.

I. Phylum Mollusca: Molluscans

Habitat:	aquatic or terrestrial, mostly marine
Body Arrangement:	bilateral symmetry, coelomate, head and unique body plan, gut (complete)
Level of Organization:	organ-system (tissues derived from all three primary germ layers), nervous system with brain, excretory system (nephridium or nephridia), circulatory system (open), respiratory system (most have gills or lungs), no skeletal system
Body Support:	shells, floats (aquatic)
Life Cycle:	sexual (monoecious or dioecious with external or internal fertilization), larvae in some
Special Characteristics:	coelomate; open circulatory system; respiratory system; external body divided into head, mantle, and foot

This is the second largest animal phylum and includes snails, slugs, nudibranchs, clams, mussels, oysters, octopuses, and squids. The body is typically soft (*molluscus* in Latin means "soft"). Apart from the squids and octopuses (and the land snails and slugs, which have adopted a terrestrial existence), *mollusks inhabit shallow marine and fresh waters,* where they crawl along or burrow into the soft substrate. Many species are valued as food for humans.

Two characteristics distinguish the mollusks from other animals. They possess a ventral, muscular **foot**

315

for movement and a dorsal integument, the **mantle,** which secretes the shell and functions in gaseous exchange. Even forms without a shell usually have a mantle.

MATERIALS

Per student:

• scalpel

• blunt probe or dissecting needle

• compound microscope, lens paper, a bottle of lens-cleaning solution (optional), a lint-free cloth (optional)

Per student pair:

• preserved freshwater clam or mussel

• dissection pan

• prepared slide of a section of clam gill

• dissection microscope

• glass petri dish

• prepared slide of *glochidia*

Per lab room:

• labeled demonstration dissection of a freshwater clam or mussel (section A.1.b, c, e) (optional)

• labeled demonstration dissection of a freshwater clam or mussel (section A.1.f.)

• collection of pelecypod shells and preserved specimens

• freshwater aquarium with snails (for example, *Physa*)

• fluorescent or gooseneck lamp

• black construction paper

• aquarium lid

• collection of gastropod shells and preserved specimens

• several preserved squids

• several preserved octopuses and chambered nautiluses

• large plastic bag for disposal of dissected specimens

PROCEDURE

A. Class Pelecypoda: Bivalves

Bivalves are so named because they have shells composed of two halves or valves. Included in this class are the clams, mussels, oysters, and scallops. They have a hatchet-shaped (laterally flattened) body and a foot used for burrowing in soft substrates. *Growth ridges* on the shell's outer surface represent periods of restricted winter growth and thus the age of the organism.

1. Examine a freshwater clam or mussel.

a. Examine a freshwater clam or mussel in a dissection pan. Each valve has a hump on the dorsal surface

(back), the *umbo,* which points toward the front of the organism. The umbo is the oldest part of the valve. Encircling the umbo are concentric rings of annual growth, some of which form the growth ridges mentioned previously. Determine the age of your mussel by counting the number of ridges formed by the growth rings and record below.

_____

b. *The following dissection of a freshwater clam or mussel may be done as a demonstration exercise, prepared by your instructor. If it is done as a demonstration, locate the structures in boldface and italics in the following description.* Remove the left valve by cutting the *anterior* and *posterior adductor muscles* that hold the valves together (fig. 24-1). To do this, slip a scalpel between the valves and cut back and forth in the areas where the two valves come together at the dorsal surface. You should be able to feel the scalpel cutting through the adductor muscles. Once these muscles are severed, slide the scalpel between the valve to be removed and the tissues that line its inner surface. With the flat side of the scalpel, hold the tissues down while you remove the valve.

The relatively simple body plan includes a ventral **foot** and a dorsal *visceral mass.* The membrane that lines the inner surface of each valve is the **mantle,** which secretes the shell. The shell is composed of three layers, each formed of calcium carbonate. Pearls are formed as a result of irritations (usually grains of sand or small pebbles) in the mantle and are sometimes found embedded in the *pearly layer* of the shell. Note the *incurrent* and *excurrent siphons* at the rear end. These represent extensions of the mantle. What are their functions?

_____

_____

_____

c. Remove the mantle on the left side by cutting it free with a scalpel (fig. 24-2). Identify the *gills,* one on each side of the *visceral mass.* The gills function in gaseous exchange. Water enters the incurrent siphon and flows over the gills, ultimately leaving through the excurrent siphon. Gills also trap food particles contained in incoming water; these are transported to the mouth by the cilia on the gills.

d. Examine the demonstration slide of a transverse section of the gill. Identify the structure labeled in figure 24-3.

e. The circulatory system is *open,* so the blood passes from arteries leading from the heart through body spaces called *sinuses.* Blood is returned to the heart by veins draining the mantle and gills. Find the *heart* (fig. 24-4) in the pericardial sac between the visceral mass and the hinge between the valves. Dispose of your dissected specimen in the large plastic bag provided for this purpose.

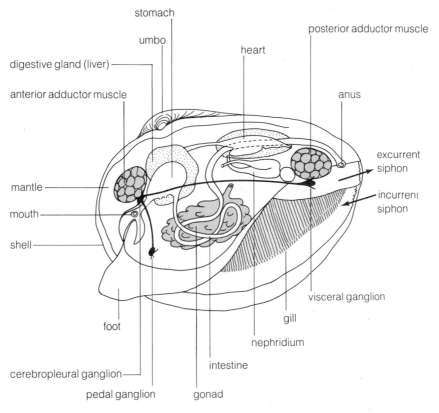

Figure 24-1 Internal anatomy of a freshwater clam or mussel. (After Boolootian and Stiles, 1981.)

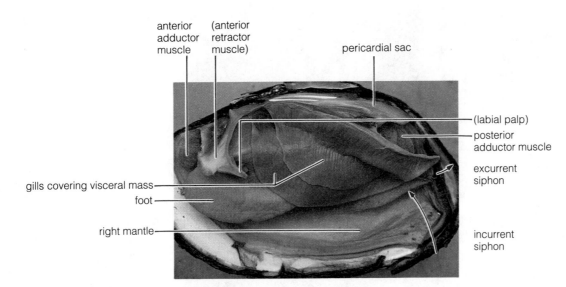

Figure 24-2 Dissection of freshwater clam. The left valve and mantle have been removed. (Photo by D. Morton.)

f. Examine the labeled dissection prepared by your instructor. Note the *mouth, esophagus, stomach, intestine, digestive gland* (liver), *anus, nephridium* (kidney), and *gonad* (fig. 24-4).

g. Fertilization is internal in bivalves. Sperm are taken in through the female's incurrent siphon and fertilize eggs in the gills. Larvae (*glochidia*) develop from the fertilized eggs and after several months leave the gill area via the excurrent siphons to parasitize fish by clamping onto the gills or fins with their jawlike valves. The fish grows tissue over the parasite, which remains attached for two to three months. The glochidium then breaks out of the fish to develop into an adult on the lake or river bottom. Look at glochidia in a prepared slide (fig. 24-5).

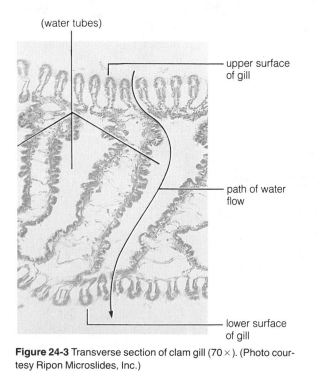

(water tubes)

upper surface of gill

path of water flow

lower surface of gill

Figure 24-3 Transverse section of clam gill (70 ×). (Photo courtesy Ripon Microslides, Inc.)

2. *Assorted pelecypods.* Examine an assortment of pelecypod shells and preserved specimens. List the features they have in common.

_____

_____

_____

List the differences.

_____

_____

_____

_____

How can you tell which live on sandy bottoms?

_____

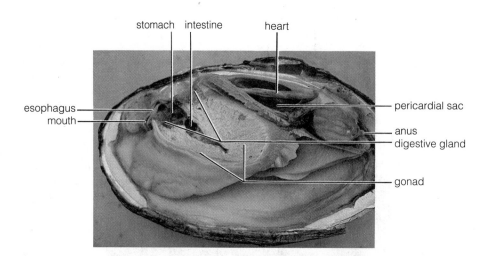

stomach   intestine   heart

esophagus
mouth

pericardial sac

anus
digestive gland

gonad

Figure 24-4 Further dissection of freshwater clam. The left gills and part of the foot have been removed. (Photo by D. Morton.)

Figure 24-5 Whole mount of *glochidia* (81 ×). (Photo courtesy Ripon Microslides, Inc.)

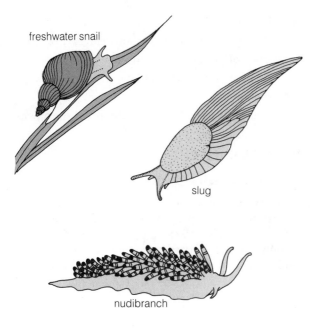
freshwater snail

slug

nudibranch

B. Class Gastropoda: Snails, Slugs, and Nudibranchs (fig. 24-6)

Slugs and nudribranchs have no shell; the former are terrestrial, the latter marine. Nudibranchs are very ornate gastropods, often brightly colored and frequently adorned with numerous fleshy appendages.

1. Place a live freshwater snail (for example, *Physa*) in a glass petri dish. Once the snail has attached itself to the glass, invert the petri dish and use the dissection microscope to observe it moving along a slimy path. A *slime gland* in the front of the foot secretes a mucus. Note the muscular contractions of the foot that allow the animal to glide over this mucus. You can see the *mouth* in this position.

2. In snails with spirally coiled shells, the shell either spirals to the snail's right or to the snail's left as it grows. Which way does your snail's shell spiral?

_____

3. Observe the feeding behavior of snails in a freshwater aquarium that has sides coated with algae. Are most of the snails oriented with their heads directed upward or downward?

_____

4. *Phototaxis.* At this time, your instructor will place a light over one half of the aquarium and cover the other half with dark paper or a lid. Count how many snails are in each half of the aquarium and record below.

Number of snails in light half = _____

Number of snails in dark half = _____

At the end of the lab period check to see if there has been any movement of the snails toward one side or the other. Again count the snails in the light and dark sides of the aquarium.

Number of snails in light half = _____

Number of snails in dark half = _____

What is their response to light?

_____

_____

5. Gastropods, especially snails, are important intermediate hosts for trematode parasites of vertebrates (see Exercise 23). Look at the variety of gastropod shells and preserved specimens on display.

C. Class Cephalopoda: Squids, Octopuses, and Nautiluses (fig. 24-7)

Cephalopods have several significant modifications of the basic molluskan body plan. They are adapted for swimming and a carnivorous habit. Octopuses have no shell.

1. Examine the external morphology of a squid. The foot of the squid has become divided into four pairs of arms and two tentacles, each with suckers, and a *siphon (funnel).* The tentacles, with their terminal suckers, shoot out to grasp prey. The siphon functions in jet propulsion by forcing jets of water outward through this opening to propel the animal through the water. Two large eyes superficially resemble those of vertebrates. The shell of the squid lies dorsally beneath the mantle. Located posteriorly is a pair of *lateral fins.* What might be the function of these fins?

_____

_____

2. Look at the variety of cephalopod shells and preserved specimens on display.

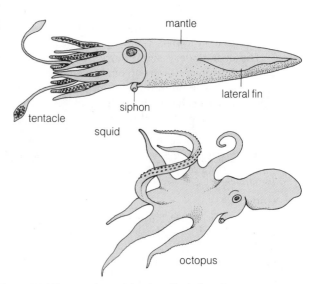
squid — mantle, lateral fin, siphon, tentacle

octopus

Figure 24-7 Two members of the class Cephalopoda.

II. Phylum Annelida: Annelids

Habitat:	aquatic or moist terrestrial, some parasitic
Body Arrangement:	bilateral symmetry, coelomate, head, body segmented, gut (complete)
Level of Organization:	organ-system (tissues derived from all three primary germ layers), nervous system with brain, circulatory system (closed), excretory system (nephridia), respiratory system (gills in some), no skeletal system
Body Support:	hydrostatic (coelom)
Life Cycle:	sexual reproduction (monoecious or dioecious with mostly internal fertilization), larvae in some, asexual reproduction (mostly transverse fission)
Special Characteristics:	segmented, closed circulatory system

The most striking annelid characteristic is the division of the cylindrical trunk into a series of similar segments. *Annelids* were the first animals to evolve the condition of **segmentation.** Segmentation is internal as well as external, with segmentally arranged components of various organ systems and the body cavity. The coelom is more or less divided by *septa* into compartments within each segment, each lined by **peritoneum.** Unlike the proglottids of the cestodes, segments cannot function independently of each other. The *circulatory system* is *closed,* with the blood entirely contained in vessels. There is a well-developed *nervous system* with a central nervous system composed of two fused dorsal ganglia (brain) and ventral nerve cords. The circulatory system is not segmented.

The evolutionary significance of segmentation is that each segment or group of segments can become specialized. Although this is an important characteristic of most of the phyla we have yet to study, it is barely evident in the earthworm.

MATERIALS

Per student:

• pins
• blunt probe or dissecting needle

Per student pair:

• preserved earthworm
• dissection microscope
• dissection pan
• fine dissecting scissors

Per lab room:

• labeled demonstration dissection of an earthworm
• collection of preserved polychaetes
• collection of preserved leeches
• live freshwater leeches in an aquarium
• large plastic bag for disposal of dissected specimens

PROCEDURE

A. Class Oligochaeta: Oligochaetes

1. You may know several characteristics of oligochaetes because of past associations with the earthworm (*Lumbricus terrestris*), especially if you go fishing. Recalling past experiences with this organism, list as many features of the earthworm as you can.

Terrestrial & Fresh water; few setae per segment; lack well developed head; hermaphroditic, Segmented

2. Let's see how you did! Obtain a preserved earthworm and have a dissection microscope handy. Is its body bilaterally or radially symmetrical?

Bilateral

3. Examine the external anatomy (fig. 24-8). Note the segments. Count them and record the number.

140

Very few segments are added after hatching. Note the *prostomium,* a fleshy lobe that hangs over the *mouth;* both structures are part of the first segment. Take the specimen between your fingers and feel the **setae,** tiny, bristlelike structures. What is their function?

Movement

4. Although you will not separate the following layers of the body wall of the earthworm, it is important that you be aware of their relationship to each other. A thin *cuticle* covers the body. The cuticle is secreted by the epidermis, which lies beneath it, and internal to this are an outer circular layer and an inner longitudinal layer of muscle (fig. 24-9).

5. Use the dissection microscope to help you find the following. A pair of small *excretory pores* is found on the lateral or ventral surfaces of each segment, except the first few and the last. On the sides of segment 14,

320

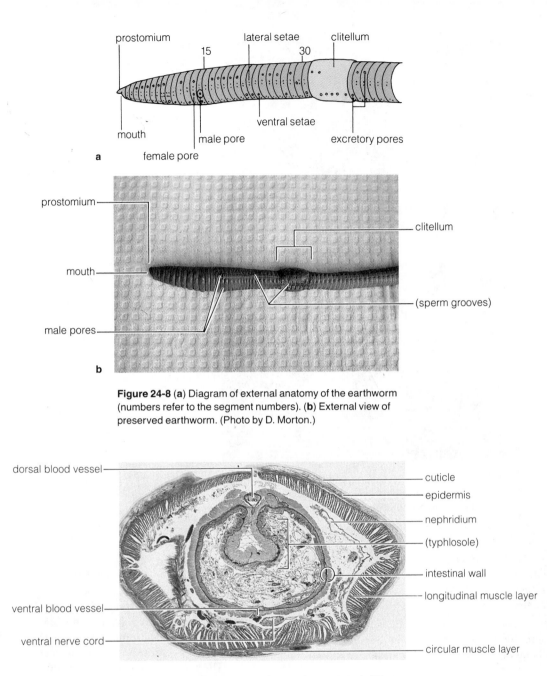

Figure 24-8 (a) Diagram of external anatomy of the earthworm (numbers refer to the segment numbers). (b) External view of preserved earthworm. (Photo by D. Morton.)

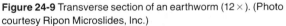

Figure 24-9 Transverse section of an earthworm (12 ×). (Photo courtesy Ripon Microslides, Inc.)

the openings (*female pores*) of the oviducts can be seen. The openings (*male pores*) of the sperm ducts, with their swollen lips, can be found on segment 15. The **clitellum** is the enlarged ring beginning at segment 31 or 32 and ending with 37. This glandular structure secretes a slimy band around two copulating individuals. The *anus* is a vertical slit in the terminal segment.

6. Place your worm in a dissection pan and stick a pin through each end of the specimen. With fine scissors, make an incision just to the right of the dorsal black line (blood vessel), ten segments anterior to the anus, and cut superficially to the mouth. (You can use the prostomium as an indicator of the dorsal side of the animal if the black line cannot be seen). Gently pull

the incision apart and note that the coelom is divided into a series of compartments by septa. Carefully cut the septa free of the body wall with a scalpel and pin both sides to the pan at 5- to 10-segment intervals. Cover the specimen with about 0.5–1.0 cm of tap water and examine the internal anatomy (fig. 24-10).

7. Use your blunt probe or dissection needle to point out the male reproductive organs, the light-colored *seminal vesicles*. These can be found in segments 9–12. They are composed of three sacs, within which are the *testes*. A sperm duct leads from the seminal vesicles to the male pore.

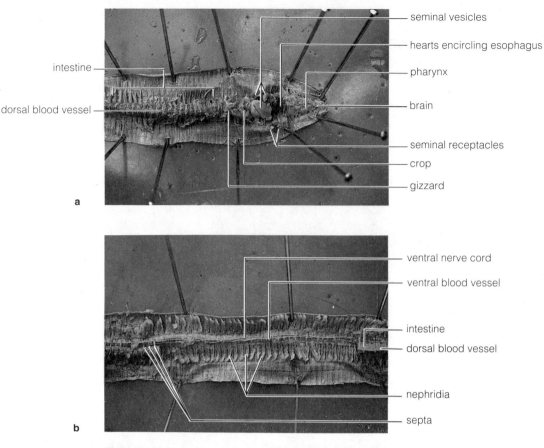

Figure 24-10 Internal anatomy of the (**a**) anterior and (**b**) posterior ends of an earthworm. The intestine has been partially removed in (**b**) to better view the ventral nerve cord. (Photos by D. Morton.)

8. Find two pairs of female reproductive organs — the white, spherical *seminal receptacles* in segments 9 and 10. There are ovaries in segment 13. An oviduct runs from each ovary to the female pore.

The earthworm is monoecious — individuals have both male and female reproductive tracts. Copulation in earthworms usually takes two to three hours. Individuals face belly to belly in opposite directions and come together at their clitellums. These organs secrete a slimy substance that forms a band around the worms. Sperm are transferred between the participants, and the sperm are stored temporarily in the seminal receptacles. A few days after the individuals separate, each worm secretes a **cocoon**. The cocoon receives eggs and stored sperm from the seminal receptacles. The eggs are fertilized, and the cocoon moves forward as the worm backs out. The cocoon then slips off the front end of the worm and is deposited in the soil. In this species, a single young earthworm completes development using the other fertilized eggs as food, eventually breaks free of the cocoon, and becomes an adult in several weeks. As long as the stored sperm lasts, the earthworm continues to form cocoons.

9. Locate the organs of the digestive tract. The *mouth (buccal) cavity* occupies the first three segments. The muscular *pharynx* is the swelling in the digestive tract just behind the mouth. This organ is responsible for sucking soil particles containing food into the mouth. The *esophagus* extends from the pharynx through segment 14. The *crop* is a large, thin-walled sac in segments 15 and 16 that serves as a temporary storage organ for food. The *gizzard*, a grinding organ, is posterior to the crop. The *intestine*, the site of digestion and absorption, extends from the gizzard to the anus.

10. Trace the parts of the circulatory system. The *dorsal blood vessel* runs along the dorsal side of the digestive tract and contains blood that flows toward the head. Five pairs of so-called *hearts* encircle the esophagus in segments 7–11, connecting the dorsal blood vessel with the *ventral blood vessel*. The ventral vessel runs along the ventral side of the digestive tract, carrying blood backward. The primary pumps of the circulatory system are the dorsal and ventral vessels, not the hearts. The dorsal vessel contains valves to prevent blood from flowing backward.

11. Find the two-lobed *brain* (cerebral ganglia) on the anterior, dorsal surface of the pharynx in segment 3. The *ventral nerve cord* arises from the brain and extends along the floor of the body cavity to the last segment. Although they are difficult to locate, paired

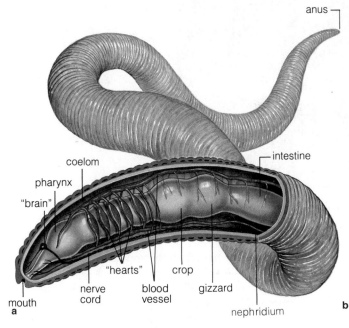

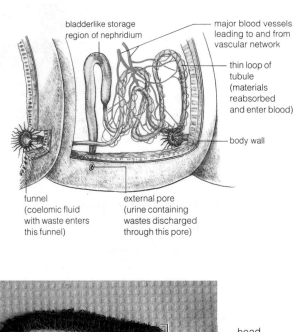

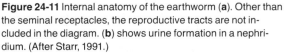

Figure 24-11 Internal anatomy of the earthworm (**a**). Other than the seminal receptacles, the reproductive tracts are not included in the diagram. (**b**) shows urine formation in a nephridium. (After Starr, 1991.)

lateral nerves branch from the segmental ganglia along the nerve cord.

12. Can you find a pair of *nephridia*—coiled white tubes—at the base of every segment but the first three and last one? These structures constitute the "kidneys" of the earthworm (fig. 24-11). Dispose of your dissected specimen in the large plastic bag provided for this purpose.

B. Class Polychaeta: Polychaetes

The polychaetes are abundant marine annelids with dorsoventrally flattened bodies (fig. 24-12). They are most common at shallow depths in the intertidal zone at the seashore. They are prey for a variety of marine invertebrates.

Examine the demonstration of preserved polychaetes. Although they are similar to oligochaetes in many ways, notice that the polychaetes have fleshy, segmental outgrowths called *parapodia* (fig. 24-12). These structures are equipped with numerous setae, the characteristic for which they get their class name, Polychaeta, meaning "many bristles." Notice also the definite head.

C. Class Hirudinea: Leeches

The leeches are dorsoventrally flattened annelids that are usually found in fresh water.

1. Examine the demonstration of preserved specimens of leeches (fig. 24-13). They lack parapodia and setae. They are predaceous or parasitic; parasitic species suck blood with the aid of a muscular pharynx.

Figure 24-12 External anatomy of the clamworm, *Neanthes*. (Photo by D. Morton.)

The anterior segments are modified into a small *sucker* surrounding the mouth. The mouth is supplied with jaws, made hard with the substance chitin, used to wound the host and supply the leech with blood. Parasitic leeches secrete an anticoagulant into the wound to prevent the blood from clotting. A *posterior sucker* helps the leech attach to the host or prey.

2. If living freshwater leeches are available, observe the action of their oral sucker by allowing them to *briefly* attach to and hang from your hand or arm.

Observe their movement as they swim through the water in an aquarium. Explain this movement in terms of the contraction and relaxation of circular and longitudinal layers of muscle that comprise part of their body wall.

_____

_____

_____

_____

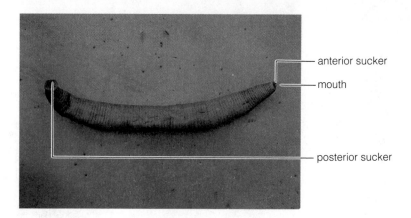

anterior sucker

mouth

posterior sucker

Figure 24-13 A leech (class Hirudinea). (Photo by D. Morton.)

III. Phylum Arthropoda: Arthropods

Habitat:	aquatic, terrestrial (land and air), some parasitic
Body Arrangement:	bilateral symmetry, coelomate, body regions (head, thorax, and abdomen), segmented (specialized segments form mouthparts and jointed appendages), gut (complete)
Level of Organization:	organ-system (tissues derived from all three primary germ layers), nervous system with brain, excretory system (green glands or Malphigian tubules), circulatory system (open), respiratory system (gills, tracheae, or book lungs)
Body Support:	exoskeleton of chitin
Life Cycle:	sexual reproduction (mostly dioecious with internal fertilization), larvae, metamorphosis, some parthenogenesis
Special Characteristics:	specialized segments, exoskeleton of chitin, metamorphosis, flight in some, complex behavior.

In numbers of both species and individuals, this is the largest animal phylum. There are more than 1 million species of arthropods, more in fact than in all the other animal phyla combined. Arthropods are believed to be evolved from annelid ancestors, the annelid cuticle having become thicker and harder, to form an **exoskeleton** (external skeleton) of **chitin** (chitin is a structural polysaccharide). Like the annelids, the arthropods are segmented, with a pair of jointed appendages per body segment.

The body is typically divided into three regions: the *head, thorax,* and *abdomen.* The mouthparts are modified appendages (legs). The coelom is filled with

324

the blood, or **hemolymph,** and connected to an *open circulatory system.* Arthropods possess a highly developed central nervous system and complex sense organs and behavior, features that are partly responsible for their evolutionary success and current abundance.

MATERIALS

Per student:
- dissecting scissors
- blunt probe or dissecting needle
- scalpel

Per student pair:
- dissection microscope
- preserved crayfish
- dissection pan
- grasshoppper (*Romalea*)

Per student group:
- living crayfish
- 5-gallon aquarium filled with pond or aquarium water
- black construction paper
- gooseneck lamp with dim light bulb
- several live millipedes (optional)

Per lab room:
- collection of preserved and mounted crustaceans
- collection of preserved and mounted insects
- demonstration of prepared slide of grasshopper mouth parts, whole mount
- collection of preserved centipedes
- collection of preserved millipedes
- collection of preserved and mounted arachnids
- large plastic bag for disposal of dissected specimens

PROCEDURE

A. Class Crustacea: Crustaceans

Some of our most prized food items are animals that belong to the class Crustacea. This group contains

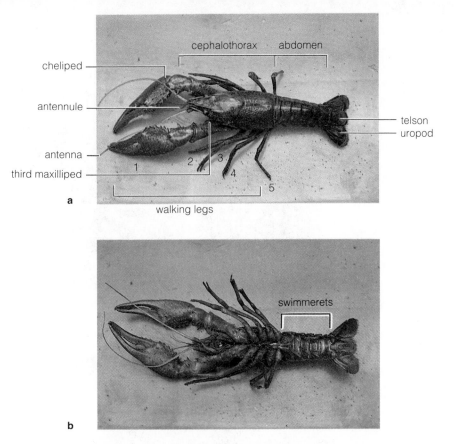

cheliped

antennule

antenna

third maxilliped

cephalothorax   abdomen

telson
uropod

1   2   3
4
5

a

walking legs

swimmerets

b

Figure 24-14 (**a**) Dorsal and (**b**) ventral views of the external anatomy of the crayfish. The five walking legs are numbered in (**a**). (Photos by D. Morton.)

shrimps, lobsters, crabs, and crayfish — and of considerably less culinary interest, water fleas, sand fleas, isopods (sow or pill bugs), ostracods, and barnacles. Most species are marine, but some live in fresh water. Most isopods occupy moist areas on land. For the most part, crustaceans are carnivorous, scavenging, or parasitic.

The head of a crustacean has two pairs of antennae and three pairs of mouth parts — one pair of mandibles and two pairs of maxillae. The exoskeleton is hardened dorsally and sometimes laterally to form a **carapace.** The appendages of the body are specialized for a wide variety of functions, and the eyes may be simple or compound.

Larger crustaceans often can adapt to the color of their background because of special pigment-containing cells, the *chromatophores,* in the outer body wall.

1. *Dissection of a crayfish.* The crayfish is a relatively large representative of the phylum Arthropoda whose internal anatomy is readily apparent in preserved specimens.

a. Examine the demonstration specimens of crustaceans. You will study in detail the external features of the crayfish (*Cambarus* or *Procambarus*) as a representative of the class Crustacea. Obtain a specimen and rinse it in water at the sink before you begin your examination. Place it in a dissection pan.

b. Notice that the body is divided into two major regions (fig. 24-14): an anterior *cephalothorax,* composed of a fused head and thorax, and a posterior *abdomen.* The hard outer cover of the *cephalothorax* is the protective **carapace.**

c. Note the large *compound eyes* on movable stalks, two large *antennae,* and two pairs of smaller *antennules.* The *mouth* is ventral (fig. 24-15c), surrounded by several pairs of appendages modified for handling food. There are five pairs of large *walking legs;* the first, called the *chelipeds,* bear large pincers. There are five pairs of abdominal appendages called *swimmerets.* The *uropods* are a pair of large, flattened lateral appendages near the end of the abdomen. The *telson* is an extension of the last abdominal segment, bearing the *anus* ventrally.

d. Crayfish exhibit **sexual dimorphism** (morphological differences between male and female). Females have a broad abdomen, five pairs of swimmerets of roughly the same size on the abdominal segments, and a *seminal receptacle* ventrally between the bases of the fourth and fifth pairs of walking legs. Males have a narrower abdomen, the front two pairs of swimmerets enlarged for copulation and transferring sperm to the female, and openings of the *vasa deferentia* (sperm ducts) at the base of each fifth walking leg. Observe both sexes.

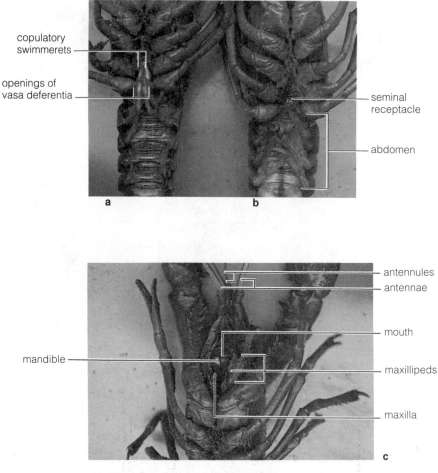

Figure 24-15 Ventral views of (**a**) a male, (**b**) a female, and (**c**) mouthparts of the crayfish. (Photos by D. Morton.)

e. Examine the appendages on one side of the body, listed (in the next column) from anterior to posterior with their functions.

f. Locate the **gills** within the branchial chambers by carefully cutting away the lateral flaps of the carapace with your scissors (fig. 24-16a). The gills are feathery structures containing blood channels that function in gaseous exchange. The second maxilla creates a current of water that flows past the gills, bringing oxygen in contact with the blood supply and carrying away carbon dioxide.

g. With scissors, superficially cut forward from the rear of the carapace to just behind the eyes. With your scalpel, carefully separate the hard carapace from the thin, soft, underlying tissue. Next, remove the gills to reveal the internal organs (fig. 24-16b).

i. *antennule* — contains a statocyst, an organ of balance, in the basal segment.

ii. *antenna* — for chemoreception.

iii. *mandible* — modified as a functional jaw.

iv. *maxillae* (first and second) — modified for handling food, the second for creating a current for gaseous exchange in the gills.

v. *maxillipeds* (first, second, and third) — modified for handling food.

vi. *walking legs* (one through five) — for defense (the cheliped), movement.

vii. *swimmerets* (one through five) — the first two are modified in males for copulation and transferring sperm to the female; young are brooded among these structures on the female.

viii. *uropod* — modified to form (with the *telson*) the tail fan, used for swimming backward.

a

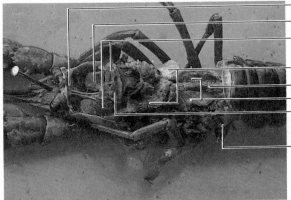

brain
green gland
(mandibular muscle)

digestive gland
intestine
flexor muscles
stomach

extensor muscle

b

Figure 24-16 Internal anatomy of the crayfish with (**a**) lateral flap of carapace removed and (**b**) carapace, gills, and heart removed. (Photos by D. Morton.)

h. Note the two longitudinal bands of *extensor muscles* that run dorsally through the thorax and abdomen. In the abdomen, find the large *flexor muscles* lying below the extensor muscles. What is the function of these two sets of muscles?

_____

_____

i. Locate the small, membranous heart or its location just posterior to the stomach. Remember, the circulatory system is open; from the arteries, blood flows into open spaces, or *sinuses*, before returning to the heart through openings in the wall of this organ.

j. Identify the organs of the upper digestive tract. The *mouth*, hidden by several oral appendages, leads to the tubular *esophagus*, which leads to the *cardiac stomach*. At the border between the cardiac stomach and the *pyloric stomach* is the *gastric mill*, a grinding apparatus composed of three chitinous teeth. What role in digestion do you think this structure plays?

_____

_____

_____

k. Note the large *digestive gland*. This organ secretes digestive enzymes into the cardiac stomach and takes up nutrients from the pyloric stomach. Absorption of nutrients from the tract continues in the intestine, which runs from the abdomen to the *anus*.

l. Locate the *green glands,* the excretory organs situated ventrally in the head region near the base of the antennae. A duct leads to the outside from each green gland.

m. The gonads lie beneath the heart. They will be obvious if your specimen was obtained during the reproductive season. If not, they will be small and difficult to locate. In a female, try to find the *ovaries* just beneath the heart. In the male, two white *testes* will occupy a similar location.

n. Remove the organs in the cephalothorax to expose the *ventral nerve cord* (fig. 24-16b). Observe the segmental ganglia and their paired lateral nerves. Trace the nerve cord forward to locate the *brain*. Note the nerves leading from the brain toward the eyes, antennules, antennae, and mouth parts. Dispose of your dissected specimen in the large plastic bag provided for this purpose.

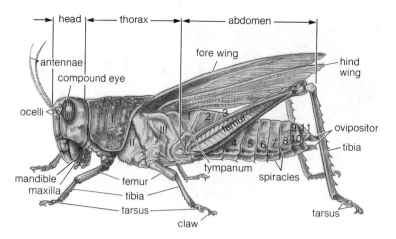

Figure 24-17 External anatomy and body form of a grasshopper, a typical insect. (After Jensen, Heinrich, Wake, and Wake, 1979.)

2. *Crayfish behavior.*

a. Place a crayfish in a small aquarium containing pond or aquarium water. Cover the aquarium to prevent light from entering, except for one end. Shine a dim light toward the uncovered end of the aquarium and describe the crayfish's response.

_____

_____

b. Remove the crayfish from the aquarium. Threaten the animal with a large object (for example, a notebook) and describe its behavior.

_____

_____

Approach the animal from a variety of angles, including from the side and behind it. Based on the animal's response, would you say that its eyes are effective in detecting movement from all directions? Explain.

_____

_____

c. Repeatedly place the crayfish on a moderate incline and observe its response and the orientation or movement it exhibits. Is its response positive (down) or negative (up) to gravity?

_____

_____

_____

B. Class Insecta: Insects

Of the more than 1 million species of arthropods, more than 850,000 of them are insects. These animals occupy virtually every kind of terrestrial habitat, and they have invaded fresh waters as well. A number of the typical arthropod adaptations have contributed to their success, including the chitinous exoskeleton that physically protects the internal organs, keeps foreign substances out of the body, and prevents loss of water. Moreover, many have an additional adaptation that many biologists believe has been especially responsible for their proliferation — flight. A currently popular hypothesis suggests that wings became adapted for flight secondarily, with their initial function as heat-absorbing devices. Whatever the initial function of wings, insects were the first organisms to fly and thus were able to exploit a variety of opportunities not available to other animals.

1. Examine members of the class Insecta on demonstration in the lab.

2. Examine the grasshopper, *Romalea,* described here as a representative insect. The general structure of insects is relatively uniform. The body consists of a *head, thorax,* and *abdomen* (fig. 24-17).

An insect's body is covered by an exoskeleton. The thorax is composed of three segments, each of which bears a pair of legs. The middle and back segments can each bear a pair of wings. The abdomen, unlike that in crustaceans, beans no appendages. Evidence of the respiratory system is located on the sides of the thorax and abdomen in the form of *spiracles,* tiny openings into the **tracheae,** or breathing tubes, which course throughout the body and connect directly with the tissues.

3. Note that the head of the grasshopper contains a pair of compound eyes between which are three simple eyes, the *ocelli,* light-sensitive organs that do not form images.

4. Locate the single pair of antennae that distinguishes the insects from the crustaceans (which have two) and the arachnids (which have none). Feeding appendages consist of a pair of *mandibles* and two pairs of *maxillae,* the second pair fused together to form the lower lip, the *labium.* The upper lip, the *labrum,* covers the mandibles. Near the base of the labium is a

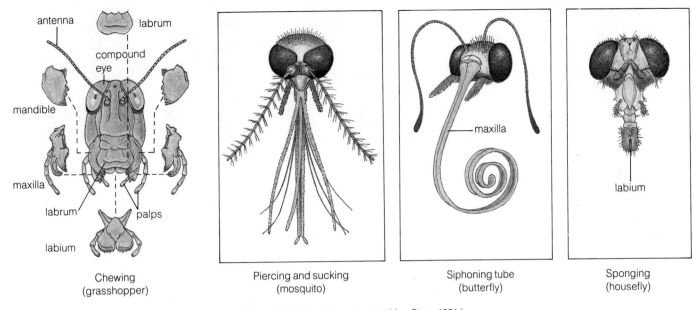

antenna

labrum

compound eye

mandible

maxilla

labrum

palps

labium

Chewing (grasshopper)

Piercing and sucking (mosquito)

maxilla

Siphoning tube (butterfly)

labium

Sponging (housefly)

Figure 24-18 Insect headparts. (After Starr, 1991.)

tonguelike process called the *hypopharynx*. Mouthparts are extremely variable in insects, with modifications for chewing, piercing, siphoning, and sponging (fig. 24-18).

5. Look at the demonstration slide of a whole mount of grasshopper mouthparts. Locate and label a mandible, a maxilla, the labium, and the labrum on the illustration below (photo courtesy Ripon Microslides Inc.). The middle mouthpart is the hypopharynx.

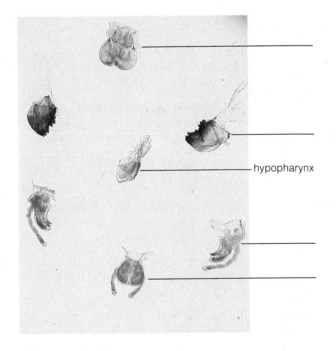

hypopharynx

C. Class Chilopoda: Centipedes

Centipedes are predaceous arthropods adapted for running. The body is flattened dorsoventrally, with one pair of legs per segment, except for the head and the rear two segments (fig. 24-19a).

The first segment of the body bears a pair of legs modified as *poison claws* for seizing and killing prey. In some tropical species as long as 20 cm, the poison can be dangerous to humans.

Centipedes are swift arthropods and live under stones or the bark of logs. They may be abundant on the forest floor in your area. Their prey consists of other arthropods, worms, and mollusks. The common house centipede eats roaches, bedbugs, and other insects.

Examine several different species of centipedes.

D. Class Diplopoda: Millipedes

Although superficially similar, the millipedes and centipedes differ in many ways. The millipedes are typically cylindrical in body form. All segments, except those of the short thorax, bear two pairs of legs. There are no poison claws (fig. 24-19b).

Millipedes are very slow arthropods that live in dark, moist places. You may find them in the same habitat as centipedes. They scavenge on decaying organic matter. When handled or disturbed, they often curl up into a ball and secrete a noxious fluid from their scent glands as a means of defense.

1. Examine several different species of millipedes.

2. If your instructor has provided live specimens, hold one to your nose to see if you can detect a smelly secretion.

a

b

c

Figure 24-19 Preserved representatives of a centipede (**a**), millipede (**b**), and a live spider (**c**). (Photos by D. Morton [a, b] and J. H. Howard [c].)

E. Class Arachnida: Spiders, Scorpions, Ticks, and Mites

The body of spiders and scorpions consists of a *cephalothorax* and *abdomen*. In ticks and mites, these parts are fused. The cephalothorax bears six pairs of appendages, the rear four of which are walking legs (fig. 24-19c). How many pairs of walking legs do insects have?

_____

Arachnids have no true mandibles or antennae, and the eyes are simple. In spiders, the first pair of appendages bear fangs. Each fang has a duct that is connected to a poison gland. The second pair of appendages are used to chew and squeeze their prey.

They are also sensory and are used by males to store and transfer to females their sperm.

Most, but not all, spiders spin webs to catch their prey. They secrete digestive enzymes into the prey's (usually an insect's) body, and the liquified remains are sucked up.

Examine the assortment of arachnids on display in the lab. How do ticks and mites obtain nourishment?

_____

_____

_____

_____

PRE-LAB QUESTIONS

_____ 1. The protostomes are animals whose (a) stomach is in front of the crop, (b) mouth is covered by a fleshy lip, (c) blastopore becomes a mouth, (d) digestive tract is lined by mesoderm.

_____ 2. One of the two major distinguishing characteristics of mollusks is (a) the presence of three body regions, (b) the mantle, (c) segmentation of the body, (d) jointed appendages.

_____ 3. Earthworms exhibit segmentation, defined as the (a) division of the body into a series of similar segments, (b) presence of a true coelom, (c) difference in size of the male and female, (d) presence of a "head" equipped with sensory organs.

_____ 4. The copulatory organ of the earthworm is the (a) penis, (b) clitellum, (c) gonopodium, (d) vestibule.

_____ 5. Leeches belong to the phylum (a) Arthropoda, (b) Mollusca, (c) Annelida, (d) Cnidaria.

_____ 6. The animal phylum with the most species is (a) Mollusca, (b) Annelida, (c) Arthropoda, (d) Platyhelminthes.

_____ 7. Sexual dimorphism, as seen in the crayfish, is (a) the presence of male and female individuals, (b) the production of eggs and sperm by the same individual, (c) another term for copulation, (d) the presence of observable differences between males and females.

_____ 8. The grinding apparatus of the digestive system of the crayfish is the (a) oral teeth, (b) gizzard, (c) pharyngeal jaw, (d) gastric mill.

_____ 9. The insects were the first organisms to (a) show bilateral symmetry, (b) exhibit segmentation, (c) fly, (d) develop lungs.

_____ 10. Like the insects, the arachnids (spiders and so on) have (a) three pairs of walking legs, (b) one pair of antennae, (c) true mandibles, (d) an abdomen.

EXERCISE 24
_____

Mollusks, Segmented Worms, and Joint-Legged Animals

POST-LAB QUESTIONS

1. Indicate the differences between protostomes and deuterostomes, and list the phyla of animals in each group.

2. Explain what causes the growth ridges found on the shells of bivalves.

3. Define *segmentation* and indicate how this phenomenon is exhibited by the annelids, arthropods, and *your* body.

4. Define *sexual dimorphism,* describe it in the crayfish, and list two other animals that exhibit this phenomenon.

5. How has flight been at least partly responsible for the success of the insects?

6. List several differences between insects and arachnids.

7. Identify and label the following structures.

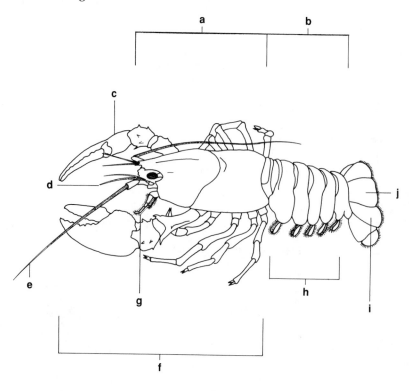

8. Write the phylum and choose the appropriate description for each animal (or group of animals) listed in the following table.

Symmetry: none, radial, or bilateral

Level of Organization: cell-specialization, tissue, or organ-system

Body Cavity: acoelomate, pseudocoelomate, or coelomate

Gut: none, incomplete, complete

Animal	Phylum	Symmetry	Level of Organization	Body Cavity	Gut
sponge				none	
Hydra or jellyfish				none	
flatworm					
roundworm					
rotifer					
clam, snail, or octopus					
segmented worm					
crayfish, insect, or spider					

Echinoderms, Hemichordates, and Invertebrate Chordates

OBJECTIVES

After completing this exercise you will be able to:

1. define *deuterostomes, endoskeleton, water-vascular system, madreporite, tube foot, notochord, gill slits, dorsal hollow nerve cord, invertebrate;*

2. describe the natural history of members of phylum Echinodermata, phylum Hemichordata, and the invertebrate members of phylum Chordata;

3. identify representatives of the echinoderm classes (Asteroidea, Ophiuroidea, Echinoidea, Holothuroidea, and Crinoidea);

4. compare and contrast acorn worms (phylum Hemichordata) and the invertebrate chordates (subphyla Cephalochordata and Urochordata);

5. identify structures (and indicate associated functions) of the representatives of these phyla and subphyla.

INTRODUCTION

The echinoderms, hemichordates, and chordates are **deuterostomes,** animals whose blastopore develops into an anus. Although echinoderms and chordates look quite different, chordates are thought to have evolved from the bilaterally symmetrical larvae of ancestral echinoderms some 600 million years ago. The hemichordates are a small phylum whose members have both echinoderm and chordate characteristics.

> ### CAUTION
> Preserved specimens are kept in a formalin-based or other preservative solution. Wash any part of your body exposed to this solution with copious amounts of water. If preservative solution is splashed into your eyes, wash them with the safety eyewash bottle for fifteen minutes.

I. Phylum Echinodermata: Echinoderms

Habitat:	aquatic, marine
Body Arrangement:	radial symmetry (larvae are bilaterally symmetrical), coelomate, no well-defined head, no segmentation, gut (usually complete)
Level of Organization:	organ-system (tissues derived from all three primary germ layers), circular nervous system with radial nerves, reduced circulatory system, no respiratory and excretory systems, unique water-vascular system, endoskeleton (mesoderm)
Body Support:	endoskeleton of calcium carbonate
Life Cycle:	sexual reproduction (mostly dioecious with external fertilization), larvae, metamorphosis, high regenerative potential
Special Characteristics:	radial symmetry, no segmentation, water-vascular system, endoskeleton

Echinoderm means "spiny skin." Members of this phylum include the sea stars, brittle stars, sea urchins, sea cucumbers, sea lilies, and feather stars. These animals are all *marine*, living on the bottom of both shallow and deep seas. Their feeding methods range from trapping organic particles and plankton (sea lilies and feather stars) to scavenging (sea urchins) and predatory behavior (sea stars).

The echinoderms exhibit five-part *radial symmetry* and a calcareous (containing calcium carbonate) **endoskeleton** (internal skeleton) composed of many small plates. Much of the *coelom* is taken up by the **water-vascular system,** important to movement, attachment, respiration, excretion, food handling, and sensory perception. We will examine the sea star as a representative echinoderm.

MATERIALS

Per student:
- dissecting scissors
- blunt probe or dissecting needle

Per student pair:
- dissection microscope
- preserved sea star
- dissection pan
- dissection pins
- preserved slide of a cross section of a sea star arm (ray)
- compound microscope, lens paper, a bottle of lens-cleaning solution (optional), and a lint-free cloth (optional)

Per lab room:

- collection of preserved or mounted brittle stars
- collection of sea urchins, sea biscuits, and sand dollars (preserved specimens and skeletons)
- collection of preserved sea cucumbers
- collection of preserved or mounted feather stars and sea lilies
- large plastic bag for disposal of dissected specimens

PROCEDURE

A. Class Asteroidea: Sea Stars

The sea stars are familiar occupants of the sea along shores and coral reefs. They are slow-moving animals that sometimes gather in large numbers on rocky substrata. Many are brightly colored. Large sea stars may be more than 1 meter in size from the tip of one arm to the tip of the farthest arm from it.

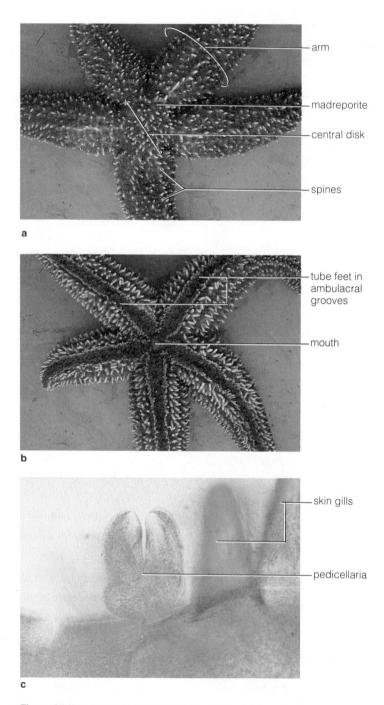

Figure 25-1 (**a**) Aboral and (**b**) oral views of a sea star. (**c**) shows a pedicellaria and skin gills (186 ×). (Photos by D. Morton.)

1. Obtain a preserved specimen of a sea star and keep your specimen moist with water in a dissection pan. Note the central disk on the *aboral* side — the side without the mouth — the five *arms,* and the **madreporite,** a light-colored calcareous sieve near the edge of the disk between two arms (fig. 25-1a). The madreporite is the opening of the water-vascular system.

2. With a dissection microscope, note the many *spines* scattered over the surface of the body. Near the base of the spines are many small pincerlike structures, the *pedicellariae* (fig. 25-1c). These structures grasp objects that land on the surface of the body. For what might these be an adaptation?

_____

_____

_____

Also among the spines are many soft, hollow skin gills that communicate directly with the coelom and function in the exchange of gases and excretion of ammonia (a nitrogen-containing metabolic waste).

3. Locate the *mouth* on the *oral* side (fig. 25-1b). Note that an *ambulacral groove* extends from the mouth down the middle of the oral side of each arm. Numerous **tube feet** extend from the water-vascular system and occupy this groove. Each tube foot consists of a bulblike structure attached to a sucker (fig. 25-2). The amount of water in the bulb of a tube foot determines whether it applies suction to or releases suction from the substratum. The animal moves by alternating the suction and release mechanisms of the tube feet. The suction created by the tube feet also is used to adhere to the shells of bivalves as the sea star uses muscular action to pry the shells open to get at their soft insides. The tube feet, along with the skin gills, also function in the exchange of gases and excretion of ammonia.

4. With your dissecting scissors, cut across the top of an arm about 1 cm from the tip. Next, cut out a rectangle of spiny skin by carefully cutting along each side of the arm to the central disk and then across the top of the arm at the edge of the disk. Observe the hard, calcareous plates of the endoskeleton as you cut. Remove the rectangle of skin to uncover the large *coelom,* which contains the internal organs (fig. 25-3).

5. Cut around the madreporite to remove the upper portion of the body wall of the central disk. The *mouth* connects with an extremely short esophagus, which leads to the pouchlike *cardiac stomach.* The cardiac stomach opens into the upper *pyloric stomach.* A slender, short *intestine* leads from the upper side of the stomach to the *anus.* Find the two green, fingerlike *digestive glands* in each arm, which produce digestive enzymes and deliver them to the pyloric stomach.

6. Identify the dark *gonads* that are located near the base of each arm. The sexes are separate but are difficult to distinguish, except by microscopic examination.

7. The water-vascular system is unique to echinoderms. The madreporite leads to a short *stone canal,* which in turn leads to the *circular canal* surrounding the mouth. Five *radial canals* lead from the circular canal into the ambulacral grooves. Each radial canal connects by short side branches with many pairs of tube feet. Locate the stone canal.

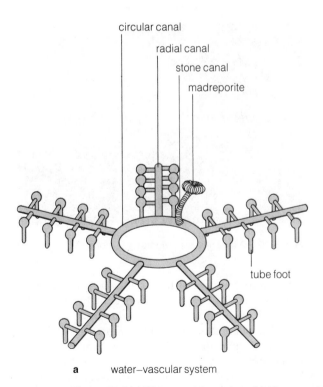

circular canal
radial canal
stone canal
madreporite

tube foot

a   water–vascular system

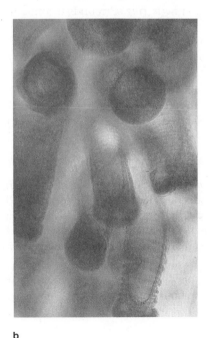

b

Figure 25-2 (a) Water-vascular system. **(b)** Close-up of tube feet (74×). (Photo by D. Morton.)

Figure 25-3 Internal anatomy of a sea star. (Photo by D. Morton.)

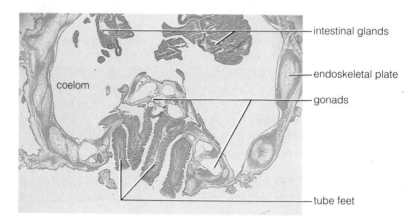

Figure 25-4 Cross section of sea star arm (9 ×). (Photo courtesy Ripon Microslides, Inc.)

8. The nervous system (not shown in fig. 25-3) is simple. A circular *nerve ring* surrounds the mouth, and a *radial nerve* extends from this into each arm, ending at a light-sensitive eyespot. There are no specific excretory organs. Dispose of your dissected specimen in the large plastic bag provided for this purpose.

9. Examine a prepared slide of a cross section of an arm with the compound microscope. Identify the structures labeled in figure 25-4.

B. Other Echinoderms

1. *Class Ophiuroidea* includes the brittle stars (fig. 25-5). These echinoderms have slender arms and move relatively rapidly, for echinoderms! In life, the rays are much more flexible than those of the sea stars. Examine preserved or plastic-mounted specimens of brittle stars.

2. *Class Echinoidea* contains the sea urchins, animals that lack arms but that contain long, movable spines on a compact skeleton (*test*) (fig. 25-6).

Figure 25-5 Dried brittle star. (Photo by D. Morton.)

The sea urchins ingest food by means of a complex structure, *Aristotle's lantern,* which contains teeth. The sand dollars are also members of this class. They have flat, disk-shaped bodies. Both sea urchins and sand dollars, despite their lack of arms, exhibit five-part radial symmetry and move by means of tube feet. Study preserved specimens and the tests (fig. 25-7) of members of this class.

Figure 25-6 (**a**) Aboral and (**b**) oral views of a sea urchin. (Photo by D. Morton.)

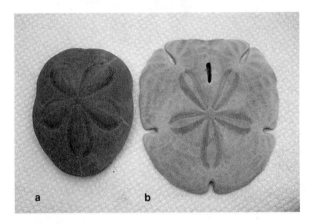

Figure 25-7 (**a**) Sea biscuit and (**b**) sand dollar tests. (Photo by D. Morton.)

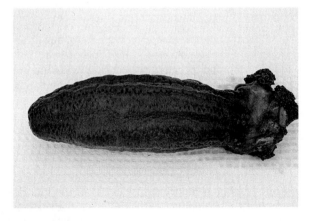

Figure 25-8 Preserved sea cucumber. (Photo by D. Morton.)

3. *Class Holothuroidea* are the sea cucumbers (fig. 25-8). No spines are present. A collection of tentacles at one end surrounds the mouth. In some forms, the tube feet occur over the entire surface of the body. Sea cucumbers move in wormlike fashion as a result of the contractions of two layers of circular and longitudinal muscles. Examine preserved sea cucumbers.

Figure 25-9 Sea lily.

4. *Class Crinoidea* contains the graceful, flowerlike feather stars and sea lilies (fig. 25-9). These stationary animals attach to the substratum by a long stalk. The arms are featherlike; each contains a ciliated groove and tentacles (derived from tube feet) to direct food to the mouth. There can be from five to over two hundred arms, some of which can reach 35 cm long. Examine preserved or plastic-mounted specimens of this class.

II. Phylum Hemichordata: Acorn Worms

Hemichordates like the acorn worms are related to members of the phylum Chordata, but just what the relationship is and how close it is are not known. The key chordate characteristics are poorly developed, and the very presence of one of them — the notochord — has been a subject of hot debate. Chordate characteristics found in hemichordates are a *dorsal hollow nerve cord* and *gill slits.* There is also a ventral nerve cord, a structure common in many of the animal phyla you have already studied.

Acorn worms are *long, slender, and wormlike. Most have been found in shallow seawater.* They obtain food by burrowing in mud and sand or *filter feeding.* Evidence of the former activity can be seen on exposed tidal flats where mud that has passed through their bodies is left in numerous coiled ropelike piles.

MATERIALS

Per student:
• blunt probe

Per student pair:
• dissection microscope
• preserved acorn worm

PROCEDURE

1. Using a dissection microscope, study the preserved acorn worm, *Balanoglossus*. Anteriorly, the acorn worms have a tough, flexible, muscular snout or *proboscis* (fig. 25-10), which is used as either a burrowing organ or a sticky trap for suspended particles of organic matter and microscopic organisms.

2. Behind the proboscis find the thick *collar,* the "cap" of the acorn, which contains the *dorsal hollow nerve cord.* Behind the collar is the trunk. Locate in the most anterior portion of this region of the body the *gill slits.* Identify the *anus,* which opens at the end of the trunk.

III. Phylum Chordata: Chordates

Habitat:	aquatic, terrestrial (land and air)
Body Arrangement:	bilateral symmetry, coelomate, head and postanal tail, segmentation, gut (complete)
Level of Organization:	organ-system (tissues derived from all three primary germ layers), nervous system with brain, circulatory system (closed with a ventral heart), organ systems highly developed, many have an endoskeleton (mesoderm)
Body Support:	most have an endoskeleton of cartilage or bone
Life Cycle:	sexual reproduction (mostly dioecious with internal or external fertilization), some larvae, some metamorphosis
Special Characteristics:	notochord, gill slits, and dorsal hollow nerve cord at some stage of life cycle

The word *chordate* refers to one of the three major diagnostic features of members of this phylum, the **notochord.** This structure is a dorsal, flexible rod that provides support for most of the length of the body during at least some portion of the life cycle. In addition to the notochord, chordates possess **gill slits.** Gill slits persist through adulthood in some chordates (for example, fish), and they are present during some portion of the life cycle in all chordates. The third chordate feature is a **dorsal hollow nerve cord.**

MATERIALS

Per student:

• blunt probe

Per student pair:

• dissection microscope

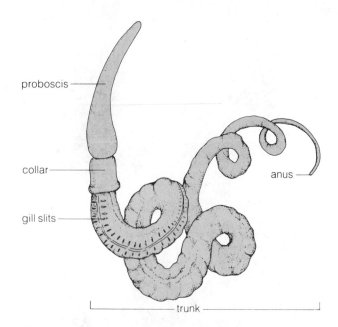

Figure 25-10 General body structure of an acorn worm.

• compound microscope, lens paper, a bottle of lens-cleaning solution (optional), and a lint-free cloth (optional)
• dissection pan
• preserved and mounted sea tunicates (sea squirts)
• preserved and mounted Amphioxus specimens
• whole mount of Amphioxus
• cross section of Amphioxus in region of the pharynx

PROCEDURE

A. Subphylum Urochordata: Tunicates

The invertebrate chordates include members of the subphlya Urochordata and Cephalochordata. They are **invertebrates** because they lack a vertebral column. All of the animal phyla you have studied so far are also invertebrates.

Tunicates are also called sea squirts. Both common names are descriptive of obvious features of these animals. A *leatherlike "tunic"* covers the adult, and water is squirted out of an excurrent siphon (fig. 25-11).

Tunicate larvae resemble tadpoles in general body form and have a notochord confined to the tail; the notochord degenerates when the larva becomes an adult tunicate. The larva also has a dorsal hollow nerve cord with a rudimentary brain and sense organs. These structures undergo reorganization into a nerve ganglion and nerve net in the adult.

1. Examine preserved specimens and a plastic mount of the sea squirt with a dissection microscope. These animals are small, inactive, and very common marine organisms that inhabit coastal areas of all oceans. The adult is a filter-feeder, capturing organic particles into

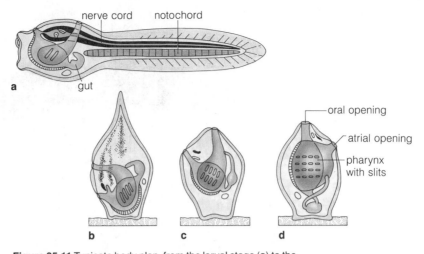

Figure 25-11 Tunicate body plan, from the larval stage (**a**) to the adult (**d,e**). The tadpole-like larva swims for only a few minutes or days until it locates a suitable living site. It attaches its head to a substrate (**b**), and metamorphosis begins; the tail, notochord, and most of the nervous system are resorbed (recycled to form new tissues). The slits in the pharynx multiply (**c**). Organs become rotated until the openings through which water enters and leaves are directed away from the substrate (**d**). (After Starr, 1991.)

an incurrent siphon by ciliary action. Individuals either float freely in the water (singly or in groups) or are sessile, attached to the bottom as branching individuals or colonies.

2. Water enters a tunicate through the *incurrent siphon*. It travels into the *pharynx*, where it seeps through *gill slits* to reach a chamber, the *atrium*. The water eventually exits through the *excurrent siphon*. Back in the pharynx, food particles are trapped in sticky mucus and then passed to the digestive tract. Undigested materials are discharged from the anus into the atrium, to be expelled with water out the excurrent siphon. Identify as many of these structures as possible.

B. Subphylum Cephalochordata: Lancelets

The *lancelets* are distributed worldwide and are especially abundant in coastal areas with *warm, shallow waters*. Amphioxus (fig. 25-12) is the commonly studied representative of this subphylum. The word *Amphioxus* means "sharp at both ends," and this is descriptive of the *elongate, fishlike body form* of cephalochordates.

1. Examine a preserved Amphioxus with the dissection microscope. The lancelets are translucent, with a low, continuous *dorsal* and *caudal* (tail) *fin*. They reach 50–75 mm in length as adults. Despite their streamlined appearance, these organisms are not very active. They spend most of their time buried in the sand with their heads projecting while they filter organic particles from the water.

2. Feeding is similar to that described above for tunicates. Obtain a plastic mount of Amphioxus. Referring to figures 25-12 and 25-13a, trace the flow of water from

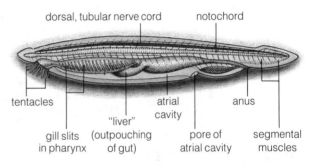

Figure 25-12 Amphioxus, showing internal features. (After Starr, 1991.)

the *mouth* (surrounded by *tentacles*), through the roughly 150 *gill slits* in the *pharynx* of the *gut*, into the *atrium*, and out of the *pore* of the atrial cavity.

3. Now describe the capture of food and the path it takes through the digestive system, as was done previously for the tunicates. Unlike the tunicates, the *anus* in Amphioxus opens externally.

_____

_____

_____

4. Find the flexible notochord extending nearly the full length of the individual; this structure persists into the adult stage. There is a small brain, with a *dorsal hollow nerve cord* that bears light-sensitive cells (the *eyespot*) at its anterior end. Locate the *segmental muscles*.

5. Examine a cross section of Amphioxus with a compound microscope and identify the structures indicated in figure 25-13b.

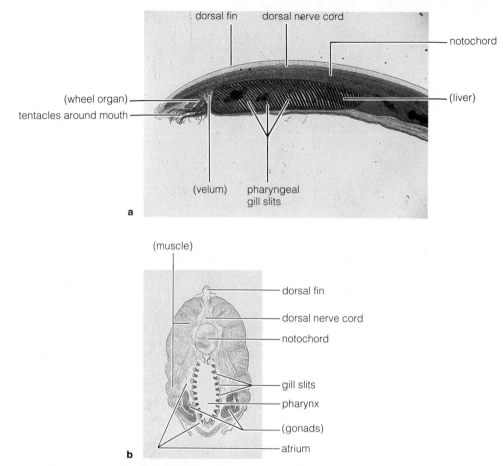

dorsal fin   dorsal nerve cord

notochord

(wheel organ)

(liver)

tentacles around mouth

(velum)   pharyngeal
gill slits

a

(muscle)

dorsal fin

dorsal nerve cord

notochord

gill slits

pharynx

(gonads)

atrium

b

Figure 25-13 (**a**) Whole mount (9 ×) and (**b**) cross section of
Amphioxus in region of pharynx (12 ×). (Photos courtesy
Ripon Microslides, Inc.)

PRE-LAB QUESTIONS

_____ 1. The deuterostomes are animals whose
(a) blastopore becomes a mouth, (b) blastophore becomes an anus, (c) digestive system opens into a
cloaca, (d) circulatory systems are open.

_____ 2. The skin gills of sea stars are organs (a) for
exchange of gases and excretion of ammonia, (b) of
defense, (c) for movement, (d) that produce the skeletal elements.

_____ 3. The tube feet of a sea star (a) bear tiny toes,
(b) are located only at the tips of the arms, (c) function in movement, (d) protect the organism from
predatory fish.

_____ 4. Aristotle's lantern is a tooth-bearing structure
of the (a) sea stars, (b) tunicates, (c) acorn worms,
(d) sea urchins.

_____ 5. The madreporite, stone canal, circular canal,
and radial canals of a sea star are structures of the
(a) nervous system, (b) water-vascular system,
(c) digestive system, (d) excretory system.

_____ 6. The endoskeleton of echinoderms is made
of (a) cartilage, (b) bone, (c) calcium carbonate,
(d) chitin.

_____ 7. Which of the following is _not_ a major diagnostic feature of the chordates? (a) dorsal hollow
nerve cord, (b) notochord, (c) gill slits, (d) vertebral
column.

_____ 8. The tunicates and lancelets have an inner
chamber from which water is expelled. This chamber
is called the (a) intestine, (b) bladder, (c) atrium,
(d) nephridium.

_____ 9. Invertebrates (a) do not have a vertebral column, (b) have a vertebral column, (c) are all members
of the phylum Chordata, (d) do not include animals
that are members of the phylum Chordata.

_____ 10. Which one of the following animals is a
chordate? (a) sea star, (b) sea cucumber, (c) acorn
worm, (d) lancelet.

EXERCISE 25

Echinoderms, Hemichordates, and Invertebrate Chordates

POST-LAB QUESTIONS

1. Compare echinoderms and arthropods in terms of their similarities and differences.

2. List the structures that comprise the water-vascular system of a sea star.

3. Describe the various functions of tube feet.

4. Identify the following structures on this aboral view of the tip of a sea star arm.

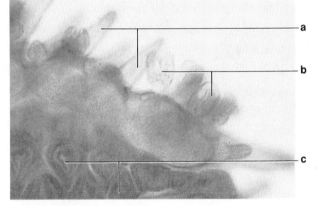

a _____

b _____

c _____

(74×). (Photo by D. Morton.)

5. What chordate structure is missing in hemichordates?

6. List the three major characteristics of chordates and tell how these are exhibited in adult tunicates and lancelets.

EXERCISE 26
Vertebrates

OBJECTIVES

After completing this exercise you will be able to:

1. define *vertebrate, vertebral column, cranium, vertebrae, cloaca, lateral line, placoid scale, operculum, atrium, ventricle, artery, vein, ectothermic, endothermic, viviparous;*

2. describe the basic characteristics of members of the subphylum Vertebrata;

3. identify representatives of the vertebrate classes, Agnatha, Chondrichthyes, Osteichthyes, Amphibia, Reptilia, Aves, and Mammalia;

4. identify structures (and indicate associated functions) of representatives of the vertebrate classes;

5. construct a dichotomous key to the animals (optional).

INTRODUCTION

Vertebrata is by far the largest chordate subphylum. The **vertebrates** have the *three basic characteristics of chordates* listed on page 338, plus a **vertebral column.** The dorsal hollow nerve cord has differentiated into a *brain* and a *spinal cord.* Both of these structures are protected by bones, the brain by the bones of the **cranium** (braincase) and the spinal cord by the **vertebrae,** the bones that make up the vertebral column.

Like other chordates, vertebrates are *bilaterally symmetrical* and *segmented.* In adults, segmentation is most easily seen in the musculature, vertebrae, and ribs. The body is typically divided into a *head, neck, trunk,* and *tail.* If appendages are present, they are paired, lateral *thoracic appendages* (pectoral fins, forelimbs, wings, and arms) and *pelvic appendages* (pelvic fins, hindlimbs, and legs), which support the body and serve in movement.

> ### CAUTION
> Preserved specimens are kept in a formalin-based or other preservative solution. Wash any part of your body exposed to this solution with copious amounts of water. If preservative solution is splashed into your eyes, wash them with a safety eyewash bottle for fifteen minutes.

MATERIALS

Per student:

- scalpel
- compound microscope, lens paper, a bottle of lens-cleaning solution (optional), and a lint-free cloth (optional)
- clean microscope slide and coverslip
- distilled water in dropping bottle
- forceps
- dissecting scissors
- blunt probe or dissecting needle
- dissection pins

Per student pair:

- dissection microscope
- preserved leopard frog
- dissection pan

Per student group:

- whole mount of a lamprey larva or ammocoete
- preserved sea lamprey
- preserved dogfish (shark)
- preserved yellow perch

Per lab room:

- collection of preserved cartilaginous fishes
- collection of preserved bony fishes
- collection of preserved amphibians
- skeleton of frog
- skeleton of human
- collection of preserved reptiles
- skeleton of snake
- turtle skeleton or shell
- an assortment of feathers
- collection of stuffed birds
- several field guides to the birds
- collection of stuffed mammals
- collection of mammalian placentas and embryos (in utero)
- large plastic bag for the disposal of dissected specimens

PROCEDURE

A. Class Agnatha: Jawless Fishes

The *agnathans* were the first vertebrates to evolve. They include the present-day lampreys and hagfishes. *Agnatha* means "without jaws," a condition characteristic of these *jawless fishes.* They have a cartilaginous (made of cartilage), primitive skeleton without a cranium and with incomplete vertebrae.

The hagfishes are marine, and the lampreys are represented by both marine and freshwater species.

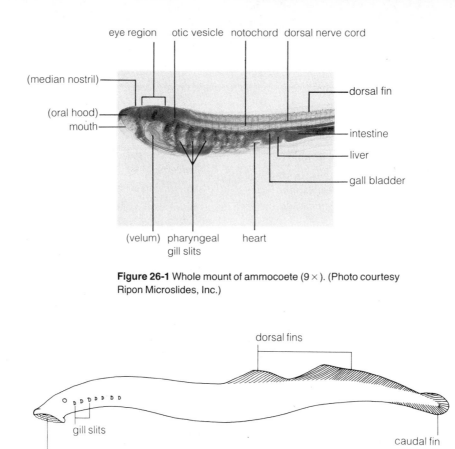

eye region   otic vesicle   notochord   dorsal nerve cord

(median nostril)

(oral hood)

mouth

dorsal fin

intestine

liver

gall bladder

(velum)   pharyngeal
gill slits

heart

Figure 26-1 Whole mount of ammocoete (9 ×). (Photo courtesy Ripon Microslides, Inc.)

dorsal fins

gill slits

caudal fin

mouth

Figure 26-2 Side view of a sea lamprey to show external features.

Both hagfishes and lampreys feed on the blood and tissue of fishes, rasping wounds in their sides. Hagfish actually burrow into and often through the bodies of their prey.

A landlocked population of the sea lamprey is infamous for having nearly decimated the commercial fish industry in the Great Lakes. This industry was restored recently by a vigorous control program that targeted lamprey larvae.

1. With a dissection microscope observe a whole mount of a lamprey larva or ammocoete. Sea lampreys spawn in freshwater streams. Ammocoetes hatch from their eggs and, after a period of development, burrow into the sand and mud. They are long-lived, metamorphosing into adults after as long as seven years. Because of their longevity and dissimilar appearance, adult lampreys and ammocoetes were long considered to be separate species.

The ammocoete is considered by many to be the closest living form to ancestral chordates. Unlike *Amphioxus*, it has an eye, an ear (otic vesicle), and a heart, as well as several other typical vertebrate organs. There are actually two median eyes. Identify the structures labeled in figure 26-1.

2. Examine a preserved sea lamprey (fig. 26-2). Note its slender, rounded body. The skin of the lamprey is soft and lacks scales. Look at the *round, suckerlike*

mouth, inside of which are circular rows of *horny, rasping teeth* and a deep, *rasping tongue.* Although there are no lateral, paired appendages, there are two *dorsal fins* and a *caudal* (tail) *fin.* Note the *gill slits.*

B. Class Chondrichthyes: Cartilaginous Fishes

The first vertebrates to evolve jaws and paired appendages, according to the fossil record, were the heavily armored *placoderms.* These fishes arose about 420 million years ago. The evolution of jaws and paired appendages was a critical event in the history of vertebrate evolution, allowing predation on larger and more active prey. Current hypotheses suggest that the placoderms ultimately gave rise to the cartilaginous and bony fishes.

Members of the class *Chondrichthyes* are *jawed fishes* with *cartilaginous skeletons.* This mostly marine group includes the carnivorous skates, rays, and sharks. *Paired appendages* are present in these animals.

1. Examine the assortment of cartilaginous fishes on display.

2. Examine a preserved dogfish (shark) (fig. 26-3). How many dorsal fins are there?

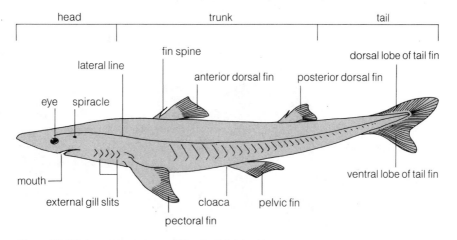

Figure 26-3 External appearance of the dogfish (shark), *Squalus acanthias*. (After S. Wischnitzer, *Atlas and Dissection Guide for Comparative Anatomy*, Third Edition. Copyright © 1967, 1972, 1979 W.H. Freeman and Company. Used by permission.)

3. Notice the front paired appendages, the *pectoral fins,* and the back pair, the *pelvic fins.* There are five to seven pairs of naked gill slits in cartilaginous fishes, six in the dogfish. The most anterior gill slit is called a *spiracle* and is located just behind the eye. The *tail fin* has a dorsal lobe larger than the ventral one. In the male, the pelvic fins bear *claspers,* thin processes for transferring sperm to the oviducts of the female. Examine the pelvic fins of a female and a male. Also examine the nearby *cloaca.* The **cloaca** is the terminal chamber that receives the products of the digestive, excretory, and reproductive systems.

4. Locate the *nostrils,* which open into blind olfactory sacs; they do not connect with the pharynx, as your nostrils do. They function solely in olfaction (smell) in the cartilaginous fishes. The *eyes* are effective visual organs at short range and in dim light. There are no eyelids.

5. Find the dashed line that runs along each side of the body. This is called the **lateral line** and functions in the detection of vibrations in the water. It consists of a series of minute canals perpendicular to the surface that contain sensory hair cells. When the hairs are disturbed, a nerve impulse is initiated that ultimately results in the detection and interpretation of the disturbance.

6. The body is covered by **placoid scales,** toothlike outgrowths of the skin. Run your hand from head to tail along the length of the animal. How does it feel?

_____

Now run your hand in the opposite direction along the animal. How does it feel this time?

_____

Cut out a small piece of skin with your scalpel and examine it with the dissection microscope. Draw what you see in figure 26-4.

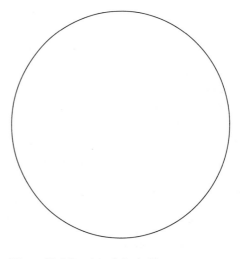

Figure 26-4 Drawing of shark skin.

C. Class Osteichthyes: Bony Fishes

The bony fishes, the fishes with which you are most familiar, inhabit virtually all the waters of the world and are the largest vertebrate group. They are economically important, commercially and as game species. The *skeleton* is at least *partly ossified (bony),* and the flat *scales* that cover at least some of the surface of most bony fishes are *bony* as well. Gill slits are housed in a common chamber covered by a bony movable flap, the **operculum.**

1. Examine the assortment of bony fishes on display.

2. Examine a yellow perch as a representative advanced bony fish (fig. 26-5). Notice that the *nostrils* are double, leading into and out of *olfactory sacs.* The eyes are similar to those in sharks, with no eyelids. Pry open the operculum to see the organs of respiration, the *gills.* A spiny *dorsal fin* is in front of a soft *dorsal*

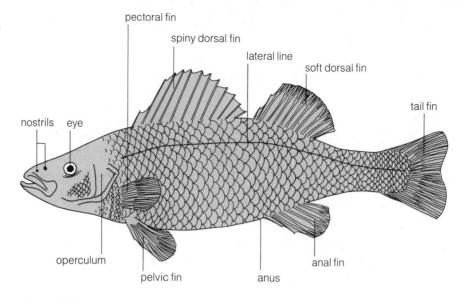

Figure 26-5 External anatomy of a bony fish.

fin. Paired *pectoral* and *pelvic fins* are present, and behind the *anus* is an unpaired *anal fin.* The *tail fin* consists of dorsal and ventral lobes of approximately equal size. The fins, as in the cartilaginous fishes, are used for braking, steering, and maintaining an upright position in the water.

3. Find the *lateral line.* This functions similarly to that of the shark.

4. Bony scales cover the body. Remove a scale, make a wet mount, and examine it with the compound microscope. Locate the *annual rings,* which indicate the age of the fish. How are the annual rings of the fish scale analogous to the annual ridges of the mussel or clam shell? (See page 316.)

_____

_____

Draw the scale in figure 26-6.

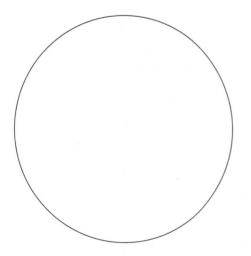

Figure 26-6 Drawing of the scale of a bony fish.

D. Class Amphibia: Amphibians

The *amphibians* were the first vertebrates to assume a terrestrial existence, having evolved from a group of lobe-finned fishes. The paired appendages are modified as legs, which support the individual during movement on land. Respiration is by lungs, gills, and the highly vascularized skin and lining of the mouth. There is a *three-chambered heart,* with a *double circulation* through it. Reproduction requires water, or at least moist conditions on land. The larvae generally live in water. The skeleton is more bony than that of the bony fishes, but a considerable proportion of it remains cartilaginous. The *skin* is usually *smooth and moist,* with mucous glands; scales are usually absent. This group of vertebrates includes the frogs, toads, and salamanders.

1. Examine preserved specimens of a variety of amphibians.

2. The leopard frog illustrates well the general features of the vertebrates and the specific characteristics of the amphibians. Study a preserved specimen after rinsing it in fresh water. Examine the external anatomy of the frog (fig. 26-7).

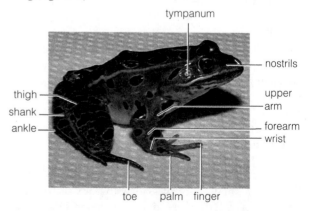

Figure 26-7 External anatomy of the frog. (Photo by D. Morton.)

3. Find the two nostrils at the tip of the head. These are used for inspiration and expiration of air. Just behind the eye is a disklike structure, the *tympanum*, the outer wall of the middle ear. There is no external ear. The tympanum is larger in the male than in the female. Examine the frogs of other students in your lab. Is your frog a male or female?

4. At the back end of the body locate the *cloacal opening*.

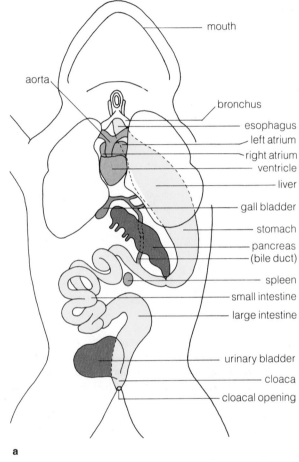

- mouth
- aorta
- bronchus
- esophagus
- left atrium
- right atrium
- ventricle
- liver
- gall bladder
- stomach
- pancreas (bile duct)
- spleen
- small intestine
- large intestine
- urinary bladder
- cloaca
- cloacal opening

a

5. The forelimbs are divided into three main parts: the *upper arm, forearm,* and *hand.* The hand is divided into a *wrist, palm,* and *fingers* (digits). The three divisions of the hindlimbs are the *thigh, shank* (lower leg), and *foot.* The foot is further divided into three parts: the *ankle, sole,* and *toes* (digits).

6. Fasten the frog, ventral side up, with pins to the wax of a dissection pan. Lift the skin with forceps. Then make a superficial cut with your scissors from the end of the trunk forward, and just left or right of center, to the tip of the lower jaw. Pin back the skin on both sides.

7. Note the white line along the midventral line and the large abdominal muscles you have exposed. Lift these muscles with your forceps, and cut through the body wall with your scissors from the end of the trunk to the tip of the lower jaw, cutting through the *sternum* (breastbone) but not damaging the internal organs. Pin back the body wall as you did the skin to expose the internal organs. Refer to figure 26-8 to study the internal anatomy.

8. Locate the *spleen* and the following organs of the digestive system: *mouth, tongue, pharynx* (throat), *esophagus, liver, gall bladder, stomach, pancreas, small intestine, large intestine* (colon), and *cloaca.* Go back to the esophagus and find the *bronchi* (the singular is *bronchus*), which lead toward the *lungs.* Are the bronchi dorsal or ventral to the esophagus?

c

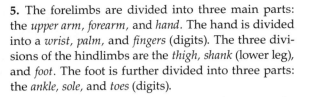

- (internal nostrils)
- (vomerine teeth)
- (eustachian tubes)
- esophagus
- (glottis)
- tongue
- pharynx

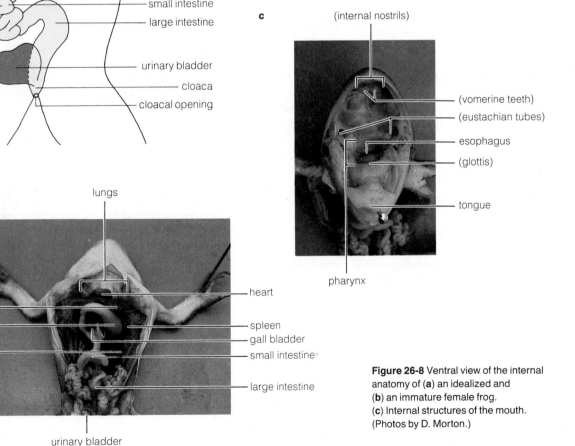

- lungs
- liver
- stomach
- kidney
- heart
- spleen
- gall bladder
- small intestine
- large intestine
- urinary bladder

b

Figure 26-8 Ventral view of the internal anatomy of (**a**) an idealized and (**b**) an immature female frog. (**c**) Internal structures of the mouth. (Photos by D. Morton.)

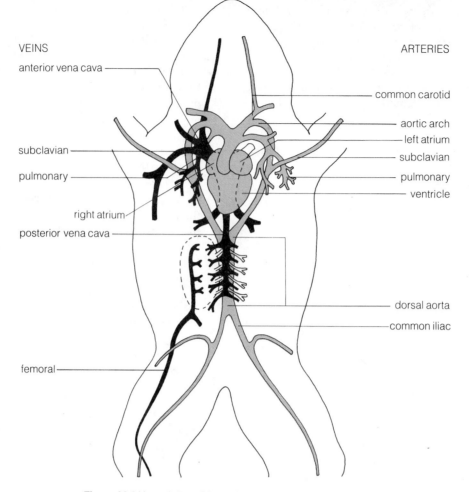

VEINS

anterior vena cava

subclavian

pulmonary

right atrium

posterior vena cava

femoral

ARTERIES

common carotid

aortic arch
left atrium
subclavian
pulmonary
ventricle

dorsal aorta
common iliac

Figure 26-9 Ventral view of the major blood vessels of the frog.

9. Amphibians have a three-chambered *heart*, with two thin-walled **atria** and a thick-walled **ventricle.** There is a *pulmocutaneous circulation*, involving the lungs and the skin, and a *systemic circulation*, involving the rest of the body. Even though there is only one ventricle and because of the structure of the heart, most of the venous (deoxygenated) blood moves from the heart to the *lungs* and back before it is pumped to the other organs of the body. Blood is carried from the ventricle to the systemic circulation by a large artery, the *aorta*.

10. Arteries carry blood (usually oxygenated) away from the heart, and **veins** carry blood (usually deoxygenated) toward the heart from the tissues. Refer to figure 26-9 and find the following major arteries: *pulmonary, aortic arch, subclavian, dorsal aorta*, and *common iliac*.

Where does blood in the dorsal aorta go?

_____

_____

Does the pulmonary artery contain oxygenated or deoxygenated blood?

_____

11. Locate the following major veins: *anterior* and *posterior vena cavae, subclavian, femoral*, and *pulmonary*. Find the lungs and describe the path of the blood from the tissues of the body, through the major veins to the heart, and through the major arteries ultimately back to the tissues.

_____

_____

_____

_____

12. The excretory and reproductive organs together comprise the *urogenital system* (fig. 26-10). Locate the urine-producing excretory organs, the pair of *kidneys* located dorsally in the body cavity. A duct, the *ureter*, leads from the kidney to the *urinary bladder*, an organ that stores urine for resorption of water from the urine into the circulatory system. The bladder empties into the cloaca.

13. In the female, find the *ovaries*; these organs expel eggs into the *oviducts*. The oviducts lead into the *uterus*. The reproductive tract ends in the cloaca.

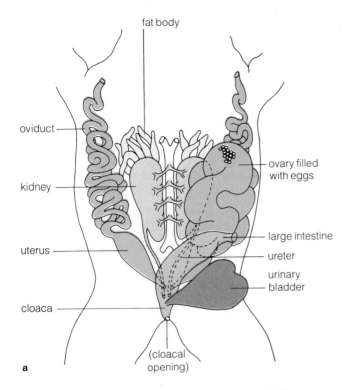

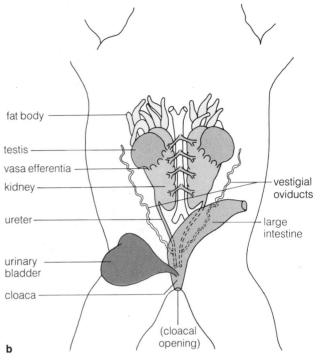

Figure 26-10 Ventral views of the urogenital systems of (**a**) the female and (**b**) the male frog.

14. In the male, locate the *testes*; these organs produce sperm that are carried to the kidneys through tiny tubules, the *vasa efferentia*. The ureters serve a dual function in male frogs, transporting both urine and sperm to the cloaca. Find the vestigial female reproductive tract, which is located lateral to the urogenital system.

15. Locate the *fat bodies*, many yellowish branched structures just above the kidneys. Their function is to store food reserves for hibernation and reproduction.

16. The nervous system of vertebrates is composed of two divisions: (1) the *central nervous system*, the brain and spinal cord, and (2) the *peripheral nervous system*, nerves extending from the central nervous system. Turn your frog over and remove the skin from the dorsal surface of the head between the eyes and along the vertebral column. With your scalpel, shave thin sections of bone from the skull, noting the shape and size of the *cranium*, until you expose the *brain*. Pick away small pieces of bone with your forceps to expose the entire brain. Use the same procedure to expose the vertebrae and *spinal cord*. Note the *cranial nerves* coming from the brain and the *spinal nerves* coming from the spinal cord. Dispose of your dissected specimen in the large plastic bag provided for this purpose.

17. The skeleton of vertebrates consists of the following: (1) the *axial skeleton* and its *skull, sternum,* and *vertebral column;* and (2) the *appendicular skeleton* and its *girdles* (support structures) and their appendages. Compare the skeleton of the frog and human (fig. 26-11).

There are differences between the two skeletons, but their basic plan is remarkably similar. To a large extent the size and shape of the different parts of the skeleton of a vertebrate correlate with body specializations and behavior. List some of the reasons for differences in the girdles, appendages, and cranium between the frog and human skeletons.

_____

_____

_____

E. Class Reptilia: Reptiles

Reptiles are the first group of vertebrates *adapted to living primarily on land*, although many do live in fresh or sea water. Their *skeleton* is more *bony* than that of amphibians. The *skin* is *dry* and covered by *epidermal scales*. There are virtually no skin glands. Reptiles lay *amniotic eggs* (fig. 26-12). Inside these eggs, embryos are suspended by fetal membranes in an internal aquatic environment surrounded by a shell to prevent their drying out.

There is no larval stage. The amniotic egg (characteristic of birds and some mammals as well as reptiles) and lack of larvae are adaptations to living an entire life cycle on land. The heart consists of two atria and a partially divided ventricle or two ventricles (in crocodiles). The nervous system, especially the brain, is more highly developed than that of amphibians. Reptiles, like the fishes and amphibians, are largely **ectothermic**, without the capability of maintaining their body temperatures physiologically.

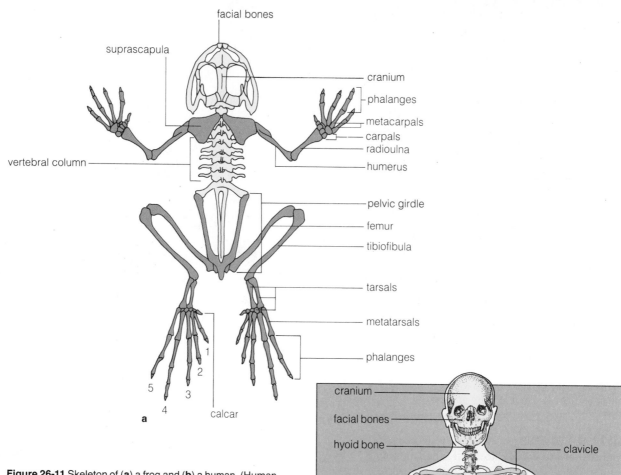

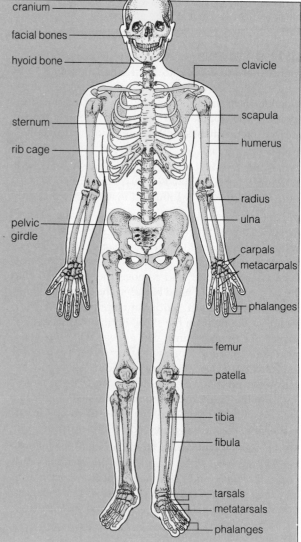

Figure 26-11 Skeleton of (**a**) a frog and (**b**) a human. (Human skeleton after Starr, 1991.)

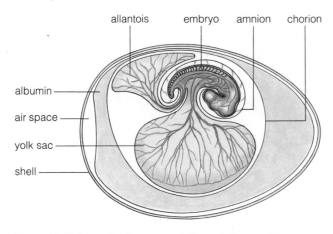

Figure 26-12 Generalized structure of the amniotic egg. (After Starr, 1991.)

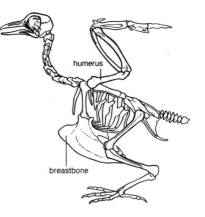

Figure 26-13 Generalized bird skeleton. (After Starr, 1991.)

1. Examine an assortment of preserved reptiles and note their diversity as a group. This diverse group of vertebrates includes the turtles, lizards, snakes, crocodiles, and alligators.

2. Compare the skeleton of a snake with that of the frog. Describe any differences you see.

_____

_____

_____

3. Look at the inside of the *carapace* (the upper portion) of a turtle shell. What portions of the skeleton are incorporated into the shell?

_____

F. Class Aves: Birds

The *birds* are **endothermic** vertebrates, capable of maintaining their body temperatures physiologically. Their body is covered with *feathers;* and *scales,* reminiscent of their reptilian heritage, are present on the feet. The front limbs in most birds are modified as *wings* for flight. An additional internal adaptation for flight is the *well-developed sternum* (breastbone), *with a keel*

for the attachment of powerful muscles for flight (fig. 26-13).

The major *bones* of birds are *hollow* and contain *air sacs connected to the lungs.*

Birds have a *four-chambered heart,* with two atria and two ventricles (fig. 26-14). This permits the complete separation of oxygenated and deoxygenated blood. Why is this circulatory arrangement an advantage for birds, as opposed to that found in amphibians and reptiles? (*Hint:* Recall that birds are endothermic and can fly.)

_____

_____

_____

The nervous system is similar to that found in reptiles. However, the brain is larger, permitting more sophisticated behavior and muscular coordination; the optic lobes are especially well developed in association with a keen sense of sight.

1. Examine a feather with the dissection microscope and note the *rachis* (shaft), *vane,* and *barbs* and *barbules* that comprise the vane (fig. 26-15). What else does the bird use its feathers for besides flight?

_____

_____

_____

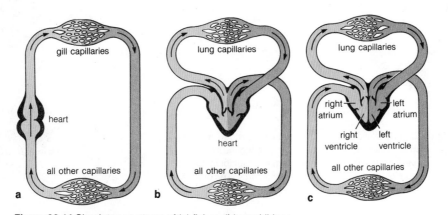

Figure 26-14 Circulatory systems of (**a**) fishes, (**b**) amphibians, and (**c**) birds and mammals. (After Starr, 1991.)

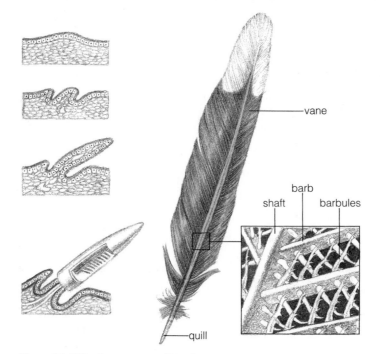

Figure 26-15 Feather structure. (After Storer *et al.*, 1979.)

2. The upper and lower jaws are modified as variously shaped *beaks* or *bills,* with the shapes reflecting the feeding habits of the species. No teeth are present in adults. Examine the assortment of birds and speculate on their food preferences by studying the configurations of their bills. Check your conclusions with a field guide or other suitable source that describes the food habits.

G. Class Mammalia: Mammals

Like the birds, *mammals* are *endothermic.* The term *Mammalia* stems from the *mammary glands,* which all mammals possess. The *heart* is *four-chambered,* as in birds. The *brain* is relatively *larger than that of other vertebrates,* and its surface area is increased by grooves and folds (fig. 26-16). The ear frequently has a cartilaginous outer portion called the *pinna.* Mammals are equipped with modifications and outgrowths of the skin, in addition to hair. There are four types of skin glands: *mammary, sebaceous, sweat,* and *scent glands. Hair* is present during some portion of the life cycle.

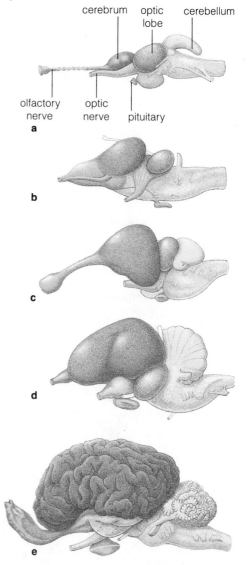

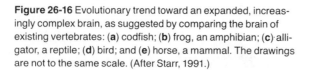

Figure 26-16 Evolutionary trend toward an expanded, increasingly complex brain, as suggested by comparing the brain of existing vertebrates: (**a**) codfish; (**b**) frog, an amphibian; (**c**) alligator, a reptile; (**d**) bird; and (**e**) horse, a mammal. The drawings are not to the same scale. (After Starr, 1991.)

1. Study the assortment of mammals on display and list below as many functions for hair as you can.

_____

_____

_____

2. Most mammals are **viviparous,** the female bearing live young after supporting them in the uterus with a *placenta,* a nutritive connection between the mother and the embryo. Examine the placentas and embryos in uteri of mammals on display in the lab. What relationship between the early development of mammals and the comparative sophistication of the nervous system and behavior of adult mammals can you suggest?

_____

_____

OPTIONAL

H. Construction of a Dichotomous Key to the Animals

Your instructor may provide you with instructions.

PRE-LAB QUESTIONS

_____ 1. Animals that have the three basic characteristics of chordates plus a vertebral column are (a) invertebrates, (b) hemichordates, (c) cephalochordates, (d) vertebrates.

_____ 2. The ammocoete is the larva of (a) lampreys, (b) sharks, (c) bony fishes, (d) amphibians.

_____ 3. Placoid scales are characteristic of class (a) Agnatha, (b) Chondrichthyes, (c) Osteichthyes, (d) Reptilia.

_____ 4. The bony movable flap that covers the gills and gill slits is called the (a) operculum, (b) cranium, (c) tympanum, (d) colon.

_____ 5. Which of the following groups of animals is endothermic? (a) fishes, (b) mammals, (c) reptiles, (d) amphibians.

_____ 6. The bones that surround and protect the brain are collectively called the (a) spinal cord, (b) vertebral column, (c) pelvic girdle, (d) cranium.

_____ 7. The lampreys and hagfishes are unusual vertebrates in that they have no (a) eyes, (b) jaws, (c) gill slits, (d) mouth.

_____ 8. The structure of the cartilaginous and bony fishes that detects vibrations in the water is the (a) anal fin, (b) operculum, (c) nostrils, (d) lateral line.

_____ 9. Amphibians have (a) a two-chambered heart, (b) a three-chambered heart, (c) a four-chambered heart, (d) none of the above.

_____ 10. The amniotic egg of reptiles, birds, and mammals is an adaptation to (a) carnivorous predators, (b) a life on land, (c) compensate for the short period of development of the young, (d) protect the young from the nitrogenous wastes of the mother during the formation of the embryo.

E X E R C I S E   2 6

Vertebrates

P O S T - L A B   Q U E S T I O N S

1. List the ways in which agnathans differ from other vertebrates.

2. List the basic characteristics of vertebrates.

3. Describe an amniotic egg. What is its evolutionary significance?

4. Describe the evolution of the heart in the vertebrates.

5. What is an ammocoete? What is its possible significance to the evolution of vertebrates?

6. Identify this vertebrate structure. What do the rings represent?

(Photo courtesy Ripon Microslides, Inc.)

(23×)

7. What characteristic of vertebrates is missing in all invertebrates?

8. Describe the difference between ectothermic and endothermic animals.

9. List the unique characteristics of mammals.

Plant Organization:

Vegetative Organs of Flowering Plants

OBJECTIVES

After completing this exercise you will be able to:

1. define *vegetative, morphology, cotyledon, dicotyledon (dicot), taproot system, node, internode, monocotyledon (monocot), adventitious root, herb (herbaceous plant), woody (woody plant), heartwood, sapwood, growth increment (annual ring), pore, vessel, sieve tube, P-protein;*

2. identify and give the function of the external structures of flowering plants (those in **boldface**);

3. identify and give the functions of the tissues and cell types of roots, stems, and leaves (those in **boldface**);

4. determine the age of woody branches.

INTRODUCTION

The color green seems to promote a feeling of well-being among humans. Perhaps that's why we find it so relaxing to stroll through the woods on a summer day. It has been suggested that we find green so pleasant because of our own evolutionary history.

It's difficult to overstate the importance of plant life. Are you sitting upon a wooden chair? If so, you're perched upon part of a tree. No, you say? Perhaps your chair is covered with fabric. If it's natural fabric other than wool, it's a plant product. If the chair is covered with plastic, that cover was made from petroleum products derived from plant material that lived millions of years ago. The energy used to create the chair was probably derived from burning petroleum products or coal, also derived from plant material.

Obviously, plants are an important part of our lives. This exercise will introduce you to the external and internal structure of the organs of flowering plants that are associated primarily with uptake and transport of water and minerals, photosynthesis, and respiration, but not sexual reproduction. The organs—the roots, stems, and leaves—are called **vegetative** organs, in contrast to the flower, which is the sexual reproductive organ.

Each organ is usually distinguished by its shape and form, its **morphology.** But there is remarkable similarity in the cells, the basic unit of life, and tissues, groups of cells functioning together, comprising these three organs. Each organ is covered by the protective dermal tissue; each organ possesses vascular tissue that transports water, minerals, and the products of photosynthesis; and each also contains ground tissue, that which is covered by the dermal tissue and in which the vascular tissue is embedded. Examine figure 27-1, illustrating these relationships.

Thus, the organs of the plant body are really much more similar than dissimilar. In fact, for this reason the differences that exist between a root and a stem or leaf are said to be *quantitative* rather than *qualitative.* That is, these differences are in the number and arrangement of cells and tissues, not the type. Consequently, the plant body is a continuous unit from one organ to the next.

I. External Structure of the Flowering Plant

MATERIALS

Per student group (table):
• mature corn plant

Per lab room:
• living bean and corn plants in flats
• potted geranium and dumbcane plants
• dishpan half-filled with water

PROCEDURE

A. Dicotyledons

The common garden bean, *Phaseolus vulgaris*, is a plant with two seed leaves—called *cotyledons*—and thus belongs to the large category of plants, called **dicotyledons.** Other familiar examples of dicotyledons (dicots) are sunflowers, roses, cucumbers, peas, maples, and oaks.

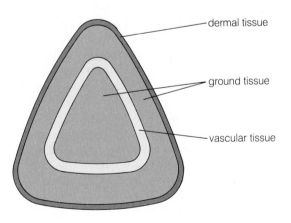

Figure 27-1 Hypothetical plant organ.

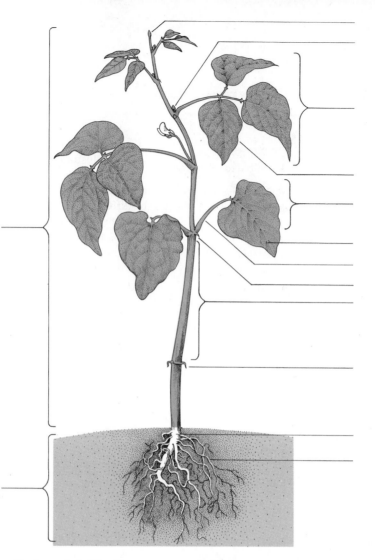

Figure 27-2 External structure of the bean plant. (After Starr and Taggart, 1984).

Labels: root system, shoot system, primary root (taproot), lateral root, node, internode, axillary bud, terminal bud, cotyledon remnant, petiole, blade, simple leaf, compound leaf, leaflet

1. Obtain a bean plant by gently removing it from the medium in which it is growing. Wash the root system in the dishpan provided, not in the sink.

Label figure 27-2 as you study the bean plant.

2. The plant consists of a **root system** and a **shoot system.** Examine the root system first. The root system of the bean is an example of a **taproot system,** that is, one consisting of one large **primary root** (the **taproot**) from which **lateral roots** arise. Identify the taproot and lateral roots.

3. Now turn your attention to the shoot system, consisting of the stem and the leaves. Identify the points of attachment of the leaves to the stem, called **nodes;** the regions between nodes are **internodes.** Look in the upper angle created by the junction of the stem and leaf stalk for the **axillary bud.** These buds give rise to branches and/or flowers.

4. Find the **terminal bud** at the very tip of the shoot system. The terminal bud contains an apical meristem

that accounts for increases in length of the shoot system.

If lateral branches are produced from axillary buds, each lateral branch is terminated by a terminal bud and possesses nodes, internodes, and leaves, complete with axillary buds. As you see, the shoot system can be a highly branched structure.

5. Look several centimeters above the soil line for the lowermost node on the stem. If the plant is relatively young, you should find the **cotyledons** attached to this node. The cotyledons shrivel as food stored in them is used for the early growth of the seedling. Eventually the cotyledons fall off.

The cotyledons are sometimes called *seed leaves* because they are fully formed (although unexpanded) in the seed. By contrast, most of the leaves you are observing on the bean plant were immature or not present at all in the seed.

6. Now let's examine the other component of the shoot system, the foliage leaves (also called *true leaves* in contrast to the cotyledons — the seed leaves). Identify the **petiole** (leaf stalk) and **blade** on each leaf. The first-formed foliage leaves are **simple leaves,** each leaf having one undivided blade. Find the simple leaves. In bean plants, subsequently formed leaves are **compound leaves,** consisting of three **leaflets** per petiole. Each leaflet has its own short stalk and blade.

7. Note the netted arrangement of veins in the blades. Veins contain vascular tissues — the xylem and phloem. The **midvein** is largest and runs down the center of the blade, giving rise to numerous lateral veins.

Now that you have a general idea of the external structure of a typical dicot, let's look at another, transferring your knowledge from a particular plant to make some generalizations.

8. Obtain a potted geranium plant and examine the external structure of the shoot. Identify the following parts, checking off in the blank before each as you go along.

_____ node

_____ internode

_____ axillary bud

_____ terminal bud

_____ leaf petiole

_____ leaf blade

What color is the stem of all of the plants you have examined?

_____

What structure in the cytoplasm of the cells making up the stem is responsible for this color?

_____

What is the function of this structure?

_____

What, then, is one function of the stem?

_____

9. Compare the leaves of the bean with that of the geranium. Are the geranium leaves simple or compound? (You may wish to refer to some of the figures in Exercise 14 ["Taxonomy: Classifying and Naming Organisms"] if you have difficulty deciding.)

_____

Is there a single midvein in the geranium, or are there many large veins?

_____

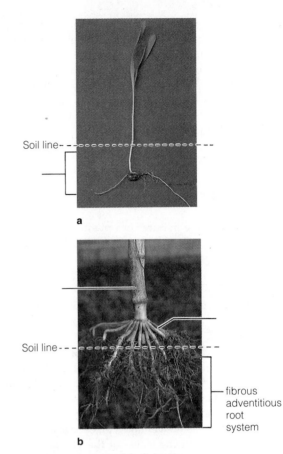

Labels: root system, prop root, leaf sheath

Figure 27-3 External structure of a corn plant. (**a**) Seedling (0.5 ×). (**b**) Lower portion of mature plant (0.1 ×). (Photos by J. W. Perry.)

List all the features shared by the leaves of beans and geraniums.

_____

_____

List the differences that you observe in the leaves.

_____

_____

B. Monocotyledons

Corn (*Zea mays*) is a **monocotyledon,** or monocot. These plants have only one cotyledon (seed leaf). You're probably familiar with a number of monocots: lilies, onions, orchids, coconuts, bananas, and the grasses. (Did you realize that corn is actually a grass?)

1. Remove a single young plant from its growing medium and wash its root system in the dishpan.

Label figure 27-3 as you study the corn plant.

2. The seed (grain) may still be attached to the plant if you were careful in removing the plant. Identify the **root system.** Note that there is no one particularly prominent root. In most monocots the primary root is short-lived and is replaced by numerous **adventitious roots.** These are roots that arise from places other than existing roots. Trace these roots back to the corn grain. Where do they originate?

_____

As the roots branch, they develop into a **fibrous root system,** one particularly well suited to prevent soil erosion.

3. Examine the mature corn plant. Identify the large **prop roots** at the base of the plant. Where do prop roots arise from?

_____

Would you classify these as adventitious roots?

_____

4. The shoot system of a young corn plant appears somewhat less complex than that of the bean. There seem to be no nodes or internodes. Look at the mature corn plant again. You should see that nodes and internodes do indeed exist. In your young plant, elongation of the stem has not yet taken place to any appreciable extent. Strip off the leaves of the young corn plant. Keep doing so until you find the shoot apex. (It's deeply embedded, don't you agree?)

5. Examine in more detail the leaves of the corn plant. Note the absence of a petiole and the presence of a **sheath** that extends down the stem. The leaf sheath adds strength to the stem. Look at the veins, which have a parallel arrangement. (Contrast this to the petioled, netted-vein arrangement of the bean leaves.)

Later in this exercise, we will study in more detail the root and shoot systems of dicotyledons.

How representative of all monocotyledons is corn? As it turns out, it's quite representative of most grasses, but not particularly so of monocots as a whole. Let's look at another monocot, a common horticultural plant found in many homes, called dumbcane (*Diffenbachia*).*

6. Obtain a potted specimen of the dumbcane plant. Observe its external morphology, comparing it with the corn plant. Does the dumbcane have sessile leaves, or does each leaf have a petiole?

_____

_____

*The common name *dumbcane* has its origin in a use by the ancient Greeks. When they tired of long orations by their senators, Greeks sometimes ground up parts of the shoot and added it to a drink. Because certain cells contain needle-shaped crystals, consumption caused a temporary paralysis of the larynx (voice box), ending the oration. Today, dumbcane is a hazard to young children. Ingestion can cause throat swelling leading to suffocation.

Are the veins in the leaves parallel, or is netted venation present? (Look on the lower surface of the leaves, where it is most obvious.)

_____

Is there a midvein? _____

Is a leaf sheath present? _____

Is the terminal bud obvious or is it deeply embedded, as in the corn plant?

_____

Are prop roots present on the dumbcane plant?

_____

II. The Root System

MATERIALS

Per student:
• single-edged razor blade
• clean microscope slide
• coverslip
• prepared slide of buttercup (*Ranunculus*) root, cross section
• compound microscope

Per student pair:
• distilled water (dH$_2$O) in dropping bottles

Per lab room:
• germinating radish seeds in large petri dishes
• demonstration slide of Casparian strip in endodermal cell walls

PROCEDURE

A. Living Root Tip

1. Obtain a germinating radish seed. Identify the **primary root.** Its fuzzy appearance is due to the numerous tiny **root hairs** (fig. 27-4).

2. Using a razor blade, cut off and discard the seed, and make a wet mount of the primary root. (Add a

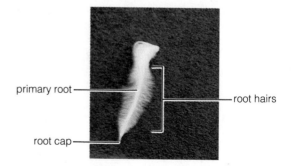

Figure 27-4 Living primary root (2×). (Photo by J. W. Perry.)

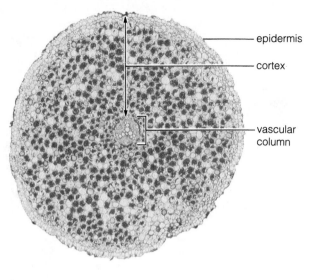

epidermis

cortex

vascular
column

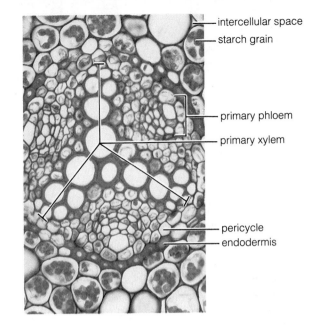

intercellular space

starch grain

primary phloem

primary xylem

pericycle

endodermis

a

Figure 27-5 Cross section of buttercup (*Ranunculus*) root. (**a**) 50 × . (**b**) Portion of cortex and vascular column (75 ×). (Photos by J. W. Perry.)

b

copious amount of water so there is no air surrounding the root. *Do not* squash the root.)

3. Examine your preparation with the low-power objective of your compound microscope. (If you are having difficulty seeing the root clearly, increase its contrast by closing the microscope's diaphragm somewhat.) Locate the conical root tip. The very end of the root tip is covered by the protective **root cap.** As a root grows through the soil, the tip is thrust between soil particles. Were it not for the root cap, the apical meristem containing the dividing cells would be damaged.

4. Find the root hairs. Do they extend all the way down to the root cap?

_____

Examine the root hairs carefully. Does their length increase from the root tip?

_____

The youngest root hairs are the shortest. What does this imply regarding their point of origin and pattern of maturation?

_____

_____

_____

Root hairs increase the absorptive surface of the root tremendously.

B. Root Anatomy

Now let's see what the internal architecture of the root looks like.

1. Obtain a prepared slide containing a mature buttercup (*Ranunculus*) root in cross section. Refer to figure 27-5 as you study this slide. Examine the slide first

with the low-power objective of your compound microscope to gain an overall impression of the organization of the tissues present.

2. Starting at the edge of the root, identify the **epidermis.** Moving inward, locate the **cortex** and the central **vascular column.** These regions represent the dermal, ground, and vascular tissue systems, respectively.

3. Switch to the medium-power objective for further study. Look at the outermost layer of cells, the **epidermis.** What function does the epidermis serve?

_____

_____

4. Beneath the epidermis find the relatively wide **cortex,** consisting of parenchyma cells that contain numerous **starch grains.** Based upon the presence of starch grains, what would you suspect one function of this root might be?

_____

_____

5. Switch to the high-dry objective. Between the cells of the cortex find numerous **intercellular spaces.**

6. The innermost layer of the cortex is given a special name. This cylinder, a single cell thick, is called the **endodermis.** Locate the endodermis on your slide and in figure 27-5b. Unlike the rest of the cortical cells, endodermal cells *do not* have intercellular spaces between them. The endodermis regulates the movement of water and dissolved substances into the vascular column. Each endodermal cell possesses a **Casparian strip** within its radial and transverse walls. To visualize this arrangement, imagine a rectangular box (the

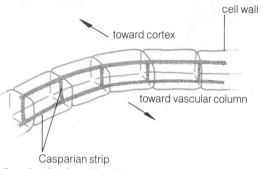

Figure 27-6 Diagrammatic representation of endodermal cells with Casparian strip.

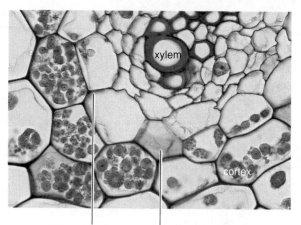

Figure 27-7 Endodermal cells showing Casparian strip (296 ×). (Photo by J. W. Perry.)

endodermal cell) that has a rubber band (the Casparian strip) around its long dimension. Now imagine that the rubber band is actually part of the wall of the box. Figure 27-6 is a diagram of this arrangement.

The Casparian strip consists of waxy material. Water and substances dissolved in the water normally move through the cell *walls* as they flow radially through the root. Because the Casparian strip consists of a waxy material, it effectively waterproofs the cell wall of the endodermal cells. Consequently, substances moving radially must flow through the cytoplasm of the endodermal cell. As you will recall, the cytoplasm is bounded by the differentially permeable plasma membrane. Thus, dissolved substances are "filtered" through endodermal cells, regulating what goes into or comes out of the vascular column.

7. In many cases, the Casparian strip is difficult to distinguish. With the highest magnification available, look for a red "dot" on the radial wall of the endodermal cells. If you cannot distinguish it on your own slide, examine the demonstration slide (fig. 27-7), which has been selected to show the Casparian strip. On this slide, you should be able to see its bandlike nature.

8. Return to your own slide, switch back to the medium-power objective, and focus your attention on the central vascular column. The cell layer immediately beneath the endodermis is the **pericycle.** Cells of the pericycle may become meristematic and produce lateral roots, like those you saw in the bean plant (fig. 27-2).

9. Finally, find the **primary xylem,** consisting of three or four ridges of thick-walled cells. (The stain used by most slide manufacturers stains the xylem cell walls red.) The xylem is the principal water-conducting tissue of the plant. Between the "arms" of the xylem find the **primary phloem,** the tissue responsible for long-distance transport of carbohydrates produced by photosynthesis (known as photosynthates).

Primary xylem and phloem make up the *primary* vascular tissues, which are those produced by the apical meristems at the tips of the root and shoot. Later, we'll examine a *secondary* vascular tissue.

III. The Shoot System

MATERIALS

Per student:

- prepared slide of herbaceous dicot stem, cross section (flax, *Linum;* or alfalfa, *Medicago*)
- prepared slide of woody stem, cross section (basswood, *Tilia*)
- prepared slide of dicot leaf, cross section (lilac, *Syringa*)
- woody twig (hickory, *Carya;* or horse chestnut, *Aesculus*)
- metric ruler or meter stick
- compound microscope

Per student pair:

- cross section of woody branch (tree trunk)
- dissecting microscope

Per lab room:

- demonstration slide of lenticel

PROCEDURE

A. Primary Structure of Stems

Remember that the root and shoot are basically similar in structure; only the arrangement of tissues differs. Dicot stems have their vascular tissues arranged in a more or less complete ring of individual bundles of vascular tissue (called vascular bundles). Moreover, the ground tissue of dicots can be differentiated into two regions: **pith** and **cortex.**

You've probably heard the term *herb.* An **herb** is an **herbaceous plant,** one that develops very little wood. Beans, flax, and alfalfa are examples of herbaceous

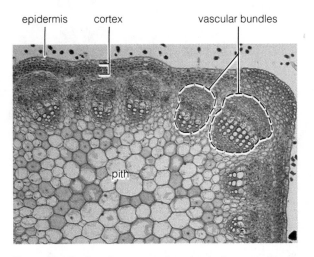

Figure 27-8 Portion of a cross section of an herbaceous dicot stem (75×). (Photo by J. W. Perry.)

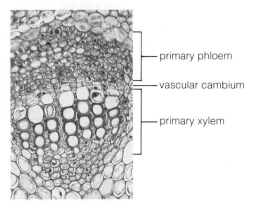

Figure 27-9 Vascular bundle of an herbaceous dicot stem (192×). (Photo by J. W. Perry.)

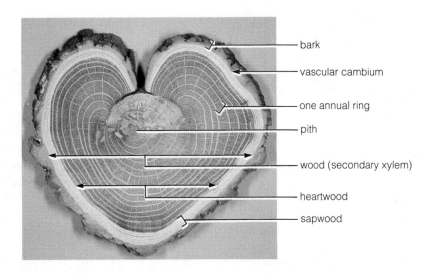

Figure 27-10 Cross section of a woody locust stem. This stem was injured several years into its growth, causing the unusual shape. (Photo by J. W. Perry.)

dicots. Maples and oaks are woody dicots. Let's look at an herbaceous stem first.

1. Obtain a slide of an herbaceous dicot stem (flax or alfalfa). This slide may be labeled "herbaceous dicot stem." As you study this slide, refer to figure 27-8, a photograph of a partial section of an herbaceous dicot stem.

2. Using the low- and medium-power objectives, identify the single-layered **epidermis** covering the stem, a multilayered **cortex** between the epidermis and **vascular bundles,** and the **pith** in the center of the stem.

3. Observe a vascular bundle with the high-dry objective (fig. 27-9). Adjacent to the cortex, find the **primary phloem.** Just to the inside of the primary phloem will be located a **vascular cambium,** which for the most part is inactive. (The vascular cambium is the lateral meristem that produces wood and secondary phloem, both secondary tissues. Despite your specimen being an herbaceous plant, a vascular cambium may be pres-

ent. Generally, this meristem does not produce enough secondary tissue to result in the plant's being considered woody.) Locate thick-walled cells of the **primary xylem.** The primary xylem is that vascular tissue closest to the pith. (The wall of the xylem cells is probably stained red.)

B. Secondary Growth: Woody Stems

Woody plants are those that undergo secondary growth. Both roots and shoots may have secondary growth. This growth occurs because of activity of the two meristems near the edge of the plant—the vascular cambium and the cork cambium. The vascular cambium produces secondary xylem (wood) and secondary phloem, while the cork cambium produces the periderm.

1. Examine a cross section of a tree trunk (fig. 27-10; *trunk* is the nonscientific term for any large, woody stem). A tiny region in the center of the stem is the

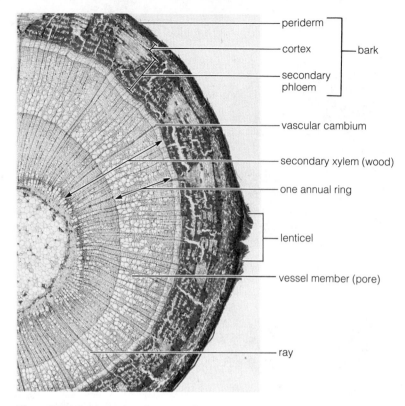

Figure 27-11 Cross section of a woody dicot stem (15 ×). (Photo by J. W. Perry.)

pith. Most of the trunk is made up of **wood** (also called *secondary xylem*).

2. As a woody stem grows larger, it requires additional structural support and more tissue for transport of water. These functions are accomplished by the wood. However, it is only the outer few years' growth of wood that is actually involved in water conduction. In some trees, like that pictured in figure 27-10, the distinction between conducting and nonconducting wood is very obvious because of the color differences in each. The nonconducting wood, very dark in figure 27-10, is called the **heartwood.** The heartwood of many species, like black walnut or cherry, is highly prized for furniture making. Although virtually all trees develop heartwood, not all species have heartwood that is *visibly* distinct from the conducting wood. Attempt to distinguish it in your stem.

3. The conducting wood, significantly lighter in color in figure 27-10, is the **sapwood.** Be careful in examining the photograph, because the color difference between the sapwood and the inner layers of the bark is subtle. Attempt to identify the sapwood on the stem you are examining.

4. Count the **growth increments (annual rings)** within the wood to estimate the age of the stem when the section was cut. (In most woods, a growth increment includes *both* a light and a dark layer of cells.) How old would you estimate your section to be?

_____ years

5. Now find the **vascular cambium** located *between* the most recently formed wood (secondary xylem) and the **bark.** The bark is everything *external* to the vascular cambium.

6. Identify the bark, consisting of **secondary phloem** and **periderm.** (As the vascular cambium produces secondary phloem to the outside, the primary phloem, cortex, and epidermis are sloughed off, much as dead skin on your body is shed.) The periderm performs the same function as did the epidermis before the epidermis ruptured as a result of increase in girth of the stem. What is the function of the periderm?

_____

7. Within the wood, find the **rays,** which appear as lines running from the center toward the edge of the stem.

8. Examine the wood with a dissecting microscope and locate the numerous holes in the wood. These are the cut ends of the water-conducting cells, often called **pores.**

Now that you've got an idea of the composition of a woody stem, let's examine one with the microscope.

9. Obtain a prepared slide of a cross section of a woody stem (basswood, *Tilia,* or another stem). Examine it with the various magnifications available on the *dissecting* microscope. Use figure 27-11 as a reference for your study.

10. Starting at the edge, identify the **periderm** (darkly stained cells). Depending upon the age of the stem, **cortex** (thin-walled cells with few contents) may be present just beneath the periderm. Now find the broad band of **secondary phloem** (consisting of cells in pie-shaped wedges). Identify the **vascular cambium** (a narrow band of cells separating the secondary phloem from secondary xylem), **secondary xylem (wood),** and pith (large, thin-walled cells in the center). Count the number of growth increments (annual rings). How old is this section?

_____ years

Note the largest, thick-walled cells in the wood. These are the pores.

11. Find the **rays** running through the wood. Rays are parenchyma cells that carry water and photosynthates *laterally* in the stem. (For the most part, the xylem and phloem carry substances *vertically* in the plant.)

Recall that the periderm replaces the epidermis as secondary growth takes place. The epidermis had stomata, which allowed the exchanging of gases between the plant and the environment. When the epidermis was shed, so were the stomata. But the need for exchange of gases still exists because the living cells require oxygen for respiration (Exercise 8). The plant has solved this problem by having special regions, **lenticels,** in the periderm. Lenticels are groups of cells with lots of intercellular space, in contrast to the tightly packed cells in the rest of the periderm.

12. Identify a lenticel on your slide. If none is found, examine the *demonstration* slide, specifically chosen to demonstrate this feature.

C. External Features of Woody Stems

1. Examine a twig of hickory (*Carya*) or buckeye (*Aesculus*) that has lost its leaves. Label figure 27-12 as you study the twig.

2. Find the large **terminal bud** at the tip of the twig. If your twig is branched, each branch has its own terminal bud.

3. Identify the shield-shaped **leaf scars** at each **node.** (Remember, a node is the region where a leaf attaches to the stem.) Leaf scars represent the point at which the leaf petiole was attached on the stem.

4. Within each leaf scar, note the numerous dots. These are the **vascular bundle scars.** Immediately above and adjacent to most leaf scars should be an **axillary bud.**

5. In the **internode** regions of the twig locate the small raised bumps on the surface; these are **lenticels,** the regions of the periderm that allow for exchange of gases.

6. Return to the terminal bud. Note that the bud is surrounded by **bud scales.** When these scales fall off during spring growth, they leave **terminal bud scale scars.** Because a terminal bud is produced at the end of each growing season, the groups of terminal bud scale scars can be used to determine the age of a twig. If the most recent growth took place during the last growing season (summer), when was the portion of the twig immediately adjacent to the cut end produced?

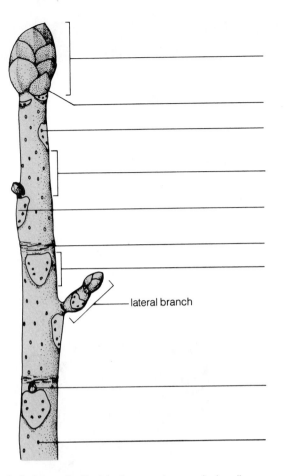

— lateral branch

Labels: terminal bud, leaf scar, node, vascular bundle scar, axillary bud, internode, lenticel, bud scale, terminal bud scale scars

Figure 27-12 External structure of a woody stem.

(Remember to date all regions between successive bud scale scars.)

Now that you've got some idea of the structure of a particular woody plant, let's get an impression of the amount of growth that may take place in several different species of plants.

7. Measure the distance between a number of successive terminal bud scale scars in your specimen. Average the results to obtain an idea of about how much growth is produced on an annual basis. Record your results in table 27-1 on page 366.

8. Now obtain twigs of several other species, including the tree of heaven (*Ailanthus*), and do the same.

Would you say the growth rate is similar or quite variable among species that grow in your area?

_____

D. Leaf Anatomy

1. Obtain a prepared slide of a cross section of a dicot leaf (lilac or other). Refer to figure 27-13 as you examine the leaf.

Table 27-1 Average Annual Terminal Growth in Woody Stems

Species	Average Distance Between Terminal Bud Scale Scars (cm)
1. *Carya* (or *Aesculus*)	
2. *Ailanthus*	
3.	
4.	

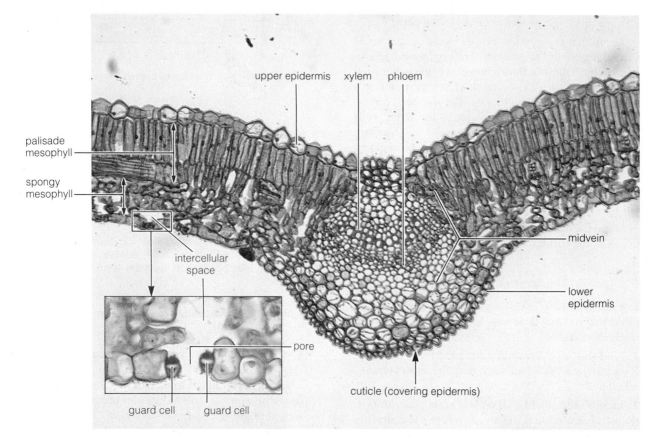

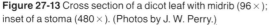

Figure 27-13 Cross section of a dicot leaf with midrib (96 ×); inset of a stoma (480 ×). (Photos by J. W. Perry.)

2. Examine the leaf first with the low-power objective to gain an overall impression of its morphology. Note the size and orientation of the **veins** within the leaf. The veins contain the xylem and phloem. Find the centrally located **midvein** within the **midrib,** the midvein-supporting tissue. You might think of the midvein as the major pipeline of the leaf, carrying water and minerals to the leaf and materials produced during photosynthesis to sites where they will be used during respiration.

3. Use the high-dry objective to examine a portion of the blade to one side of the midvein.

4. Starting at the top surface of the leaf, find the **cuti-**

cle, a waxy, water-impervious substance covering the **upper epidermis.** The epidermis is a single layer of tightly appressed cells.

5. The ground tissue of the leaf is represented by the **mesophyll** (literally "middle leaf"). In dicot leaves the mesophyll is usually divided into two distinct regions; immediately below the upper epidermis find the two layers of **palisade mesophyll.** These columnar-shaped cells are rich in chloroplasts. Below the palisade mesophyll find the loosely arranged **spongy mesophyll.** Note the large volume of intercellular space within the spongy mesophyll. Does the spongy mesophyll contain any chloroplasts?

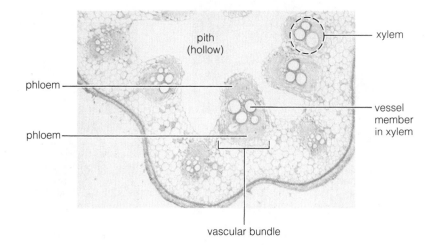

Figure 27-14 Squash stem (*Cucurbita maxima*) stem, c.s.
(12 ×). (Photo by J. W. Perry.)

What then is one function that occurs within the spongy mesophyll?

_____

6. In the **lower epidermis** find a **stoma** (plural: *stomata*) with its **guard cells** and the **pore** (inset, fig. 27-13). Large **epidermal hairs** shaped somewhat like mushrooms are usually found on the lower epidermis. Is the lower epidermal layer covered by a cuticle?

_____

7. Compare the abundance of stomata within the lower epidermis with that in the upper epidermis. Which epidermal surface has more stomata?

_____

8. Examine the midvein in greater detail. The thick-walled cells (often stained red) are part of the **xylem** tissue. Below the xylem locate the **phloem** (usually stained green). Now identify and examine the smaller veins within the lamina (blade). Note that these, too, contain both xylem and phloem.

IV. Structure of the Vascular Tissues

During the evolution of land plants from green algae, structural changes took place that allowed these new organisms to colonize the hostile environment of dry land. The algal ancestors were small and bathed in water. Any cell was only a short distance from the water needed for metabolism, and thus they relied upon diffusion for water transport. Moreover, most probably all cells of the algal ancestors were photosynthetic, so a supply of carbohydrates necessary for respiration was readily available.

As you know, land plants have an aerial shoot system responsible for photosynthesis and a root system anchored in the soil, the source of water and minerals. Consequently, a transport system is necessary to carry water to the shoot system and carbohydrates to the nonphotosynthetic root system. These functions take place in the xylem and phloem tissues, respectively. The transition from water to land could not have occurred without these tissues, just as the evolution of complex animals such as humans could not have taken place without their elaborate vascular systems. Let's study more thoroughly the structure of these important transport tissues.

MATERIALS

Per student:

• prepared slide of squash (*Cucurbita maxima*) stem, cross and longitudinal sections (on same slide)
• compound microscope

PROCEDURE

A. Xylem Tissue

The primary function of **xylem tissue** is the transportation of water and dissolved minerals throughout the plant body. Because some of the cells within the xylem have thick, rigid walls, the xylem also offers considerable structural support that assists the plant in standing erect.

Xylem is a *complex tissue,* which means that a number of different cell types make up the tissue. For the purposes of understanding the function of the xylem, the most important cell type in flowering plants is the **vessel member,** a cell that is *dead* at maturity! Lacking cytoplasm, vessel members serve as pipelike conduits for the flow of water.

1. Obtain a prepared slide of a squash (*Cucurbita maxima*) stem. This slide has both a cross section (c.s.) and longitudinal section (l.s.). Use figure 27-14 as a guide

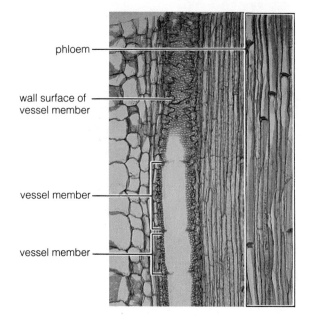

Figure 27-15 Longitudinal section through xylem and phloem of squash (25 ×). (Photo by J. W. Perry.)

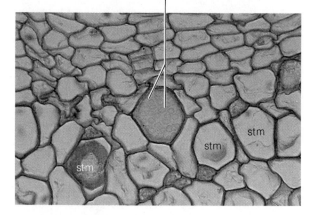

Figure 27-16 Detail of phloem of *Cucurbita*, showing several sieve tube members (stm) (465 ×). (Photo by J. W. Perry.)

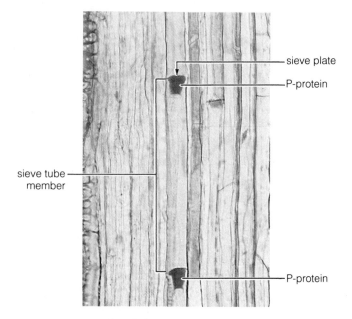

Figure 27-17 Longitudinal section through phloem tissue (296 ×). (Photo by J. W. Perry.)

to find a **vascular bundle** on the cross section with the low-power objective. The largest cells with a wide, clear lumen and thick walls in the vascular bundle are the **vessel members.** Their thick walls are impregnated with a substance called *lignin,* a polymer that waterproofs the wall. (The stain used by most slide manufacturers colors lignified walls red.)

A series of vessel members united in a vertical file comprise a **vessel.** You might think of a vessel as a long pipeline consisting of a series of individual pipes (the vessel members) connected end to end. In some plants, a vessel may be several meters in length!

2. Now turn your attention to the longitudinal section (fig. 27-15). Study this section with the low-power objective. Find the **vessel members** stacked into a longitudinal file forming a **vessel.** Notice that the lumen of each vessel member is empty, making for a pipelike conduit for the flow of water.

OPTIONAL
Experiments: Xylem Transport

Your instructor may provide you with descriptions of two optional experiments demonstrating the movement of water through the xylem.

B. Phloem Tissue

The phloem tissue is also a complex tissue with more than one cell type. Here the principal conducting cell is the **sieve tube member,** a cell that conducts the materials produced during photosynthesis to growing plant parts or storage organs such as roots. Sieve tube members are alive at maturity and when functioning, although they have lost some of their cellular constituents, most notably the nucleus. The remaining con-

tents become distributed in a thin layer along the cell wall so that sugars moving through the cell have a relatively unobstructed journey.

1. The vascular bundle of *Cucurbita* contains phloem tissue on both sides of the xylem. Locate it in figure 27-14.

2. On the cross section, find the phloem tissue, using the low-power objective. Then switch to the high-dry objective for more detailed study. The largest cells in the phloem are the **sieve tube members** (figure 27-16). As in the xylem, sieve tube members are united in longitudinal stacks, forming **sieve tubes.**

3. Next identify an end wall of a sieve tube member. This wall contains numerous pores; consequently, the end wall is called a **sieve plate** (fig. 27-16). Sugars produced during photosynthesis move from sieve element to sieve element through the pores in the sieve plate.

4. Finally, examine the longitudinal section of the phloem. Use figure 27-15 to locate the tissue with the low- or medium-power objective and then switch to the high-dry objective for further study. Find a sieve tube member by locating the plug of material (often stained red) aggregated at the end of the cell against a sieve plate (fig. 27-17). This plug is *P-protein*, a substance that probably serves to seal off injured sieve tube members, preventing excessive loss of sugars from the plant.

When the sieve tube member is alive and functioning, the P-protein is dispersed within the cell, and there is no plug. When the stem of the squash plant was cut prior to processing for slide preparation, the sieve tube members were damaged, and the plug formed.

OPTIONAL

Experiments: Phloem Function

Your instructor may provide you with directions for two optional experiments dealing with the function of the phloem tissue.

PRE-LAB QUESTIONS

_____ 1. The study of a plant's structure is (a) physiology, (b) morphology, (c) taxonomy, (d) botany.

_____ 2. A plant with two seed leaves is (a) a monocotyledon, (b) a dicotyledon, (c) exemplified by corn, (d) a dihybrid.

_____ 3. A taproot system lacks (a) lateral roots, (b) a taproot, (c) both of the above, (d) none of the above.

_____ 4. Which of the following is not part of the shoot system? (a) stems, (b) leaves, (c) lateral roots, (d) axillary buds.

_____ 5. An axillary bud (a) would be found along internodes, (b) produces new roots, (c) is the structure from which branches and flowers arise, (d) is the same as a terminal bud.

_____ 6. The endodermis (a) is the outer covering of the root, (b) is part of the vascular tissue, (c) contains the Casparian strip, which regulates the movement of substances, (d) none of the above.

_____ 7. Meristems are (a) located at the tips of stems, (b) located at the tips of roots, (c) regions of active growth, (d) all of the above.

_____ 8. To determine the age of a woody twig, you would count the number of (a) nodes, (b) leaf scars, (c) lenticels, (d) regions between sets of terminal bud scale scars.

_____ 9. The midrib of a leaf (a) contains the midvein, (b) contains only xylem, (c) is part of the spongy mesophyll, (d) contains only phloem.

_____ 10. P-protein would be found in (a) sieve tube members, (b) vessels, (c) xylem, (d) vessel members.

EXERCISE 27

Plant Organization: Vegetative Organs of Flowering Plants

POST-LAB QUESTIONS

1. A major problem for land plants is water conservation. Most water is lost through stomata due to evaporation at the surface of the leaf. Many plants, including lilac (the leaf section you examined), orient their leaves perpendicular to the drying force of the sun's rays. What did you observe about the relative abundance of stomata in the lower epidermis versus the upper epidermis? Why do you think this distribution has evolved?

2. On the figure below, identify structures **a**, **b** and **c**.

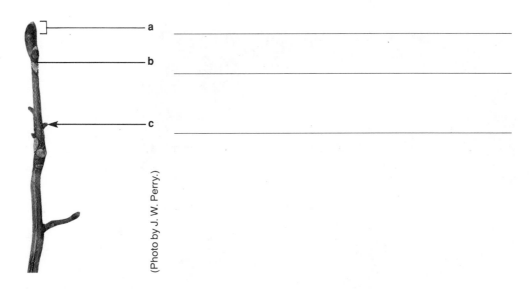

(Photo by J. W. Perry.)

a _____

b _____

c _____

3. Identify the structures labeled **a** and **b** on the figure below.

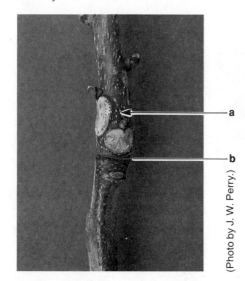

(Photo by J. W. Perry.)

a _____

b _____

4. What feature(s) would you use to determine the age of a woody twig?

5. Axillary buds grow into _____ .

6. Describe the location, structure, and importance of the Casparian strip.

7. The figure below illustrates a section cut from an ash branch.
   a. Identify region **a**.
   b. Which meristem is located at **b**?
   c. Within ± 3 years, how old was this branch when cut?

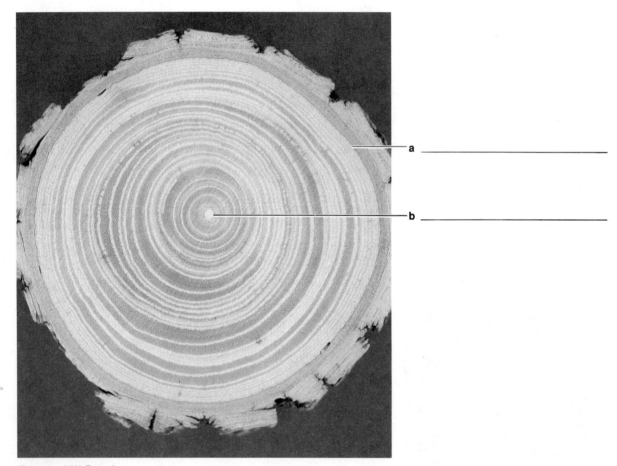

a _____

b _____

(Photo by J. W. Perry.)

8. The figure below shows the microscopic appearance of maple wood. Basing your answer on the knowledge gained from the study of the woody stem section, identify cell type **a** and the "line" of cells at **b**. (Note: the outside of the tree from which this section was taken is toward the bottom of the page.)

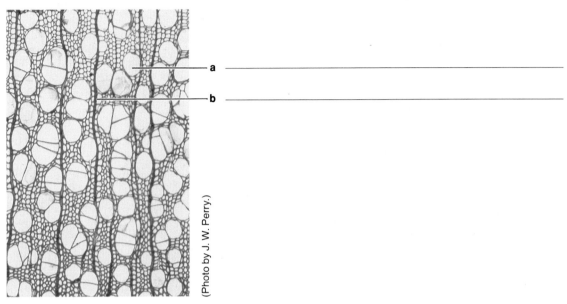

a _____

b _____

(58×).

(Photo by J. W. Perry.)

9. The figure below shows a section of a leaf from a plant that is adapted to a dry environment. Its lower epidermis has depressions, and the stomata are located in the cavities. Even though it's different from the leaf you studied in lab, identify the regions labeled **a**, **b**, and **c**.

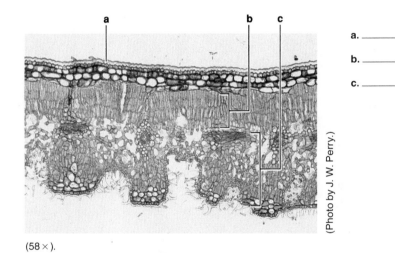

a. _____

b. _____

c. _____

(58×).

(Photo by J. W. Perry.)

10. Even though woody stems are not photosynthetic, they do contain cells that are alive and undergoing respiration. What structure present in the periderm allows oxygen to reach these cells?

Animal Organization

OBJECTIVES

After completing this exercise you will be able to:

1. define *tissue, organ, system, organism, histology, basement membrane, goblet cell, cilia, brush border of microvilli, keratinization, keratin, collagen fiber, elastic fiber, fibroblast, fat cell, lumen, chondrocyte, lacuna, Haversian canal, lamella, osteocyte, actin filaments, myosin filaments, intercalated disks, neuron, cell body, dendrites, axon, neuroglia;*

2. discuss the high degree of organization present in animal structure and explain its significance;

3. recognize the four basic tissues and their common mammalian subtypes;

4. list the functions of the four basic tissues and their common mammalian subtypes;

5. explain how the four basic tissues are combined to make organs.

6. list each system of a mammal and its vital functions;

7. describe the basic plan of the mammalian body;

8. locate the major organs in a mammal's body.

INTRODUCTION

Unicellular eukaryotic organisms contain organelles that carry on all the functions vital to life. In multicellular organisms, each cell is specialized to emphasize certain functions, although most carry on basic activities such as cellular respiration.

Collections of similar cells that interact as a structural, functional unit are called **tissues.** A tissue includes any extracellular material produced and maintained by its cells. There are four basic tissue types in animals, and they are combined much like a quadruple-decker sandwich to make all of the **organs.** Organs perform specific functions and are strung together functionally, and usually structurally, to form **systems.** Each system carries on a vital function, and considered together, the systems equal the animal **organism.** Tissues, organs, and systems are present in most, but not all, animals.

I. Tissues

The study of tissues is called **histology.** The four basic tissues types are **epithelial tissue, connective tissue, muscle tissue,** and **nervous tissue.** Each type has subtypes. In order to study these subtypes, you will examine a variety of tissues and organs, all permanently mounted on glass microscope slides. Most slides have sections of tissues and organs, usually 6 to 10 μm thick. Other slides have whole pieces of organs that are either transparent or are teased (gently pulled apart) and spread on the surface of the slide until they are thin enough to see through. Some organs are simply smeared onto the surface of slides.

These prepared slides have been chosen to show not only the characteristics of each tissue but also how variations in the basic structure of each allows for the related but distinct functions of its subtypes. For the sake of simplicity, the following exercise will consider only the common subtypes of adult mammalian tissues.

MATERIALS

Per student:

• compound microscope, lens paper, a bottle of lens-cleaning solution (optional), a lint-free cloth (optional), a dropper bottle of immersion oil (optional)

• prepared slides of the following:
— whole mount of mesentery (simple squamous epithelium)
— section of the cortex of the mammalian kidney
— section of trachea
— cross section of small intestine (preferably of the ileum)
— section of esophagus
— section of mammalian skin
— sections of contracted and distended urinary bladders
— teased spread of loose (areolar) connective tissue
— longitudinal section of tendon
— section of white adipose tissue
— ground cross section of compact bone
— sections of the three muscle types
— smear of the spinal cord of an ox (neurons)

Per lab room

• demonstration of intercalated disks in cardiac muscle tissue
• 50-mL beaker three-fourths full of water
• 50-mL beaker three-fourths full of immersion oil
• 2 small glass rods

PROCEDURE

A. Epithelial Tissues

Epithelial tissues are widespread throughout the body, covering both the body's outer (epidermis of

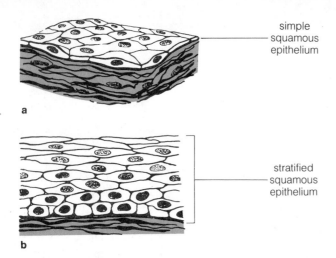

Figure 28-1 Squamous epithelia. (a) Simple. (b) Stratified.

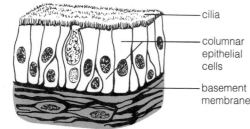

Figure 28-2 Pseudostratified columnar epithelium. It is columnar because of the shape of the cells that reach the free surface.

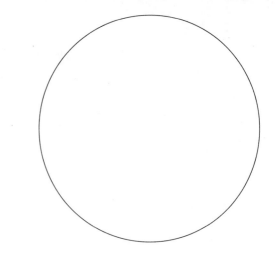

Figure 28-3 Drawing of a surface view of simple squamous epithelium (_____ ×).

skin) and inner surfaces (ventral body cavities), and lining the inner surfaces of tubular organs (the small intestine, for example). Their main functions are *protection* and *transport* (secretion and absorption, for example). Specialized functions include sensory reception and the maintenance of the body's gametes (egg and developing sperm).

Epithelial cells carry on rapid cell division in the adult, and various stages of mitosis (see Exercise 9) are often seen in this tissue type. As a consequence, epithelia have the highest rates of cell turnover among tissues. For example, the lining of the small intestine replaces itself every three to four days. Epithelial tissues don't have blood vessels and are attached to the underlying connective tissue by an extracellular **basement membrane** that is difficult to see if it is not specially stained. The following rules are used to name the subtypes of epithelial tissue:

RULE 1 What is the shape of the outermost cells? There are three choices: *squamous* (scalelike or flat), *cuboidal,* or *columnar.* These choices describe the shape of the cells when viewed in a section perpendicular to the surface. In surface view, all three types are polygonal.

RULE 2 How many layers of cells are there? Epithelial tissues are *simple* if there is one layer of cells; they are *stratified* if there are two or more.

These principles are illustrated for squamous epithelia in figure 28-1.

There is a further complication. If an epithelium appears stratified because it has more than one layer of nuclei, but electron-microscopic examination shows that all cells reach the basement membrane, it is called *pseudostratified* (fig. 28-2).

Also, the epithelium lining the inside of the urinary bladder and of some of the other urinary system organs changes its subtype as it fills with urine. This epithelium is called *transitional* to avoid confusion.

There are six common subtypes of epithelia: *simple squamous, simple cuboidal, simple columnar, pseudostratified columnar, stratified squamous,* and *transitional.* Find and examine these subtypes in the following slides:

1. *Whole mount of mesentery.* The mesentery is a fold of the abdominal wall and holds the intestines in place. This slide provides a surface view of the **simple squamous epithelium,** which lines the ventral body cavities. What is the shape of the surface cells?

_____

Draw in figure 28-3 what you see. In the space provided at the end of the figure title, note the total magnification of the compound microscope you used to make this drawing. Repeat this procedure for each subsequent drawing.

2. *Kidney.* With the help of figure 28-4, find a *renal corpuscle.* A renal corpuscle is composed of a tuft of capillaries called the *glomerulus,* which lies in *Bowman's capsule,* the cup-shaped end of one of the kidney's functional units, the nephron. The walls of Bowman's capsule and the capillaries of the glomerulus are simple squamous epithelia. A sectional view of simple squamous epithelium is easiest to see in the outer wall of Bowman's capsule. Around the renal corpuscle you can see a number of transverse and oblique sections of the tubular portion of the nephron. Their walls are composed of **simple cuboidal epithelium.** The main

376

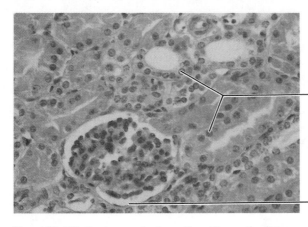

simple cuboidal epithelium

simple squamous epithelium

Figure 28-4 Photomicrograph of a section of the cortex of the kidney (297 ×). (Photo by D. Morton.)

cilia

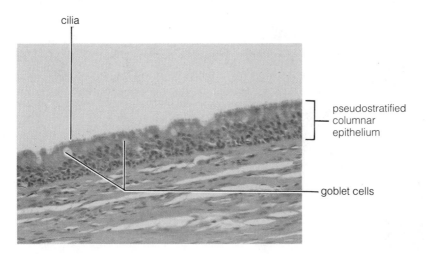

pseudostratified columnar epithelium

goblet cells

Figure 28-5 Photomicrograph of a section of the inner surface of the trachea (297 ×). (Photo by D. Morton.)

function of the nephrons and the entire kidney is the production of urine.

3. *Trachea.* The trachea is a tubular organ that conveys air to and from the lungs. Note that its inner surface is lined by **pseudostratified columnar epithelium** (fig. 28-5). Locate unicellular glands called **goblet cells** because of their shape. They secrete mucus that is difficult to see unless it is specifically stained. Using high power, do you see the numerous hairlike structures that project from the surface of the columnar epithelial cells? These are **cilia,** which in life move synchronously, sweeping mucus and trapped bacteria and debris up the trachea to the throat. When you clear your throat, you collect this mucus and swallow it.

4. *Small intestine.* The small intestine is a tubular organ that connects the stomach to the large intestine. Its inner surface is lined with **simple columnar epithelium** (fig. 28-6). Note that *goblet cells* are present. Don't confuse the **brush border of microvilli** with cilia. Compare the height of this border with that of the cilia from the previous slide. The size of the nucleus is a good reference point.

The microvilli are primarily responsible for the large surface area of the small intestine. Individual microvilli can best be seen at the electron-microscopic level (see fig. 28-7).

5. *Esophagus.* The tubular esophagus connects the throat to the stomach. Examine the epithelium lining the inner surface of the esophagus (fig. 28-8). Is the shape of the outermost cells squamous, cuboidal, or columnar?

_____

Are there one or many layers of cells in this tissue?

_____

Name this subtype of epithelial tissue.

_____

6. *Skin.* The skin is divided into three layers (fig. 28-9). The *epidermis* is composed of stratified squamous epithelium, while the *dermis* and *hypodermis* (or subcutaneous layer) are connective tissue and will be

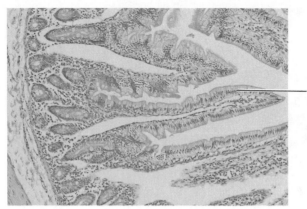

simple columnar
epithelium

a

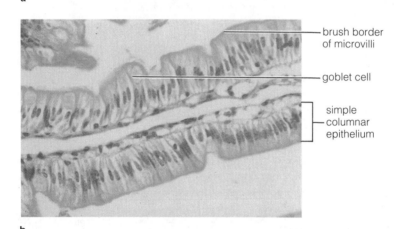

brush border
of microvilli

goblet cell

simple
columnar
epithelium

b

Figure 28-6 Photomicrographs of a section of the inner surface of the small intestine. (**a**) 74×. (**b**) 297×. (Photos by D. Morton.)

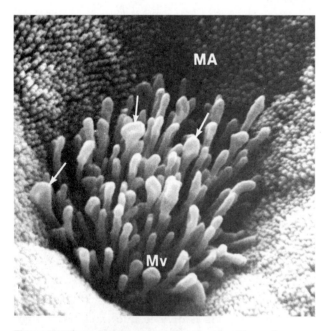

MA

Mv

Figure 28-7 Scanning electron microscopic view of the surface of the small intestine: MA—Small microvilli on surface of absorptive cells and Mv—Longer, larger microvilli on surface of a goblet cell (12,220×). (Photo from R. Kessel and R. Kardon, *Tissues and Organs.* Copyright © 1979 W. H. Freeman and Company. Used by permission.)

studied later. The epidermis is the most extreme example of a protective epithelium. Strata in the epidermis are caused by the process of **keratinization,** whereby the cells transform themselves into bags of **keratin.** It is this protein that gives skin its tough, flexible, and water-resistant surface.

At one of the free edges of your section of mammalian skin, locate the **keratinized stratified squamous epithelium.** It will be stained bluer than the predominately pink connective tissue layers. Hair follicles and multicellular sweat glands may be present in the connective tissue layers. These structures grow into the connective tissue layers from the epidermis during the development of the skin.

7. *Urinary bladder.* There are two sections on this slide. One is from a contracted bladder, the other from a distended bladder. Locate the **transitional epithelium** at the surface of one of these sections (fig. 28-10). The transitional epithelium from the contracted bladder looks like stratified cuboidal epithelium. The transitional epithelium of the distended bladder is thinner and looks like stratified squamous epithelium. Draw a high-power view of these two extremes in figure 28-11.

direction of surface

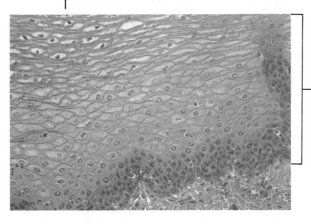

stratified squamous epithelium

Figure 28-8 Photomicrograph of a section of the epithelial tissue lining the inner surface of the esophagus (197 ×). (Photo courtesy Ripon Microslides, Inc.)

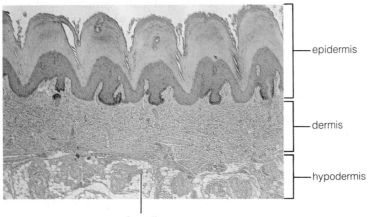

epidermis

dermis

hypodermis

fat cell

Figure 28-9 Photomicrograph of a section of skin (35 ×). (Photo courtesy Ripon Microslides, Inc.)

contracted transitional epithelium

distended transitional epithelium

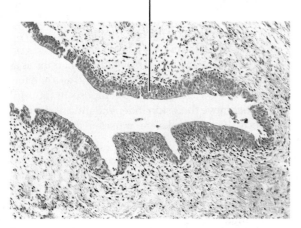

a

b

Figure 28-10 Photomicrographs of low-power views of (**a**) contracted and (**b**) distended urinary bladder (89 ×). (Photo courtesy Ripon Microslides, Inc.)

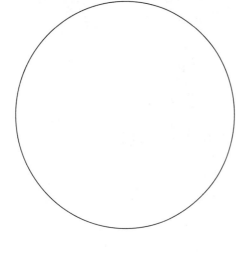

a   b

Figure 28-11 Drawings of high-power views of transitional epithelia from (**a**) contracted and (**b**) distended urinary bladders (_____ ×).

B. Connective Tissue

Connective tissues occur in all parts of the body. They contain a large amount of material external to the cells, called the *extracellular matrix*. This matrix consists of *fibers* embedded in *ground substance*. There are two basic kinds of fibers: collagen and elastic. Collagen fibers are tough, flexible, and inelastic, whereas elastic fibers stretch when pulled, returning to their original length when the pull is removed. Except for cartilage, connective tissues contain blood vessels. Similar to epithelia, connective tissue cells are capable of cell division in the adult, but at a reduced rate. Connective tissue subtypes can be classified as follows:

GROUP 1   Soft (connective tissue proper)
a. loose (few fibers that run in all directions; also called areolar)
b. dense (many fibers)
   i. regular (fibers run in the same direction)
   ii. irregular (fibers run in all directions)
c. special (for example, white adipose)

GROUP 2   Hard
a. cartilage (ground substance is polymerized or jellylike)
b. bone (ground substance is mineralized)

GROUP 3   Blood (although blood is a connective tissue, we will cover it in a later exercise)

Find and examine these subtypes in the following slides.

1. *Teased spread of loose connective tissue.* **Loose connective tissue** forms much of the packing material of the body and fills in the spaces between other tissues. Many other kinds of cells live in this tissue and play important roles in the body's immune system.

Find **collagen fibers** and **elastic fibers** in your slide

of loose connective tissue (fig. 28-12). Collagen fibers are stained light pink and are variable in diameter, but they are wider than the elastic fibers. Elastic fibers are darkly stained, thin, and branched. Areolar connective tissue contains a number of different cell types. To see a **fibroblast**—the cell that produces the matrix of loose and other soft connective tissues—look for an elongated, oval-shaped nucleus associated with a fiber. The amorphous ground substance of all soft connective tissues consists of a viscous soup of carbohydrate-protein molecules and is usually extracted from sections of these tissues during processing.

2. *Skin.* Reexamine the section of skin. Look at the second layer of the skin, the dermis, which is primarily composed of **dense fibrous irregular connective tissue** (fig. 28-9). The term *fibrous* refers to the high concentration of collagen fibers produced by the resident fibroblasts. As the name of this subtype indicates, there is a large number of apparently randomly oriented collagen fibers in the matrix. The dermis cushions the body from everyday stresses and strains.

Note that the looser fibrous irregular connective tissue of the hypodermis has a lower concentration of collagen fibers and islands of fat cells. It functions as a shock absorber, as an insulating layer, and as a site for storing water and energy (white adipose tissue). In animals that move their skin independently of the rest of the body (like cats), skeletal muscle tissue is found in the hypodermis.

3. *Tendon.* A tendon connects a skeletal muscle organ to a bone organ. Tendons are composed predominately of **dense regular fibrous connective tissue.** Examine a longitudinal section of a tendon. How are the fibers arranged (fig. 28-13)?

_____

_____

_____

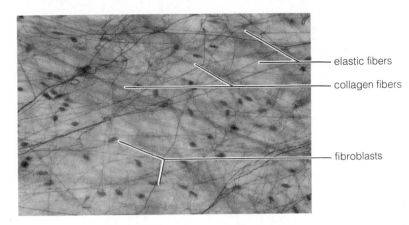

Figure 28-12 Photomicrograph of a spread of loose connective tissue (297 ×). (Photo by D. Morton.)

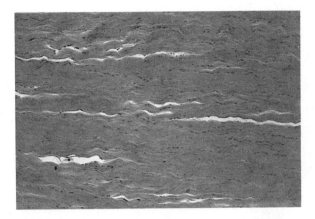

Figure 28-13 Photomicrograph of a section of dense regular fibrous connective tissue (70 ×). Photo courtesy Ripon Micro-slides, Inc.)

This design makes tendons very strong, much like a rope composed of braided strings, which in turn are made of even smaller fibers. How are the fibroblasts oriented relative to the arrangement of the fibers?

_____

_____

_____

4. *White adipose tissue.* As you see in figure 28-14, **white adipose tissue** consists mainly of **fat cells.** However, careful examination of the section using high power shows them to be surrounded by delicate collagen fibers, fibroblasts, and capillaries. Because the fat has been lost during the slide preparation, the fat cells look empty. The primary function of this tissue is energy storage.

5. *Trachea.* Reexamine the section of the trachea. Locate a portion of one of the rings of **hyaline cartilage** in the wall of the trachea (fig. 28-15). The rings of cartilage prevent the wall of the trachea from collapsing and closing the lumen. A **lumen** is the space within a hollow organ. Look for **chondrocytes** (cartilage cells) within the matrix. Chondrocytes are located

in **lacunae** (the singular is *lacuna*), small holes in the matrix.

Although invisible, many collagen fibers are embedded in the polymerized ground substance. They cannot be seen because the indices of refraction of these two matrix components are similar. Your instructor has set up a demonstration of this phenomenon. Observe the two labeled beakers, one filled with immersion oil and the other with water. Look at the glass rod that has been placed in each of them. The index of refraction of glass is about 1.58, that of water about 1.33, and that of immersion oil about 1.52. In which fluid is it easier to see the glass rod?

_____

In locations where cartilage has to be more durable (intervertebral disks, for example), the collagen content is higher and the fibers are visible. In other sites (such as outer ear flaps), large numbers of elastic fibers are present. Elastic cartilage is deformed by a small force and returns to its original shape when the force is removed.

6. *Bone.* Living bones are amazingly strong. **Bone** is a hard yet flexible tissue. Its hardness is due to a mineral (predominantly a calcium-phosphate salt called hydroxyapatite) deposited in the matrix. Its flexibility comes from having the highest collagen content of all connective tissues.

Examine the cross section of compact bone tissue (fig. 28-16). This preparation has been produced by grinding a piece of the shaft of a long bone with coarse and then finer stones until a thin wafer remains. Although only the mineral part of the matrix is present, the basic architecture has been preserved. In living bone, blood vessels and nerves are present in the large **Haversian canals.** These are surrounded by concentric layers of matrix called **lamellae** (the singular is *lamella*). There are intervening rings of **lacunae** (smaller holes), which in living bone contain cells called **osteocytes.** Find a *Haversian canal* and identify *lamellae* and *lacunae* within it.

ANIMAL ORGANIZATION

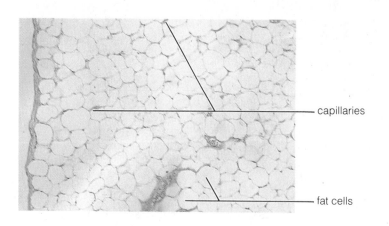

capillaries

fat cells

Figure 28-14 Photomicrographs of a section of white adipose tissue (70 ×). (Photo courtesy Ripon Microslides, Inc.)

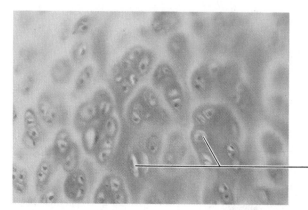

chondrocytes in lacunae

Figure 28-15 Photomicrograph of hyaline cartilage in the wall of the trachea (297 ×). (Photo by D. Morton.)

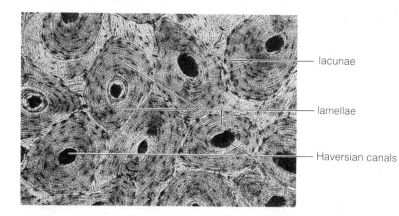

lacunae

lamellae

Haversian canals

Figure 28-16 Transverse section of ground compact bone (70 ×). (Photo courtesy Ripon Microslides, Inc.)

In young living bone, the lines that you see connecting lacunae with each other and with the Haversian canal are little canals that contain the cytoplasmic processes of osteocytes. Thus, the osteocytes can easily exchange nutrients, wastes, and other molecules with the blood. By comparison, substances in cartilage have to diffuse across the matrix between chondrocytes and blood vessels in the surrounding soft connective tissue. Which tissue, bone or cartilage, will heal quicker?

382

Table 28-1 Characteristics of Muscle Fibers

Muscle Tissue	Nucleus		Transverse Striations	Special Features
	Number	Position		
skeletal	many	peripheral	yes	——
cardiac	one	central	yes	intercalated disks
smooth	one	central	no	——

C. Muscle Tissue

Muscle tissue is contractile. Its cells (fibers) can shorten and produce changes in the position of body parts, or they try to shorten and produce changes in tension. Contraction results from interactions between two types of protein filaments: **actin** and **myosin.** Like epithelial tissue, muscle tissue is primarily cellular; but unlike both epithelial and connective tissues, its cells do not normally divide in the adult. Therefore, dead fibers usually cannot be replaced. There are three subtypes of muscle tissue: *skeletal, cardiac,* and *smooth.* The main characteristics of their fibers are summarized in table 28-1.

Skeletal muscle tissue is found in skeletal muscle organs and is under voluntary control. This means that your conscious mind can order it to contract, but not all of its contractions are voluntary. In fact, most are not.

Cardiac muscle (located in the heart) and smooth muscle (found in the walls of tubular organs like the small intestine) are both involuntary. *Involuntary* means that these muscle types are normally controlled at the unconscious level and cannot be directly controlled by the conscious mind.

Examine your slide with sections of all three muscle subtypes (fig. 28-17). Identify their characteristics (see table 28-1). In skeletal and cardiac muscle fibers, actin and myosin filaments overlap to produce the alternating pattern of light and dark bands (transverse striations) seen in these tissues. Only cardiac muscle cells are branched. Where the branch of one fiber joins another, the cells are stuck together and in direct communication through a complex of cell-to-cell junctions. The complex is called an **intercalated disk.** They are present in your section, but if you have trouble seeing them, look at the demonstration of intercalated disks set up by your instructor.

D. Nervous Tissue

Nervous tissue is found in the brain, spinal cord, nerves, and all of the body's organs. Its function is the point-to-point transmission of information. A mes-

nuclei of cardiac muscle cells   intercalated disks

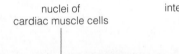

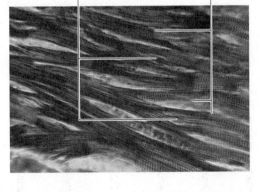

a

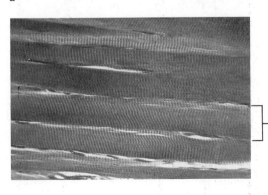

portion of a longitudinally sectioned skeletal muscle cell

b

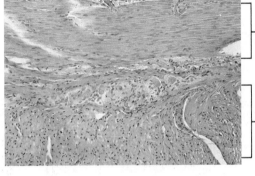

longitudinally sectioned smooth muscle cells

transversely sectioned smooth muscle cells

c

Figure 28-17 Photomicrographs of sections of muscle tissue. (**a**) Cardiac (400×). (**b**) Skeletal (250×). (**c**) Smooth (70×). (Photos courtesy Ripon Microslides, Inc.)

ANIMAL ORGANIZATION

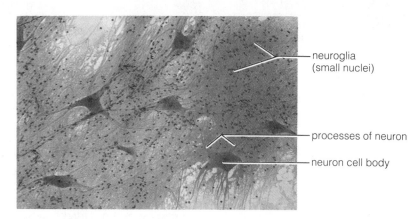

Figure 28-18 Photomicrograph of motor neurons in smear of spinal cord (70 ×). (Photo courtesy Ripon Microslides, Inc.)

sage is carried by impulses that travel along the functional unit of the nervous system, the **neuron.** Like muscle cells, neurons do not divide in the adult, and therefore replacement is impossible.

The largest cells in figure 28-18 are *motor neurons,* which connect the spinal cord to muscle fibers or glands. Find a similar cell in your smear of an ox spinal cord. The motor neuron has a **cell body** and a number of slender cytoplasmic extensions called neuron *processes,* including one long **axon** and several shorter **dendrites.** The dendrites and cell body are stimulated within the spinal cord, and the axon conducts impulses out of it. The cells with smaller nuclei are accessory cells of the spinal cord and brain and are called **neuroglia.** Accessory cells help neurons function and make up about half the mass of nervous tissue.

II. Analysis of an Organ

As part of your study of tissues, you already have examined some aspects of the microscopic structure of a number of organs, including the kidney, trachea, small intestine, and skin. To repeat an important point, each of these organs is composed of various subtypes of the four basic tissues. The subtypes present in any organ contribute to its specific function. Let us examine more closely the function of the tissues of the small intestine (fig. 28-19).

The specific functions of the small intestine are digestion of food, absorption of the end products of digestion, and transportation of indigestible material (fiber and so on) to the large intestine.

MATERIALS

Per student:

• compound microscope, lens paper, a bottle of lens-cleaning solution (optional), a lint-free cloth (optional), a dropper bottle of immersion oil (optional)

• prepared slide with a cross section of small intestine (preferably of the ileum)

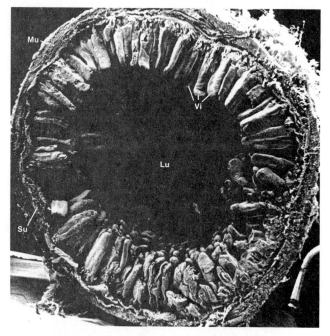

Figure 28-19 Scanning electron micrograph of a section of the small intestine: Lu—Lumen, Mu—Muscle layer, Su—Submucosa, and Vi—Villi (14 ×). (Photo from R. Kessel and R. Kardon, *Tissues and Organs.* Copyright © 1979 W. H. Freeman and Company. Used by permission.)

PROCEDURE

Find and examine the following tissue subtypes in your section of the small intestine.

A. Simple Columnar Epithelium

This tissue is ideally suited for absorption of molecules from a hostile environment. The epithelium is one cell thick to facilitate transport and contains goblet cells that secrete mucus, which is thought to protect its surface from being digested. The simple columnar epithelium is continuous with the ducts of multicellular glands that deliver enzymes and other molecules important in digestion to the lumen of the small intestine.

B. Loose Connective Tissue

The simple columnar epithelium combines with loose connective tissue, which in turn combines with a thin layer of smooth muscle called the *muscularis mucosae* to form the *mucosa*. The muscularis mucosae separates the rest of the mucosa from a denser connective tissue layer, the *submucosa*.

The epithelial and connective tissue layers of the mucosa form fingerlike projections that protrude into the lumen. These projections are called *villi*. After the microvilli discussed earlier in this exercise, the villi are the second most important contributor to the large surface area of the small intestine. The connective tissue core of a villus contains a rich network of capillaries and a lymphatic capillary to receive the absorbed molecules. The walls of these blood vessels are porous, simple squamous epithelium.

C. Smooth Muscle of the Muscle Layer

The muscle layer contains two sublayers of smooth muscle arranged perpendicular to each other. In the inner sublayer the fibers are circular, while in the outer sublayer they are parallel to the longitudinal axis of the small intestine. These sublayers produce both *segmental contractions*, which mix the contents of the small intestine, and waves of contraction known as *peristalsis*, which sweep the contents of the small intestine along its length.

D. Nervous Tissue

Two nets of neurons in the wall of the small intestine control the contractions of the smooth muscle. One is in the submucosa, while the other is between the circular and longitudinal smooth muscle sublayers. The latter one is easier to find. Look for neuron cell bodies between the sublayers of the muscle layer.

E. Simple Squamous Epithelium of Serosa

The outer wall of the small intestine is lined with simple squamous epithelium that is continuous with and similar to the lining of the body cavity. This serosal membrane provides a moist, low-friction surface that allows for the free movement of abdominal organs.

As you see, the specific function of an organ is the sum of the individual functions of its tissues. Each organ in an animal can be analyzed in a similar fashion.

III. Systems

There are eleven systems in mammals, and each of them contains a number of organs. In this portion of

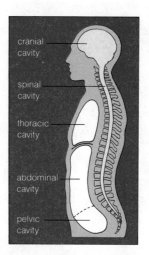

Figure 28-20 Major body cavities of humans. White signifies ventral body cavities; yellow, dorsal cavities. (After Starr and Taggart, 1989.)

the exercise, you will identify the vital functions of these systems and many of their constituent organs. Many of these organs are located in or near the major body cavities. The body cavities of humans, which are typical of mammals, are illustrated in figure 28-20.

MATERIALS

Per student:

- colored pencils — green, yellow, black, red, brown, pink, and blue

Per lab room:

- labeled demonstration dissection of a mouse (optional)
- demonstration dissection of a sheep brain

PROCEDURE

A. Vital Functions of Organ Systems

Complete table 28-2 on page 389 with the vital functions of each system.

B. Major Organs and Systems of the Ventral Body Cavities

Look at the following photographs of dissected mice (figs. 28-21, 28-22, and 28-23) showing the organs of the *ventral body cavities* — the *thoracic cavity* in the *thorax* and the *abdominal* and *pelvic cavities* (or *abdominopelvic cavity*) in the *abdomen*. In addition, your instructor may have prepared a labeled demonstration dissection of a mouse. Use your text to identify the systems to which the labeled organs belong. Indicate the system to which each organ belongs by underlining its name with a colored pencil — green for the respiratory

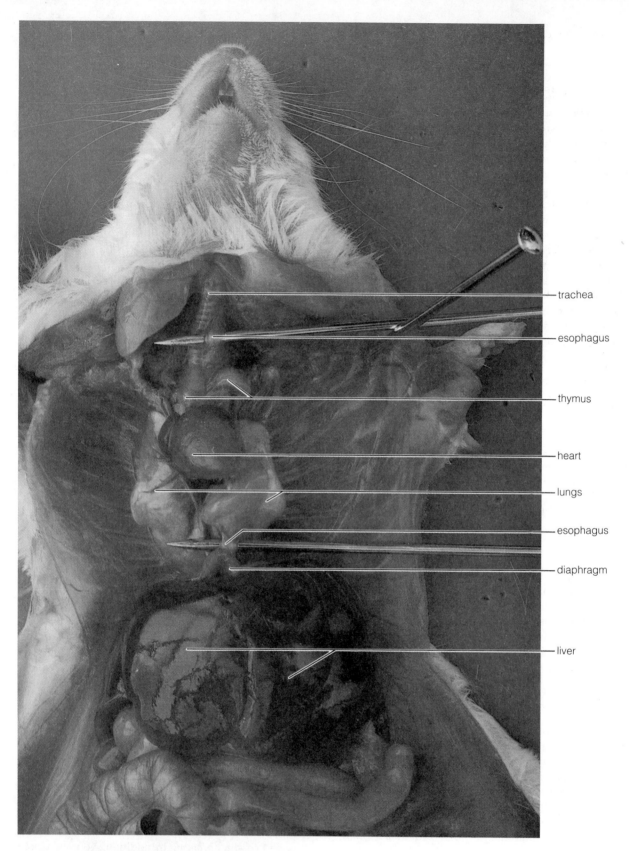

—— trachea

—— esophagus

—— thymus

—— heart

—— lungs

—— esophagus

—— diaphragm

—— liver

Figure 28-21 Ventral view of organs in thoracic cavity. (Photo by D. Morton.)

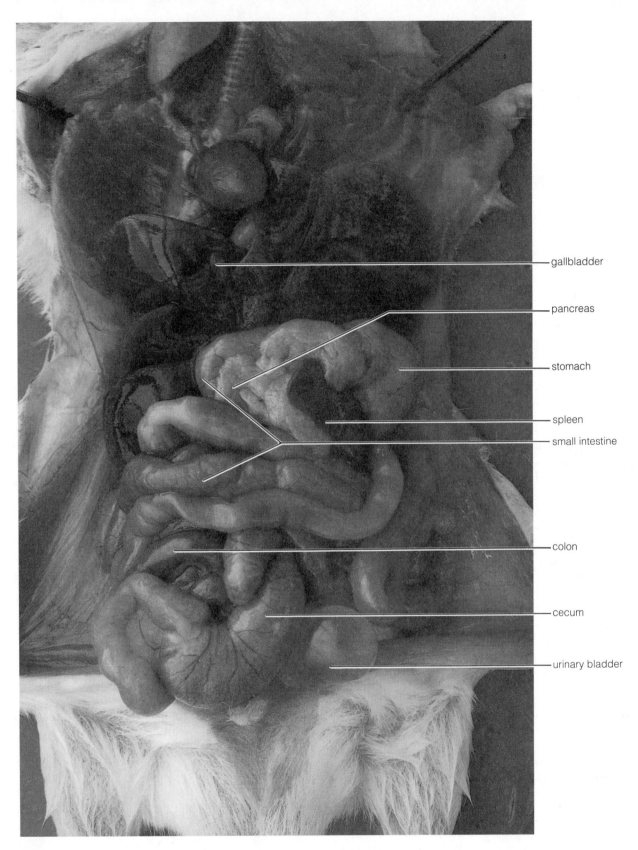

gallbladder

pancreas

stomach

spleen

small intestine

colon

cecum

urinary bladder

Figure 28-22 Ventral view of organs of abdominopelvic cavity.
(Photo by D. Morton.)

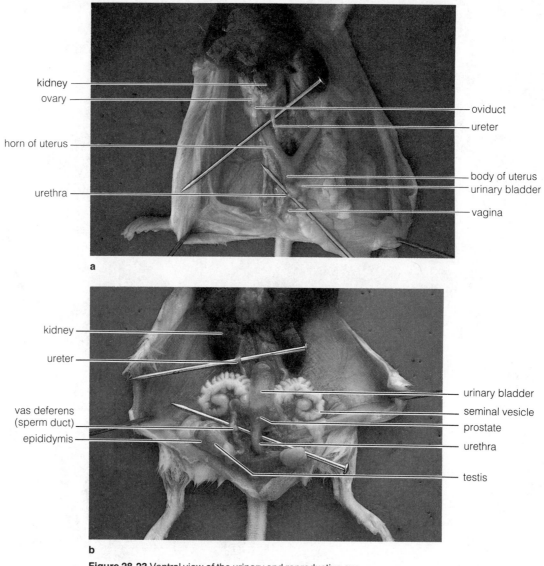

Figure 28-23 Ventral view of the urinary and reproductive systems of (**a**) a female and (**b**) a male mouse. The digestive system has been removed from each mouse. (Photos by D. Morton.)

system, yellow for the digestive system, black for the lymphatic system, red for the circulatory system, brown for the urinary system, pink for the female reproductive system, and blue for the male reproductive system.

C. Major Organs and Systems of the Dorsal Body Cavities

The organs of the nervous system are located in the *dorsal body cavities*—the brain in the *cranial cavity* and the spinal cord in the *spinal cavity.* Examine the demonstration dissection of a sheep brain and identify the structures indicated in figure 28-24 on page 390.

O P T I O N A L
Mouse Dissection

Your instructor may provide you with directions for a mouse dissection.

Table 28-2 The Organ Systems of Mammals

Systems	Vital Functions
Integumentary	
Nervous	
Endocrine	
Skeletal	
Muscular	
Circulatory	
Lymphatic	
Respiratory	
Digestive	
Urinary	
Reproductive	

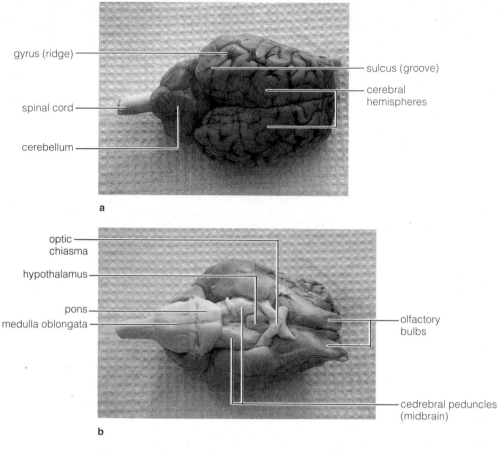

gyrus (ridge)

spinal cord

cerebellum

sulcus (groove)

cerebral
hemispheres

a

optic
chiasma

hypothalamus

pons

medulla oblongata

olfactory
bulbs

cedrebral peduncles
(midbrain)

b

Figure 28-24 (**a**) Dorsal and (**b**) ventral views of sheep brain.
(Photos by D. Morton.)

PRE-LAB QUESTIONS

_____ 1. Histology is the study of (a) cells, (b) organelles, (c) tissues, (d) organisms.

_____ 2. Collections of similar cells and their products that interact as a structural, functional whole are called (a) organs, (b) systems, (c) tissues, (d) organelles.

_____ 3. Organs strung together functionally and usually structurally form (a) organs, (b) systems, (c) tissues, (d) organelles.

_____ 4. The structures that carry on all the functions vital to life in unicellular eukaryotic organisms are called (a) organs, (b) systems, (c) tissues, (d) organelles.

_____ 5. Structures formed when the four tissues are combined are called (a) organs, (b) systems, (c) tissues, (d) organelles.

_____ 6. An epithelial tissue formed by more than one layer of cells and with columnlike cells at the surface would be called (a) simple squamous, (b) stratified squamous, (c) simple columnar, (d) stratified columnar.

_____ 7. The middle layer of the skin (a) is called the dermis, (b) is connective tissue, (c) contains collagen fibers, (d) is all of the above.

_____ 8. Connective tissues are composed of (a) cells, (b) extracellular matrix, (c) axons, (d) a and b.

_____ 9. To which subtype of muscle tissue does a fiber with transverse striations and many peripherally located nuclei belong? (a) skeletal, (b) cardiac, (c) smooth, (d) none of the above.

_____ 10. In which tissue would you look for cells that function in point-to-point communication? (a) connective, (b) epithelial, (c) muscle, (d) nervous.

E X E R C I S E   2 8

Animal Organization

P O S T - L A B   Q U E S T I O N S

1. In the correct order from smallest to largest, list the levels of organization present in most animals.

2. Describe the main structural characteristics of the four basic tissues.
   a. epithelial tissue

   b. connective tissue

   c. muscle tissue

   d. nervous tissue

3. Describe the main functions of the four basic tissues.
   a. epithelial tissue

   b. connective tissue

   c. muscle tissue

   d. nervous tissue

4. Choose any organ. What is its specific function? Describe its functional histology.

5. Identify the following tissues (all photos by D. Morton).

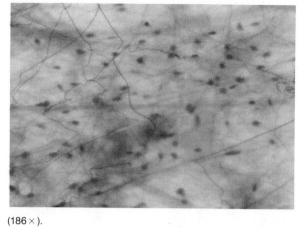

(186×).

a. _____

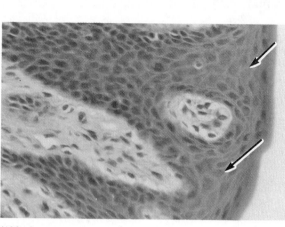

(186×).

b. _____

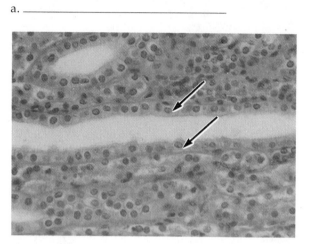

(186×).

c. _____

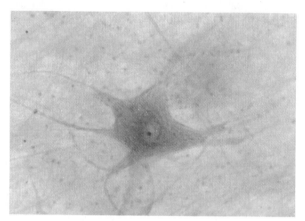

(297×).

d. _____

6. Identify the following structures.

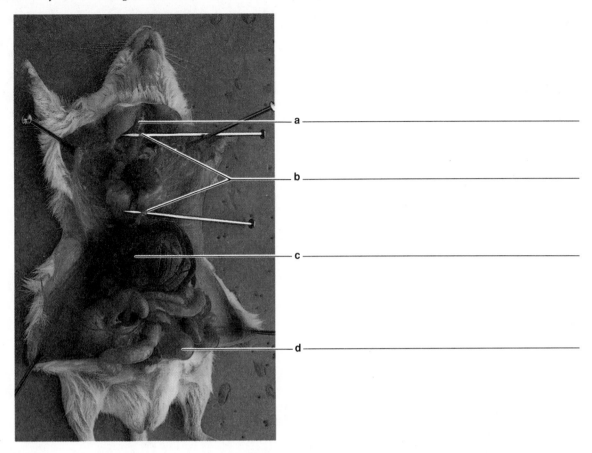

a _____

b _____

c _____

d _____

7. Jacob Bronowski referred to the different levels of organization present in living things as "stratified stability." What do you think this phrase means relative to the structural evolution of animals?

Dissection of the Fetal Pig:

Introduction, External Anatomy, and the Muscular System

OBJECTIVES

After completing this exercise you will be able to:

1. define *fetus, digitigrade locomotion, plantigrade locomotion, antagonistic muscles*;

2. locate and describe the external features of a fetal pig;

3. determine the sex of a fetal pig;

4. describe the function of the umbilical cord;

5. define the origin, insertion, and action of a skeletal muscle;

6. identify some of the major skeletal muscles of a mammal.

INTRODUCTION

In this and the following two exercises, you will examine in some detail the external and internal anatomy of a fetal (from **fetus,** an unborn mammal) pig. As the pig is a mammal, many aspects of its structural and functional organization are identical with those of other mammals, including humans. Thus, a study of the fetal pig is in a very real sense a study of ourselves.

The specimens you will use in the laboratory were purchased from a biological supply house, which obtained them from a plant where pregnant sows were slaughtered for food. On the average, a sow may produce seven to twelve offspring per litter. The period of development in the uterus (*gestation period*) is approximately 112–115 days. Generally, lab specimens are approximately 20–30 cm (8–12 inches) long, and their age is between 100 days and nearly full term.

At the slaughterhouse the fetuses are quickly removed from the sow, cooled, and embalmed with formalin or another preservative, which is injected through one of the umbilical arteries. Following this, the arterial and venous systems are injected under pressure with latex or a rubberlike compound. The arteries are injected with red latex through the umbilical artery, and the venous system is injected with blue latex through one of the jugular veins at the base of the throat.

During the fetal pig exercises, keep several points in mind. First, be aware that *to dissect* does not mean "to cut up," but rather primarily "to expose to view." Thus, proceed carefully and never cut or move more than is necessary to expose a given part. Second, for each structure or organ that you identify in the pig, ask yourself if an equivalent one is present in your body. If so, where is it located, and is its function similar to that in the fetal pig? Finally, pay particular attention to the spatial relationships of organs, glands, and other structures as you expose them. Realize that their positions in the body are not random. Carefully identify each structure and determine to what organ system it belongs, what its relationship to that organ system is, and what its general function is. Then determine how it is related to the other organ systems in the body. By proceeding in this manner, you will greatly enhance your understanding of the structure and function of the mammalian body.

To understand the dissection directions, you will need to become familiar with the terms used in virtually all anatomical work (see the appendix on p. 514). Spend a few minutes relating each of these terms to the fetal pig body and to your body as well. Refer to figure 29-1 to aid you in this exercise.

CAUTION

Preserved fetal pigs are kept in a formalin-based or other preservative solution. Wash any part of your body exposed to this solution with copious amounts of water. If the formalin solution is splashed into your eyes, wash them with the safety eyewash bottle for fifteen minutes.

I. External Anatomy of the Fetal Pig

Before you begin your examination of the internal structure of the fetal pig, you will examine the external features of its body. This will give you an opportunity to compare the body of the pig with other mammals. Remember, it is the external surface of an organism that has the greatest amount of contact with the environment. Thus, the greatest differences between two organisms may be their external features rather than their internal features.

MATERIALS

Per student pair:
• one preserved fetal pig injected with red and blue latex
• plastic bag to store fetal pig
• dissection pan
• plastic gloves and/or lanolin hand cream
• goggles if you wear contact lenses
• paper towels

Per lab room:
• liquid waste disposal bottle

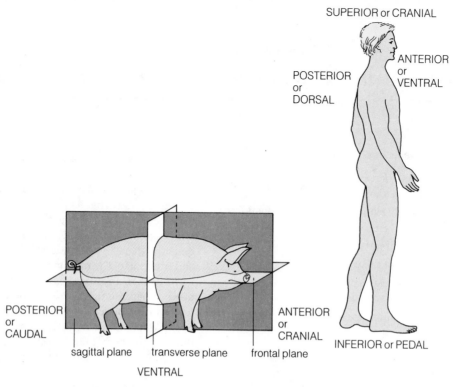

Figure 29-1 Lateral view of pig and human.

PROCEDURE

A. Preparing the Fetal Pig for Observation

- box of name tags (tags may be provided with fetal pigs)
- marking pens or pencils

1. The fetal pigs have been preserved in formalin or a phenol-based chemical, which may be irritating to hands and eyes. Therefore, wear protective plastic gloves or apply lanolin cream to your hands before proceeding with the dissection. Students wearing contact lenses should wear protective goggles.

2. Obtain a fetal pig and place it on a dissection pan lined with paper towels. If the plastic bag that may be supplied with your pig contains any excess preservative, pour the liquid into the waste bottle provided by your instructor, and save the bag. As you will be using the same fetal pig for several days, you should place it in the plastic bag at the end of each day's exercise so it does not dry out.

For easy identification, tie a name tag to a hindleg, the bag, or both, *according to the instructor's wishes.* Use a marking pen or pencil to fill in the tag.

B. General Observations

Identify the four regions of the fetal pig body: the large, compact **head;** the **neck;** the **trunk** with four **appendages** (the *limbs*) and the *tail* (see fig. 29-2). The trunk may be divided further into the **thoracic region,**

lumbar region, and **sacral region.** These regions, along with the cervical region in the neck, also describe the corresponding regions of the vertebral column or spine. We will return to this region during the study of the nervous system.

C. Head and Neck Region

1. Examine the head in more detail (fig. 29-3) and identify the **eyes** with **upper** and **lower lids,** the **external ears,** the **mouth,** and the characteristic **nose** or *snout.* Note the position of the **nostrils** or *external nares* on the snout. Feel the texture of the snout. It is composed of bone, cartilage, and other tough connective tissue and as such allows the pig to root and push soil and debris in its search for food.

2. Open the pig's mouth and note the **tongue** with its covering of **papillae,** which contain *taste buds.* Papillae are especially concentrated and prominent along the posterior edges and tip of the tongue. Also notice if any *baby teeth* are present. Like humans, pigs are omnivores; that is, they eat both animal and plant matter. We will return to the structure and placement of teeth in the section on the mouth.

D. Trunk

1. Place the pig on its back (dorsal surface) and examine its *abdomen* (belly). The most prominent feature of the underside (ventral surface) of the fetal pig is the **umbilical cord** seen near the posterior end of the abdomen (see fig. 29-4). Is this structure present in the adult pig?

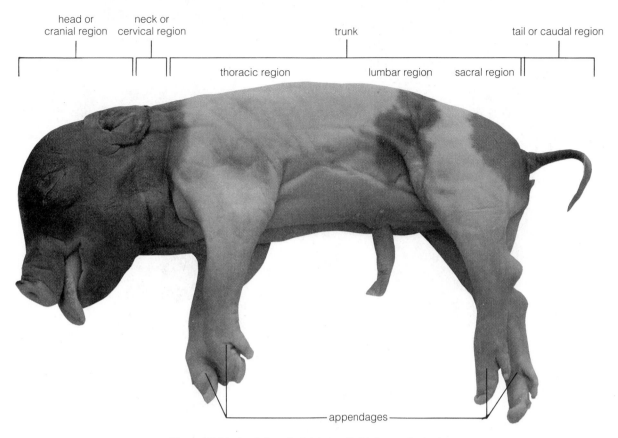

Figure 29-2 Lateral view of a fetal pig with the four major body regions indicated. (Photo by D. Morton.)

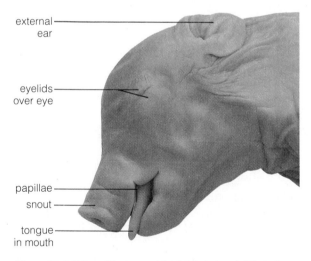
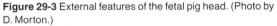

Figure 29-3 External features of the fetal pig head. (Photo by D. Morton.)

(Yes or no) _____

Adult human?

(Yes or no) _____

During its development, the fetus was connected to the placenta on the uterine wall of its mother's reproductive system via the umbilical cord. The cord contains two *arteries* (red), a large *vein* (blue), and a fourth vessel, usually collapsed, the *allantoic duct*. The blood in the umbilical vein carries nutrients and oxygen from the mother to the fetus, and blood in the umbilical arteries carries waste materials and carbon dioxide from the fetus to the mother. We will study the umbilical cord in more detail during the exercises on the urogenital (excretory and reproductive) and circulatory systems.

2. Note on the ventral surface of the pig the pairs of **nipples** or *teats*. Both male and female pigs may have from five to eight pairs of these structures situated in two parallel rows on the **thoracic region** (chest) and **abdominal region** of the body. Finally, locate the **anus,** the posterior opening of the digestive tract. The anus is situated immediately under (ventral to) the tail.

E. Determining the Sex of Your Fetal Pig

1. The male is identified by (1) the presence of a single *urogenital opening* to the urinary and reproductive systems just behind the umbilical cord (see fig. 29-4b), and (2) the presence of a swelling on the posterior portion of the abdomen between the upper ends of the

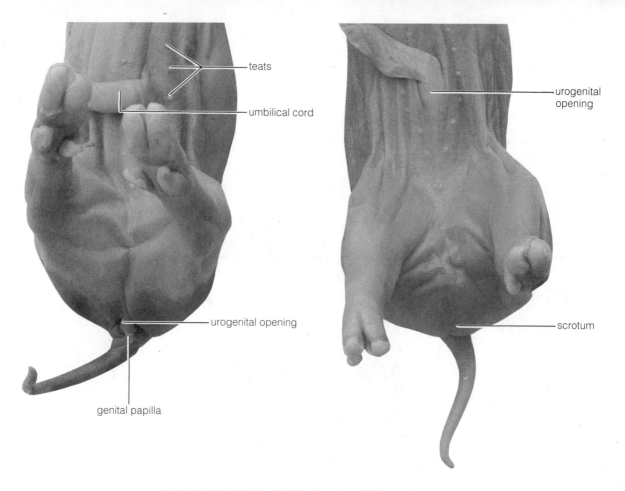

Figure 29-4 Ventral body of (**a**) the female and (**b**) the male fetal pig. (Photos by D. Morton.)

hindlimbs. The swelling is the **scrotum,** which contains the *testes,* a pair of small, oval structures that are part of the male reproductive system. These are generally easy to locate in older fetuses. Identify the *penis,* a large, tubular structure immediately under the skin posterior to the urogenital opening.

2. A female fetal pig can be identified by the presence of a single urogenital opening immediately ventral to the anus (see fig. 29-4a). A small fleshy piece of tissue, the *genital papilla,* projects from the urogenital opening.

3. Note that in both male and female fetal pigs, there is a common urogenital opening shared by the urinary system and the genital (reproductive) system. Both adult male pigs and human males have a similar structure. In adult female pigs and humans, however, there are separate opening to the urinary and reproductive systems.

F. Appendages

1. Examine carefully the feet and legs of your pig. The first *toe,* or *digit,* which corresponds to your big toe or thumb, is absent in both forelimbs and hindlimbs of the pig. Furthermore, the second and fifth digits are reduced in size, and the middle two digits, the third

and fourth, are flattened or *hoofed.* Pigs and other hoofed animals, referred to as **ungulates,** walk with the weight of their body borne on the tips of the digits. This type of walking is referred to as **digitigrade locomotion.** By contrast, humans use the entire foot for walking and have **plantigrade locomotion.** Compare the structure of your hands and feet with the foot of a pig.

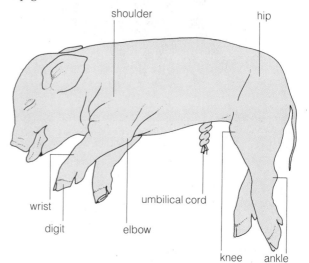

Figure 29-5 Lateral view of external features of the fetal pig.

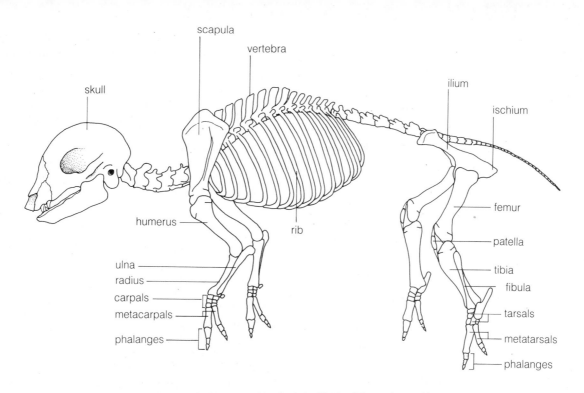

Figure 29-6 Skeleton of the fetal pig. The hyoid bone, located in the upper neck, and the sternum, to which many of the ribs are attached midventrally, are not included in this illustration. (After Gilbert, 1966).

2. Complete your study of the appendages of the fetal pig by using figure 29-5 to locate and identify the following structures and joints: **wrist, elbow, shoulder, ankle, knee,** and **hip.**

II. Muscular System

The contractions of *skeletal muscle* organs enable the body to move. Through the voluntary contractions of skeletal muscles, you may wink your eye, wave at a friend, or tap your foot in time to the beat of music. In this portion of the exercise, you will examine the structure of skeletal muscles and learn to identify some of them and describe their functions. The movement produced by a skeletal muscle is called its **action.**

Skeletal muscles are attached to the various parts of the skeleton by tough strips of dense fibrous connective tissue called *tendons.* The three parts of a typical muscle are the **origin** (the end attached to the less mobile portion of the skeleton), the **insertion** (the end attached to the portion of the skeleton that moves when the muscle contracts), and the *belly* (the middle portion between the points of attachment).

A skeletal muscle that moves an appendage one way usually has an opposing muscle that moves it in the opposite direction. Muscles with such opposite actions are called **antagonistic muscles.** For example, the biceps brachii is responsible for flexing the forelimb (in the pig) or the forearm (in humans), and the triceps brachii (its opposing muscle) straightens or extends them.

Realize that the attachments of the muscles can only be understood with reference to the skeleton, and that the shapes of the bones are meaningless when considered apart from the leverage they provide for muscles. Although we will not examine the skeleton in detail, it will be necessary to refer to its various parts during this portion of the exercise. Thus, during your dissection, use the illustration of the fetal pig skeleton (fig. 29-6) as a reference.

MATERIALS

Per student pair:

- one preserved fetal pig injected with red and blue latex
- dissection pan
- paper towels
- one dissecting kit including the following: scalpel, blunt probe, dissecting needle, scissors, forceps
- 4 large rubber bands *or* 2 pieces of string, each 60 cm long
- plastic gloves and/or lanolin hand cream
- goggles if you wear contact lenses

PROCEDURE

A. Directions for Dissection of the Skeletal Muscles

1. Place your specimen on its dorsal side in a dissection pan lined with paper towels and secure the feet with rubber bands or string as follows:

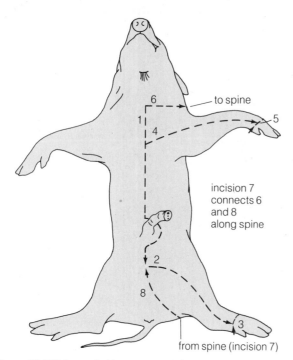

6 — to spine

1
4
5

incision 7
connects 6
and 8
along spine

2

8

3

from spine (incision 7)

Figure 29-7 Cuts needed to expose muscles.

Tie one end of a string to the left forelimb at the wrist, pass the string underneath the tray, and then tie the other end of the string to the right wrist so that the legs are spread apart under tension. Repeat with the hindlimbs, using the second piece of string. If rubber bands are available, tie two rubber bands together to make them longer. Then loop one end of the band around the right forelimb close to the foot. Bring the rubber band under the dissecting tray and loop it around the other forelimb to anchor the feet securely. Repeat this procedure with another set of rubber bands and anchor the hindlimbs to the tray.

2. For the following dissection, refer to the drawing in figure 29-7. The numbers in the drawing refer to the incisions to be made in the following dissection.

Using a scalpel, make an incision on the ventral side of the pig at the base of the neck (1). *Make sure you cut only through the skin and not the underlying tissues.* With your scissors, continue the incision posteriorly to the umbilical cord, then around the cord on the left side (the pig's left side) and to the region between the hindlegs. From this point cut down the medial surface of the left hindleg toward the foot (2) and around the pig's ankle (3). Make a similar cut from the midventral incision in the thorax down the medial surface of the left forelimb (4) and around the pig's wrist (5).

3. Return to the original incision on the ventral surface of the neck and extend it around the left side of the pig's neck dorsally toward the spine (6). To do this you will need to remove the rubber bands or string and place the pig on its right side in the dissecting pan. When you reach the spine, continue the incision posteriorly along the spine to the base of the tail (7). Finally, connect the posterior ends of incisions 7 and 1

(8). After completing the incisions, place the pig on its back and secure the legs with rubber bands or string as you did earlier.

4. To skin your specimen, grasp the cut edge of the skin at the base of the throat and begin easing the skin loose from the underlying tissues. Use your blunt probe between the skin and underlying connective tissues, working slowly until you have removed the skin from the ventral portion of the trunk and the limbs. Then, turn your pig on its right side and remove the skin along the lateral (left side) and dorsal surfaces to the spine.

5. You may notice in the region of the neck, shoulder, and trunk, a layer of light brown muscle fibers adhering to the skin. These fibers comprise the *cutaneous maximus* muscle, which is responsible for the twitching of the skin that gets rid of insects and other irritants. Humans do not have this layer of muscle.

6. After your specimen is skinned, the muscles will not appear as clearly defined as in the illustrations. This is because they are covered by adipose tissue (fat) and two layers of relatively loose connective tissue or *fascia*. The first layer or *superficial fascia* connects the skin to the muscles and is relatively easy to remove. The second layer or *deep fascia* connects one muscle to another and maintains them in their proper position to one another. It is considerably tougher than the superficial layer. It will be necessary to break through the deep fascia as you proceed through the exercise. Remember, however, that you are working with a fetus and as such the structures are immature and can be easily torn. Therefore, proceed with care.

7. As you attempt to identify the various muscles, you may find that the boundary of a given muscle is readily apparent, whereas in others it seems to blend with those around it. In order to define the limits of a muscle, use a blunt probe to tease away the overlying adipose and connective tissues until you can see the direction of the muscle fibers. Look for changes in the direction of the muscle fibers and attempt to slip the blunt probe or flat edge of your scalpel handle between the two separate layers at this point. If the two layers separate readily from one another, you have located two different muscles.

Now proceed with the exercise and attempt to identify some of the major muscles in the fetal pig *as directed by your instructor.* Obviously, not all of the skeletal muscles have been included, but only those that are relatively easy to identify and that will illustrate the general principles of skeletal muscular action.

B. Muscles of the Shoulder and Back

1. The **latissimus dorsi** (fig. 29-8) is a broad muscle wrapped around the sides of the thoracic region and chest. Carefully pick away the adipose tissue and connective layer from the sides of the chest until its fibers are apparent. The origin of this muscle is the lumbar

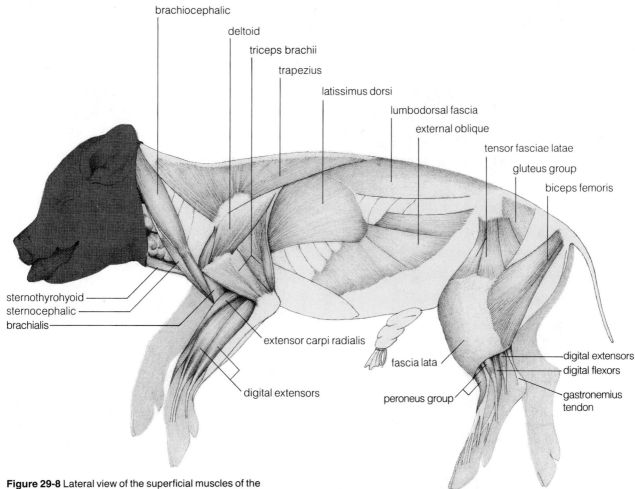

Figure 29-8 Lateral view of the superficial muscles of the fetal pig.

vertebrae, some of the posterior thoracic vertebrae, and the **lumbodorsal fascia** (fig. 29-8). It is inserted by a tendon into the proximal end of the humerus (the major bone of the upper forelimb; see fig. 29-6). The action of the latissimus dorsi is to move the forelimb dorsally and posteriorly.

2. The **trapezius** (fig. 29-8) is a broad muscle anterior to the latissimus dorsi. Its origin is from the base of the skull and from the first ten thoracic vertebrae, and it is inserted on the scapula (the shoulder blade). Its action is to draw or pull the shoulder medially.

3. The **deltoid** (fig. 29-8) is a relatively broad muscle over the shoulder region. It originates from the scapula and is inserted into the humerus. Its action flexes the forelimb.

4. If you have not done so, carefully remove the cutaneous muscle and gelatinous connective tissues covering the side of the neck. You should now be able to observe a broad, flat strip of muscle, the **brachiocephalic** (figs. 29-8 and 29-9), extending from the back of the skull to the foreleg. This muscle is inserted in the distal end of the humerus, and its action flexes the forelimb.

C. Muscles of the Forelimb

1. The **triceps brachii** (figs. 29-8 and 29-9) is a large muscle that virtually covers the entire outer and posterior surface of the upper forelimb. Its origin is from the scapula and proximal third of the humerus, from which it extends posteriorly, and it is inserted on the proximal end of the ulna. Its action is to extend the forelimb.

2. To locate the **biceps brachii** (fig. 29-9), carefully place your pig on its dorsal side and secure the legs with string or rubber bands. The biceps are a rather small, spindle-shaped muscle extending along the anterior surface of the humerus. This muscle originates on the scapula and inserts on the radius and ulna. Its action is to flex the forelimb and act antagonistically with the triceps. In order to see it clearly, you will need to cut through the overlying muscle (the **brachialis;** see fig. 29-9) and pull the cut edges back. The brachialis is a small muscle that also flexes the forelimb. Its origin is the proximal humerus, and it inserts on the ulna.

3. There are numerous other muscles in the lower forelimb, most of which are concerned with extending or flexing the foot and digits; (see figs. 29-8 and 29-9).

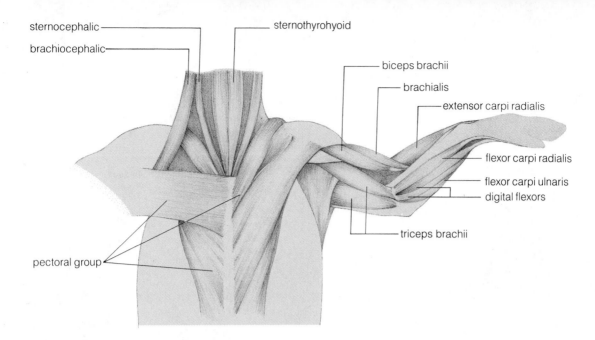

Figure 29-9 Muscles of the ventral thoracic region and forelimb. Some of the superficial muscles of the pig's left side have been removed. (After Gilbert, 1966.)

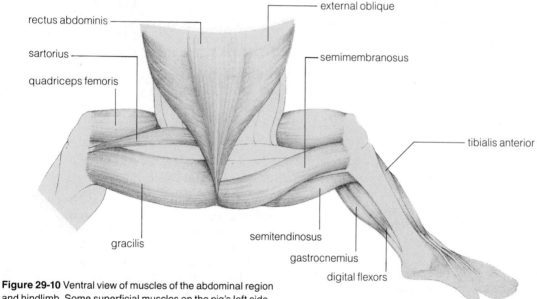

Figure 29-10 Ventral view of muscles of the abdominal region and hindlimb. Some superficial muscles on the pig's left side have been removed. (After Gilbert, 1966.)

D. Muscles of the Throat and Chest

1. Continue working from the ventral side of your specimen and locate the **sternocephalic** (figs. 29-8, 29-9) and **sternothyrohyoid** (figs. 29-8, 29-9). The former is a long muscle lying ventral to the brachiocephalic. Its origin is at the sternum, and it is inserted into the lateral portion of the skull just behind the external ear known as the mastoid process. Its action flexes the head.

2. The sternothyrohyoid consists of two long flat muscles extending from the sternum (origin) to the hyoid (insertion). The action of these muscles is to retract and depress the base of the tongue and the larynx, as, for example, when swallowing.

3. The **pectoral group** (fig. 29-9) originates on the ventral side of the sternum and inserts on the humerus. The pectoral group draws, or adducts (moves the appendage medially), and retracts the forelimb toward the chest. If you wish, you may carefully cut the belly of the superficial pectoral and bend it back to examine the underlying muscles more closely.

E. Muscles of the Abdominal Region

The major lateral abdominal muscles consist of the outer **external oblique,** the **internal oblique** (the midlayer), and the **transversus** (the inner layer). To locate these muscles, turn your specimen on its side. Now

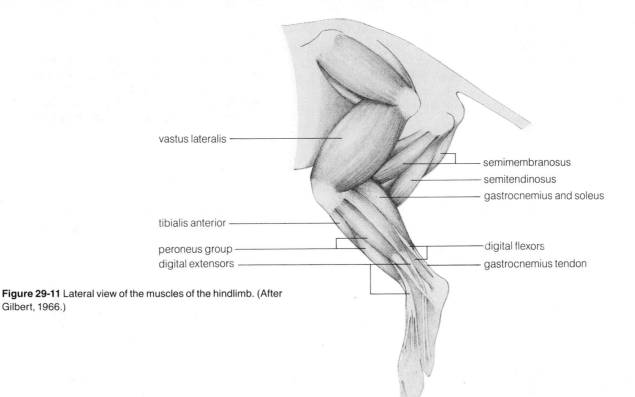

vastus lateralis

semimembranosus

semitendinosus

gastrocnemius and soleus

tibialis anterior

peroneus group

digital extensors

digital flexors

gastrocnemius tendon

Figure 29-11 Lateral view of the muscles of the hindlimb. (After Gilbert, 1966.)

find the fibers of the external oblique (figs. 29-8 and 29-10) and observe that they run diagonally, so that their ventral ends are posterior to their dorsal ends. By carefully cutting a "window" approximately 1 cm square in the external oblique, you will expose a portion of the internal oblique. (Remember that the muscles of the fetus are extremely thin. Use care not to cut into the body cavity at this time, as you will release a large amount of fluid.) Notice that the fibers of the internal oblique run at nearly right angles to the direction of those of the external oblique. Now, using the same careful technique as above, remove a small portion of the internal oblique and attempt to reveal the fibers of the transversus, which as the name suggests run transversely. Collectively, the actions of these abdominal muscles, together with the **rectus abdominus** (fig. 29-10), is to flex the trunk and compress the abdominal viscera to aid expiration or defecation.

F. Muscles of the Hip and Thigh

1. Begin your dissection of the muscles of the hip and thigh by locating the **biceps femoris** (fig. 29-8). This conspicuous superficial muscle covers most of the caudal half of the lateral surface of the thigh and originates from the ischium of the pelvic girdle. It is inserted in the femur and the upper part of the tibia. Its action extends and abducts (moves the appendage laterally) the hindlimb.

2. Next, locate the **tensor fasciae latae** (fig. 29-8), the most anterior of the thigh muscles. It is short and triangular-shaped and continues as a sheet of connec-

tive tissue, the *fascia lata*, which attaches to the tibia. The action of the tensor fasciae latae is to tense the fascia lata, flex the hip, and extend the knee.

3. Now carefully free the tensor fasciae latae at its insertion, but leave the origin and its medial portion intact. Cut the biceps femoris at its insertion near the tibia, and peel it back to expose the deeper muscles of the thigh and foot. Attempt to identify the *vastus lateralis*, one of the four **quadriceps femoris** muscles, and the **semitendinosus** (figs. 29-10 and 29-11), which extend the knee and hip, respectively. Between these two muscles are several arteries and nerves that must be carefully removed to expose the *semimembranosus* (fig. 29-11).

4. To locate other muscles of the thigh, place your pig on its dorsal side and tie back the legs with string or rubber bands. Referring to figures 29-10 and 29-11, locate the **sartorius, gracilis,** and **semimembranosus.** The actions of these muscles are to adduct the hindlimb, flex the hindlimb (sartorius), and extend the hindlimb (semimembranosus). The biceps femoris, semitendinosus, and semimembranosus are collectively called the hamstring muscles.

G. Muscles of the Hindlimb

1. The **gastrocnemius** (figs. 29-10 and 29-11) and *soleus* (fig. 29-11) originate, respectively, from the distal end of the femur and the head of the fibula. They are both inserted by the Achilles tendon to the calcaneus (heel bone). These muscles extend the foot, and the gastrocnemius helps flex the knee.

2. Other muscles of the lower hindlimb flex the foot (*tibialis anterior*, figs. 29-10 and 29-11; *peroneus group*, figs. 29-8 and 29-11), while others flex and extend the digits (*digital flexors*, figs. 29-8 and 29-11; *digital extensors*, figs. 29-8 and 29-11). Spend a few minutes examining the lower hindlimb and attempt to identify several of these muscles.

> **NOTE**
>
> When you have completed the laboratory, place your pig in a plastic bag, tying the bag shut to prevent it from drying out. Dispose of any paper towels that contain formalin or preservative as directed by your instructor. Clean your dissecting tools and the laboratory table.

PRE-LAB QUESTIONS

_____ 1. A fetus is (a) a newborn pig, (b) a newborn human, (c) an unborn mammal, (d) all of the above.

_____ 2. *To dissect* means primarily (a) to cut open, (b) to remove all internal organs, (c) to expose to view, (d) all of the above

_____ 3. When the directions for a fetal pig dissection refer to the left, they are referring to (a) your left, (b) the pig's left, (c) the pig's right, (d) a and c.

_____ 4. In a fetal pig, *dorsal* and *ventral* refer to (a) the head and tail regions of the body, respectively; (b) the tail and the head regions of the body, respectively; (c) the upper (that is, back) portion and the lower (that is, underside) portion of the body, respectively; (d) the lower (that is, underside) portion and the upper (that is, back) portion of the body, respectively.

_____ 5. The umbilical cord functions to (a) carry waste products in the blood from the fetus to the mother, (b) carry waste products in the blood from the mother to the fetus, (c) carry oxygen in the blood from the mother to the fetus, (d) a and c.

_____ 6. The female *fetal* pig is similar to the male *fetal* pig in that its body (a) has separate openings for the urinary system and the reproductive system, (b) has a common opening for the urinary system and the reproductive system, (c) has an opening for the urinary system but none for the reproductive system, (d) has an opening for the reproductive system but none for the excretory system.

_____ 7. The pig and the horse are called ungulates because they walk on (a) their ankles, (b) the soles of their feet, (c) the tips of their toes, which are modified as hooves, (d) their hands and knees.

_____ 8. Pigs and humans are similar in that they are omnivores. That is, they eat (a) only animal matter, (b) only plant matter, (c) both plant and animal matter, (d) neither plant nor animal matter.

_____ 9. The *insertion* of a muscle is (a) attached to the less mobile portion of the skeleton, (b) attached to the portion of the skeleton that moves when the muscle contracts, (c) never attached to the skeleton, (d) its belly.

_____ 10. The biceps brachii is responsible for flexing the forearm while the triceps brachii extends the forearm. Collectively, these opposing muscles are called (a) cooperative, (b) antagonistic, (c) involuntary, (d) sensory.

EXERCISE 29

Dissection of the Fetal Pig: Introduction, External Anatomy, and the Muscular System

POST-LAB QUESTIONS

1. What is the function of the umbilical cord?

2. Using external features, briefly describe how you can determine the difference between a male and a female fetal pig.

3. What does the term *ungulate* mean?

4. What is meant by the *origin, insertion,* and *action* of a muscle?

5. What are antagonistic muscles?

6. What is the difference between a sagittal plane, a frontal plane, and a transverse plane of the body?

7 a. What is meant by the type of locomotion referred to as *plantigrade?*

   b. *Digitigrade?*

8. Identify these external features of a fetal pig.

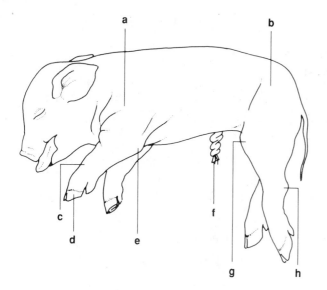

a. _____

b. _____

c. _____

d. _____

e. _____

f. _____

g. _____

h. _____

EXERCISE 30

Dissection of the Fetal Pig:

Digestive, Respiratory, and Circulatory Systems

OBJECTIVES

After completing this exercise you will be able to:

1. define *diaphragm, thoracic cavity, abdominopelvic cavity, exocrine gland, endocrine gland;*

2. locate the organs of the digestive, respiratory, and circulatory systems in a fetal pig;

3. describe and give the functions of the organs of the digestive, respiratory, and circulatory systems;

4. explain the importance of the digestive, respiratory, and circulatory systems to a living mammal;

5. trace the pathway of oxygen and carbon dioxide into and out of the lungs of a mammal;

6. locate, name, and describe the functions of the internal structures of the heart.

INTRODUCTION

You will begin your study of the mammalian digestive, respiratory, and circulatory systems by carefully opening the ventral body cavities of your fetal pig to expose the internal organs for further examination. Remember that *dissecting* does not primarily mean "cutting up" but rather "exposing to view." Thus, proceed carefully, as the internal organs are fragile. And work closely with your partner, making sure that each structure is fully identified and studied before proceeding to the next step.

> ### CAUTION
> Preserved fetal pigs are kept in a formalin-based or other preservative solution. Wash any part of your body exposed to this solution with copious amounts of water. If the formalin solution is splashed into your eyes, wash them with the safety eyewash bottle for 15 minutes.

I. Directions for Dissection

MATERIALS

Per student pair:
- one preserved fetal pig injected with red and blue latex
- dissection pan
- paper towels
- dissecting kit
- dissecting pins
- bone shears
- 4 large rubber bands *or* 2 pieces of string, each 60 cm long
- piece of string 20 cm long
- plastic gloves and/or lanolin hand cream
- goggles if you wear contact lenses

Per lab room:
- liquid waste disposal bottle

PROCEDURE

For the following dissection, use figure 30-1 as a guide for making the various incisions. The numbers in the figure correspond to the numbers in the following directions.

1. Place the pig, ventral side up, in the dissection pan and restrain it using rubber bands or string as you did in the previous exercise. Begin your incision at the small tuft of hair on the upper portion of the throat (1), and continue the incision posteriorly to approximately 1.5 cm anterior to the umbilical cord. You should cut through the muscle layer but not too deeply, or the internal organs may be damaged.

2. You determined the sex of your specimen in the previous exercise. If it is a *female,* make the second incision (2F) completely around the umbilical cord and proceed with a single incision posteriorly for approximately 3 cm between the hindlimbs.

3. If your fetal pig is a *male,* make the second incision (2M) as a half circle anterior to the umbilical cord and then proceed with *two incisions* posteriorly to the region between the hindlimbs. The two incisions are necessary to avoid cutting the *penis,* which lies under the skin just posterior to the umbilical cord. The incisions made in the region of the *scrotum* should be made carefully so as not to damage the testes, lying just under the skin.

4. Deepen incisions 1 and 2 until the body cavity is exposed. Proceed carefully, however, as the body cavity may be filled with a dark fluid. In order to make lateral flaps of the muscle tissue, which can be folded out of the way, make a third (3) and fourth (4) incision as illustrated in figure 30-1. Now, carefully open the body cavity. If it is filled with fluid, pour the fluid into the waste container provided (not into the sink!) and carefully rinse out the cavity with a little water.

5. Use your fingers to locate the lower margin of the *rib cage.* Just below it, make a fifth (5) incision laterally in both directions from the first incision (1). In this

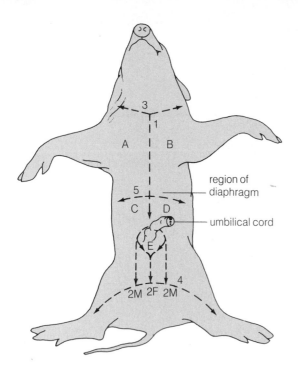

Figure 30-1 Ventral view of the fetal pig, with the position of incisions indicated. Specific incisions for male and female specimens are marked M and F, respectively.

region is the **diaphragm,** a sheet of tissue connected to the body wall and separating the two major ventral body cavities: the **thoracic cavity** and the **abdomino-pelvic cavity.** By using your scalpel, free the diaphragm (do not remove it, however) where it is in contact with the body wall.

6. Carefully peel back flaps A, B, C, and D (see fig. 30-1) and pin them beneath your pig. It may be necessary to cut through the ventral part of the rib cage with a pair of scissors or bone shears to separate body wall flaps A and B. Do so carefully, so as not to damage the heart and lungs, which lie in this region.

7. To free the umbilical cord and the flesh immediately surrounding it (flap E), locate the umbilical vein—a dark, tubular structure extending from the umbilical cord forward (anteriorly) to the liver. Tie small pieces of string around the vein in two places (approximately 1.5 cm apart) and with your scissors cut through the vein between the pieces of string. The umbilical cord and attached tissue (flap E) can now be laid back between the hindlegs and pinned to the body. The pieces of string around the umbilical vein will aid in identifying this structure during the exercise on the circulatory system.

8. When the body cavities are fully exposed, carefully remove any excess red or blue latex that may be present. (This occurs when some veins and arteries burst when injected with latex.) Remove large pieces with forceps. Smaller pieces should be rinsed out at a sink equipped to screen out any debris.

II. Digestive System

The digestive system of a vertebrate consists of the **digestive tract** or *alimentary canal* (mouth, oral cavity, pharynx, esophagus, stomach, small intestine, large intestine or colon, rectum, and anus) and associated structures and glands (salivary glands, gall bladder, liver, and pancreas). In this exercise you will locate and identify all of these structures. In addition, you will identify the thymus and thyroid, two endocrine system glands.

MATERIALS

Per student pair:
- one preserved fetal pig injected with red and blue latex
- dissection pan
- paper towels
- dissecting kit
- dissecting pins
- bone shears
- 4 large rubber bands *or* 2 pieces of string, each 60 cm long
- plastic gloves and/or lanolin hand cream
- goggles if you wear contact lenses
- dissection microscope

Per lab room:
- meter stick
- liquid waste disposal bottle

PROCEDURE

A. Thymus and Thyroid Glands

1. Work from the ventral side of your pig with the legs secured by string or rubber bands and the body wall flaps pinned to the sides of the body.

Identify the **thymus gland,** a whitish structure that is divided into two lobes and is in the neck and upper thoracic cavity (fig. 30-2). This gland, which partially covers the anterior portion of the heart and extends along the trachea to the larynx, plays important roles in the development and maintenance of the body's immune system.

2. Immediately beneath the thymus in the neck region is the **thyroid gland** (fig. 30-2). This is a small reddish, oval mass with a consistency more solid than the thymus. Thyroid hormones function in the regulation of metabolism, growth, and development.

B. Salivary Glands

There are three pairs of salivary glands: the *parotid*, the *mandibular (submaxillary)*, and the *sublingual* (see fig. 30-3). Together they produce *saliva*. To expose

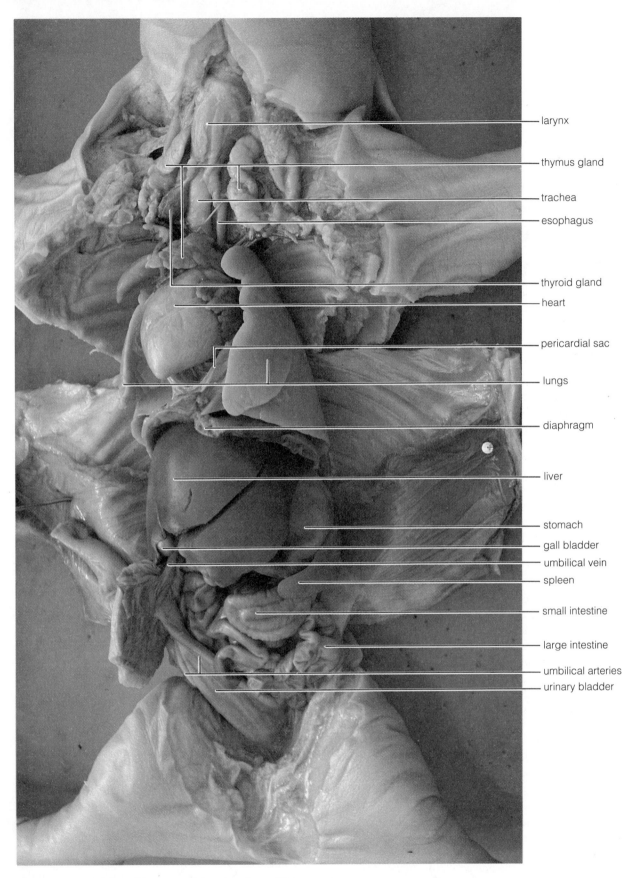

larynx

thymus gland

trachea

esophagus

thyroid gland

heart

pericardial sac

lungs

diaphragm

liver

stomach

gall bladder

umbilical vein

spleen

small intestine

large intestine

umbilical arteries

urinary bladder

Figure 30-2 Ventral view of the general internal anatomy of the fetal pig. (Photo by D. Morton.)

DISSECTION OF THE FETAL PIG: DIGESTIVE, RESPIRATORY, AND CIRCULATORY SYSTEMS

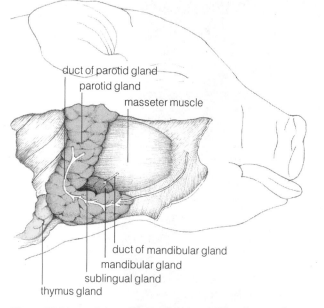

Figure 30-3 Lateral view of the fetal pig head with salivary and thymus glands exposed. (After Gilbert, 1966.)

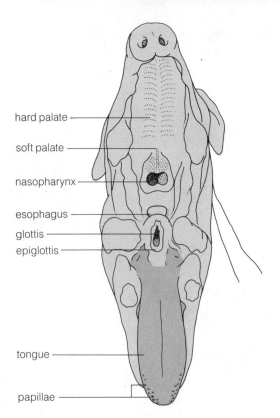

Figure 30-4 Structures of the oral cavity and pharynx.

these glands, it will be necessary to remove the skin and muscle tissue from one side of the face and neck of your specimen. Place your pig on its right side and, proceeding from the base of the ear, carefully cut through the skin to the corner of the eye, then ventrally toward the chin, and, finally, continue the incision posteriorly toward the forelimb. Carefully remove the skin as described in the first fetal pig exercise.

1. Parotid gland. After removing the skin, tease away the muscle tissue below the ear to reveal a large, relatively dark, triangular gland, the *parotid*. This gland extends from the edge of the ear posteriorly to the midregion of the neck (fig. 30-3). Can you distinguish the parotid from the large masseter muscle? The difference is that the parotid appears to be composed of many small nodules, while the muscle tissue is fibrous.

2. Mandibular (submaxillary) gland. The *mandibular gland* is relatively large and somewhat lobed. It lies just below the parotid (fig. 30-3). You will need to cut through the middle of the parotid gland to expose the mandibular gland. Do not confuse the mandibular gland with the small, oval lymph nodes present in the head and neck region.

3. Sublingual gland. The third salivary gland, the *sublingual*, is long and rather slender. It is difficult to locate.

The fluids secreted by the mandibular and sublingual glands are more viscous than that secreted by the parotid. Collectively, the secretions by the three salivary glands maintain the oral cavity in a moist condition, ease the mixing and swallowing of food, and contain enzymes that begin the breakdown of starch to sugars.

C. Mouth

1. The mouth is the beginning of the digestive tract, a long tubelike structure where food digestion and absorption of nutrients occur. Observe the area between the **lips** and **gums;** this is called the **vestibule.** The larger area behind the gums is referred to as the **oral cavity.**

2. With the scissors or bone shears, carefully cut through the corners of the mouth and back toward the ears until the lower jaw can be dropped and the oral cavity exposed (see fig. 30-4).

3. If teeth are present, carefully extract a **tooth** and examine it. A tooth consists of the *crown*, the *neck* (surrounded by the gum), and the *root* (embedded in the jawbone). If your specimen does not have exposed teeth, cut into the gums and determine whether developing teeth are present.

4. Feel the roof of the oral cavity and determine the position of the **hard palate** and **soft palate** (fig. 30-4). What is the difference between the two regions?

_____

_____

_____

5. Posterior to the soft palate is the **pharynx** (in humans the portion of the pharynx just posterior to the oral cavity is also referred to as the *throat*). Note that unlike humans, the pig does not have a fingerlike piece of tissue, the **uvula,** projecting from the posterior region of the soft palate. Confirm its presence in humans by looking into the throat of your lab partner.

6. Carefully close the pig's jaws. Now, in the neck locate the **trachea** (fig. 30-2), a tube that is supported throughout its length by a series of cartilaginous rings. Although the trachea is actually a part of the respiratory system, its identification will aid in finding the **esophagus,** which lies on its dorsal surface. Carefully slit the esophagus and insert a blunt probe into it and run it back toward the mouth. Open the mouth and note where the probe emerges. This is the opening of the esophagus (fig. 30-4). If you would run your probe posteriorly through the esophagus, where would it emerge?

_____

7. Continue your study of the oral cavity by locating the opening to the _larynx,_ the **glottis.** It can be identified by the presence of a small white cartilaginous flap, the **epiglottis,** on its ventral surface (fig. 30-4). The epiglottis covers the glottis when a mammal swallows.

D. Liver, Gallbladder, and Pancreas

1. The largest organ of the abdominopelvic cavity is the **liver** (fig. 30-2), a brownish four-lobed gland that produces _bile._ Count and carefully determine the extent of the four lobes.

In addition to producing bile, the liver plays a very important role in maintaining a stable composition of the blood. When the nutrients from a digested meal are absorbed by the blood capillaries of the small intestine, they contain high concentrations of such compounds as sugars like glucose and amino acids. This blood is transported from the small intestine to the liver (via the hepatic portal vein), and there the excess glucose is converted to glycogen for storage. If the liver has stored a full capacity of glycogen, it converts the glucose into fat, which is stored in other parts of the body. The liver also removes excess amino acids from the blood by converting them to carbohydrates and fats. During this process, an amino group ($-NH_2$) is removed from the amino acid and converted into ammonia (NH_3). Ammonia is a very toxic substance, and the liver combines it with carbon dioxide to form urea. The urea, which is less toxic than ammonia, will be eliminated from the body in urine.

2. Lift the right central lobe of the liver to locate the **gall bladder.** This structure is a sac for storing bile secreted by the liver.

The _cystic duct_ from the gallbladder unites with the _hepatic duct_ from the liver to form the common _bile duct._ The latter empties into the first portion of the _small intestine_ (fig. 30-2). _If your instructor indicates,_ attempt to locate the hepatic and common bile ducts and trace them from the liver to the small intestine. Be careful not to injure the _hepatic portal vein,_ which parallels these ducts.

3. Carefully move the small intestine and locate the **pancreas,** an elongated granular mass lying between the _stomach_ and small intestine. The _pancreatic duct_ carries digestive enzymes and other substances produced by the pancreas to the duodenum. (Do not attempt to find the pancreatic duct, however, as it is too small be dissected satisfactorily.)

The pancreas is both an **exocrine gland** (whose secretions are released into a duct) and an **endocrine gland** (whose hormones are released into the blood). The endocrine portion of the pancreas secretes insulin and other hormones involved with controlling the levels of glucose in the blood of mammals.

E. Stomach, Small Intestine, Large Intestine or Colon, Rectum, and Anus

1. Earlier in the exercise, you made a small slit in the esophagus. Return to this incision and trace the esophagus from the oral cavity to the **stomach.** You may be aided in this by inserting a blunt probe through the slit in the esophagus and pushing it (posteriorly) toward the stomach, a bean-shaped organ dorsal to the liver. (You will need to carefully push the lobes of the liver to one side to fully expose the stomach.) Note that the esophagus penetrates the diaphragm before joining the _cardiac end_ (near the heart) of the stomach. The other end of the stomach, which empties into the small intestine, is called the _pyloric end._

Two muscular rings, the _cardiac sphincter_ and the _pyloric sphincter,_ control the movement of food through the stomach. Feel for these sphincters by gently squeezing the entrance and exit of the stomach between your index finger and thumb. Unlike skeletal muscles, the cardiac and pyloric sphincters are involuntary smooth muscle tissue.

2. Cut the stomach lengthwise with your scissors. Describe the contents of the stomach.

_____

_____

_____

The contents of the fetal pig's digestive tract are called _meconium_ and are composed of a variety of substances, including amniotic fluid swallowed by the fetus, epithelial cells sloughed off from the digestive tract, and hair.

3. Clean out the stomach and note the folds or _rugae_ on its internal surface. What role might the rugae play in digestion?

_____

_____

_____

4. The **small intestine** (fig. 30-2) is divided into three regions: the *duodenum*, the *jejunum*, and the *ileum*. The first portion, the duodenum, leaves the pyloric end of the stomach and runs along the edge of the pancreas. The junctions of the duodenum and ileum with the jejunum cannot easily be distinguished.

5. The coils of the small intestine are held together by thin membranes called **mesenteries.** *At the direction of your instructor,* cut the mesenteries and uncoil the small intestine. Measure its length and record below.

_____

A rule of thumb is that the small intestine in both pigs and humans is about five times the length of the individual.

6. Using your scissors, cut a 0.5-cm section of the intestine, slit it lengthwise, and place it in a clear, shallow dish filled with water. Now examine it using a dissection microscope. How does the inner surface appear?

_____

_____

_____

Locate the *villi.* Most of the nutrients provided by the digestive process are absorbed by these small projections from the wall of the small intestine.

7. Locate the juncture of the **large intestine** (fig. 30-2), or *colon,* and the ileum. This may be more difficult in a pig than in a human because in the former there is not such a noticeable difference in the size of the small and large intestines. However, it is marked by the presence of a blind pouch, the *cecum,* which in the pig is relatively large. In humans, the cecum is very short and bears a small fingerlike projection known as the *appendix.*

8. As with other junctures in the alimentary canal, the region where the ileum joins the large intestine is the site of a muscular sphincter, the *ileocecal valve.* Feel for it by rolling the junction between your index finger and thumb.

9. The coiled large intestine stretches from the cecum to the straight **rectum,** which opens to the outside at the **anus.** The anus is the site of the final muscle in the alimentary canal, the *anal sphincter.* Locate the rectum, anus, and anal sphincter, but do not dissect these structures at this time. You may, however, *at the direction of your instructor,* remove a piece of the colon and examine it with a dissecting microscope as you did earlier with the small intestine. How do their internal surfaces compare?

_____

_____

_____

III. Respiratory System

The respiratory system of a mammal consists of various organs and structures associated with (1) the intake (*inhalation*) of air rich in oxygen, (2) the exchange of oxygen and carbon dioxide between the blood in the lung capillaries and the air sacs of the lungs, and (3) the release (*exhalation*) of gases rich in carbon dioxide.

MATERIALS

Per student pair:
- one preserved fetal pig injected with red and blue latex
- dissection pan
- paper towels
- dissecting kit
- dissecting pins
- 4 large rubber bands *or* 2 pieces of string, each 60 cm long
- plastic gloves and/or lanolin hand cream
- goggles if you wear contact lenses

PROCEDURE

A. Structures of the Head and Oral Cavity

1. In the pig and other mammals, molecules of air enter the body through the nostrils and pass through a pair of **nasal cavities** dorsal to the hard palate and into the *nasopharynx* (fig. 30-4). Examine the nostrils and hard and soft palates, and then carefully cut the soft palate longitudinally to examine the nasopharynx of your specimen.

2. From the nasopharynx, air passes through the glottis into the **larynx** (figs. 30-2 and 30-5a and b). In humans, the front of the larynx is often referred to as the *Adam's apple* or *voice box*. Slit the larynx longitudinally to expose the *vocal cords* (fig. 30-5c).

B. Trachea, Bronchial Tubes, and Lungs

1. Locate the trachea and, as a review, distinguish it from the esophagus (fig. 30-5). The trachea extends from the larynx and divides into two major branches, the **bronchi** (singular *bronchus*), to the lungs. Note again the series of cartilaginous rings that prevent the trachea from collapsing. These rings are actually incomplete on their dorsal side.

2. Direct your attention to the thoracic cavity. It is divided into two **pleural cavities,** which contain the lungs, and the **pericardial sac,** which contains the heart. The latter is located between the pleural cavities. Carefully examine the lungs and note the two **pleural membranes:** thin, transparent tissues, one of which lines the inner surface of the thoracic cavity and the other the outer surface of the lungs. The right lung

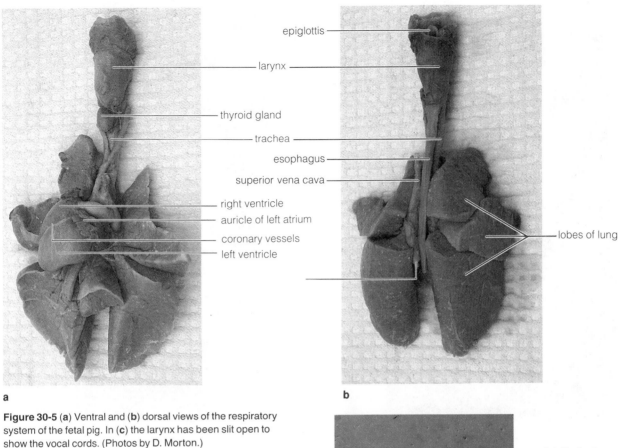

epiglottis

larynx

thyroid gland

trachea

esophagus

superior vena cava

right ventricle

auricle of left atrium

coronary vessels

left ventricle

lobes of lung

a

b

Figure 30-5 (a) Ventral and **(b)** dorsal views of the respiratory system of the fetal pig. In **(c)** the larynx has been slit open to show the vocal cords. (Photos by D. Morton.)

consists of four lobes and the left of two or three. Are the lungs of the fetal pig filled with air?

(Yes or no) _____

The pericardial sac is likewise lined by a **pericardial membrane,** as is the surface of the heart.

3. Carefully push the heart to one side and gently tease away some of the lung tissue to expose the bronchi. Attempt to see that the bronchi divide into smaller and smaller branches. These are called *bronchioles;* they continue to divide and branch into finer and finer structures, eventually ending as microscopic air sacs called **alveoli** (singular *alveolus*). The thin walls of the alveoli are extensively supplied with blood capillaries, and it is here that the exchange of carbon dioxide and oxygen occurs in an adult. Where does this exchange occur in the fetus?

_____

4. Now relocate the diaphragm and note the position of this thin sheet of muscular tissue in relation to the lungs.

5. Complete your study of the respiratory system of the fetal pig by tracing the pathway of a carbon dioxide molecule from an alveolus to the nostrils.

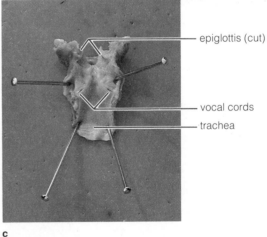

epiglottis (cut)

vocal cords

trachea

c

IV. Circulatory System

The circulatory system of the fetal pig consists of a vast network of **vessels** (*arteries, arterioles, capillaries, veins,* and *venules*), which contain blood and transport water, oxygen, carbon dioxide, nutrients, metabolic wastes, hormones, and other substances to and from every living cell in the body. In mammals, blood is propelled through the arteries, arterioles, and capillaries by a muscular four-chambered **heart.**

Through this extensive system, oxygen is added to the blood in the capillaries of the alveoli of the lungs, while carbon dioxide is removed for exhalation from the body. In the capillaries of the small intestine, various nutrients (for example, glucose and amino acids)

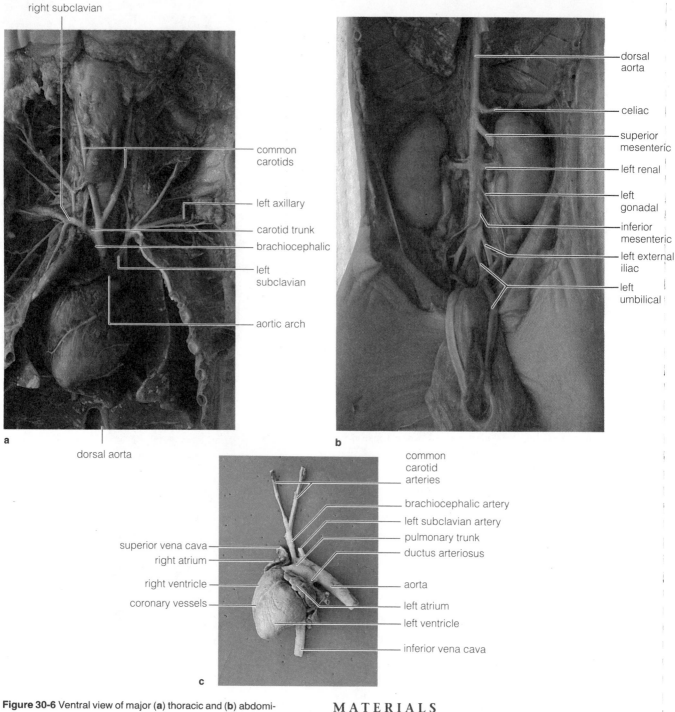

right subclavian

common
carotids

left axillary

carotid trunk

brachiocephalic

left
subclavian

aortic arch

dorsal aorta

a

dorsal
aorta

celiac

superior
mesenteric

left renal

left
gonadal

inferior
mesenteric

left external
iliac

left
umbilical

b

common
carotid
arteries

brachiocephalic artery

left subclavian artery

pulmonary trunk

ductus arteriosus

superior vena cava

right atrium

right ventricle

coronary vessels

aorta

left atrium

left ventricle

inferior vena cava

c

Figure 30-6 Ventral view of major (**a**) thoracic and (**b**) abdominopelvic arteries and (**c**) the heart of a fetal pig. (Photos by D. Morton.)

are added to the blood, while in the capillaries of the kidneys, the blood is cleansed of various metabolic wastes and excess ions.

The circulatory system is divided into two circuits: the *pulmonary circuit*, which involves blood flow to and from the lungs, and the *systemic circuit*, which is concerned with the flow of blood to and from the rest of the body. In this section, you will study these two circuits and examine how the heart directs the flow of blood through them both in a fetus and in an adult.

MATERIALS

Per student pair:

• one preserved fetal pig injected with red and blue latex

• dissection pan

• paper towels

• dissecting kit

• dissecting pins

• 4 large rubber bands *or* 2 pieces of string, each 60 cm long

• plastic gloves and/or lanolin hand cream

• goggles if you wear contact lenses

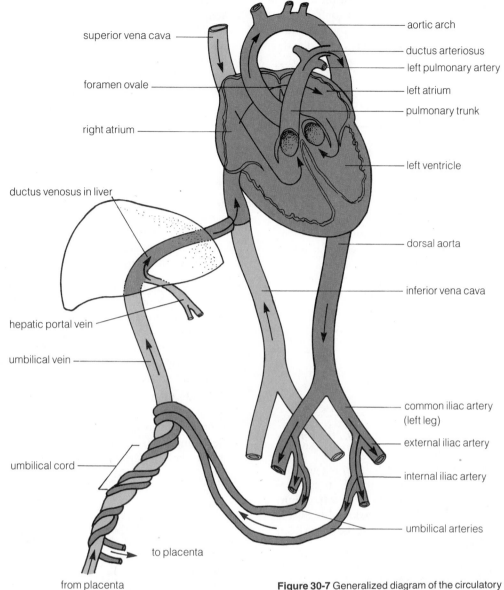

superior vena cava

foramen ovale

right atrium

ductus venosus in liver

hepatic portal vein

umbilical vein

umbilical cord

to placenta

from placenta

aortic arch

ductus arteriosus

left pulmonary artery

left atrium

pulmonary trunk

left ventricle

dorsal aorta

inferior vena cava

common iliac artery (left leg)

external iliac artery

internal iliac artery

umbilical arteries

Figure 30-7 Generalized diagram of the circulatory system of a fetal mammal. Arrows indicate the flow of blood. Pink represents fully oxygenated blood. The blue indicates oxygen-depleted blood. (After Weller and Wiley, 1985.)

PROCEDURE

For the following dissections, refer to figure 30-6 for the major arteries of the fetal pig and to figure 30-7, which is a diagrammatic representation of the heart and the major arteries and veins of a fetal mammal.

A. Pulmonary Circuit and Surface Anatomy of the Heart

1. Continue working from the ventral side of your specimen as you did in the preceding section. Make sure that the legs of the pig are secured with string or rubber bands and that the skin flaps are pinned to the sides, or dorsal portion, of the body.

2. Locate the heart in the thoracic cavity and carefully remove the pericardial sac that surrounds it. Identify

the four chambers of the heart: the thin-walled **right atrium** (fig. 30-6) and **left atrium** (fig. 30-5), and the larger **right** and **left ventricles** (see fig. 30-5). You should also be able to locate the **coronary artery** and **coronary vein** lying in the diagonal groove between the two ventricles (fig. 30-5). The coronary artery and its branches supply blood directly to the heart. (The heart is a muscle and as such has the same requirements as any other organ.) When these vessels become severely occluded, a heart attack may occur. It is the coronary arteries and their branches that are replaced or "bypassed" in coronary bypass surgery.

3. In adult pigs, oxygen-poor (or carbon-dioxide-rich) blood returning to the right atrium of the heart from the systemic circulation does so through the large veins known as the **superior vena cava** (or anterior vena cava) and the **inferior vena cava** (or posterior vena cava). Gently push the heart to the right and identify these relatively large blue veins. Describe the difference in diameter between the superior and inferior vena cava.

_____

_____

_____

Trace the inferior vena cava a short distance from the heart. Where does it lead?

_____

4. The blood that enters the right atrium passes to the right ventricle and then to the **pulmonary trunk,** a large vessel that branches to the lungs. This vessel, which lies between the left and right atria and extends dorsally and to the pig's left, branches to form the **left** and **right pulmonary arteries.** Do these arteries contain red or blue latex?

_____

In the adult, do they carry oxygen-rich or oxygen-poor blood?

_____

Carefully move the heart aside and follow the pulmonary arteries to the lungs.

5. In the adult, once the blood has been oxygenated (and the carbon dioxide removed) in the lungs, it returns to the left atrium of the heart via the **left** or **right pulmonary veins.** Carefully move the lungs and heart and locate these large vessels that carry the blood from the lungs to the left atrium of the heart.

6. From the left atrium of the adult, the oxygenated blood passes to the left ventricle and from there passes into the **aorta** and into the systemic circulatory system. Locate the aorta, which leads dorsally out of the left ventricle of the heart. Note that its base is partially covered by the pulmonary trunk coming from the right ventricle.

7. Blood circulation is different in fetal mammals.

a. The preceding description of blood flow through the heart and lungs is only representative of a pig or other mammal following birth. In the fetus, most of the pulmonary circuit is bypassed twice. First, most blood from the right ventricle enters into the aorta directly from the pulmonary trunk through the **ductus arteriosus,** a large but short vessel connecting the pulmonary trunk directly to the aorta. With the first breath of the fetus, the ductus arteriosus contracts and circulation is established with the pulmonary system. Then, during the eight weeks following birth, the ductus arteriosus forms a fibrous strand of connective tissue, the _ligamentum arteriosum._ Locate the ductus arteriosus in your fetal pig.

b. The second bypass occurs when most of the blood arriving from the posterior portions of the body via the inferior vena cava and entering the right atrium passes directly into the left atrium via a temporary opening (_foramen ovale_) in the wall separating the right and left atria. Thus, this blood bypasses the pulmonary circuit. As with the ductus arteriosus, the foramen ovale closes at birth, and pulmonary circulation is established. Why is it not necessary for large quantities of blood to enter the pulmonary system of a fetus?

_____

_____

B. Systemic Circuit — Major Arteries and Veins Anterior to the Heart
(see figs. 30-6 and 30-8)

1. The systemic circuit begins with the aorta. This large vessel leads anteriorly out of the left ventricle of the heart and makes a sharp turn to the left (the so-called **aortic arch**) and proceeds posteriorly through the body as the **dorsal aorta.** All of the major arteries of the body arise from these two regions (the aortic arch and dorsal aorta) of the aorta.

2. Locate the first visible vessel, the **brachiocephalic artery,** to branch from the aortic arch. The first vessels to branch from the aorta are the coronary arteries; these cannot be seen without dissecting the heart. The brachiocephalic artery branches to give rise to the

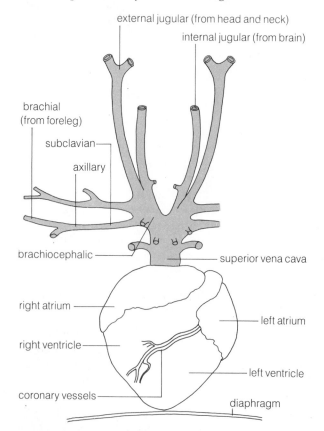

Figure 30-8 Diagram of veins anterior to the heart.

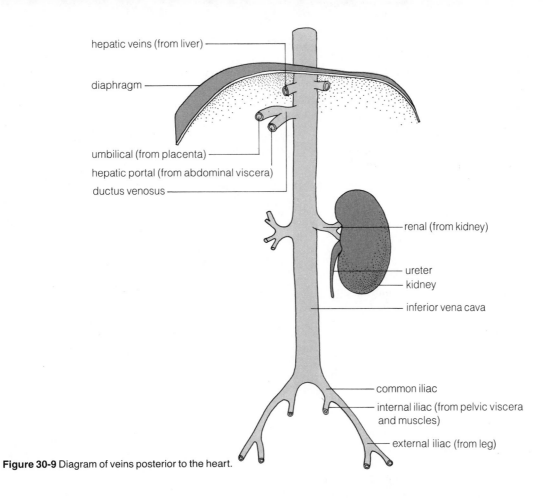

hepatic veins (from liver)

diaphragm

umbilical (from placenta)

hepatic portal (from abdominal viscera)

ductus venosus

renal (from kidney)

ureter

kidney

inferior vena cava

common iliac

internal iliac (from pelvic viscera and muscles)

external iliac (from leg)

Figure 30-9 Diagram of veins posterior to the heart.

right subclavian artery, going to the right forelimb, and the **carotid trunk,** whose branches course anteriorly through the neck and head. Trace the right subclavian artery and its branches through the shoulder region to the right forelimb. As it passes through the shoulder region, the name of the right subclavian changes to the *axillary artery* and then to the *brachial artery* when it enters the upper forelimb.

3. Return to the aorta and locate the second visible vessel to branch from the aortic arch, the **left subclavian artery.** The left subclavian artery and its branches pass through the shoulder and left forelimb or arm in the same manner as the right subclavian artery, described above. As you trace the course of the left subclavian, notice that some of its branches feed the muscles of the chest and back.

4. Return to the right forelimb and locate the venous system that passes through this appendage. Because the veins are relatively thin-walled, this may be very difficult. Also, some of them may not be injected with blue latex and will appear a brownish color. If possible, follow the *brachial vein* to the *axillary* and the **subclavian vein** until the latter becomes the **brachiocephalic vein.** It should be relatively easy to follow the brachiocephalic to its juncture with the superior vena cava (it forms a prominent "V"), which returns to the right atrium of the heart.

5. In order to examine the arterial system that serves the throat and head, locate the carotid trunk (a branch of the brachiocephalic artery; see above). This short branch of the brachiocephalic artery immediately splits into the **left** and **right common carotid arteries.** Each of these vessels divides into the *internal* and *external common carotid arteries.* Remove the thymus and thyroid glands and considerable muscle tissue in the throat to locate the anterior portions of the common carotid arteries.

As you locate and trace the carotid arteries, look for a white "fiber" that parallels them. This is the *vagus nerve.*

6. On either side of the neck are the major veins that drain the head and throat region. The **internal** and **external jugular veins** join the subclavian veins (from the forelimbs) to form the **brachiocephalic vein.** The latter leads into the superior vena cava, which returns to the right atrium of the heart.

C. Systemic Circuit—Major Arteries and Veins Posterior to the Heart (see figs. 30-6 and 30-9)

1. The posterior extension of the aortic arch is the dorsal aorta. As the name implies, the dorsal aorta lies in a middorsal position along the spine. From this large vessel arise all of the arterial branches that feed the organs, glands, and muscles of the abdominal region and the muscles of the hindlimbs and tail.

2. Follow the dorsal aorta posteriorly. Carefully move the liver and stomach of the pig and use a dissection needle to scrape away the sheet of tissue that connects the dorsal aorta to the pig's back. Locate the **celiac artery,** whose branches deliver oxygenated blood to the stomach, spleen, and liver. Continue to follow the dorsal aorta posteriorly and locate the **superior mesenteric artery.** This vessel, just posterior to the celiac artery, branches to the pancreas and duodenum of the small intestine.

3. Posterior to the superior mesenteric artery are the **renal arteries,** relatively short vessels that connect the dorsal aorta and the kidneys. At this time it is easy to locate the **renal veins,** which drain blood from the kidneys to the inferior vena cava.

4. As you follow the dorsal aorta posteriorly beyond the kidneys, the **external iliac arteries** branch, one into each hindlimb. Each leg is also supplied with a major vein, the **iliac vein,** which joins the inferior vena cava.

5. Follow the branches of the dorsal aorta into the tail region, being careful not to cut the two intervening branches. The small extension toward the tail region is called the *sacral artery* as it leaves the dorsal aorta and the *caudal artery* when it enters the tail.

6. Just anterior to the sacral artery, the **internal iliac arteries** branch from the dorsal aorta. These enlarge and form the two **umbilical arteries,** which run through the umbilical cord to the placenta. Cut the umbilical cord transversely and note the arrangement of the umbilical arteries within it. Consider the composition of the blood as it travels through the umbilical arteries to the placenta. Is it rich in oxygen or carbon dioxide?

---

7. Locate the two pieces of vein that you tied with string in Exercise 29. This is the **umbilical vein,** through which blood rich in nutrients and oxygen flows from the placenta of the mother back to the fetus. Locate the umbilical vein in the umbilical cord and follow it anteriorly toward the liver. When the umbilical vein reaches the liver, it becomes the **ductus venosus** (fig 30-7), which continues anteriorly within the substance of the liver and joins the inferior vena cava. The umbilical arteries, the umbilical veins, and ductus venosus become modified into ligaments following the birth of the fetus. What is the relationship between the navel and the umbilical cord?

---

8. The *hepatic portal system* consists of a network of veins that collects blood from the lower digestive tract and associated organs (stomach, small intestine, pancreas, and spleen) and carries it to the liver via the **hepatic portal vein.** In general, a portal vein is one that collects blood from the capillaries of one organ and transfers it to the capillaries of another organ. Locate the hepatic portal vein.

9. In the adult, the blood of the liver is drained by the **hepatic veins.** These join the inferior vena cava just anterior to the point where the ductus venosus joins the inferior vena cava (fig. 30-9).

10. Complete your dissection of the systemic circulatory system by tracing the inferior vena cava from the abdominal cavity through the diaphragm and to the thoracic cavity, where it joins the superior vena cava before entering the right atrium of the heart.

D. Internal Structure of the Heart

Refer to figure 30-10 for the dissection of the heart and the study of its internal structure.

1. Carefully free the heart from the fetal body by cutting through the superior and inferior venae cavae, the subclavians, the common carotids, and the dorsal aorta just posterior to the heart. Cut through the left and right pulmonary veins and the pulmonary arteries at their juncture with the lungs. Remove the heart from the fetus and place it on paper towels with its ventral surface facing you, as it was in the thoracic cavity. If any of the whitish membranous pericardium is still present, carefully remove it from around the heart.

2. Review the location of the four heart chambers: the left and right atria and the left and right ventricles. Locate the coronary artery and vein in the longitudinal groove running between the left and right ventricles. Carefully follow these vessels to the dorsal side of the heart. Although you do not need to confirm this, the coronary arteries are the first vessels to branch from the aortic arch.

3. Place the heart dorsal-side up and locate the inferior and superior venae cavae. Cut through these vessels with your scissors and expose the interior chamber of the right atrium (see fig. 30-10, incision 1). Carefully remove any latex and coagulated blood in the right atrium. Between the atrium and the right ventricular cavity are three membranous cusps attached to the wall of the right atrium. This is the *tricuspid valve.* The open ends of the cusps face downward into the cavity of the right ventricle.

4. Continue working from the dorsal side and cut into the right ventricle as indicated by incision 2 in figure 30-10. With your forceps and needle, remove any latex that obstructs your view. You should be able to see the three cusps of the *semilunar valve* at the juncture of the pulmonary artery and the right ventricle. The open ends of the cusps face into the pulmonary trunk and thus prevent a backward flow of blood into the ventricle.

5. Examine the internal walls of the ventricle. If you wish, you may extend incision 2 to the ventral side of

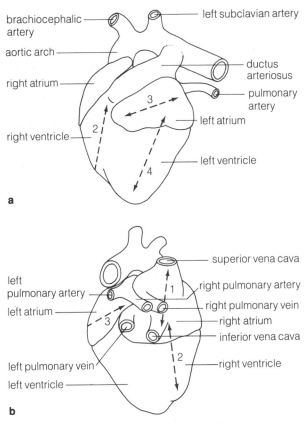

brachiocephalic artery
left subclavian artery
aortic arch
ductus arteriosus
right atrium
pulmonary artery
left atrium
right ventricle
left ventricle

a

superior vena cava
left pulmonary artery
right pulmonary artery
right pulmonary vein
left atrium
right atrium
inferior vena cava
left pulmonary vein
right ventricle
left ventricle

b

Figure 30-10 (**a**) Ventral and (**b**) dorsal views of fetal pig heart, showing numbered incisions for dissection.

the right ventricle. Can you see the muscular ridges? These are the *papillary muscles,* and arising from them are relatively fine fibers, the *chordae tendinae.* The chordae tendinae are attached to the edges of the tricuspid valve.

6. Next, with the heart's ventral surface facing you, locate the ductus arteriosus and the aorta. (Remember, the ductus arteriosus is a shunt between the pulmonary trunk and the aorta.)

Cut open the left atrium (incision 3) and the left ventricle (incision 4). Remove the latex. On the dorsal wall of the heart, find the pulmonary veins from the inside of the left atrium. Next, locate the *bicuspid valve* (consisting of two cusps) between the left atrium and left ventricle. Do the cusps appear similar to the tricuspid valve?

(Yes or no) _____

7. Turn the heart so that the ventral surface is facing you and examine the cavity of the left ventricle. Note the papillary muscles and the chordae tendinae in the left ventricle. The chordae tendinae are commonly called heart strings. Do they appear similar to those in the right ventricle?

(Yes or no) _____

8. Insert a probe into the aorta from the exterior of the heart and note where it enters the cavity of the left ventricle. At this point there is another valve, the *aortic semilunar valve,* with three cusps. Is the orientation of the aortic semilunar valve similar to that of the semilunar valve between the pulmonary trunk and the right ventricle?

(Yes or no) _____

9. Recall that in the fetus a temporary opening, the foramen ovale, exists between the right and left atria. It is not necessary to locate this opening in the fetal heart, but you should review its function.

10. Complete your study by looking for differences in the thickness of the walls of the atria and those of the ventricles. Describe any differences you find.

_____

_____

_____

Explain why the wall of the left ventricle is thicker than the wall of the right ventricle.

_____

_____

_____

NOTE

When you have completed this exercise, return the body wall flaps to their original positions and place the pig in the plastic bag. Dispose of any paper towels that contain formalin or preservative as directed by your instructor. Clean your dissecting tools and laboratory table.

_____ 1. The two *major* body cavities of a fetal pig are the (a) thoracic and pleural, (b) thoracic and pericardial, (c) abdominopelvic and thoracic, (d) abdominopelvic and pericardial.

_____ 2. The diaphragm is a muscular sheet of tissue that separates the (a) thoracic and pleural cavities, (b) thoracic and pericardial cavities, (c) thoracic and abdominopelvic cavities, (d) pleural and pericardial cavities.

_____ 3. The digestive system is concerned with (a) blood circulation, (b) digestion and the absorption of nutrients, (c) reproduction, (d) excretion of urine.

_____ 4. The liver functions to (a) produce bile, (b) pump blood, (c) form urea, (d) a and c.

_____ 5. A vein is a blood vessel that always carries (a) blood toward the heart, (b) blood away from the heart, (c) oxygen-rich blood, (d) oxygen-poor blood.

_____ 6. As a general rule, the small intestine of a pig or human is (a) about 60 cm long, (b) about 1.5 m long, (c) about as long as the individual is tall (or long in the case of the pig), (d) about five times the height of the individual.

_____ 7. In humans, the front of the larynx is commonly referred to as the (a) voice box, (b) Adam's apple, (c) food pipe, (d) a and b.

_____ 8. The microscopic air sacs or alveoli are the sites where blood (a) picks up oxygen, (b) gives up carbon dioxide, (c) gives up oxygen, (d) a and b.

_____ 9. The hearts of a fetal pig and a human are similar in that they are (a) the primary pump of the circulatory system of the body, (b) both four-chambered, (c) composed of cardiac muscle tissue, (d) all of the above.

_____ 10. The cardiac, pyloric, anal, and ileocecal sphincters are all part of the (a) digestive tract, (b) respiratory tract, (c) circulatory system, (d) muscular system.

EXERCISE 30

Dissection of the Fetal Pig: Digestive, Respiratory, and Circulatory Systems

POST-LAB QUESTIONS

1. What are the two *major* ventral body cavities?

2. What is the digestive tract?

3. What is the difference in the structure and function of the trachea and esophagus?

4. What is the function of the villi in the small intestine?

5. Identify the following structures.

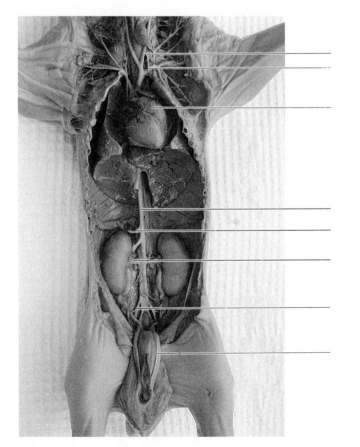

a _____

b _____

c _____

d _____

e _____

f _____

g _____

h _____

6. What is the difference between the pulmonary and the systemic circuits of the circulatory system?

7. What is the foramen ovale?

8. With regard to blood circulation, what is the difference between an artery and a vein?

9. Briefly describe the function of a portal vein system (for example, the hepatic portal vein).

EXERCISE 31

Dissection of the Fetal Pig:

Urogenital and Nervous Systems

OBJECTIVES

After completing this exercise you will be able to:

1. define *ovulation, homologous, inguinal hernia, vasectomy, semen, nephron, urea, urine, meningitis;*

2. locate the organs of the urinary, reproductive, and nervous systems in a fetal pig;

3. describe and give the functions of the organs of the urinary, reproductive, and nervous systems;

4. explain the importance of the urinary, reproductive, and nervous systems to a living mammal;

5. locate, name, and describe the function of the internal structures of the kidney.

INTRODUCTION

In today's exercise you will complete your study of the internal anatomy of the fetal pig. In previous exercises, you have dissected the digestive system, whose function is to break down the large complex organic compounds present in food to smaller molecules that can be absorbed by the body; and you have dissected the respiratory system, which brings oxygen into the body and exchanges it for carbon dioxide. In addition, you have examined the circulatory system, which transports nutrients and gases dissolved in the blood throughout the body.

Now you will examine the system that removes metabolic wastes from the bloodstream (the *urinary system*) and the system largely responsible for integration and control in the organism (the *nervous system*). In addition, you will study both male and female specimens in order to examine the system responsible for the production of new individuals or offspring (the *reproductive system*). The urinary and reproductive systems are traditionally studied together (as the *urogenital system*) because they share several anatomical features.

I. Urogenital System

A. Urinary System

Like humans, the pig is a terrestrial organism and, as such, must conserve body fluids or water. At the same time, metabolic wastes must be continuously removed from the blood. Furthermore, the composition of the blood must be constantly monitored and adjusted so that the cells of the body are bathed in a fluid of constant composition.

Much of the potentially poisonous waste occurs in the form of *urea* and results from the metabolism of amino acids in the liver. Urea is filtered from the bloodstream in the kidneys, which also regulate water and salt balance.

MATERIALS

Per student pair:

- one preserved fetal pig injected with red and blue latex
- dissection pan
- paper towels
- dissecting kit
- dissecting pins
- 4 large rubber bands *or* 2 pieces of string, each 60 cm long
- plastic gloves and/or lanolin hand cream
- goggles if you wear contact lenses

PROCEDURE

1. Place your pig on its back in the dissection pan and use string or rubber bands to secure the legs, as you did in the preceding exercises. Pin the lateral body-wall flaps to the dorsal side of your specimen and pull the umbilical cord and surrounding tissue back between the hindlimbs.

2. The **kidneys** are situated in the lumbar region of the body cavity against the dorsal body wall (fig. 31-1). In addition to the various abdominal viscera, they are covered by the **peritoneum,** the smooth, rather shiny membrane that lines the abdominopelvic cavity. (You may have already removed much of this during the dissection of the circulatory system in the previous exercise.) To expose the right kidney and its *ureter,* carefully lift up the abdominal organs and move them anteriorly and to the pig's left. Using a dissecting needle, carefully scrape away the peritoneum so that

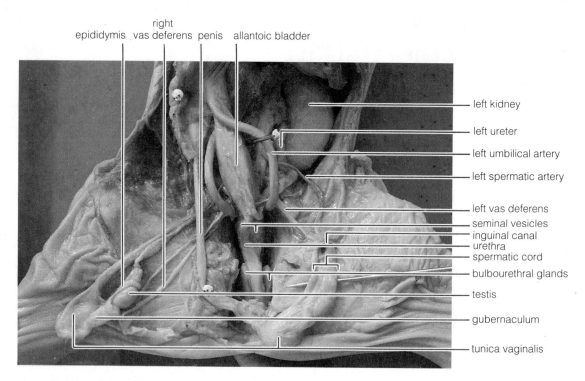

right
epididymis vas deferens penis allantoic bladder

left kidney

left ureter

left umbilical artery

left spermatic artery

left vas deferens
seminal vesicles
inguinal canal
urethra
spermatic cord
bulbourethral glands

testis

gubernaculum

tunica vaginalis

Figure 31-1 Ventral view of the male urogenital system of the fetal pig. (Photo by D. Morton.)

the kidney, a bean-shaped structure, and its ureter are easily seen. Note the central depression in the surface of the kidney. This is the *hilus,* the region where the ureters and the *renal vein* leave and the *renal artery* enters the kidney.

3. Carefully follow the ureters from the hilus to the **allantoic bladder.** Then lift the bladder and find the **urethra.** The latter is the structure through which urine passes from the bladder to the outside of the animal. In the male, the urethra is very long and passes through the penis to the outside of the body (fig. 31-1). Notice that the urethra passes posteriorly for a distance of approximately 2 cm and then turns sharply anteriorly and ventrally before entering the penis. In the female fetal pig, the urethra is short and passes posteriorly to join with the *vagina* to form the **urogenital sinus** (see fig. 31-2).

4. Locate the *allantoic duct,* which leads from the allantoic bladder into the umbilical cord. The allantoic duct is largely a vestigial structure, for most of the wastes produced in the kidneys of the fetus are carried in the bloodstream through the umbilical vein to the placenta, where they are removed in the body of the mother. Following the birth of the fetus, the allantoic duct collapses, and the allantoic bladder is incorporated into the **urinary bladder.**

5. We will examine the urinary system and the internal structure of the kidneys more closely following the study of the male and female reproductive systems.

B. Female Reproductive System

In terrestrial organisms, fertilization (the fusion of the nuclei of male and female gametes) occurs internally, where a relatively stable aquatic environment is maintained. Once fertilization has occurred, the zygote divides to form an embryo and eventually a fetus. In mammals, all of this growth and development occurs within the female's uterus, which nourishes the developing offspring until it can pass through the birth canal and exist on its own in the outside world.

In this portion of the exercise, you will study the structures (external genitalia, uterus, oviducts, ovaries, and associated ducts) of the mammalian female reproductive system and examine how each contributes to the process of internal fertilization and to the growth and development of a viable fetus.

1. Examine the **vulva,** the collective term for the external genitalia of the female. In the pig, the vulva includes the *genital papilla* on the outside of the body, the **labia** or lips found on either side of the urogenital sinus, and the **clitoris,** a small body of erectile tissue on the ventral portion of the urogenital sinus. Also included in the vulva is the opening of the urogenital sinus itself (fig. 31-2).

2. In the female fetal pig, the urogenital sinus is the common passage for the urethra and the **vagina.** To locate these structures, carefully insert your scissors into the opening of the urogenital sinus and cut this structure from the side. Locate where the ducts of the vagina and urethra enter to form the urogenital sinus.

The urogenital sinus is not present in the adult female pig. During the subsequent development of the

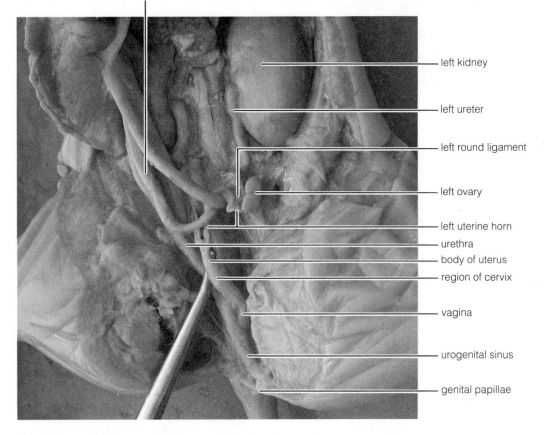

allantoic bladder

left kidney

left ureter

left round ligament

left ovary

left uterine horn
urethra
body of uterus
region of cervix

vagina

urogenital sinus

genital papillae

Figure 31-2 Ventral view of the female urogenital system of the fetal pig. (Photo by D. Morton.)

fetus, the sinus is reduced in size until the vagina and the urethra each develop their own, separate opening to the outside. Thus, in the adult female pig, urine exits through the urinary opening. This is the situation in most adult female mammals. How does this compare with the structure of the reproductive system of most male mammals?

---

3. Again, locate the clitoris. This small, rounded region on the inner ventral surface of the urogenital sinus is **homologous** (that is, similar in structure and origin) with the male *penis*. In the male, the tissues of the penis develop around and enclose the urethra, while in the female, the urethra opens posteriorly to the clitoris.

4. Follow the urogenital sinus anteriorly and identify the thick-walled muscular vagina, which is continuous with the **uterus**. In the pig, the uterus consists of three structures or regions: the **cervix** at the entrance to the uterus, the **uterine body,** and the two **uterine horns** (fig. 31-2). Determine in your dissection that the uterine horns unite to form the body of the uterus.

The pig has a *bicornuate uterus,* in which the fetuses develop in the uterine horns. In the human female there are no uterine horns, and the fetus develops within the body of the *simplex uterus.*

5. From the uterine horns, follow the **oviducts** to the **ovaries** (female gonads). The ovaries are small, yellowish kidney-shaped structures that lie just posterior to the kidneys. They are the sites of egg production and the source of the female sex hormones, estrogen and progesterone.

All of the eggs that a female will produce during her lifetime are present in the ovaries at birth. After puberty, eggs will mature, rupture from the surface of the ovaries, and enter the oviducts. This process is referred to as **ovulation.**

If viable sperm are present in the upper third of an oviduct when it contains eggs, fertilization may occur. In this case, the fertilized egg or zygote will develop into an embryo and pass down the oviduct to become implanted in the wall of the uterine horn. In the human female, however, it will become implanted in the uterine body.

6. The ovaries, oviducts, and uterine horns are supported by a sheet of mesentery or connective tissue, the **broad ligament,** which originates from the dorsal body wall (fig. 31-2). A second mesentery, the **round**

ligament, which also supports the ovaries, extends from the lateral wall and crosses the broad ligament diagonally. Identify the broad and round ligaments.

C. Male Reproductive System

The male reproductive system of a mammal consists of the external genitalia (penis, scrotum, and testes) and various internal structures: the urethra, sex accessory glands, and associated ducts. In this section, you will locate and identify these structures and glands in the fetal pig and examine how they contribute to the reproductive process.

1. Begin your dissection of the male reproductive system by locating the **testes** (male gonads), the site of *sperm* production and source of *testosterone,* the male sex hromone. In older fetuses they are located in the *scrotum,* but in younger fetuses they may be found anywhere between the kidneys and the scrotum.

In order for viable sperm to be produced in adult males, the testes must be situated outside of the abdominal cavity, where body temperatures are slightly lower than within. Thus, during normal development, the testes undergo a posterior migration, or descent, into the scrotum.

The following discussion and dissection apply to an older male fetus with testes fully descended into the scrotum.

2. Locate the scrotum. Make a midline incision through this structure, cutting through the muscle tissue. Pull out the two elongated bulbous structures covered with a transparent membrane. This membrane is the *tunica vaginalis* and is actually an outpocketing of the abdominal wall. Notice the tough white cord that connects the posterior end of the testes to the inner face of the sac (see fig. 31-1). This cord, the **gubernaculum,** is homologous to the round ligament in the female reproductive system. It grows more slowly than the surrounding tissues and thus aids in pulling the testes posteriorly into the scrotal sacs.

3. Cut through the tunica vaginalis to expose a single testis and locate the **epididymis.** This is a tightly coiled tube that lies along one side of the testis. Sperm produced in the testes are stored in the epididymides until ejaculation moves them out of the body in the *semen.*

4. The slender, elongated structure that emerges from each testis is the **spermatic cord.** Gently pull the cord and note that it moves through an opening, the **inguinal canal,** which is actually an opening in the abdominal wall between the abdominopelvic cavity and the cavity of the scrotum. It is through this opening that the testes descend during their migration into the scrotum.

Some human males develop an **inguinal hernia,** a condition in which part of the intestine drops through the inguinal canal into the scrotum. Pigs and other quadripeds (hint) do not develop inguinal hernias. Explain why.

_____

_____

_____

5. The spermatic cord consists of the **vas deferens** (plural is *vasa deferentia*), the spermatic vein, the spermatic artery, and the spermatic nerve. It is the vas deferens that is severed when a human male has a **vasectomy.** Follow the vas deferens to the base of the bladder, where it loops up and over the ureter and then continues posteriorly to enter the urethra. During ejaculation, sperm stored in the epididymis move through the vas deferens into the urethra for transport out of the body.

6. Before the sex accessory glands can be identified, it is necessary to expose the full length of the **penis** and its juncture with the urethra. Make an incision with a scalpel through the muscles in the midventral line between the hindlegs until they lie flat. Carefully remove the muscle tissue and pubic bone on each side of the cut until the urethra is exposed. With a blunt probe, tear the connective tissue connecting the urethra to the rectum, which lies dorsal to it.

7. Locate a pair of small glands, the **seminal vesicles,** on the dorsal surface of the urethra where the two vasa deferentia enter. Situated between the bases of the seminal vesicles is the **prostate gland.** The other sex accessory glands are the **bulbourethral glands,** two elongate structures lying on either side of the juncture of the penis and urethra.

The seminal vesicles, the prostate gland, and the bulbourethral glands all secrete fluids that, together with sperm, form **semen,** which is ejaculated during sexual intercourse. In addition to sperm, semen is mostly water, sugar, and other molecules that provide the correct aquatic environment for the flagellated sperm.

II. Internal Anatomy of the Kidney

In the first portion of today's laboratory, you located the kidneys, a pair of bean-shaped structures lying on either side of the spine in the lumbar region of the body. Although the kidneys are situated below the diaphragm, they are actually located outside of the peritoneum, the membrane that lines the abdominal cavity. During this procedure, you will examine in greater detail the structure of the kidney, including its internal anatomy, and the activity of its functional unit, the **nephron.**

MATERIALS

Per student pair:

• one preserved fetal pig injected with red and blue latex

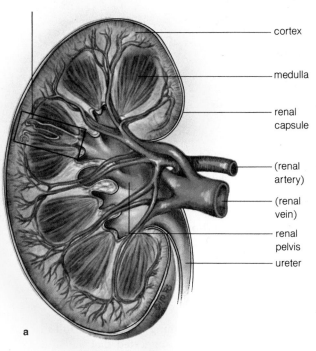

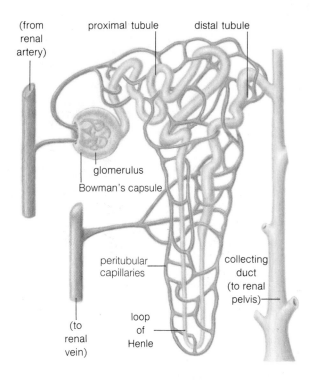

Figure 31-3 (**a**) Longitudinal section of kidney. (After Starr, 1991.) (**b**) A nephron, the functional unit of the kidney.

- dissection pan
- paper towels
- dissecting kit
- dissecting pins
- 4 large rubber bands *or* 2 pieces of string, each 60 cm long
- plastic gloves and/or lanolin hand cream
- goggles if you wear contact lenses
- compound microscope
- prepared section of kidney

PROCEDURE

A. Gross Internal Anatomy of the Kidney (fig. 31-3a)

1. Locate one of the kidneys and free it by severing the renal vein, renal artery, and ureter. Remove the kidney from the body cavity and place it on a paper towel with the central depression to the right. Attempt to identify the *adrenal gland,* a tiny, cream colored, comma-shaped body located on the medial, anterior side of the kidney.

2. With your scalpel, carefully cut the kidney in half lengthwise, as you would separate the two halves of a bean or peanut. Examine the cut surface of one of the halves and locate the three major regions of the kidney: the outer *cortex,* the *medulla,* and the *renal pelvis* (see fig. 31-3). The cortex and medulla contain the functional units of the kidney: the *nephrons* and their

associated blood vessels. Nephrons remove **urea** (produced in the liver), excess water, salts, and other wastes from the blood to form **urine.** Urine then collects in the space of the renal pelvis, travels through the ureters to the bladder, and eventually leaves the body through the urethra. Before proceeding, review the location of the ureters, bladder, and urethra (see figs. 31-1 and 31-2).

B. Microscopic Structure of the Nephrons

1. Examine a prepared section of the kidney with your compound microscope. Identify the *cortex* and *medulla.* In the cortex locate *glomeruli, Bowman's capsules,* and sections of nephrons (fig. 31-4).

2. Blood enters the kidney via branches of the renal artery (from the dorsal aorta) and is filtered out of knots of capillaries in the glomeruli (singular is glomerulus). The filtrate is literally forced out of the capillaries by the relatively high blood pressure and collects in the spaces within Bowman's capsules. As the filtrate travels through the rest of the nephron (fig. 31-4b) (the *proximal tubule,* the *loop of Henle,* and the *distal tubule*), much of the water, ions, sugars, and other useful substances are reabsorbed into the blood in the *peritubular capillaries.* While the reabsorption process carries many useful materials back into the blood, such substances as ammonia, potassium, and hydrogen ions are actually secreted from the blood to the convoluted tubules, where they join urea and other substances, forming urine.

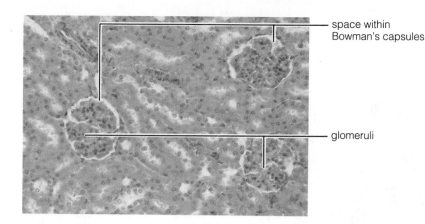

Figure 31-4 Section of the cortex of a kidney (186 ×). (Photo by D. Morton.)

The urine moves out of the nephrons into the *collecting ducts,* where it drains into the renal pelvis. It is in the collecting ducts that the final urine concentration is determined. The blood from the peritubular capillaries returns to vessels that join to form the renal vein, which returns the filtered blood to the inferior vena cava. Thus, the nephron carries out its excretory and osmoregulatory functions in three steps: filtration, reabsorption, and tubular secretion.

Although all of the activities of the nephron are extremely vital to the health of a mammal, the importance of the reabsorption function is especially easy to understand. For example, the glomeruli of the kidneys produce approximately 180 L (approximately 180 quarts) of filtrate each day in adult humans. However, about 99% of this filtrate is reabsorbed as water, primarily by the nephrons. If they were not so efficient, we would have to drink constantly just to replenish the fluid lost through filtration.

III. Nervous System

The nervous system of the pig and, in fact, all vertebrates can be divided into two major components: the *central nervous system* (CNS) and the *peripheral nervous system* (PNS). The CNS consists of the brain and the spinal cord, which serves as the primary link between the brain and much of the PNS. The PNS includes the large network of nerves outside of the central system. Through the cranial and spinal nerves of the peripheral system, impulses enter and leave the central nervous system.

MATERIALS

Per student pair:

• one preserved fetal pig injected with red and blue latex
• dissection pan
• paper towels
• dissecting kit

• plastic gloves and/or lanolin hand cream
• goggles if you wear contact lenses
• dissection microscope

PROCEDURE

As it is very time-consuming to expose the full length of the spinal cord, it is suggested that the students work in groups of four during the first portion of the exercise (section A). One pair in the group could expose the anterior portion of the spinal cord and the other expose the posterior region. In any case, each pair of students should carry out section B, the dissection of the brain up to its juncture with the spinal cord.

A. Spinal Cord

1. Turn the body wall flaps of the fetal pig inward and place your specimen ventral-side down on the dissection pan. Proceed carefully with this portion of the dissection, as the nervous tissues of a fetus are extremely delicate and may be easily destroyed.

2. Carefully remove a strip of muscle about 1.5 cm wide from the base of the neck posteriorly along the spinal column to the tail. This will expose the *spines* and the *vertebral arches* of the vertebrae (see fig. 31-5). With a sharp scalpel, gradually cut away the spines and the neural arches of the vertebrae until the **spinal cord** is exposed (fig. 31-6 on p. 430).

3. Note the enlargements of the spinal cord at the level of the forelimbs and hindlimbs. These are the *cervical* and *lumbar enlargements,* respectively; they result from the large number of nerve cells, *neurons,* supplying the appendages in these regions.

4. At the anterior end of the body, the spinal cord widens to become the **medulla oblongata,** the most posterior portion of the brain. At its caudal end, the spinal cord narrows to a relatively thin strand of tissue called the *filum terminale.*

5. Surrounding the spinal cord and the brain are a set of three membranes, the **meninges.** The outermost layer, the **dura mater,** is the most apparent and adheres

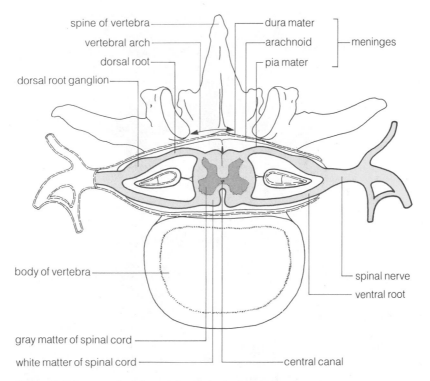

Figure 31-5 Cross section of a vertebra and spinal cord of a mammal.

to the underside of the cranial and spinal bones. The dura mater is a tough, fibrous sheath that must be slit in order to expose the spinal cord. The middle layer, the **anachnoid,** will not be apparent. The innermost layer, the **pia mater,** adheres closely to the surface of the spinal cord and brain. If you cannot identify the outer and inner meninges, attempt to locate them when you dissect the brain in section B.3 of this exercise.

6. Note the origin of the **spinal nerves** on either side of the spinal cord (see fig. 33-6 to determine the relationships of the spinal nerves to the spinal cord). There are thirty-three pairs of spinal nerves associated with the spinal cord: eight cervical, fourteen thoracic, seven lumbar, and four sacral. (Do not attempt to locate all of these, however.) Determine that a spinal nerve is composed of a **dorsal** and a **ventral root.** The dorsal root, carrying sensory impulses into the spinal cord, can be easily identified by the presence of a distinct swelling called the **dorsal root ganglion.** The ventral root, which carries motor impulses from the cord to some type of effector (a muscle, for example), has no ganglion and is not as easily identified as is the dorsal root.

7. Remove a short cross section (0.5 cm long) of the spinal cord and examine it with a dissection microscope. Notice the tissue that forms the prominent "H" in the transverse plane. This is the **gray matter,** composed of the cell bodies of motor neurons. The **white matter** around the "H" is made up of neuron fibers that conduct messages to and from the brain. Many of these fibers are insulated by myelin sheaths, which are responsible for the white color.

B. Brain

1. Using your scalpel and scissors, make a longitudinal cut through the skin and muscle tissue of the dorsal portion of the head, beginning at the base of the snout and ending at the base of the skull. From the anterior portion of this incision, make a transverse cut to the angle of the jaws and another transverse incision from the base of the skull to a level just ventral to the ears. You should now be able to remove the skin and muscle to expose the skull.

2. To remove the skull, make a longitudinal cut along the middorsal line of the skull. Do not cut too deeply and damage the brain, however. Now make two cuts, about 2 cm apart, at right angles to the longitudinal incision. To expose the brain, carefully break off pieces of the skull until the entire dorsal and lateral areas of the brain are exposed (fig. 31-5).

3. If you did not identify the meninges or membranes covering the spinal cord in section A.5, locate the dura mater and the pia mater on the surface of the brain at this time. As with the spinal cord, the arachnoid layer (middle layer) will not be apparent.

In certain severe viral or bacterial infections, the meninges around the spinal cord and/or brain may become inflamed. This serious condition is known as **meningitis.**

4. The gross features of the brain can be more easily identified by cutting the spinal cord at the base of the brain and carefully removing it from the skull. The brain is composed of the right and left *cerebral hemispheres* (collectively, the largest part of the brain, the **cerebrum**), separated by a prominent *longitudinal*

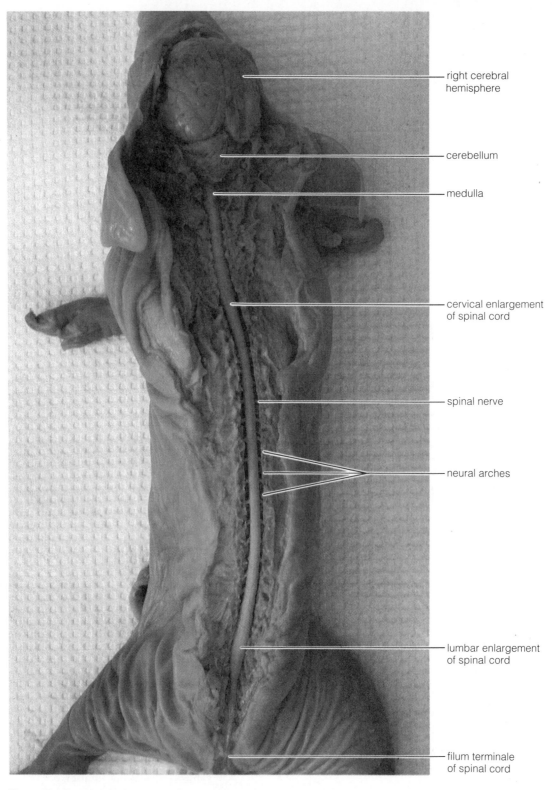

right cerebral
hemisphere

cerebellum

medulla

cervical enlargement
of spinal cord

spinal nerve

neural arches

lumbar enlargement
of spinal cord

filum terminale
of spinal cord

Figure 31-6 Dorsal view of the brain, spinal cord, and spinal
nerves of a fetal pig. (Photo by D. Morton.)

groove; a smaller mass posterior to the cerebrum, the **cerebellum;** and the medulla oblongata, or more simply, the *medulla,* under the cerebellum. The **pons,** which is not easily seen, lies just anterior to the medulla on the ventral side of the brain.

In general, most involuntary, unconscious, and mechanical processes are directed by the more posterior portions of the brain (centers in the medulla control breathing, digestion, and heartbeat). The cerebellum unconsciously controls posture and contains motor programs (like computer programs) for many complex movements. The cerebrum is responsible for such activities as reasoning, memory, conscious thought, language, and sensory decoding — activities that are generally associated with intelligence.

> **NOTE**
>
> When you have completed this exercise, return the body-wall flaps to their original positions and place the pig in the plastic bag. Dispose of any paper towels that contain formalin or preservative as directed by your instructor. Clean your dissecting tools and laboratory table.

PRE-LAB QUESTIONS

_____ 1. The urogenital system refers to the (a) urinary and reproductive systems, (b) urinary and excretory systems, (c) reproductive system, (d) external genitalia.

_____ 2. The ureters drain urine into the (a) renal pelvis, (b) cecum, (c) urinary bladder, (d) small intestine.

_____ 3. The clitoris of the female and the penis of the male are homologous structures. This means they have (a) a similar function, (b) a similar structure, (c) a similar origin, (d) a similar origin and structure.

_____ 4. The testes of a male differ from the ovaries of a female in that the testes (a) develop in the body cavity and migrate to a position outside of the body cavity, (b) require a slightly higher temperature than that of the body to produce viable gametes, (c) produce zygotes, (d) a and b.

_____ 5. When a human male has a vasectomy, the operation involves (a) removal of the male gonads or testes, (b) removal of the urethra, (c) the severing of the vas deferens, (d) removal of the prostate gland.

_____ 6. Semen contains (a) sperm, (b) male sex hormones, (c) eggs, (d) a and b.

_____ 7. The functional unit of the kidney is the (a) renal pelvis, (b) ureter, (c) cortex, (d) nephron.

_____ 8. The central nervous system of a mammal includes (a) the brain, (b) the spinal cord, (c) the brain and spinal cord, (d) the brain, spinal cord, and every major nerve in the body.

_____ 9. The brain is surrounded by a set of membranes called the (a) pleural membranes, (b) peritoneum, (c) pericardial membranes, (d) meninges.

_____ 10. The largest part of the brain of a mammal is the (a) cerebrum, (b) cerebellum, (c) pons, (d) medulla oblongata.

EXERCISE 31

Dissection of the Fetal Pig: Urogenital and Nervous Systems

POST-LAB QUESTIONS

1. What is the urogenital system?

2. Briefly describe the functions of the kidney, ureters, bladder, and urethra in the adult male pig.

3. What is the vulva?

4. What is the function of the uterine horns in the female pig?

5. What is the inguinal canal and an inguinal hernia?

6. What is the functional unit of the kidney? Briefly describe how it operates.

7. Identify the following structures.

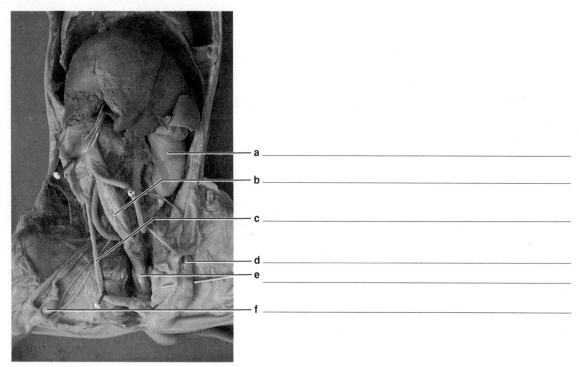

a _____

b _____

c _____

d _____

e _____

f _____

(Photo by D. Morton.)

8. What organs constitute the central nervous system of a mammal?

Human Sensations, Reflexes, and Reactions

OBJECTIVES

After completing this exercise you will be able to:

1. define *consciousness, sensory neurons, receptors, stimulus, motor neurons, somatic motor neurons, autonomic motor neurons, chemical synapse, effectors, interneurons, integration, sensations, proprioception, modality, projection, adaptation, free neuron endings, encapsulated neuron endings, phantom pain, reflex, reflex arc, stretch reflexes, patella reflex, swallowing reflex, pupillary reflex, reaction, reaction time;*

2. describe the flow of information through the nervous system;

3. state the nature and function of sensations;

4. describe a stretch reflex;

5. describe the pupillary reflex;

6. distinguish between a reflex and a reaction;

7. measure visual reaction time.

INTRODUCTION

How do you interact with the external environment? To answer this question, you first have to be able to analyze your interactions. This means you have to be conscious. **Consciousness** is the state of being aware of the things around you, your responses, and your own thoughts. Being conscious allows you to learn, to remember, and to show emotion. Second, you have to understand the flow of information through the nervous system (fig. 32-1).

Sensory neurons carry messages from **receptors** to the spinal cord and brain, which comprise the central nervous system (CNS). Receptors are located both within and body and on its surface. Receptors within the body receive information from the internal environment, while those on the surface of the body receive information from the external environment. Each piece of information received by a receptor is called a **stimulus** (the plural is *stimuli*).

Motor neurons carry messages from the CNS to **effectors.** Effectors are muscles or glands that respond to stimuli. **Somatic motor neurons** control skeletal muscles, and **autonomic motor neurons** control smooth muscles, cardiac muscle, and glands.

In the CNS, a sensory neuron can directly stimulate a motor neuron across a **chemical synapse;** more frequently, though, one or more **interneurons** connect the sensory and motor neurons.

A chemical synapse is a junction between two neurons, or between a neuron and an effector, that are separated by a small gap. A chemical transmitter substance released from the first neuron diffuses across the gap and then binds to (and produces changes in) the receiving cell.

The function of the neurons within the CNS is **integration.** At this level, integration is the processing of messages received from receptors and the activation of the appropriate motor neurons, if any, to initiate responses by effectors. Your conscious mind is located in the cerebral cortex of the brain and is aware, and indeed is a part, of some of this activity.

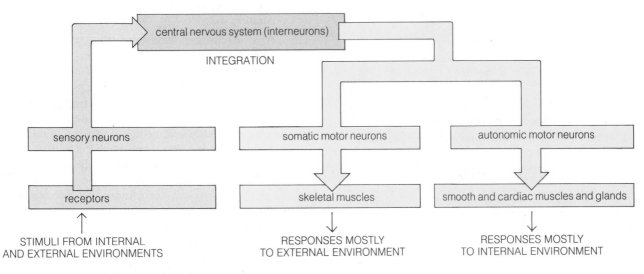

Figure 32-1 The flow of information through the nervous system.

I. Sensations

A receptor is the smallest part of a sense organ (such as skin) that can respond to a stimulus. The receptor is linked to the CNS by a single sensory neuron. Our bodies have receptors for light, sound waves, chemicals, heat, cold, tissue damage, and mechanical displacement. Senses for which we have sensations include sight, hearing, taste, smell, pain, touch, pressure, temperature, vibration, equilibrium, and **proprioception** (knowledge of the position and movement of the various body parts). **Sensations** are that portion of the sensory input to the CNS that is perceived by the conscious mind. There are also a number of complex sensations such as thirst, hunger, and nausea.

Most sensations inform the conscious mind about the state of the external environment. Sensations from the internal environment inform the conscious mind about problems such as dehydration. If you are thirsty, you will make a conscious decision to find and drink water.

Receptors and the sensations they produce have three characteristics: modality, projection, and adaptation. These characteristics can be easily demonstrated by investigating the skin's receptors.

MATERIALS

Per student pair:

- compound microscope
- prepared slide of mammalian skin stained with hematoxylin and eosin
- felt-tip, nonpermanent pen
- bristle
- dissecting needle
- 2 blunt probes in a 250-mL beaker of ice water
- 2 blunt probes in a 250-mL beaker of hot tap water (the hot water will have to be changed every 5 minutes)
- ice bag
- camel-hair brush
- reflex hammer
- 1,000-mL beaker containing ice water
- 1,000-mL beaker containing 45°C water
- 1,000-mL beaker containing room-temperature water
- tissue paper

Per lab room:

- demonstration slide showing a Pacinian corpuscle

PROCEDURE

A. Modality

Modality is the particular sensation that results from the stimulation of a particular receptor. For example, the modalities of taste—bitter, salty, sour, and sweet—are associated with four different types of taste buds. However, although every receptor has evolved to be most sensitive to one type of stimulus, modality actually depends on where in the brain the sensory neurons from the receptor (or the interneurons to which they connect) terminate. Modality cannot be encoded in the messages carried by sensory neurons, because every impulse in that message is identical. The only information the neurons can transmit is the absence or presence of a stimulus and (if one is present) its intensity—low stimulus-intensities producing a low frequency of impulses and high stimulus-intensities producing a high frequency of impulses.

1. Examine a prepared section of skin (fig. 32-2). There are two categories of receptors present: **free neuron endings** and **encapsulated neuron endings.**

Free neuron endings are almost impossible to see in typically stained sections, but note their distribution in figure 32-2. Stimulating different free neuron endings produces sensations of pain, crude touch, and perhaps cold and hot.

Encapsulated neuron endings consist of neuron endings surrounded by a connective tissue capsule. You can see the connective tissue capsule in typically stained sections. Look for Meissner's corpuscles in the dermal papillae. Meissner's corpuscles are receptors for fine touch and low-frequency vibration. Now find Pacinian corpuscles between the dermis and hypodermis. Pacinian corpuscles look like a cut onion and are receptors for pressure and high-frequency vibration. Not all skin sections will contain a Pacinian corpuscle. If you cannot locate one, look at the demonstration slide.

2. With a felt-tip, nonpermanent ink pen, have your lab partner draw a 25-cell, 0.5-cm grid (fig. 32-3) on the inside of your forearm, just above the wrist.

3. You are now the subject, and your lab partner is the investigator. At this point the investigator asks the subject to close his or her eyes. Using a bristle, the investigator touches the center of each box in the grid. If the bristle bends, you are pressing too hard. Ask the subjects to announce when they feel the touch. Do not count those responses that are given when you remove the bristle. Just count those that coincide with the initial touch. Mark each positive response with a *T* in the upper left-hand corner of the corresponding box in figure 32-3.

4. Repeat the above with a clean dissecting needle. This time, if you feel a prick, mark *P* for "pain" in the upper right-hand corner of the corresponding box in figure 32-3.

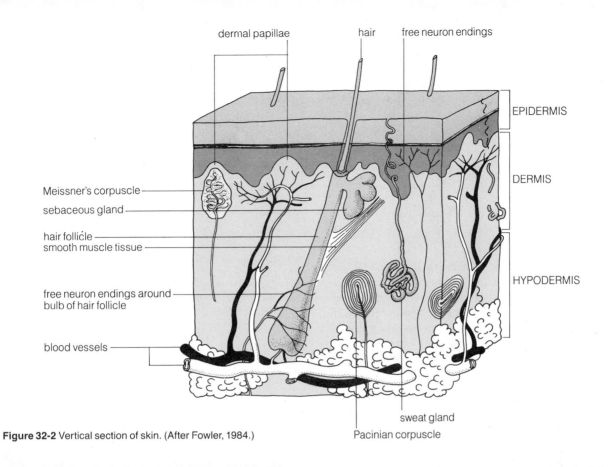

Figure 32-2 Vertical section of skin. (After Fowler, 1984.)

Labels on figure: dermal papillae, hair, free neuron endings, EPIDERMIS, DERMIS, HYPODERMIS, Meissner's corpuscle, sebaceous gland, hair follicle, smooth muscle tissue, free neuron endings around bulb of hair follicle, blood vessels, sweat gland, Pacinian corpuscle

CAUTION

Do not press; simply let the tip of the dissecting needle rest on the surface of the skin

5. Repeat the above with a chilled blunt probe. Before using the blunt probe, dry it with tissue paper. The blunt probe will warm up over time, so switch it with the second chilled blunt probe every five trials. This time, mark each positive response with a *C* for "cold" in the lower left-hand corner of the corresponding box in figure 32-3.

6. Repeat the above with a heated blunt probe. Before using the blunt probe, dry it with tissue paper. Use the two blunt probes alternately every five trials. This time, mark each positive response with an *H* for "hot" in the lower right-hand corner of the corresponding box in figure 32-3.

7. What is the total number of positive responses for each stimulus?

touch _____ /25 trials

pain _____ /25 trials

cold _____ /25 trials

hot _____ /25 trials

8. Can you see a pattern in the distribution of positive responses marked in figure 32-3? (yes or no)

_____

9. What can you conclude about the modality of skin receptors?

_____

_____

_____

B. Projection

All sensations are felt in the brain. However, before the conscious mind receives a sensation it is assigned back to its source, the receptor. This phenomenon is called **projection.** This is a very important characteristic of sensations because it allows the conscious mind to perceive the body as part of the world around it. You have probably experienced projection. A common example is the "pins and needles" you feel in your hand and forearm when you accidentally jar the nerve that passes over the inside of the elbow (so-called

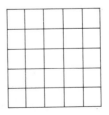

Figure 32-3 Grid for testing skin stimuli and recording modality data.

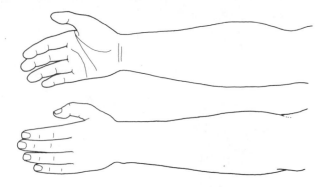

Figure 32-4 Front and back views of forearm and hand for recording projection data.

funny bone). The sensory neurons in the nerve are stimulated, and your brain projects the sensation back to the receptors. Another example is the **phantom pain** and other sensations that recent amputees seem to feel in missing limbs. This occurs because the sensory neurons that once served the missing body part are activated by the trauma of the amputation.

1. Obtain an ice bag from the freezer.

2. The investigator holds the ice bag against the inside of the subject's elbow for 2 to 5 minutes.

3. The subject describes any sensations felt in the hand or forearm, and the investigator notes them on figure 32-4.

4. While the ice bag is applied to the elbow, the investigator checks for any loss of sensation by gently stroking the arm with a camel-hair brush.

5. Sensations may also be felt after the ice bag is removed.

6. If no results are obtained, try tapping the inside of the elbow with the reflex hammer.

7. What can you conclude about projection and the receptors on the surface of the hand and forearm?

_____

_____

_____

C. Adaptation

The intensity of the signal produced by a receptor depends in part on the strength of the stimulus and in part on the degree to which the receptor was stimulated before the stimulus. Receptors undergo adaptation to a constant stimulus over time. When you enter a dark room after having been in bright light, you cannot see. After a while your photoreceptors adapt to the new light conditions, and your vision improves.

1. Partially fill each of three 1,000-mL beakers with ice water, water at room temperature, and water at 45°C.

2. The subject places one hand in the ice water and the other in the warm water. After one minute, the subject places both hands simultaneously in the water at room temperature.

3. The subject describes the sensation of temperature in each hand, and the investigator notes the results.

Hand preadapted in ice water _____

_____

Hand preadapted in warm water _____

_____

4. What can you conclude about adaptation and the receptors for temperature in the skin of the hand?

_____

_____

_____

What about other kinds of receptors? Can you give an example from your own experience? Do you feel the touch of your clothes?

_____

_____

_____

_____

_____

II. Reflexes

A **reflex** is an involuntary response to the reception of a stimulus. A **reflex arc** consists of the nervous system components activated during the reflex. The simplest reflex arc consists of a receptor, sensory neuron, motor neuron, and effector.

Involuntary means that your conscious mind does not help decide the response to the stimulus. However, the conscious mind may be aware after the fact that the reflex has taken place. Reflexes of which we are not aware occur most often in the internal environment (for example, reflexes involved in adjustments of blood pressure).

MATERIALS

Per student pair:
• reflex hammer
• penlight

PROCEDURE

A. Stretch Reflexes

Stretch reflexes are the simplest type of reflex because there is no direct involvement of interneurons (fig. 32-5). The sensory neuron synapses directly with the motor neuron in the spinal cord. Stretch reflexes are important in controlling balance and complex skeletal muscular movements such as walking. They are often tested by physicians during a physical as a check for

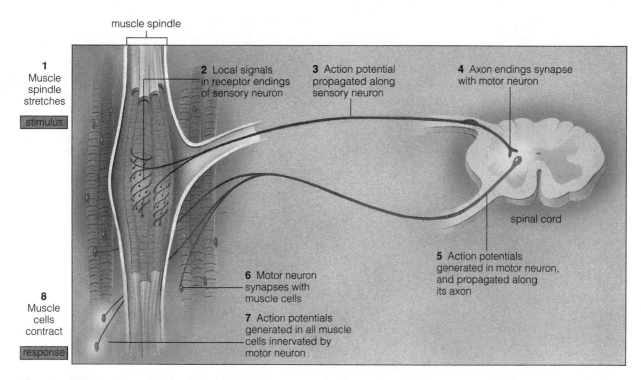

Figure 32-5 The stretch reflex. (After Starr, 1991.)

spinal nerve damage. You have probably experienced one of these tests, the **patella reflex.** In this test, the receptor is the **muscle spindle** in the quadriceps femoris muscle of the thigh, which is attached to the patella ligament. The muscle spindle detects any stretching of the muscle. The effector is the muscle itself.

1. The subject sits on a clean lab bench.

2. The investigator taps the patella ligament just below the patella bone (kneecap) with a reflex hammer (fig. 32-6). Describe the response.

_____

_____

_____

3. If you have trouble producing a response, ask the subject to shut both eyes and count backward from 10. While the subject is distracted by this task, again tap the patella ligament. Why did you obtain a response this time?

_____

_____

_____

4. Even with your eyes shut, are you aware of the stimulus or the effect, or both?

stimulus (yes or no) _____

effect (yes or no) _____

This is because of pressure receptors that sense the

tap and because of proprioceptors that sense movement of the leg.

5. Stretch reflexes are *somatic reflexes* because they involve somatic motor neurons and skeletal muscles. Can you willfully inhibit a stretch reflex?

(yes or no) _____

B. Pupillary Reflex

1. The investigator shines the penlight into one of the subject's eyes. Does the size of the pupil (the opening into the eye that is surrounded by the iris, the pigmented part of the eye) get larger or smaller?

_____

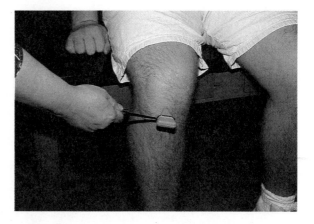

Figure 32-6 Area to tap to produce patella reflex. (Photo by D. Morton.)

HUMAN SENSATIONS, REFLEXES, AND REACTIONS

2. Now turn off the penlight. Does the size of the pupil get larger or smaller?

_____

3. Repeat steps 1 and 2 and note which is faster, constriction of the iris (which makes the pupil smaller) or dilation of the iris (which makes the pupil larger).

_____ is faster.

4. Ask if the subject is aware of the pupil's changing diameter.

(yes or no) _____

5. The pupillary reflex is an autonomic reflex because it involves an autonomic motor neuron and, in this case, smooth muscle. Can you deliberately inhibit the pupillary reflex?

(yes or no) _____

C. Complex Reflexes

Complex reflexes involve many reflex arcs and interneurons. A good example is swallowing. The stimulus in the **swallowing reflex** is the movement of saliva, food, or drink into the posterior oral cavity. The response is swallowing.

1. Cup your hand around your neck and swallow. Feel the complex skeletal muscular movements involved in swallowing. Do you consciously control all these muscles?

(yes or no) _____

2. Is it possible to swallow several times in quick succession?

(yes or no) _____

3. Explain this result. It has something to do with the stimulus.

_____

_____

_____

_____

4. What part of swallowing does your conscious mind control, and what part is a reflex?

_____

_____

_____

III. Reactions

A **reaction** is a voluntary response to the reception of a stimulus. _Voluntary_ means that your conscious mind initiates the reaction. An example is swatting a fly once it has landed in an accessible spot. Because neurons must carry the sensory message to the cerebral cortex and the message to the motor neuron to react, a reaction takes more time than a reflex. **Reaction time** has the following components:

1. The time it takes for the stimulus to reach the receptive unit.
2. The time it takes for the receptor to process the message.
3. The time it takes for a sensory neuron to carry the message to the integration center.
4. The time it takes for the integration center to process the information.
5. The time it takes for a motor neuron to carry the response to the effector.
6. The time it takes for the effector to respond.

Visual reaction time can easily be measured with a reaction-time ruler. This device makes use of the principle of progressive acceleration of a falling object.

MATERIALS

Per student group (4):
• Reaction Time Kit (Carolina Biological Supply Company)
• chair or stool
• calculator (optional)

PROCEDURE

The following instructions are modified from the _Reaction Time Kit Instructions_ booklet.

1. The subject sits on a chair or stool (fig. 32-7).

2. The investigator stands facing the subject and holds the _release end_ of the reaction-time ruler with the thumb and forefinger of the dominant hand, at eye level or higher.

3. The subject positions the thumb and forefinger of the dominant hand around the _thumb line_ on the ruler. The space between the subject's thumb and forefinger should be about one inch.

4. The subject tells the investigator when he or she is ready to be tested.

5. Once the investigator is told the subject is ready, at any time during the next 10 seconds, the investigator lets go of the ruler.

6. The subject catches the ruler between the thumb and forefinger as soon as it starts to fall. The line under his or her thumb represents visual reaction time in milliseconds.

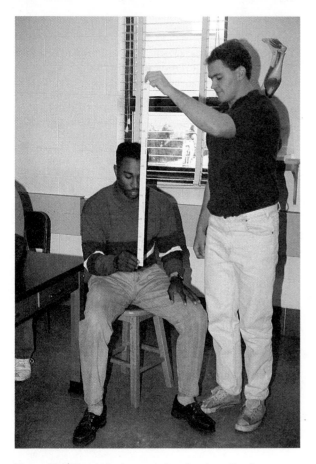

Figure 32-7 Two students measuring visual reaction time. (Photo by D. Morton.)

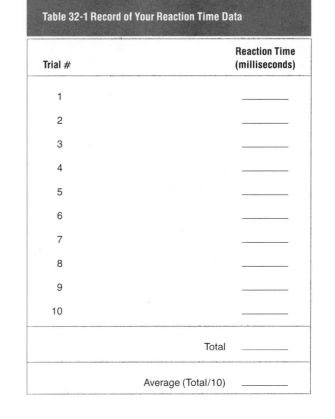

Trial #	Reaction Time (milliseconds)
1	_____
2	_____
3	_____
4	_____
5	_____
6	_____
7	_____
8	_____
9	_____
10	_____
Total	_____
Average (Total/10)	_____

7. The subject reads the reaction time from the ruler out loud, and the investigator records the data in table 32-1.

8. Repeat steps 1 through 7 ten times and calculate the average reaction time from the ten trials.

9. Repeat steps 1 through 8 for each member of the group.

10. The reaction times of most of the ten trials should be similar, but perhaps the first few or one at random may be relatively different from the others. If this is true for your data, suggest some reasons for this variability.

_____

_____

_____

_____

_____

11. If opportunity and interest allow, the *Reaction Time Kit Instructions* booklet has a number of suggestions for other experiments you can easily do with the reaction-time ruler.

_____ 1. Neurons that carry messages from receptors to the CNS are (a) sensory, (b) motor, (c) interneurons, (d) autonomic.

_____ 2. Neurons that carry messages from the CNS to effectors are (a) sensory, (b) motor, (c) interneurons, (d) a and b.

_____ 3. Neurons that carry messages within the CNS are (a) sensory, (b) motor, (c) interneurons, (d) autonomic.

_____ 4. Knowledge of the position and movement of the various body parts is (a) modality, (b) projection, (c) adaptation, (d) proprioception.

_____ 5. Skin contains (a) free neuron endings, (b) encapsulated neuron endings, (c) a and b, (d) no nervous tissue.

_____ 6. Which characteristic of receptors does phantom pain illustrate? (a) modality, (b) projection, (c) adaptation, (d) proprioception.

_____ 7. A simple reflex arc is made up of a receptor and (a) a sensory neuron, (b) a motor neuron, (c) an effector, (d) all of the above.

_____ 8. A stretch reflex is (a) somatic, (b) autonomic, (c) a and b, (d) none of the above.

_____ 9. A pupillary reflex is (a) somatic, (b) autonomic, (c) a and b, (d) none of the above.

_____ 10. A reaction is (a) a reflex, (b) involuntary, (c) voluntary, (d) a and b.

EXERCISE 32

Human Sensations, Reflexes, and Reactions

POST-LAB QUESTIONS

1. a. Where in the brain does your consciousness reside?

  b. Are organisms aware of all the activity in their nervous systems? (yes or no)

  c. Why is this an advantage to an organism?

2. In your own words, define:
  a. *modality*

  b. *projection*

  c. *adaptation*

3. What are the advantages and disadvantages of receptor adaptation to an organism?

4. List the basic steps of a simple reflex arc like the stretch reflex.

5. How does the patella reflex differ from the pupillary reflex?

6. Indicate whether the following are reactions or reflexes.
   a. a baby wetting a diaper

   b. braking a car to avoid an accident

   c. withdrawing your hand from a hot stove surface

   d. sneezing

   e. waving to a friend across the street

7. All animals do not perceive the external environment in exactly the same way. List some examples from your own knowledge and readings in the textbook.

8. To survive, an animal needs all its receptors working, and even then it cannot fully sense the external environment. What extra receptors do you think would be an advantage to the survival of humans in this modern world?

Human Skeletal and Muscular Systems

OBJECTIVES

After completing this exercise you will be able to:

1. define *bones, ligaments, joints, skeletal muscles, tendons, muscle tone, posture, sutures, synovial joints, diaphysis, epiphyses, compact bone, spongy bone, marrow cavity, lever;*

2. identify the major bones of the human skeleton;

3. describe the structure of a typical bone;

4. define *origin, insertion,* and *action* as these terms apply to skeletal muscles and their tendons;

5. distinguish between isometric and isotonic contractions of skeletal muscles;

6. describe and give everyday and anatomical examples of the three classes of levers;

7. present a simple biomechanical analysis of walking.

INTRODUCTION

The skeletal system and muscular system are often considered together to stress their close structural and functional ties. They are often referred to as the musculoskeletal system. These two systems determine the basic shape of the body, support the other systems, and provide the means by which an organism moves in the external environment.

Bones are the main organs of the skeletal system. They are primarily bone tissue, although all of the four basic tissue types are present. The places in the body where two or more bones are connected are called **joints.** The joints you are most familiar with are the shoulder, elbow, wrist, hip, knee, and ankle. However, there are many others. Around many joints, bones are held together by straplike structures called **ligaments.** Ligaments are primarily dense connective tissue that is more or less elastic. Elastic ligaments around mobile joints stretch to allow movement.

Skeletal muscles are the main organs of the muscular system and are composed primarily of skeletal muscle tissue. Skeletal muscles are connected to bones by dense fibrous connective tissue structures called **tendons.** Tendons are inelastic, so all of the force of skeletal muscle contraction is transferred to the skeleton.

When a skeletal muscle contracts, movement may or may not occur. If the skeletal muscle is allowed to shorten, the bone moves, and in doing so it moves some body part. On the other hand, if the skeletal muscle does not shorten, the tension in that muscle and in its tendons increases. All skeletal muscles exhibit tension or **muscle tone** except when you are asleep. This tension maintains the position of the body and its parts against the pull of gravity. The ability to hold the body erect is called **posture.**

The organs of the skeletal and muscular systems have other functions. Bones protect internal organs (for example, the skull protects the brain, eyes, and ears). Bones also store minerals and produce blood cells in the bone marrow. When body temperature drops below a certain level, skeletal muscles produce heat by shivering.

I. Adult Human Skeleton

An articulated human skeleton is prepared by joining together the degreased and bleached bones of an individual so that many of the bones may be moved as they were in life. Sometimes, plastic casts of the original bones are used.

Although bone tissue predominates, fresh bones are composed of all four basic tissue types. However, when they are prepared for study, their organic portion is lost. These bones consist only of the mineral portion of bone tissue. Original details remain, but the bones are brittle. *Therefore, you must handle bones gently. Use a pipe cleaner to point out details and never use a pencil or pen because it is very difficult to remove marks.*

MATERIALS

Per student:

• pipe cleaner

• compound microscope, lens paper, a bottle of lens-cleaning solution, a lint-free cloth (optional)

• prepared slide of a ground transverse section of compact bone (optional)

Per student group:

• articulated adult human skeleton (natural bone or plastic)

• femur

• femur that has been sawed into two halves lengthwise

Per lab room:

• labeled chart and illustrations of the adult human skeleton

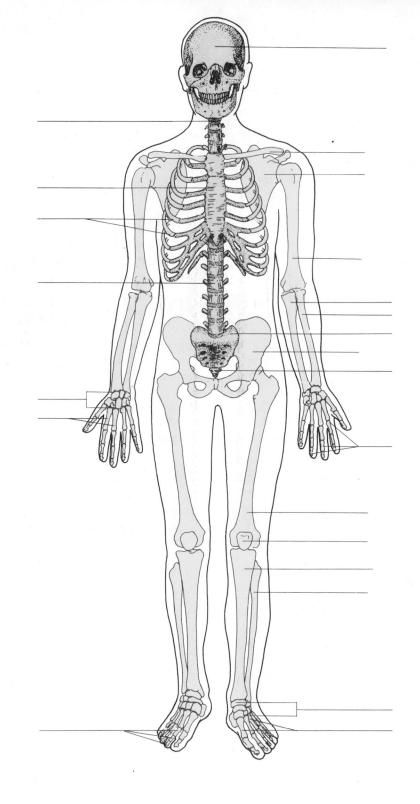

Figure 33-1 Label this front view of adult human skeleton (axial skeleton shaded gray). (After Fowler, 1984.)

Labels: bones listed in section A, wrist, elbow, shoulder, hip, knee, ankle

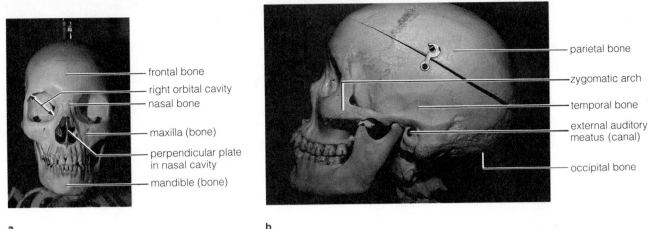

Figure 33-2 (a) Front and **(b)** side views of the skull. (Photos by D. Morton.)

PROCEDURE

A. Identification of Some Bones

There are 206 separate bones in the adult human skeleton. Using the labeled chart and illustrations of the human skeleton, identify the following bones on the articulated human skeleton and label them in figure 33-1.

1. Axial skeleton
   a. **skull** (28 separate bones, including middle ear bones) (fig. 33-2)
   b. **vertebrae** (singular is *vertebra*; 26 separate bones, including the **sacrum,** which is composed of five fused vertebrae, and the **coccyx,** which is usually composed of four fused vertebrae) (fig. 33-3)
   c. **ribs** (24 separate bones)
   d. **sternum** (three fused bones)
   e. **hyoid** (only bone that does not form a joint with another bone)

2. Appendicular skeleton (these bones are all found on each side of the body)
   pectoral girdle (shoulder)
   a. **scapula**
   b. **clavicle**
   upper appendage (arm)
   c. **humerus**
   d. **radius**
   e. **ulna**
   f. **carpals** (8)
   g. **metacarpals** (5)
   h. **phalanges** (14)
   pelvic girdle (hip)
   i. **coxal bone** (three fused bones — pubis, ischium, ilium)
   lower appendage (leg)
   j. **femur**
   k. **tibia**
   l. **fibula**
   m. **patella**
   n. **tarsals** (7)
   o. **metatarsals** (5)
   p. **phalanges** (14)

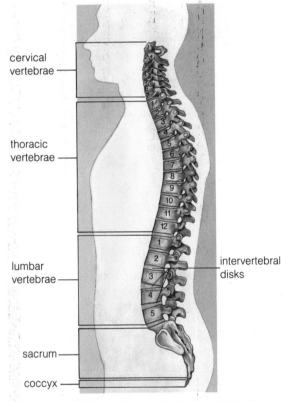

Figure 33-3 Side view of the vertebral column. (After Starr, 1991.)

B. Joints

Joints can be immobile, like the **sutures** that connect the bones of the roof of the skull of young adults; they can be freely movable, like **synovial joints** (such as the elbow and knee); or they can fall between these two extremes in the degree of motion they allow (fig. 33-4). The *fibrous capsule* of synovial joints is lined by a *synovial membrane,* which secretes lubricating *synovial fluid.* The fibrous capsule and ligaments function to stabilize synovial joints. Ligaments can be located outside and inside the capsule, and they may be thickenings of its wall. Identify the synovial joints listed in figure 33-1 and list the adjacent bones that form them in table 33-1 on p. 448.

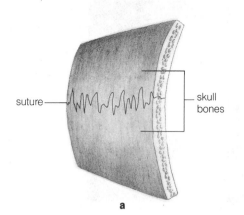

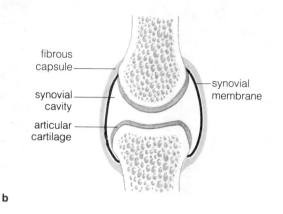

fibrous
capsule

synovial
cavity

articular
cartilage

synovial
membrane

suture

skull
bones

a

b

Figure 33-4 (**a**) A suture and (**b**) a diagrammatic synovial joint. (After Fowler, 1984.)

Table 33-1 Bones That Form the Major Synovial Joints	
Joint	**Adjacent Bones**
wrist	_____
elbow	_____
shoulder	_____
hip	_____
knee	_____
ankle	_____

Table 33-2 Surface Features and Bones	
Surface Feature	**Bone**
knuckles	_____
bump next to the wrist and on the same side of the upper appendage as the little finger	_____
smaller bump next to the wrist and on the same side of the upper appendage as the thumb	_____
bump next to and outside the ankle	_____
bump next to and inside the ankle	_____

C. Surface Features

There are many places on your body surface where bones can be felt. However, it is often difficult to tell specifically which bone you are feeling. Some are easy. For example, feel the bone of the lower jaw, the *mandible*. This is the only bone of the skull that forms a synovial joint with another skull bone.

Let us try a harder example, the process that projects from the point of the elbow joint. Touch it and alternately extend and flex the forearm, increasing and decreasing the angle between the forearm and upper arm, respectively. Which part of the arm does the process move with, forearm or upper arm?

_____

While still touching this process, alternately turn the hand palm down and up. Does the process move?

(yes or no) _____

Identify the bone to which this process belongs.

_____

In general, to identify a portion of a bone near a joint, move the body parts adjacent to the joint while touching the bone.

Identify the bones to which the surface features listed in table 33-2 belong.

D. Structure of a Bone (fig. 33-5)

1. Look at a femur, the longest bone of the skeleton. It consists of a shaft, or **diaphysis,** with two knobby ends, or **epiphyses** (the singular is *epiphysis*). One of the ends has a narrow neck and a round head. To which bone does it join?

_____

To which bone of the skeleton does the other end join?

_____

Note that there are other surface features on the femur, such as projections of various sizes and lines. These surface features are attachment sites for tendons and ligaments. Are there small tunnels opening onto the surface of the femur?

(yes or no) _____

In life, these nutrient canals serve as routes for blood vessels and nerves.

2. Examine a femur that has been sawed in half lengthwise. There are two kinds of adult bone tissue: **compact bone** and **spongy bone.** Compact bone is

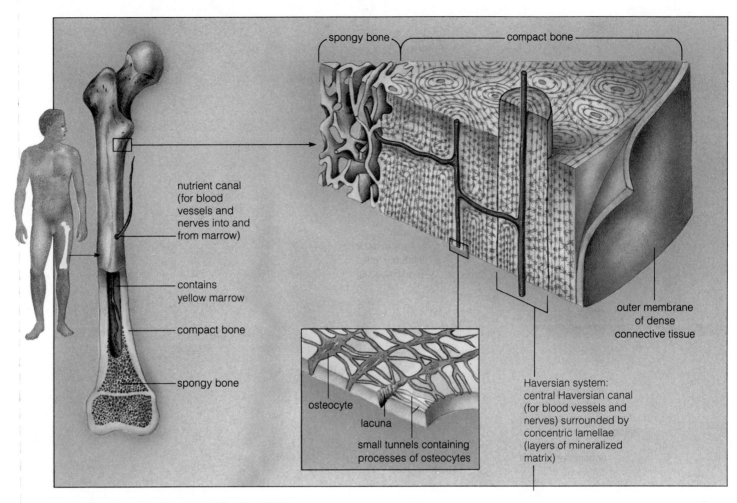

Figure 33-5 Structure of the femur. (After Starr, 1991.)

solid and dense and is found on the surface of the femur. Spongy bone is latticelike and is found on the inside of the femur, primarily in the epiphyses and surrounding the **marrow cavity.** Which kind of bone tissue looks denser?

_____

Comparing pieces of equal size, which kind of bone tissue looks lighter?

_____

3. *Optional.* Instructions for the study of a transverse section of ground compact bone are located on page 38 in Exercise 28.

E. Structure of a Skeletal Muscle (fig. 33-6)

Like bones, skeletal muscles are composed of all four basic tissue types. Skeletal muscles are mostly skeletal muscle tissue with the individual skeletal muscle fibers arranged parallel to the axis along which the muscle shortens when contracting. There is also a lot of connective tissue that surrounds the fibers and connects them to the tendons.

II. Leverage and Movement

Much of the skeletal system can be considered a system of levers, in which each bone is a lever and the joints are fulcrums. **Levers** are simple machines. When a pulling force or effort is applied to a lever, it moves about its fulcrum, overcoming a resistance or moving a load.

During a typical movement, one end of a skeletal muscle, the **origin,** remains stationary. The other end, the **insertion,** moves along with the bone and surrounding body part. The movement produced by the contraction is the **action** of the skeletal muscle. Most insertions are close to their joints, and the advantage gained by this is that the muscle has to shorten a small distance to produce a corresponding large movement of a body part.

MATERIALS

Per student pair:
• pair of scissors
• toggle switch mounted on a board (Alternatively, you can use any light switches present in the room.)

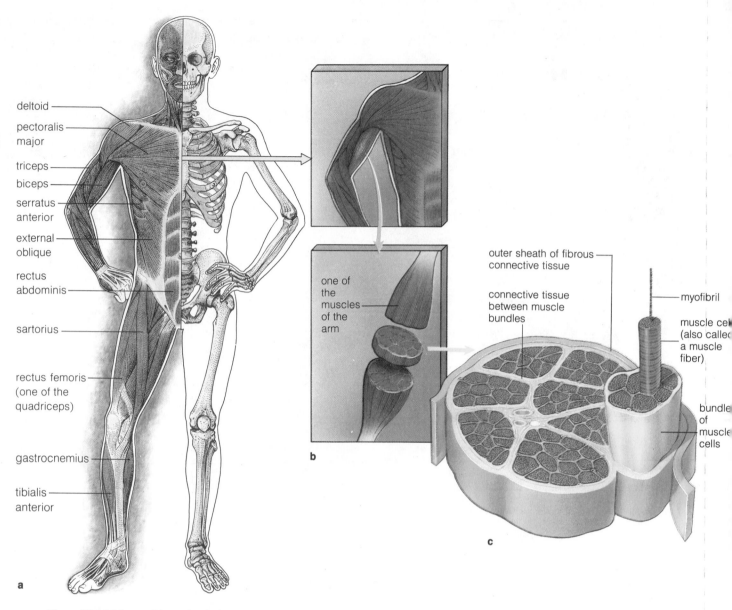

Labels for figure (a):
- deltoid
- pectoralis major
- triceps
- biceps
- serratus anterior
- external oblique
- rectus abdominis
- sartorius
- rectus femoris (one of the quadriceps)
- gastrocnemius
- tibialis anterior

a

b

one of the muscles of the arm

Labels for figure (c):
- outer sheath of fibrous connective tissue
- connective tissue between muscle bundles
- myofibril
- muscle cell (also called a muscle fiber)
- bundle of muscle cells

c

Figure 33-6 (**a**) Some of the major skeletal muscles of the human muscular system. (**b**) Closer look at the structure of a skeletal muscle organ. (**c**) Transverse section of a skeletal muscle organ. (After Starr, 1991.)

- pair of forceps
- pencil
- textbook

PROCEDURE

A. Classes of Levers

There are three classes of levers:

1. Class I. The fulcrum is located between the effort and the load (fig. 33-7).

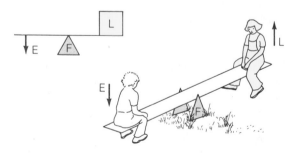

Figure 33-7 A seesaw is an example of a first-class lever; E—effort, F—Fulcrum, and L—load.

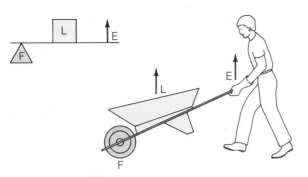

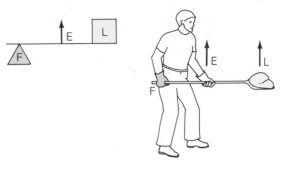

Figure 33-8 A wheelbarrow is an example of a second-class lever; E—effort, F—fulcrum, and L—load.

Figure 33-9 Lifting a spade with one hand while holding the handle stationary with the other hand is an example of a third-class lever; E—effort, F—fulcrum, and L—load.

2. Class II. The load is located between the fulcrum and the effort (fig. 33-8).

3. Class III. The effort is located between the fulcrum and the load (fig. 33-9). Third-class levers are the most common in the skeletal system.

4. Test your understanding of the three classes of levers by examining the following objects and completing the following matching question.

class of lever	objects
_____ **1.** class I	**a.** scissors
_____ **2.** class II	**b.** toggle switch or light switch
_____ **3.** class III	**c.** forceps

B. Analysis of Simple Movements

Let us analyze three simple movements: flexion of the forearm, extension of the forearm, and plantar flexion of the foot (fig. 33-10).

1. *Flexion of forearm.* While sitting, turn your hand so the palm is up and place it under the lab bench. Try to flex the forearm (decrease the angle between the forearm and upper arm). Because the skeletal muscle that is attempting to flex the forearm cannot shorten, the tension in it will increase. A contraction of a skeletal muscle in which tension increases but no movement results is called an **isometric contraction.** Feel with your other hand the front surface of the upper arm. The large tense muscle is the biceps brachii. Its origin is the scapula, and its insertion is the radius. Which joint is the fulcrum?

Now place a pencil in the palm of your hand and flex the forearm. A contraction of a skeletal muscle that results in movement is called an **isotonic contraction.** There is no increase in tension during the movement. Feel the tension in the biceps brachii as you

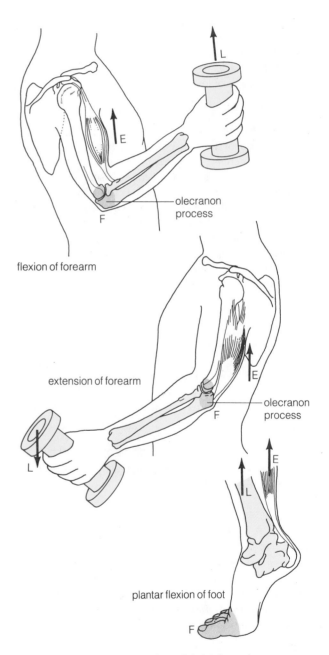

flexion of forearm

extension of forearm

plantar flexion of foot

Figure 33-10 Some simple actions of skeletal muscles.

make this movement. Repeat this procedure, but replace the pencil with a textbook. Both the pencil and the book are adding to the load being lifted, the forearm. In which case—lifting the pencil or the textbook—was the tension in the biceps brachii the greatest?

lifting the _____

When you lift any object, the tension in the muscle must equal the weight of that object before movement can occur. Therefore, normal movements have an isometric phase followed by an isotonic phase.

Where is the pulling force applied? (insertion, origin, or both the insertion and origin)

_____

_____

Even simple movements require the coordination of a group of muscles. For example, the origin does not move because other skeletal muscles hold the scapula stationary.

What class of lever is illustrated by this example?

(I, II, or III) _____

2. *Extension of forearm.* Place the hand, still palm up, on the top of the lab bench and try to extend the forearm (increase the angle between the forearm and the upper arm). Feel for a tense muscle on the back surface of the upper arm. This is the triceps brachii. The origin of the triceps brachii is the scapula and the upper humerus; its insertion is the *olecranon process* of the ulna (fig. 35-10). The fulcrum is the same as the previous example, except that it has shifted its position relative to the effort and the load. What class of lever is illustrated by this example?

_____

Extension of the forearm is the opposite movement to flexion of the forearm. Hold the textbook, palm still up, halfway between full flexion and full extension. Feel the tension in the biceps brachii and triceps brachii. Repeat this procedure without the book. Is the tension in the biceps brachii greater with or without the book?

_____

Is the tension in the triceps brachii greater with or without the book?

_____

The state of contraction of a group of skeletal muscles has to be coordinated to accomplish a particular movement or element of posture. Both the tendons of the biceps brachii and the triceps brachii are pulling on their insertions on the bones of the forearm to keep the forearm stationary. Other muscles are keeping the shoulder stationary.

3. *Plantar flexion of foot.* You need to stand up for the last example. A lab partner should stand behind you and watch that you do not fall during this procedure. With one hand on the lab bench to steady your balance, stand on the tips of your toes. With your other hand feel one of the very large tense muscles on the back of each calf. This is the gastrocnemius. The origin of the gastrocnemius is the femur, and its insertion is a tarsal—the calcaneus (the so-called heel bone). The fulcrum is the metatarsal-phalangeal joints, and the weight is the weight of the body transmitted through the tibia. Of what class of lever is this an example?

_____

III. Walking

Walking is a complex activity that requires many movements and the coordinated contractions of several groups of skeletal muscles. For each leg, walking involves two phases, which together make up the *step cycle*. The *stance phase* is the time when the leg is bearing weight, and the *swing phase* is when the leg is in the air.

MATERIALS

Per lab room:
• safe place to walk

PROCEDURE

1. *Your instructor will tell you where you can walk safely.* Walk a few normal steps, concentrating on one leg. What part of the foot strikes the ground first?

(toe or heel) _____

What part of your foot leaves the ground last?

_____

Does it leave passively, or does it push off?

_____

2. Now put your hands on your hips and concentrate on what your pelvic girdle is doing while you walk. First take short strides and then long ones. Does the pelvic girdle rotate more during short or long strides?

_____

Rotation of the pelvic girdle can be demonstrated in a different way. Find a lab partner of about equal height. Walk right next to each other but out of step, that is, with opposite feet leading. First take short steps and then long ones. What happens?

_____

_____

This sideways movement is called *lateral displacement*. Incidentally, females in general have to rotate their pelvic girdles a little more than males for a given length of stride. This is due to differences in the proportions of the female and male pelvic girdles.

3. *Vertical displacement* also occurs during walking. From the side observe two individuals of equal height walking out of step and next to each other. Do their heads remain at the same level, or do they bob up and down?

_____

PRE-LAB QUESTIONS

____ 1. Ligaments connect (a) bones to bones, (b) skeletal muscles to bones, (c) tendons to bones, (d) skeletal muscles to tendons.

____ 2. Tendons connect (a) bones to bones, (b) skeletal muscles to bones, (c) ligaments to bones, (d) skeletal muscles to tendons.

____ 3. Which of the following bones is part of the axial skeleton? (a) clavicle, (b) radius, (c) coxal bone, (d) sternum.

____ 4. The two kinds of bone tissue are (a) compact and loose, (b) compact and spongy, (c) dense and spongy, (d) loose and dense.

____ 5. There are _____ classes of levers. (a) two, (b) three, (c) four, (d) more than four.

____ 6. The class of lever in which the effort is located between the fulcrum and the load is called (a) first class, (b) second class, (c) third class, (d) fourth class.

____ 7. The end of the skeletal muscle that remains stationary during a movement is the (a) action, (b) origin, (c) insertion, (d) none of the above.

____ 8. In an isotonic contraction of a skeletal muscle (a) the tension in the muscle increases, (b) movement occurs, (c) no movement occurs, (d) a and c.

____ 9. In an isometric contraction of a skeletal muscle (a) the tension in the muscle increases, (b) movement occurs, (c) no movement occurs, (d) a and c.

____ 10. The step cycle of walking consists of a (a) stance phase, (b) swing phase, (c) a and b, (d) none of the above.

EXERCISE 33

Human Skeletal and Muscular Systems

POST-LAB QUESTIONS

1. Match the following bones to their location in the body.

    _____ **a.** radius   **i.** pectoral girdle

    _____ **b.** coxal bone   **ii.** lower appendage

    _____ **c.** ribs   **iii.** axial skeleton

    _____ **d.** scapula   **iv.** upper appendage

    _____ **e.** fibula   **v.** pelvic girdle

2. Draw a long bone (for example, femur) that has been sawed in half lengthwise. Label the diaphysis, an epiphysis, compact bone, spongy bone, and the marrow cavity.

3. Define:

    a. the *insertion* of a skeletal muscle

    b. the *origin* of a skeletal muscle

    c. the *action* of a skeletal muscle

    d. *isometric contraction*

    e. *isotonic contraction*

4. The skeletal muscles that flex the head are the sternocleidomastoids. The origins are the sternum and the clavicle, and the insertion is the mastoid process, the bump just behind the ear. Identify this class of lever and explain why you made this choice.

5. In your own words, describe one step in the walking cycle.

6. Identify the bones indicated in this illustration.

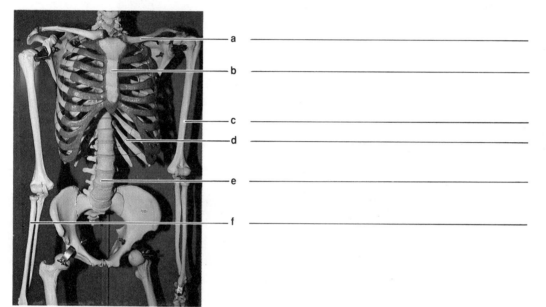

a _____

b _____

c _____

d _____

e _____

f _____

(Photo by D. Morton.)

Human Blood and Circulation

OBJECTIVES

After completing this exercise you will be able to:

1. define *blood vessels* (different types), *heart, blood, plasma, extracellular fluid, intracellular fluid, interstitial fluid, homeostatis, pulmonary circuit, systemic circuit, blood pressure, elastic membranes, valves, sinoatrial node;*

2. identify and give the functions of the different types of blood cells;

3. distinguish among an artery, capillary, and vein;

4. describe how blood flows through capillaries;

5. name the four chambers of the heart and describe the route blood takes through them;

6. describe how the heart contracts.

INTRODUCTION

Circulation—the bulk transport of fluid around the body—joins together the specialized cells of multicellular organisms, even though they are separated physically. A circulatory system is a necessary step in the evolution of complex, larger organisms.

Most coelomates (animals with a true body cavity) have a circulatory system with blood, blood vessels, and one or more hearts. The **heart** pumps the fluid **blood** around the circulatory system in pipelike **blood vessels.** Some invertebrates, like the clam, crayfish,

and insect, have an *open circulatory system*, in which the blood percolates directly through the body tissues (fig. 34-1).

Vertebrates and some invertebrates, like the earthworm, have a *closed circulatory system*, in which the blood flows solely within the blood vessels (fig. 34-2).

In mammals and birds, there are two completely separate routes leading to and from the heart: the pulmonary and systemic circuits (fig. 34-3). In each circuit, branching **arteries** convey blood to smaller and more numerous **arterioles,** which in turn deliver the blood to beds of **capillaries.** It is across the capillary walls that the exchange of dissolved gases, nutrients, wastes, and so on takes place. Within the beds the capillaries branch and merge, finally merging into **venules.** Venules merge into a smaller number of **veins,** which continue merging and which drain blood back toward the heart.

The **pulmonary circuit** carries oxygen-depleted blood to the capillary beds of the lungs, where oxygen is loaded and where excess carbon dioxide is unloaded. Pulmonary veins drain the oxygen-rich blood back to the heart. The **systemic circuit** takes the oxygen-rich blood from the heart and conveys it to the

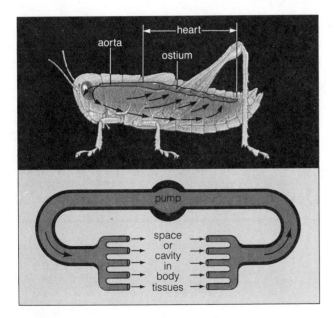

Figure 34-1 Open circulatory system. (After Starr, 1991.)

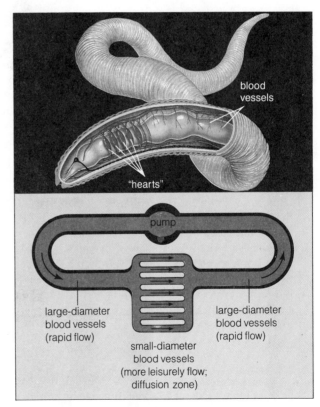

Figure 34-2 Closed circulatory system. (After Starr, 1991.)

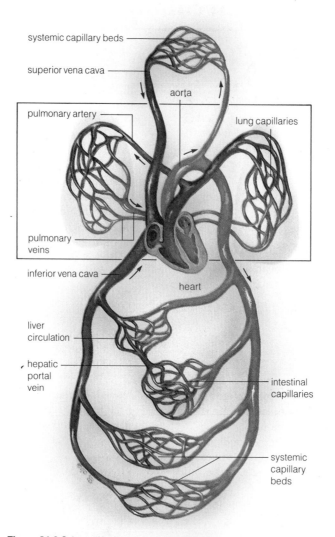

systemic capillary beds

superior vena cava

aorta

pulmonary artery

lung capillaries

pulmonary veins

inferior vena cava

heart

liver circulation

hepatic portal vein

intestinal capillaries

systemic capillary beds

Figure 34-3 Schematic of the human circulatory system. The pulmonary circuit is enclosed by the box. Red indicates oxygenated blood. The blue indicates oxygen-depleted blood. (After Starr and Taggart, 1989.)

rest of the body's capillary beds, where oxygen is unloaded and excess carbon dioxide is picked up. Systemic veins drain the oxygen-depleted blood back to the heart.

In a few cases this pattern of blood flow—heart → arteries → arterioles → capillaries → venules →veins → heart is interrupted by a **portal vein,** which connects two capillary beds. The most prominent example is the *hepatic portal vein,* which transports blood from capillary beds in the intestines, stomach, and spleen to beds of large capillaries in the liver.

I. Blood

Human **blood** is about 45% cells by volume, although it is only slightly thicker than water. Most of the cells are erythrocytes, or red blood cells. They are what gives blood its red color. Blood cells are suspended in

a straw-colored fluid called **plasma** (about 55% of the blood).

Body fluid is divided into two major components, the intracellular fluid and the extracellular fluid. The **intracellular fluid** is all of the fluid in cells. The **extracellular fluid** is the remainder of the fluid in the body—all of the fluid outside of cells. The extracellular fluid is further divided into the plasma and the interstitial fluid. The **interstitial fluid** is all of the fluid between the cells and the blood vessels. A molecule diffusing from the blood into a cell would have to move out of the plasma, pass through the interstitial fluid, and finally enter the intracellular fluid.

The circulatory system plays a central role in the maintenance of **homeostatis**—a stable internal environment. Homeostatic mechanisms throughout the body function to keep the physical and chemical properties of the blood within physiological limits. By

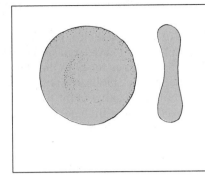

a erythrocytes

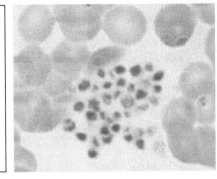

b platelets

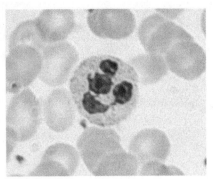

c neutrophil

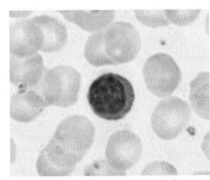

d lymphocyte

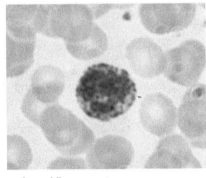

e monocyte

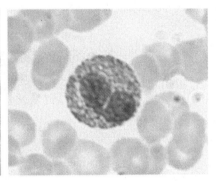

f eosinophil

g basophil

Figure 34-4 Formed elements of blood. (b–g, 1,440×). (Photos by D. Morton.)

means of the fast-circulating blood, the effects of homeostatic mechanisms are spread first to the interstitial fluid and then to the intracellular fluid throughout the body.

Plasma is mostly water but contains many dissolved substances, including gases, nutrients, wastes, ions, hormones, enzymes, antibodies, and other proteins.

MATERIALS

Per student:

• a compound microscope, lens paper, a bottle of lens-cleaning solution (optional), a lint-free cloth (optional), a dropper bottle of immersion oil (optional)

• a prepared slide of a Wright-stained smear of human blood

Per lab room:

• eosinophil on demonstration (compound microscope)

• basophil on demonstration (compound microscope)

PROCEDURE

A. Formed Elements (Cells and Platelets) of Blood

1. Use your compound microscope to examine the prepared slide of a Wright-stained smear of blood with medium power. Note the numerous pink-stained red blood cells, or **erythrocytes** (fig. 34-4a). Each erythrocyte is a biconcave disk without a nucleus. Scattered among them are a much smaller number of blue/purple-stained cells. These are white blood cells,

Table 34-1 Characteristics of Formed Elements of Blood

Cell or Fragment	Number/mm³ in Peripheral Blood	Percent of Leukocytes	Size (μm)
erythrocyte	4.5 to 5.5 million	—	7 by 2
platelets	250,000 to 300,000	—	2 to 5
neutrophils	3,000 to 6,750	65	10 to 12
eosinophils	100 to 360	3	10 to 12
basophils	25 to 90	1	8 to 10
lymphocytes	1,000 to 2,700	25	5 to 8
monocytes	150 to 750	6	9 to 15

Table 34-2 Functions of Formed Elements of Blood

Cell or Fragment	Functions
erythrocytes	contain hemoglobin, which transports oxygen, and carbonic anhydrase, which promotes transport of carbon dioxide by the blood
platelets	source of substances that aid in blood clotting
neutrophils	leave the blood early in an inflammation to become phagocytes (cells that eat bacteria and debris)
eosinophils	phagocytosis of antigen-antibody complexes; numbers are elevated during allergic reactions
basophils	granules contain a substance (histamine) that makes blood vessels leaky and a substance (heparin) that inhibits blood clotting
lymphocytes	perform many functions central to immunity
monocytes	leave the blood to form phagocytic cells called macrophages

or **leukocytes.** Center a leukocyte and rotate the nosepiece to the high-dry objective. What part of the cell is stained blue/purple?

(nucleus or cytoplasm) _____

Using high power, preferably the oil-immersion objective, move the slide slowly and look for fragments of cells between the erythrocytes and leukocytes. They usually have one small blue-stained granule in them and are often clumped together. These cell fragments are **platelets,** or thrombocytes (fig. 34-4b).

Locate at least three of the five leukocytes — **neutrophils, lymphocytes,** and **monocytes** (figs. 34-4c, 4d, and 4e). Search for them with the high-dry objective. When one is found, center and examine it. If your microscope has an oil-immersion objective (and with the permission of the instructor), use it to look at each blood cell. You may also find **eosinophils** and **baso-**

phils (figs. 34-4f and 4g), which are usually the rarest leukocyte types.

The three types of leukocytes with the suffix -*phil* (for *philic,* meaning "to like") have large *specific granules.* The prefix in their names refers to the staining characteristics of the specific granules: *neutro-* for neutral (that is, little staining by either of the two dyes in Wright stain); *eosino-* because the specific granules stain with the pink dye eosin; and *baso-* because the specific granules stain with the *basic* dye methylene blue. Eosin and methylene blue are the two dyes present in Wright stain.

3. If you have not found an eosinophil or a basophil by the time you have identified the three common leukocyte types, look at one or both of the demonstrations of these cells that have been set up by your instructor.

4. The abundance and size of the various blood cells are presented in table 34-1.

5. Table 34-2 lists the functions of the blood cells.

II. Blood Vessels

The basic structure of blood vessels is illustrated in figure 34-5.

MATERIALS

Per student:

- compound microscope
- prepared slide of a transverse section of a companion artery and vein

Per student pair:

- fish net
- small fish (3–4 cm long) in an aquarium
- 3-by-7-cm piece of absorbent cotton
- half a Petri dish
- coverslip
- dissecting needle

Per student group (4):

- container of anesthetic dissolved in dechlorinated water
- squeeze bottle of dechlorinated water

Per lab room:

- safe place to run in place
- several meter sticks taped vertically to the walls
- clock with a second hand

PROCEDURE

A. Arteries

Each contraction of the heart pumps blood into the space within the arteries. The rate of flow of blood out

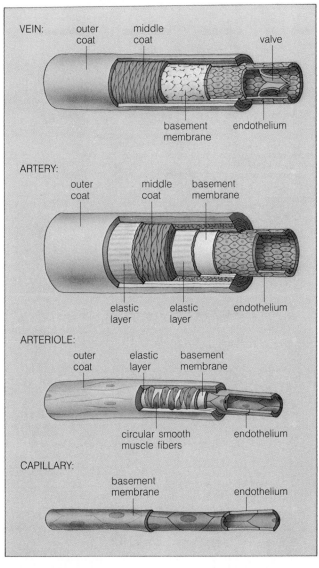

VEIN: outer coat · middle coat · valve · basement membrane · endothelium

ARTERY: outer coat · middle coat · basement membrane · elastic layer · elastic layer · endothelium

ARTERIOLE: outer coat · elastic layer · basement membrane · circular smooth muscle fibers · endothelium

CAPILLARY: basement membrane · endothelium

Figure 34-5 Structure of blood vessels. (After Spence, 1982, and Starr and Taggart, 1989.)

Location	Blood Pressure (mm/Hg)
right atrium of heart	5/0 (systolic/diastolic)
right ventricle of heart	25/5
pulmonary arteries	20
arterioles and capillaries of lung	20 to 10
pulmonary veins	10
left atrium of heart	10/0
left ventricle of heart	120/10
brachial artery	120/80
arterioles	100 to 50
capillaries	50 to 20
veins	20 to 0

Table 34-3 Blood Pressure in the Circulatory System of a Young Man at Rest

Would you expect blood flow to be more rapid in arteries or veins?

_____

Explain your answer. (*Hint:* look at the pressure differences between arteries, capillaries, and veins in table 34-3).

_____

_____

_____

The walls of the largest arteries contain many **elastic membranes,** which are stretched during contraction (*systole*) of the heart. When the heart is relaxing (*diastole*), these membranes rebound and squeeze the blood, maintaining blood pressure and flow. Valves at the point where the aorta and pulmonary arteries leave the heart prevent the backflow of blood.

When a physician takes your blood pressure, it is usually of the brachial artery of the upper arm and with the body at rest. A blood pressure of 120/80 means that the pressure during systole is 120 mm of mercury (Hg) and that the diastolic pressure is 80 mm Hg. The difference between systolic and diastolic pressures (*pulse pressure*) produces a pulse that you can feel in arteries that pass close to the skin.

Blood pressure changes with health, emotional state, activity, and other factors.

1. Get a prepared slide of a companion artery and vein. Find and examine the transverse section of an artery (fig. 34-6).

Arteries have thick walls compared to other blood vessels. They have an outer coat of connective tissue, a middle coat of smooth muscle tissue, and an inner coat of simple squamous epithelium (*endothelium*).

of the heart and into the arteries per minute is called *cardiac output* (CO). The arterial space is fairly constant, and it is somewhat difficult for blood to flow through the blood vessels, especially out of the arterioles. This resistance to the flow of blood is called *peripheral resistance* (PR). As cardiac output or peripheral resistance, or both, increase, more blood has to fit into the arterial space, which increases the force that the blood exerts on the walls of the arteries. This force is called **blood pressure** (BP). Blood pressure is directly proportional to the product of cardiac output and peripheral resistance (BP α CO × PR).

Pressure in blood vessels is highest in the arteries leaving the heart, gradually decreasing the further a vessel is located away from the heart (table 34-3). Blood or any other liquid or gas always flows from high to low pressure. So accordingly, blood flows through the circulatory system down this pressure gradient.

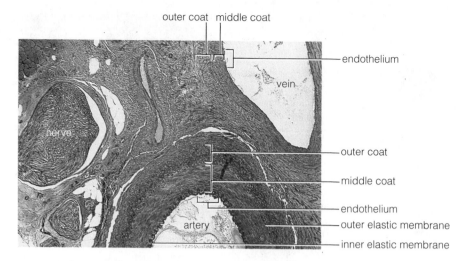

Figure 34-6 Photomicrograph of a transverse section of an artery and vein (50 ×). (Photo courtesy Ripon Microslides, Inc.)

The three coats are separated by elastic membranes. The middle coat is the thickest layer of the three coats.

2. Find your *radial pulse* in the radial artery (fig. 34-7). Use the index and middle fingers of your other hand. A pulse occurs every time the heart contracts. The strength of the pulse is a measure of the difference between the systolic and diastolic blood pressure.

3. Determine and record your heart rate by counting the number of pulses in 15 seconds and multiplying by four.

_____/minute

CAUTION

Do not do the following procedures if you have any medical problems with your lungs or heart. All subjects should be seated and should stop immediately if they feel faint.

4. Hold your breath. After 10 seconds have passed, determine your heart rate as in step 2.

_____/minute

Compared to when you were breathing normally, does the strength of the pulse increase, decrease, or remain the same when holding your breath?

_____

When you hold your breath, you decrease the return of blood to the heart. This reduces pulse pressure. Homeostatic mechanisms increase heart rate to compensate for reduced blood pressure.

5. Now run in place for 2 minutes in the area designated by your lab instructor. Immediately after sitting down, again measure your heart rate as in step 2.

_____/minute

Compared to when you are at rest, does the strength of the pulse increase, decrease, or remain the same immediately after exercise?

_____

Explain these results.

_____

_____

_____

_____

B. Capillaries

Capillaries have a very thin wall consisting of endothelium.

1. Use the net to catch a small fish from the aquarium and place it in the anesthetic fluid. Treat the fish gently, and it will not be harmed by this procedure.

2. After the fish turns belly up, wrap its body in cotton made soaking wet with dechlorinated water. Place the fish in half a Petri dish so that the tail is in the center.

3. Using dechlorinated water, make a wet mount of the posterior two-thirds of the fish's tail and examine it with the low-power, medium-power, and high-dry objectives of the compound microscope. Use the lowest illumination that still allows you to see the blood flowing in the vessels. If necessary, you can temporarily close the condenser iris diaphragm to create more contrast.

Can you see erythrocytes?

(yes or no) _____

What vessels can you identify?

_____

_____

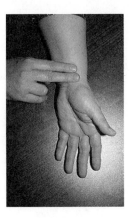

Figure 34-7 Feeling the radial pulse. (Photo by D. Morton.)

Is the blood flowing at the same speed in all of the capillaries?

(yes or no) _____

Describe blood flow.

_____

_____

_____

_____

4. Return the fish to the aquarium, wash the half Petri dish, and squeeze out the cotton into the sink before dropping it into the trash can.

C. Veins

For the blood to return to the heart after passing through capillary beds below the heart, it must overcome the force of gravity. Veins have **valves** to prevent the backflow of blood away from the heart. Blood is moved from one segment between valves to another primarily by muscular and breathing movements.

1. Again look at a prepared slide of a companion artery and vein. Find and examine the transverse section of a vein (fig. 34-6).

The vein has thinner walls and a larger lumen compared to its companion artery. Veins have an outer coat of connective tissue, a middle coat of smooth muscle tissue, and an inner coat of endothelium. Elastic membranes may be present. The outer coat is the thickest layer of the three coats. Compared to arteries and their walls, the walls of veins are more disorganized.

2. Work in pairs. Notice the veins as the subject's arm hangs down at the side of the body. You can easily see the veins because they are full of blood. This is usually best seen on the back of the hand. Now raise the arm above the head. Describe and explain any changes that take place.

_____

_____

_____

3. Using one of the meter sticks vertically taped to the wall, determine the venous pressure in the veins of the hand. Hold the subject's arm straight out at the level of the heart. Record the reading in millimeters where the hand crosses the meter stick.

measurement 1 = _____ mm

Raise the arm slowly until the veins in the hand collapse (be sure that most of the muscles in the arm are relaxed). Record the height to which the arm has been raised in millimeters.

measurement 2 = _____ mm

The difference between the two readings gives you the venous pressure expressed in millimeters of water.

_____ mm (measurement 1) − _____ mm (measurement 2) = _____ mm H_2O

Use the following formula to convert this into millimeters of mercury.

_____ mm H_2O × 0.074 mm Hg/mm H_2O

= _____ mm Hg

How does this compare to arterial pressures (table 34-3)?

_____

_____

4. Look at the veins of the subject's forearm and hand. There are swellings at various intervals. The swellings are valves. Choose a section between two swellings that does not have any side branches. Place one finger on the swelling away from the heart and with another finger press the blood forward (toward the heart) beyond the next swelling. Does the vein fill up with blood again?

(yes or no) _____

Now remove the finger and observe what happens. Try this again, but press blood in the opposite direction. Discuss your observations.

_____

_____

_____

_____

III. The Heart

Normally the heart beats over one hundred thousand times a day, pumping the blood around the circulatory system. The hearts of birds and mammals have four chambers (fig. 34-8).

The **right atrium** receives blood from the *superior vena cava, inferior vena cava* (fig. 34-1), and *coronary sinus* (which drains blood from capillary beds in the heart itself). When the right atrium contracts, blood is

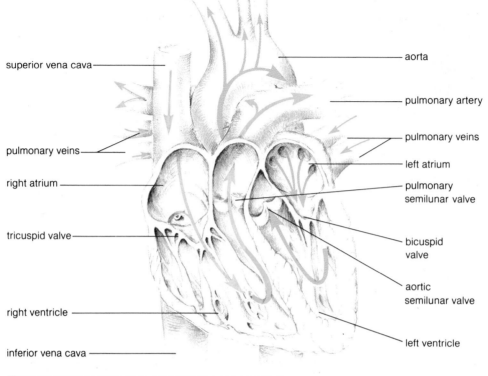

Figure 34-8 Ventral view of a human heart. The arrows indicate the direction of blood flow, and the chambers and vessels carrying blood that is rich in oxygen are shaded. (After Fowler, 1984.)

pushed through the *tricuspid valve* into the **right ventricle.** Contraction of the right ventricle pushes blood into the trunk of the *pulmonary arteries.* The **left atrium** receives blood from the *pulmonary veins,* and its contraction pushes blood through the *bicuspid valve* into the **left ventricle.** Contraction of the left ventricle pushes blood into the *aorta.* The *semilunar valves* prevent the backflow of blood from the pulmonary trunk and aorta.

MATERIALS

Per lab section:

• demonstration of the effects of acetylcholine and epinephrine on the heart of a doubly pithed frog (no brain or spinal cord) kept moist with amphibian Ringer's solution (balanced salts solution)

PROCEDURE

Your instructor has set up a demonstration of a frog heart in place in the opened thorax of a frog. Although the frog has a three-chambered heart—two atria and one ventricle—the heart's function and control are essentially the same as those of humans. The nervous system of the frog has been destroyed, so it does not feel pain or control heart action.

Is the heart beating?

(yes or no) _____

Are the contractions of the heart organized or disorganized?

_____

If you observe carefully, you can see the order in which the chambers contract. Record your observations.

_____

_____

_____

_____

The primary pacemaker of the heart, the **sinoatrial node,** is located in the right atrium. It can function independently of the nervous system, firing rhythmically. Each time the sinoatrial node fires, it initiates a message to contract. This message spreads over the atria. Then special heart cells amplify and conduct the message throughout the ventricle.

Count and record how many times the heart contracts in one minute (heart rate).

Control 1 heart rate = _____ beats/minute

Your instructor will now place several drops of an acetylcholine solution on the heart. After 1 minute, determine the heart rate.

Heart rate after acetylcholine = _____ beats/minute

After thoroughly flushing the thoracic cavity with balanced salt solution, wait 3 minutes and determine the second control heart rate.

Control 2 heart rate = _____ beats/minute

Your instructor will now place several drops of an epinephrine solution on the heart. After 1 minute, again determine the heart rate.

Heart rate after epinephrine = _____ beats/minute

Plot your results on the graph below.

Describe the effects of acetylcholine and and epinephrine on the heart rate.

_____

_____

_____

_____

In an intact frog, the heart rate is modified by input from the central nervous system. The heart rate is affected by the amount of *acetylcholine* and *norepinephrine* secreted by neurons around the sinoatrial node. During restful activities, acetylcholine slows the heart rate and thus acts as a brake on the sinoatrial node. Pain, strong emotions, the anticipation of exercise, and the fight-or-flight response all can increase the secretion of norepinephrine. Norepinephrine, which has a molecular structure and action similar to epinephrine, speeds the heart rate and thus acts as an accelerator on the sinoatrial node and can override the parasympathetic brake.

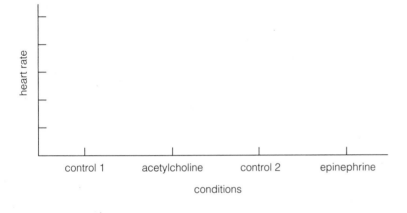

heart rate

control 1   acetylcholine   control 2   epinephrine

conditions

PRE-LAB QUESTIONS

____ 1. The extracellular fluid consists of (a) plasma, (b) interstitial fluid, (c) intracellular fluid, (d) a and b.

____ 2. Blood contains dissolved (a) gases, (b) nutrients, (c) hormones, (d) all of the above.

____ 3. Red blood cells are (a) erythrocytes, (b) leukocytes, (c) platelets, (d) all of the above.

____ 4. The most common leukocyte in the blood is a(n) (a) lymphocyte, (b) eosinophil, (c) basophil, (d) neutrophil.

____ 5. The cellular fragments in the blood that function in blood clotting are (a) erythrocytes, (b) leukocytes, (c) platelets, (d) none of the above.

____ 6. Blood vessels that return blood from capillaries back to the heart are (a) arteries, (b) veins, (c) portal veins, (d) arterioles.

____ 7. Blood vessels that connect capillary beds are (a) arteries, (b) veins, (c) portal veins, (d) b and c.

____ 8. From which chamber of the heart does the right ventricle receive blood? (a) right atrium, (b) left atrium, (c) left ventricle, (d) none of the above.

____ 9. How many chambers does the frog heart have? (a) one, (b) two, (c) three, (d) four.

____ 10. The primary pacemaker of the heart is the (a) bicuspid valve, (b) tricuspid valve, (c) aorta, (d) sinoatrial node.

EXERCISE 3 4

Human Blood and Circulation

POST-LAB QUESTIONS

1. Name and give the staining characteristics and functions of the three leukocytes with specific granules.

   a.

   b.

   c.

2. Describe the shape, content, and function of an erythrocyte.

3. Compare the structure and function of an artery, capillary, and vein.

4. Pretend you are an erythrocyte in the right atrium of the heart. Describe one trip through the human circulatory system, ending back where you started.

5. Explain how the heart of a double pithed frog can continue to contract in an organized manner after the nervous system is destroyed.

6. How is the heart rate controlled by the nervous systems in an intact organism?

7. Considering its relationship with the other systems of the body, what is the importance of the circulatory system?

EXERCISE 35
Human Respiration

OBJECTIVES

After completing this exercise you will be able to:

1. define *breathing, inspiration, expiration, ventilation, negative pressure inhalation, cohesion, positive pressure exhalation, positive pressure inhalation, tidal volume, inspiratory reserve volume, expiratory reserve volume, residual volume, vital capacity, chemoreceptor;*

2. list the skeletal muscles used in breathing and give the specific function of each;

3. explain how air moves in and out of the lungs during respiration in the human;

4. explain how air moves in and out of the lungs during respiration in the frog;

5. describe the relationship between vital capacity and lung volumes and the interrelationships among lung volumes;

6. explain the importance of CO_2 concentration in the blood and other body fluids to the control of respiration.

INTRODUCTION

Exercise 8 investigated carbohydrate metabolism and cellular respiration. Oxygen (O_2) is consumed, and carbon dioxide (CO_2) and water (H_2O) are produced during the breakdown of glucose to provide the energy (adenosine triphosphate, or ATP) to fuel cellular activities.

For cellular respiration to continue, O_2 must be replenished and CO_2 removed from cells by the process of diffusion (Exercise 5). The efficiency of diffusion to transport substances is great over short distances but decreases rapidly as distance increases. Evolution, however, has selected for organisms of different sizes—from one-celled species to the blue whale, the largest living animal. Because diffusion works well only over short distances, animals about the size of earthworms and larger have circulatory systems to move dissolved gases around the body; and animals a little larger—for example the clam and crayfish—move water or air across surfaces specialized for gas exchange (respiratory systems). O_2 is carried from the lungs to the body's cells—and CO_2 is delivered to the lungs from the body's cells—by the blood and circulatory system. Respiratory systems include the gill and associated structures (figs. 24-2 and 24-16a), the tracheal system of insects (a system of air-filled tubes reaching deep into the body), and the lung and associated structures (figs. 26-8b, 28-21, and 30-2).

In air-breathing vertebrates, O_2 uptake and CO_2 elimination occur by diffusion across the moistened thin membranes of millions of alveoli (singular is *alveolus*) and their surrounding capillaries located in the lungs (fig. 35-1). These animals are protected from excessive water loss via evaporation from the very large, moist respiratory surface by having the lungs positioned inside the body.

The main function of the rest of the respiratory system is **ventilation**—the exchange of gases between the lungs and the atmosphere. The movement of gases in and out of the respiratory system requires the contraction of skeletal muscles. The rhythm of these muscular contractions and resulting ventilation is called respiration, or **breathing.**

Before we go on with this exercise, we need to review some anatomical terms. The trunk of the body is divided into an upper *thorax*, which is supported by the rib cage and contains the *thoracic cavity*, and a lower *abdomen*. The thoracic cavity and the cavity of the abdomen are separated by a partition of skeletal muscle called the **diaphragm.** The thoracic cavity contains two *pleural sacs*, which contain the lungs and the *pericardial sac* around the heart.

The muscles of breathing include two sets of skeletal muscles between the ribs (**external** and **internal intercostal muscles**), the diaphragm, and skeletal muscles in the abdominal wall (**abdominal muscles**).

Breathing alternates between **inspiration** (inhalation) and **expiration** (exhalation). Usually, restful inspiration is accomplished by contraction of the *diaphragm*, which increases the size of the thoracic cavity by lowering its floor. Deeper inhalation requires the complementary contraction of the *external intercostal muscles* to further increase the size of the thorax and thoracic cavity by raising the rib cage (fig. 35-2a). Other muscles of the anterior regions of the neck and shoulders are also involved in the deepest inhalations.

Relaxation of the diaphragm and external intercostal muscles results in a restful or passive expiration (fig. 35-2b). When more air has to be exhaled in a shorter time, contraction of the *internal intercostal muscles* pulls down the rib cage, rapidly decreasing the size of the thoracic cavity and forcing air out of the lungs. The abdominal muscles can also be contracted to squeeze the internal organs, pushing up the diaphragm to decrease the size of the thoracic cavity.

MATERIALS

Per student:

• 2 pieces of paper (each 14×21.5 cm — half of a sheet of notebook paper)

469

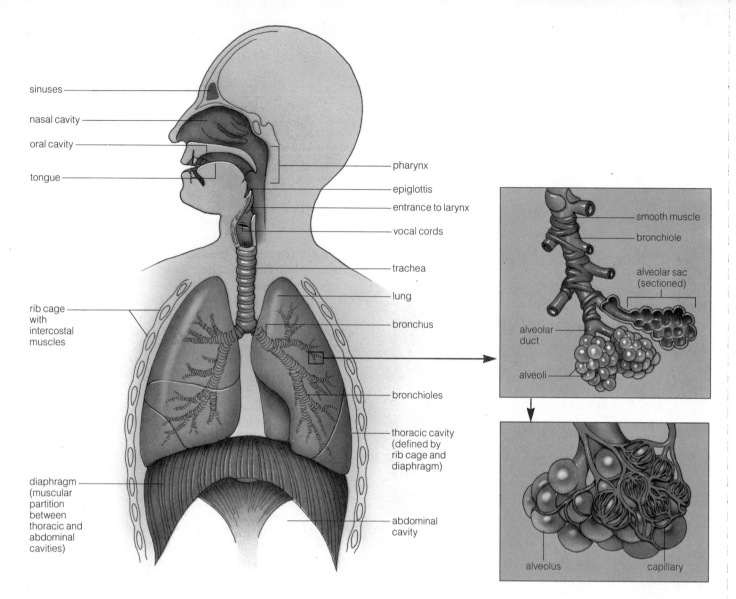

Figure 35-1 Human respiratory system. (After Starr, 1991.)

Labels on figure:
sinuses
nasal cavity
oral cavity
tongue
pharynx
epiglottis
entrance to larynx
vocal cords
trachea
lung
bronchus
rib cage with intercostal muscles
bronchioles
thoracic cavity (defined by rib cage and diaphragm)
diaphragm (muscular partition between thoracic and abdominal cavities)
abdominal cavity

smooth muscle
bronchiole
alveolar sac (sectioned)
alveolar duct
alveoli
alveolus
capillary

Per group (2):
- metric tape measure
- large caliper with linear scale (for example, Collyer pelvimeter)

Per group (4):
- noseclip (optional)
- functional model of lung
- simple spirometer or lung volume bags

Per lab room:
- frogs in terrarium or video of breathing frog
- clock with second hand that is visible to all
- designated safe area for running in place

PROCEDURE

A. Ventilation

All flow occurs down a pressure gradient. When you let go of an untied inflated balloon, it flies away, propelled by the jet of air flowing out of it. The air flows out because the pressure is higher inside than outside the balloon. The high pressure inside the balloon is maintained by the energy stored in its stretched elastic wall.

When the thoracic cavity expands during inspiration, first the pressure in the pleural sacs decreases, and then the pressure within the lungs decreases. Because the pressure outside the body is now higher than that in the lungs, and assuming the connecting *ventilatory ducts* (trachea and so on) are not blocked, air flows into the lungs (fig. 35-3b). This is called **negative pressure inhalation.**

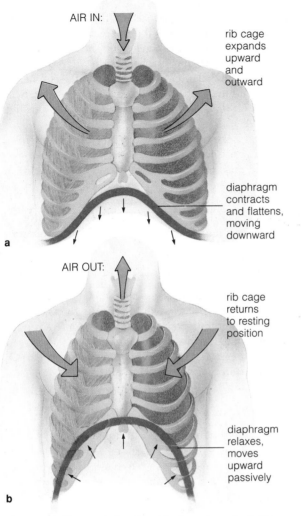

AIR IN:

rib cage
expands
upward
and
outward

diaphragm
contracts
and flattens,
moving
downward

a

AIR OUT:

rib cage
returns
to resting
position

diaphragm
relaxes,
moves
upward
passively

b

Figure 35-2 Changes in the size of the thoracic cavity during (**a**) inspiration and (**b**) expiration. (After Starr, 1991.)

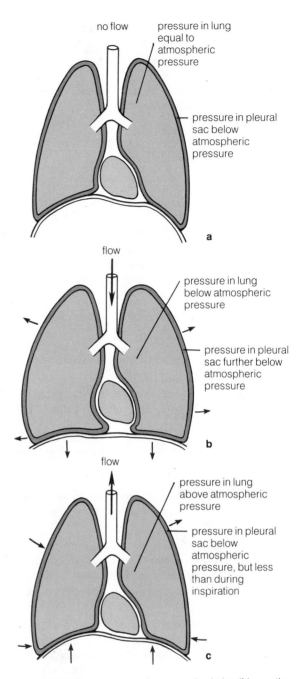

no flow

pressure in lung
equal to
atmospheric
pressure

pressure in pleural
sac below
atmospheric
pressure

a

flow

pressure in lung
below atmospheric
pressure

pressure in pleural
sac further below
atmospheric
pressure

b

flow

pressure in lung
above atmospheric
pressure

pressure in pleural
sac below
atmospheric
pressure, but less
than during
inspiration

c

Figure 35-3 Changes in the thoracic cavity during (**b**) negative pressure inhalation and (**c**) positive pressure exhalation, and corresponding movements of air. (**a**) shows the situation at the end of an expiration. (After Weller and Wiley, 1985.)

The opposite occurs during expiration. The size of the thorax and pleural sacs decreases, the pressure in the lungs increases, and air flows out of the body down its concentration gradient. This is called **positive pressure exhalation** (fig. 35-3c).

The pressure in the pleural sacs is actually always below atmospheric pressure, which means the lungs are always partially inflated after birth. Thus, a hole in a pleural sac or lung will result in a collapsed lung. Inspiration is aided by **cohesion** (sticking together) of the wet serosal membranes lining the lungs and outer walls of the pleural sacs. Expiration depends in part on the *elastic recoil* (like letting go of a stretched rubber band) of lung tissue.

1. Work in groups of four. Look at the functional lung model. The "Y" tube is analogous to the ventilatory ducts. The balloons represent the lungs. The space within the transparent chamber represents the thoracic spaces and its rubber floor (rubber "diaphragm"), the muscular diaphragm.

2. Pull down the rubber diaphragm. Describe what happens to the balloons.

As you pull down the rubber diaphragm, does the volume of the space in the container increase or decrease?

As the volume changes, is the pressure in the container increased or decreased?

_____

As the balloons inflate, does the volume of air in the balloons increase or decrease?

_____

Why do the balloons inflate?

_____

_____

_____

3. Push up on the rubber diaphragm. Describe what happens to the balloons and why it happens.

_____

_____

_____

4. Pull the rubber diaphragm down and push it up several times in succession to simulate breathing.

5. Pucker up your lips and inhale. As you inhale, place one of the pieces of paper directly over your lips. What occurs?

_____

_____

This suction is caused by the negative pressure created in your lungs by the contraction of the muscles of inspiration.

6. Fold the narrow ends of the two pieces of paper to produce 2–3 cm flaps. Open the flaps and use them as handles. Hold a piece of paper with each hand and touch their flat surfaces together in front of you. Pull them apart.

Now, thoroughly wet both pieces of paper with water and again touch their flat surfaces together in front of you. Pull them apart. What difference did the water make?

_____

_____

Inflation of the lungs of vertebrates that inhale using negative pressure is aided by the cohesion (sticking together) of the wet serosal membranes that line the lungs and the outer walls of the pleural sacs.

B. Positive Ventilation

Some vertebrates such as the frog inhale by pushing air into the lungs. This is called **positive pressure inhalation.**

Observe a frog out of water or watch a video of a breathing frog. The frog inhales by sucking in air through the nostrils by lowering the floor of the mouth. Valves in the nostrils are then closed and the floor of the mouth raised, thus increasing the pressure and forcing the air into the lungs. The upper portion of the ventilatory duct can be closed to keep the air in the lungs. Exhalation occurs by elastic recoil of the lungs with the ventilatory duct open. In the frog, both inhalation and exhalation are the result of positive pressure.

What is the frog's respiratory rate (breaths per minute)? Count and record how many times the frog lowers and raises the floor of the mouth (one breath) in 3 minutes.

_____ breaths

Divide by three to calculate the average respiratory rate.

_____/minute

Respiration in the frog is supplemented by gas exchange across the moist skin. Also, as they are ectotherms (do not maintain a high body temperature using physiological means), frogs generally have a lower metabolic rate and, therefore, a lesser demand for O_2 compared to a mammal of the same size.

C. Breathing Movements

1. Place your hands on your abdomen and take three deep breaths—three inspirations followed by three expirations. Describe and explain what you feel during:

each inspiration

_____

_____

each expiration

_____

_____

2. Place your hands on your chest and repeat step 1. Describe and explain what you feel during:

each inspiration

_____

_____

each expiration

_____

_____

D. Measurements of the Thorax

The size of the thorax can be described by three so-called diameters: the lateral diameter (LD), the anterioposterior diameter (A/PD), and the vertical diameter (fig. 35-4). The vertical diameter is the only one that cannot be measured easily.

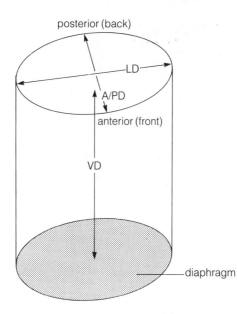

Figure 35-4 Thoracic diameters; LD—lateral diameter, A/PD—anterioposterior diameter, VD—vertical diameter.

Make the following observations and record them in table 35-1.

1. Work in pairs. Take turns measuring the circumference of each other's chest with a tape measure at two levels, under the armpits (axillae—C_{AX}) and at the lower tip of the sternum (xiphoid process—C_{XP}) for the following conditions: (a) at the end of a restful inspiration; (b) at the end of a passive expiration; (c) at the end of a forced (maximum) inspiration; and (d) at the end of a forced expiration. While the measurements are being taken, it is extremely important not to tense muscles other than those used for respiration. For example, do not raise the arms.

2. With calipers, also measure the A/PD and LD at the nipple line for these same conditions. The distance between the tips of the calipers is read off the scale in centimeters.

3. About two-thirds of the air inhaled during a restful inspiration is due to contraction of the diaphragm. Interpret the data in table 35-1 and in your own words describe changes in the size of the thorax during:

a. a restful inspiration

C_{AX} ———————————————————

C_{XP} ———————————————————

A/PD ———————————————————

LD ———————————————————

b. a passive expiration

C_{AX} ———————————————————

C_{XP} ———————————————————

A/PD ———————————————————

LD ———————————————————

Table 35-1 Chest Measurements (cm)				
Condition	C_{AX}	C_{XP}	A/PD	LD
Subject 1	———	———	———	———
Subject 2	———	———	———	———

c. a forced inspiration

C_{AX} ———————————————————

C_{XP} ———————————————————

A/PD ———————————————————

LD ———————————————————

d. a forced expiration

C_{AX} ———————————————————

C_{XP} ———————————————————

A/PD ———————————————————

LD ———————————————————

Does the size of the thorax change significantly during a restful inspiration or a passive expiration?

(yes or no) ————————————————

How does the shape of the thorax change during a forced inspiration?

————————————————————

————————————————————

————————————————————

————————————————————

How does the shape of the thorax change during a subsequent forced expiration?

————————————————————

————————————————————

————————————————————

————————————————————

E. Spirometry

Air in the lungs is divided into four mutually exclusive volumes: tidal volume (TV), inspiratory reserve volume (IRV), expiratory reserve volume (ERV), and residual volume (RV).

Tidal volume is the volume of air inhaled or exhaled during breathing. It normally varies from a minimum at rest to a maximum during strenuous exercise.

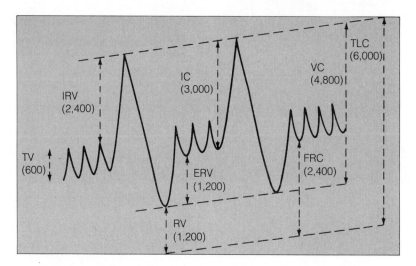

Figure 35-5 A spirogram showing the defined lung volumes and capacities. The numbers in parentheses are average values in milliliters. (After Weller and Wiley, 1985.)

Inspiratory reserve volume is the volume of air you can voluntarily inhale after inhalation of the tidal volume. **Expiratory reserve volume** is the volume of air you can voluntarily exhale after an exhalation of the tidal volume. IRV and ERV both decrease as TV increases.

Residual volume is the volume of air that cannot be exhaled from the lungs. That is, normal lungs are always partially inflated.

There are four capacities derived from the four volumes:

inspiratory capacity (IC) = TV + IRV

functional residual capacity (FRC) = ERV + RV

vital capacity (VC) = TV + IRV + ERV

total lung capacity (TLC) = total of all four lung volumes

All the lung volumes except the residual volume can be measured or calculated from measurements obtained using a simple spirometer or lung volume bag. A more sophisticated recording spirometer makes a trace of respiration over time called a spirogram. Figure 35-5 illustrates a spirogram and the relationships of the lung volumes and capacities.

CAUTION
Always use a sterile mouthpiece and do not inhale air from either a simple spirometer or a lung volume bag.

1. Work in groups of four. Sit quietly and breathe restfully. Use a noseclip or hold your nose. After you feel comfortable, start counting as you inhale. After the fourth inhalation, exhale normally into the spirometer or lung volume bag. Read the volume indicated by the spirometer or squeeze the air to the end of the lung volume bag and read the volume from the wall of the bag. Record the volume below (trial 1). Reset the spirometer or squeeze the air out of the lung volume bag. Repeat this procedure two more times (trials 2 and 3) and calculate the total and average tidal volume at rest.

trial 1 _____ mL

trial 2 _____ mL

trial 3 _____ mL

total = _____ mL

Divide the total by three = _____ mL to calculate the average TV at rest.

2. Determine the volume of air you can forcibly exhale after a restful inspiration (average of three trials).

trial 1 _____ mL

trial 2 _____ mL

trial 3 _____ mL

total = _____ mL

Divide the total by three = _____ mL. This is the average sum of expiratory reserve volume and tidal volume at rest.

Table 35-2 Vital Capacity and Lung Volumes at Rest (mL)	
Measure	**Volume (mL)**
Tidal volume	_____
Inspiratory reserve volume	_____
Expiratory reserve volume	_____
Vital capacity	_____

3. Determine the volume of air you can forcibly exhale after a forceful inspiration (average of three trials).

trial 1 _____ mL

trial 2 _____ mL

trial 3 _____ mL

total = _____ mL

Divide the total by three = _____ mL to calculate the average vital capacity.

4. Calculate the ERV at rest by subtracting the result of step 1 from the result of step 2.

_____ mL (step 2) − _____ mL (step 1)

= _____ mL.

5. Calculate the IRV at rest by subtracting the result of step 2 from the result of step 3.

_____ mL (step 3) − _____ mL (step 2)

= _____ mL.

6. Summarize your results in table 35-2.

Does **vital capacity** change as tidal volume increases or decreases?

(yes or no) _____

Measure and record your height in centimeters.

_____ cm

Write your vital capacity/height on the board—your name is not necessary.

Plot the vital capacity of each of the students in your lab section on the graph below.

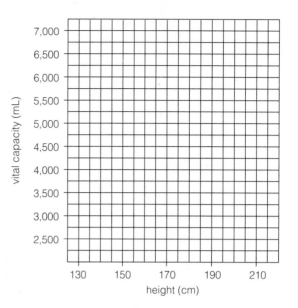

Is there a relationship between vital capacity and height? If so, describe it mathematically or with words.

_____

_____

_____

_____

_____

F. Control of Respiration

The control of respiration, both the rate and depth of breathing, is very complex. Simply stated, **chemoreceptors** (receptors for chemicals such as O_2, CO_2, and hydrogen ions, or H^+), stretch receptors in the ventilatory ducts, and centers in the brain stem (part of the brain that connects to the spinal cord) control respiration. By far the most important stimulus is the CO_2 concentration in the blood and other body fluids.

Our own experience has taught us that respiration is to some extent under the control of the conscious mind. We can decide to stop breathing or to breathe more rapidly and deeply. However, the unconscious mind can override voluntary control. The classical example of this is the inability to hold one's breath for more than a few minutes. Once the CO_2 concentration rises above a specific point, you are forced to breathe.

1. Work in pairs. The subject sits down unless instructed to do otherwise. After the subject feels comfortable, determine the respiratory rate at rest. The investigator counts and records the number of times the subject breathes in 3 minutes.

_____ breaths

Divide by three to calculate the average respiratory rate.

_____/minute

2. The subject breathes deeply, as rapidly as possible. Try to take at least ten breaths but stop as soon as you can answer this question. (In any case do not continue for more than twenty breaths.) As times goes on, does it become easier or more difficult to continue rapid deep breathing?

Forced deep breathing results in overventilation of the lungs, or **hyperventilation.** How does hyperventilation affect the CO_2 concentration of the blood? (increases CO_2, decreases CO_2, or no effect on CO_2)

_____

3. When fully recovered from step 2, measure how long the subject can hold his or her breath after a restful inspiration.

_____ seconds

Now, the subject carefully runs in place for 2 minutes in the area designated by your lab instructor. Immediately after sitting down, again measure how long the subject can hold his or her breath after a restful inspiration.

_____ seconds

How does running in place affect the CO_2 concentration of the blood? (increases CO_2, decreases CO_2)

_____

What causes the CO_2 concentration to change while you are running in place?

_____

_____

_____

OPTIONAL

G. Experiment: Physiology of Exercise

Your instructor may provide you with an experiment about the physiology of exercise.

PRE-LAB QUESTIONS

____ 1. Which of the following muscles may contract during inspiration? (a) external intercostals, (b) internal intercostals, (c) abdominal, (d) b and c.

____ 2. Which of the following muscles contract during a passive expiration? (a) external intercostals, (b) internal intercostals, (c) diaphragm, (d) none of the above.

____ 3. Which of the following muscles may contract during a more forceful expiration? (a) external intercostals, (b) diaphragm, (c) abdominal, (d) b and c.

____ 4. An untied inflated balloon flies because (a) the pressure is higher inside than outside the balloon, (b) the pressure is lower inside than outside the balloon, (c) air flows down its pressure gradient, (d) a and c.

____ 5. Human ventilation is (a) negative pressure inhalation, (b) positive pressure inhalation, (c) negative pressure exhalation, (d) b and c.

____ 6. Frog ventilation is (a) negative pressure inhalation, (b) positive pressure inhalation, (c) positive pressure exhalation, (d) b and c.

____ 7. Vital capacity is always equal to (a) tidal volume, (b) inspiratory reserve volume, (c) expiratory reserve volume, (d) a + b + c.

____ 8. An instrument that measures lung volumes is (a) a caliper, (b) a spirometer, (c) a barometer, (d) stethoscope.

____ 9. Respiration is controlled by (a) chemoreceptors, (b) stretch receptors, (c) centers in the brain stem, (d) all of the above.

____ 10. The most important stimulus in the control of respiration is the concentration in the blood and other body fluids of (a) oxygen (O_2), (b) carbon dioxide (CO_2), (c) hydrogen ions (H^+), (d) nitrogen (N_2).

EXERCISE 35

Human Respiration

POST-LAB QUESTIONS

1. Which skeletal muscles are contracted during:
   a. restful inspiration

   b. forced inspiration

   c. passive expiration

   d. forced expiration

2. Describe changes in the size of the thorax during:
   a. inspiration

   b. expiration

3. Describe changes in the potential volume of the pleural sacs during:
   a. inspiration

   b. expiration

4. Define for humans:
   a. *negative pressure inhalation*

   b. *positive pressure exhalation*

5. How does breathing in a human differ from that in a frog?

6. What substance is the most important stimulus in the control of respiration? How is its production linked to changes in metabolic rate, such as occur during exercise?

7. Explain why hyperventilation can prolong the time you can hold your breath. Can this be dangerous (for example, hyperventilation followed by swimming under water)?

Animal Development:

Gametogenesis and Fertilization

OBJECTIVES

After completing this exercise you will be able to:

1. define *sexual reproduction, fertilization, gametes, gonads, ovum, sperm, dioecious, zygote, monoecious, asexual reproduction, parthenogenesis, yolk, acrosome, blastodisc, vegetal pole, animal pole, gametogenesis, testis, ovary, seminiferous tubules, interstitial cells, testosterone, Sertoli cells, follicle, ovulation, estrogen, progesterone, corpus luteum, external fertilization, external development, internal ferterilization, internal development;*

2. draw and label a diagram of a mammalian sperm;

3. describe the structure of chicken and frog ova, and tell how they differ from a typical mammalian ovum;

4. recognize interstitial cells, seminiferous tubules, Sertoli cells, and sperm in a prepared section of a mammalian testis;

5. recognize follicles, primary oocytes, and a corpus luteum in a prepared slide of a mammalian ovary;

6. describe spermatogenesis and oogenesis in mammals;

7. describe the events and consequences of sperm penetration and fertilization.

INTRODUCTION

Most animal species reproduce sexually. **Sexual reproduction** usually involves the fusion of the nuclei of two gametes, called the ovum (the plural is *ova*) and sperm. This fusion is referred to as **fertilization.**

Gametes are produced by *meiosis* in reproductive organs called **gonads,** usually in individuals of two separate sexes. A female's gametes are **ova,** while those of a male are **sperm.** Species with male and female gonads in separate individuals are said to be **dioecious.**

Each gamete is haploid, and fertilization creates a new diploid cell, the **zygote,** whose combination of genes is unlike those of either parent. It is also very unlikely that the genes of one zygote will be identical to those of any other zygote, even those derived from the same parents. What two Mendelian principles largely account for this?

_____

_____

_____

_____

This variation is an advantage to a species in a changing and unpredictable environment. Why?

_____

_____

_____

Some animals (earthworms and snails, for example) produce both ova and sperm in the same individual, but self-fertilization is rare, occurring only in parasites with constant and predictable environments (tapeworms, for example). A species with both male and female reproductive organs in the same individual is referred to as **monoecious** or *hermaphroditic.*

Some animals reproduce asexually. **Asexual reproduction** is the production of new individuals by any mechanism that does not involve gametes (budding in sponges, for example). Also, the ova of many animals, either naturally or in the laboratory, are capable of development without fertilization. This is called **parthenogenesis.**

> **NOTE**
>
> Depending on the timing of your lab and in preparation for observation of a live frog zygote, your instructor may demonstrate the fertilization of frog eggs before starting this exercise.

I. Gametes

Most sperm have at least one flagellum. Sperm are specialized for motility and contribute little more than their chromosomes to the zygote.

Ova are specialized for storing nutrients, and they contain the molecules and organelles needed to fuel, direct, and maintain the early development of the embryo. Nutrients are stored as **yolk** in the cytoplasm of the ovum. Consequently, mammalian ova are larger than body cells and in some species reach a diameter of 0.2 mm. In the frog, additional yolk increases the diameter of the ovum to 2 mm, and in the chicken it reaches about 3 cm.

MATERIALS

Per student:

- compound microscope, lens paper, bottle of lens-cleaning solution (optional), lint-free cloth (optional), dropper bottle of immersion oil (optional)
- prepared slide with a whole mount of bull sperm
- glass microscope slide, a coverslip, and a dissecting needle

• one-piece plastic dropping pipet

Per student pair:
• unfertilized hen's egg
• several paper towels
• dissection pan
• Syracuse dish
• 2 camel-hair brushes
• dissection microscope

Per lab group (table or bench):
• a model of a frog ovum (optional)

Per lab section:
• live frog sperm in pond water
• live frog ova in pond water
• pond water

Per lab room:
• phase-contrast compound light microscope (optional)

PROCEDURE

A. Sperm

1. With your compound microscope, study a prepared slide of bull sperm and draw several in figure 36-1 as seen with the high-dry objective. Each sperm has three major segments: the *head, midpiece,* and *tail.* Label the major segments of one of the sperm in your diagram. The tail is composed primarily of a single *flagellum.*

2. *Skip this step if your compound microscope does not have an oil-immersion objective.* Using the oil-immersion objective, find the **acrosome** covering the *nucleus* in the head of a sperm and *mitochondria* in its midpiece. The acrosome contains enzymes that aid in the penetration of the egg. Considering its cell size, why does a sperm have a lot of mitochondria?

_____

_____

3. Place a drop of frog sperm suspension on a glass microscope slide and make a wet mount. Observe the movement of their flagella using either the phase-contrast compound microscope (a microscope that increases the contrast of transparent specimens) or your compound microscope with the iris diaphragm partially closed to increase contrast. Describe what you see.

_____

_____

_____

_____

Figure 36-1 Drawing of bull sperm (_____ ×).

B. Ova

1. *Unfertilized chicken egg.*

a. Obtain an unfertilized chicken egg (fig. 36-2). Crack it open as you would in the kitchen and spill the contents into a hollow made from paper towels placed in a dissection pan.

b. Only the yolk is the ovum. Look for the **blastodisc,** a small white spot just under the cell membrane. This area is free of yolk and contains the nucleus. This is where fertilization would have occurred if sperm had been present in the hen's oviduct.

c. The *albumin* (egg white) is secreted by the walls of the oviduct. Examine the *shell* and the two *shell membranes.* The shell membranes are fused except in the region of the air space at the blunt end of the egg. Note the two shock absorber-like *chalazae* (the singular is *chalaza*), which suspend the ovum between the ends of the egg. They are made of thickened albumin and may function to help rotate the ovum to keep the blastodisc always on top of the yolk.

2. *Frog egg.* Gently place a live frog egg in a Syracuse dish half filled with pond water and examine it with a dissection microscope. Use two camel-hair brushes to transfer the egg. **Do not let the egg dry out.** The ovum is enclosed in a protective jelly membrane. Does the ovum float light- or dark-side up in the pond water?

_____

Alternatively, study a model of a frog ovum. Ova from different species of animals vary in the amount and distribution of yolk. Frog ova have a moderate amount of yolk that is concentrated in the lower half of the ovum (fig. 36-3). This half of the ovum is called the **vegetal pole.** The nucleus is located in the upper, yolk-free half, the **animal pole.** Note that the animal pole is black. This is because it contains pigment granules. Why does a frog ovum have less yolk than a bird ovum?

_____

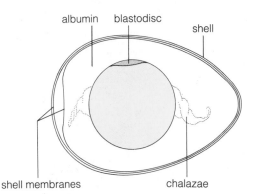

Figure 36-2 Unfertilized chicken egg.

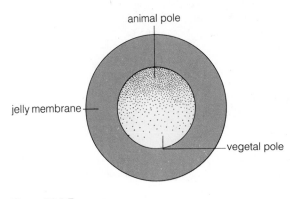

Figure 36-3 Frog egg.

Suggest one or more possible functions for the black pigment in the animal pole. (Hint: One function is the same as that for the pigment in your skin that increases when exposed to sunlight.)

_____

_____

_____

3. *Human egg.* Figure 36-4 shows a human egg as it would appear in the upper oviduct. The egg is surrounded by a membrane, the *zona pellucida,* and a capsule of follicle cells (see section II.B). Why do most mammalian ova have very little yolk?

_____

_____

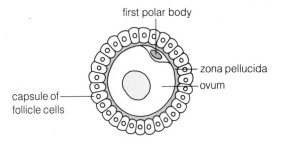

Figure 36-4 Section through a human egg in the upper oviduct.

PROCEDURE

A. Mammalian Spermatogenesis

Examine a prepared slide of the testis (fig. 36-5). Most of the interior of a testis is filled with **seminiferous tubules,** which coil to and fro. Transverse and oblique sections will be present in your slide. Look for glandular **interstitial cells** between the seminiferous tubules. Interstitial cells secrete the male sex hormone **testosterone.** Now find a transverse section of a seminiferous tubule and increase the magnification of your

II. Gametogenesis

Gametogenesis is the production of the gamates. It is called **oogenesis** in the female and **spermatogenesis** in the male. The general scheme and terminology of gametogenesis are summarized in table 36-1.

Because you have already studied the events of meiosis that bring about the haploid number of chromosomes present in gametes, we will now examine the production of gametes in the gonads of mammals. The gonads are paired organs and are called **testes** (the singular is *testis*) in males and **ovaries** in females.

MATERIALS

Per student:

• compound microscope
• prepared slide of a section of adult mammalian testis stained with iron hematoxylin
• prepared slide of a section of adult mammalian ovary with follicles
• prepared slide of a section of adult mammalian ovary with a corpus luteum

Table 36-1 Gametogenesis		
	Type of Cell	
Condition of Cell	**Male**	**Female**
mitotically active	spermatogonium	oogonium
before meiosis I	primary spermatocyte	primary oocyte
before meiosis II	secondary spermatocyte	secondary oocyte and first polar body
after meiosis II	spermatid	ovum and three polar bodies
after differentiation	sperm	—

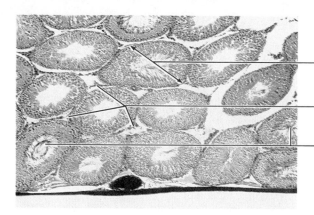

seminiferous tubule

interstitial cells

sperm

a

Figure 36-5 (a) Transverse section of the testis (58x). (Photo courtesy Ripon Microslides, Inc.) **(b)** Spermatogenesis. (After Starr, 1991.)

compound light microscope until a portion of the tubule's wall fills the field of view. The wall of the seminiferous tubule contains cells in various stages of spermatogenesis, as well as **Sertoli cells,** which function to nurture the developing sperm.

Spermatogenesis in most animal species is seasonal, its completion coinciding with mating. In humans, however, sperm production is continuous from puberty throughout a male's lifetime.

B. Mammalian Oogenesis

With the compound microscope, examine a prepared slide of a section of a mammalian ovary with follicles (fig. 36-6). After birth, in humans and most other mammals, oogonia are not present, and meiosis is suspended in prophase of the first meiotic division. The primary oocytes are the largest cells in the section and are always surrounded by a **follicle** composed of smaller cells.

There is an excess supply of primary oocytes present at birth (about 2 million in a newborn girl). Each primary oocyte is initially surrounded by a *primary follicle* whose thin walls are one cell thick. Most follicles degenerate, and in humans only about three-hundred-thousand primary oocytes remain at puberty. The majority of these oocytes will also degenerate. With each turn of the female cycle, several follicles begin to mature, and one (rarely two or more) of each batch of oocytes finally bursts from the surface of the ovary and is swept into the oviduct. The release of oocytes from the ovary is called **ovulation.**

Follicles also function to secrete the female sex hormones, **estrogen** and **progesterone.**

Find in your section of a mammalian ovary the following stages of follicular development.

1. *Primary follicle.* Look for primary follicles (fig. 36-7). Both the follicles and their primary oocytes are small compared with maturing follicles and their primary oocytes. Primary follicles tend to occur in groups lo-

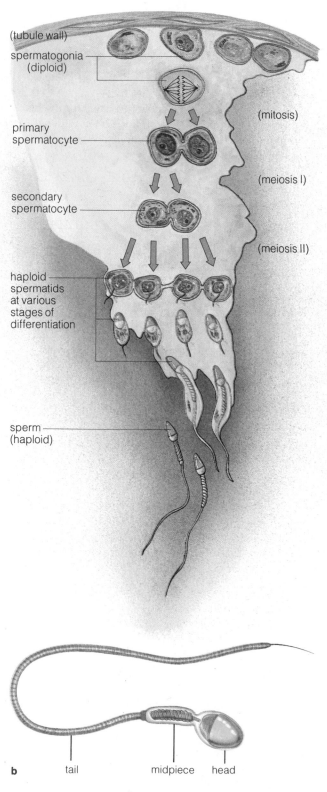

(tubule wall)

spermatogonia (diploid)

(mitosis)

primary spermatocyte

(meiosis I)

secondary spermatocyte

(meiosis II)

haploid spermatids at various stages of differentiation

sperm (haploid)

b   tail   midpiece   head

482

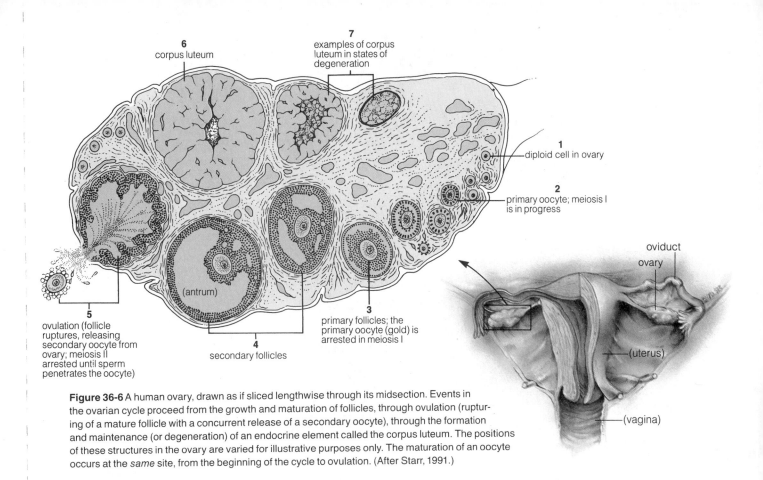

Figure 36-6 A human ovary, drawn as if sliced lengthwise through its midsection. Events in the ovarian cycle proceed from the growth and maturation of follicles, through ovulation (rupturing of a mature follicle with a concurrent release of a secondary oocyte), through the formation and maintenance (or degeneration) of an endocrine element called the corpus luteum. The positions of these structures in the ovary are varied for illustrative purposes only. The maturation of an oocyte occurs at the *same* site, from the beginning of the cycle to ovulation. (After Starr, 1991.)

Labels in Figure 36-6:

6 corpus luteum

7 examples of corpus luteum in states of degeneration

1 diploid cell in ovary

2 primary oocyte; meiosis I is in progress

3 primary follicles; the primary oocyte (gold) is arrested in meiosis I

4 secondary follicles

5 ovulation (follicle ruptures, releasing secondary oocyte from ovary; meiosis II arrested until sperm penetrates the oocyte)

oviduct

ovary

(uterus)

(vagina)

cated between maturing secondary follicles or between secondary follicles and the ovarian wall.

2. Secondary follicles. In maturing follicles, the size of the cells in the follicle wall and of the primary oocyte itself increases. Also, the follicle cells divide, causing the wall to become first two cells thick and then multilayered. As a follicle matures, a space appears between the follicle cells (fig. 36-8a). This fluid-filled space increases in size until the primary oocyte and the follicle cells immediately around the primary oocyte are suspended in it (fig. 36-8b). The mass of cells is connected to the rest of the wall by a narrow stalk of follicle cells. Just prior to ovulation, the follicle reaches its maximum size and bulges from the surface of the ovary.

3. Corpus luteum. Replace the slide on the microscope with one of a mammalian ovary with a corpus luteum. Around the time of ovulation, the first meiotic division is completed, a polar body is split off, and the secondary oocyte enters but does not complete the second meiotic division. The oocyte is surrounded by the zona pellucida and a capsule of follicle cells (fig. 36-4). A sperm must first penetrate this barrier before penetration of the egg membrane and fertilization can occur.

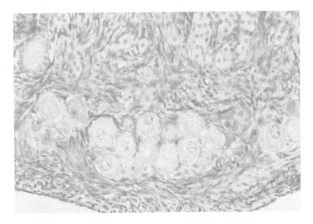

Figure 36-7 Group of primary follicles (297 ×). (Photo by D. Morton.)

After ovulation, the follicle cells that remain in the ovary develop collectively into a large roundish structure called the **corpus luteum** (yellow body) (fig. 36-9). The corpus luteum continues to secrete female sex hormones, especially progesterone.

If fertilization and implantation of the embryo in the uterus do not occur, the corpus luteum degenerates and is replaced by scar tissue. This scar is now called the *corpus albicans* (fig. 36-10).

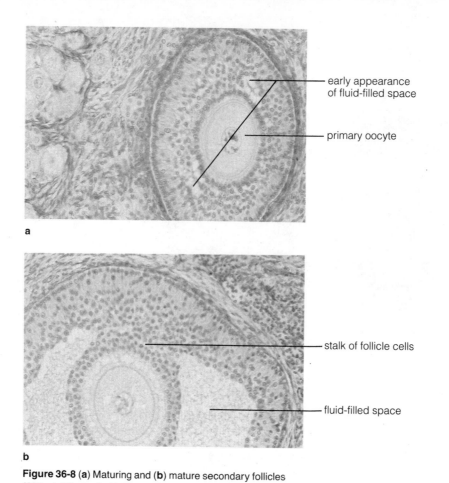

early appearance
of fluid-filled space

primary oocyte

stalk of follicle cells

fluid-filled space

a

b

Figure 36-8 (**a**) Maturing and (**b**) mature secondary follicles
(297 ×). (Photos by D. Morton.)

III. Sperm Penetration and Fertilization

Fertilization and subsequent development may be internal or external to the female's body. **External fertilization** with **external development** is common in invertebrates, fish, and amphibians. Why is external fertilization generally limited to aquatic animals?

_____

_____

_____

Land animals have **internal fertilization,** with either external development in the shell (true of most reptiles and birds) or **internal development** in the mother's *uterus* (true of most mammals). The uterus is the organ of the female reproductive system where most mammalian embryos develop until birth.

In many animals, meiosis is not complete at the time of sperm penetration. For example, in humans and most mammals sperm penetration triggers the completion of the second meiotic division and the splitting of the second polar body from the secondary oocyte to form the ovum proper. Fertilization occurs when the male and female nuclei or *pronuclei* fuse.

MATERIALS

Per student pair:

• Syracuse dish
• 2 camel-hair brushes
• dissection microscope

Per lab group (table or bench):

• model of a frog zygote (optional)

Per lab section:

• live frog zygotes in pond water
• pond water
• film or videotape of sperm penetration and fertilization in the frog (optional)

PROCEDURE

A. Frog Zygote

At the beginning of the lab, or just before, your instructor induced ovulation in a female frog by injecting pituitary gland extract into the abdominopelvic cavity. Some of these ova were fertilized by sperm obtained from the shredded testes of a male frog.

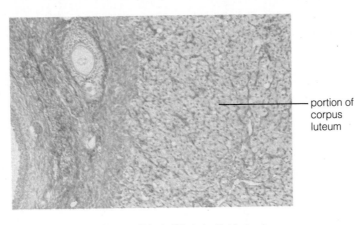

— portion of corpus luteum

Figure 36-9 Corpus luteum (74 ×). (Photo by D. Morton.)

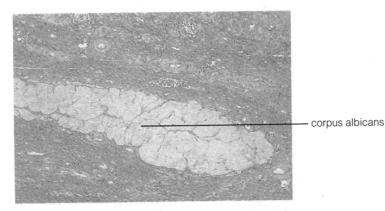

— corpus albicans

Figure 36-10 Corpus albicans (81 ×). (Photo courtesy of Ripon Microslides, Inc.)

Using two camel-hair brushes, gently transfer a live frog zygote that has not yet begun to cleave to a Syracuse dish half filled with pond water and examine it with a dissecting microscope. **Do not let the zygote dry out.** Does the zygote float animal- or vegetal-side up in the pond water?

_____

Is a fertilization membrane present or absent?

_____

Alternatively, study a model of a frog zygote or watch a film or videotape on sperm penetration and fertilization in the frog. Is there a change in the pigmentation pattern? If so, describe what you observe.

_____

_____

_____

PRE-LAB QUESTIONS

_____ 1. Most animals reproduce by (a) asexual means, (b) sexual means, (c) both of these means, (d) none of these means.

_____ 2. Most animals (a) are monoecious, (b) are dioecious, (c) have two sexes, (d) b and c.

_____ 3. Sperm (a) are male gametes, (b) are female gametes, (c) contain yolk, (d) are produced in the ovaries.

_____ 4. Ova (a) are male gametes, (b) are female gametes, (c) are specialized for motility, (d) a and c.

_____ 5. The production of gametes in the gonads is called (a) gametogenesis, (b) spermatogenesis, (c) oogenesis, (d) none of the above.

_____ 6. One primary oocyte will form (a) one ovum, (b) four ova, (c) up to three polar bodies, (d) a and c.

_____ 7. Which of the following could be found in a section of the testis? (a) secondary spermatocytes, (b) sperm, (c) Sertoli cells, (d) all of the above.

_____ 8. Oocytes are found in an ovary in (a) seminiferous tubules, (b) follicles, (c) corpora luteum, (d) none of the above.

_____ 9. Most mammals have (a) internal fertilization, (b) external fertilization, (c) internal development, (d) a and c.

_____ 10. Which process results in a zygote? (a) meiosis, (b) mitosis, (c) fertilization, (d) none of the above.

EXERCISE 36

Animal Development: Gametogenesis and Fertilization

POST-LAB QUESTIONS

1. Define and characterize:

   a. *gametes*

   b. *gonads*

   c. *gametogenesis*

2. Describe the similarities and differences between sperm and ova.

3. Why do you think four sperm cells are produced as a result of gametogenesis, but only one ovum?

4. Why is meiosis a necessary part of gametogenesis?

5. Are oogonia present in adult human females?
   (yes or no)

6. In the frog, what is the relationship of the gray crescent to sperm penetration?

7. What substances are contained in the acrosome of a sperm? What is the function of these substances during sperm penetration?

8. As various newspaper articles, books, and movies suggest, the cloning of human beings — producing new individuals from activated somatic cells, perhaps followed by uterine implant — is a distinct possibility. Can you suggest any biological advantages or disadvantages to having the earth populated with clones of a few of the best examples of our species?

EXERCISE 37

Animal Development:

Cleavage, Gastrulation, and Late Development

OBJECTIVES

After completing this exercise you will be able to:

1. define *blastomere, morula, blastula, blastocoel, blastoderm, blastocyst, inner cell mass, trophoblast, archenteron, blastopore, primitive streak, notochord, neural tube, somites, embryonic disk, implantation, amniotic cavity, placenta, umbilical cord, fetus;*

2. list and define the five stages of development;

3. compare and contrast cleavage in the sea star, frog, chicken, and human;

4. explain differences in cleavage according to (a) the amount and distribution of yolk in the ovum and (b) the evolution of mammals;

5. describe gastrulation, the formation of the primary germ layers, and their derivatives;

6. list the four extraembryonic membranes;

7. give the functions of the extraembryonic membranes in birds and mammals.

INTRODUCTION

Animal development has six stages (fig. 37-1). Having studied gametogenesis and fertilization in the previous exercise, we will continue the story through organogenesis in this exercise. During growth and specialization, the sixth stage, organs grow in size and acquire the specialized properties necessary for an independent life.

I. Cleavage

Cleavage is a special type of cell division that occurs first in the zygote and then in the cells formed by successive cleavages, the **blastomeres.** Unlike normal cell division, there is no intervening period of cytoplasmic growth between mitotic divisions. Thus, the blastomeres become smaller and smaller. After a number of cleavages, the blastomeres form a solid cluster of cells called the **morula.** The formation of a hollow ball of cells, called the **blastula** in invertebrates and amphibians, marks the end of cleavage. Because there is no cytoplasmic growth, the size of the blastula is only slightly larger than that of the zygote.

In many organisms whose ova have little yolk (the sea star and human, for example), cleavage is *complete* and nearly *equal*, resulting in separate blastomeres that are all about the same size. In amphibians like the frog and other animals with moderate amounts of

yolk, cleavage is complete but *unequal*. There is so much yolk in the ova of many animals (most fish, reptiles, birds, and the two mammals that lay eggs — the platypus and spiny anteater) that complete cleavage is impossible. This type of cleavage is incomplete and is called *discoidal* (disklike).

MATERIALS

Per student:

- compound microscope, lens paper, bottle of lens-cleaning solution (optional), lint-free cloth (optional), dropper bottle of immersion oil (optional)

- prepared slide of a whole mount of sea star development through gastrulation

Per student pair:

- 2 Syracuse dishes
- 2 blue camel-hair brushes (optional)
- 2 red camel-hair brushes
- dissection microscope

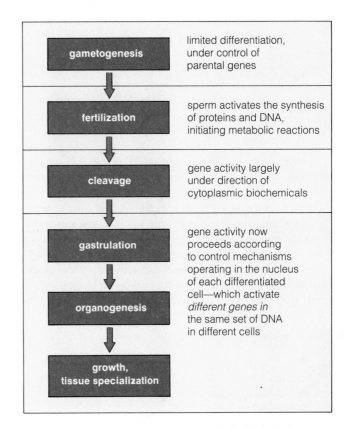

Figure 37-1 Generalized scheme and control of animal development. (After Starr and Taggart, 1989.)

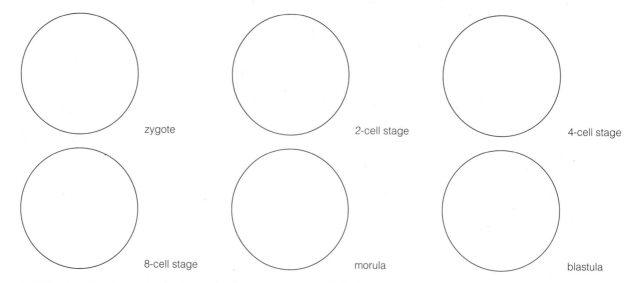

zygote · 2-cell stage · 4-cell stage

8-cell stage · morula · blastula

Figure 37-2 Drawings of early sea star developmental stages (_____ ×).

Labels: blastomeres, blastocoel

Per lab section:

• frog embryos in pond water from eggs fertilized one hour before lab (optional)

Per lab room:

• preserved specimens of two-, four-, and eight-cell cleavage stages; morulae (32-cell cleavage stage); and blastulae of frog in easily accessible screw-top containers

• source of distilled water

• pond water (optional)

• models of early frog development

• models of human development

PROCEDURE

A. Sea Star

With your compound microscope, observe a prepared slide with whole mounts of early sea star embryos. Find and draw in figure 37-2 a zygote; two-, four-, and eight-cell cleavage stages; a morula; and a blastula. Be sure to adjust the fine-focus knob to see the three-dimensional aspects of these stages. The blastula is a hollow ball of flagellated blastomeres surrounding a cavity called the **blastocoel.** Label the blastomeres and blastocoel in your diagram of a sea star blastula.

Are the first two cleavage planes parallel or perpendicular to each other?

_____

What is the orientation of the third cleavage plane compared to the first and second cleavage planes?

_____

B. Frog

1. If your instructor fertilized frog eggs one hour before lab, you will be able to watch the first two cleavage divisions during this laboratory. At room temperature about an hour after addition of the sperm to the eggs, a region of less pigmented cytoplasm, the *gray crescent,* appears opposite the site of sperm penetration (fig. 37-3). The first cleavage division occurs about two hours after fertilization, the second a half hour later, and the third after another two hours.

> **CAUTION**
>
> Preserved frog embryos are kept in a formalin preservative solution. Thoroughly wash any part of your body exposed to this solution with water. If the formalin solution is splashed into your eyes, wash them with the safety eyewash bottle for fifteen minutes.

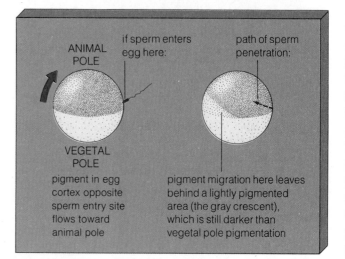

Figure 37-3 Formation of the gray crescent. (After Starr and Taggart, 1989).

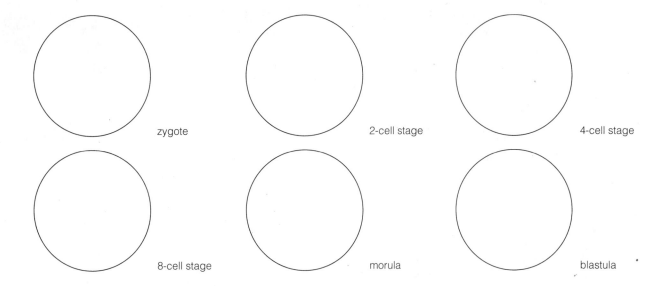

zygote   2-cell stage   4-cell stage

8-cell stage   morula   blastula

Figure 37-4 Drawings of early frog developmental stages
(_____ ×).

Gently place several live frog zygotes in a Syracuse dish half-filled with pond water and examine them with a dissection microscope. Use two blue camel-hair brushes to transfer the zygote. **Do not let it dry out.**

2. If living embryos are unavailable, and for the eight-cell, morula, and blastula stages, examine preserved specimens under the dissecting microscope. Use two red camel-hair brushes to transfer each stage in turn to a Syracuse dish half-filled with distilled water. When done, return the specimen to its container. The preserved embryos are in a preservative solution, so be careful that you **do not use the red brushes to manipulate the live embryos,** as the residual preservative will harm them.

As a supplement, study the models of the early stages of frog development.

3. In figure 37-4 draw the stages of frog development. Note that cleavage in the vegetal pole lags behind that of the animal pole. Why? (*Hint:* What is present in the vegetal pole that would hinder cleavage?)

_____

_____

_____

C. Chicken

In the chicken, cleavage of the blastodisc forms a layer of cells called the **blastoderm,** which in time becomes separated from the yolk by a cavity, the blastocoel. Further development of the embryo will occur only in the blastoderm (fig. 37-5).

D. Human

Examine the early stages of models of human development. Because the ovum has little yolk, cleavage is

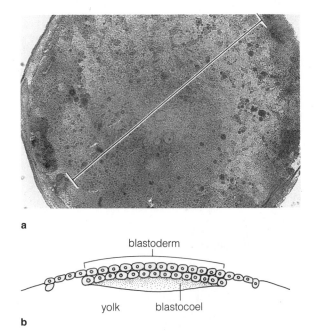

a

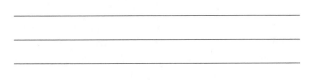

blastoderm

yolk   blastocoel

b

Figure 37-5 Late cleavage in the chicken. (**a**) surface view (37 ×) (Photo courtesy Ripon Microslides, Inc.); (**b**) transverse section along diameter indicated in (**a**).

complete and nearly equal. At the end of cleavage a hollow ball of cells is formed. However, this **blastocyst** differs from a blastula in that a group of cells aggregate at one pole of the inner surface of the blastocoel (fig. 37-6). These cells are called the **inner cell mass,** and, like the blastoderm of the chicken, further development of the embryo will proceed only here. The remaining cells that surround the blastocoel are called the **trophoblast.**

Current evolutionary thought is that modern reptiles, birds, and mammals evolved from earlier reptilian ancestors that had external development and an ovum rich in yolk. Internal development in mammals

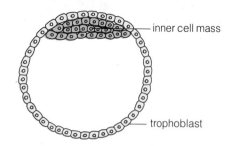

Figure 37-6 Section through a human blastocyst.

inner cell mass

trophoblast

linked to nourishment of the embryo directly by the mother made large yolky eggs redundant. Excessive yolk would have been a biological liability; therefore, by natural selection, mammals have developed ova with little yolk. However, gastrulation and subsequent developmental events in reptiles, birds, and mammals are remarkably similar. This theme will be expanded in the discussion of extraembryonic membranes later in this exercise.

II. Gastrulation

Gastrulation is a time of growth and cell migration that produces, in most animals, three **primary germ layers** and the longitudinal axis of the body. In animals having at least the organ level of organization, the primary germ layers are called **ectoderm, mesoderm,** and **endoderm** and give rise to the four tissue types (table 37-1).

MATERIALS

Per student:
- compound microscope
- prepared slide of a whole mount of sea star development through gastrulation
- prepared slides with whole mounts of chicken embryos at 18 and 24 hours incubation

Per student pair:
- 2 Syracuse dishes
- 2 blue camel-hair brushes (optional)
- 2 red camel-hair brushes
- dissection microscope
- fertile hen's eggs incubated for 18 and 24 hours (optional)

Per student group (bench or table):
- models of human gastrulation

Per lab section:
- frog embryos (early gastrulae) in pond water from eggs fertilized 21 hours before lab (optional)
- frog embryos (late gastrulae) in pond water from eggs fertilized 36 hours before lab (optional)
- pond water

Table 37-1 Tissue Derivatives of the Primary Germ Layers

Layer	Derivatives
ectoderm	nervous tissues; epidermis and its derivatives
mesoderm	muscle tissues, connective tissues, and epithelia of the urinary and reproductive systems
endoderm	epithelial lining of most of the digestive tract and respiratory tract, and associated glands (for example, liver)

Per lab room:
- preserved specimens of early and late gastrulae of frog in easily accessible screw-top containers
- a source of distilled water
- pond water (optional)
- models of early and late gastrulae of frog

PROCEDURE

A. Sea Star

The sea star has complete and nearly equal cleavage, resulting in a blastula with one layer of flagellated blastomeres that are slightly elongated at the *vegetal pole* (fig. 37-7a).

Reexamine the prepared slide bearing whole mounts of the early stages of sea star development. Find an early gastrula (fig. 37-7b). As gastrulation starts, the vegetal pole flattens and folds in like a pocket to create a new cavity, the **archenteron** ("ancient gut").

Find a late gastrula (fig. 37-7c). The hole connecting the archenteron to the outside is called the **blastopore.** Gastrulation initially forms two layers of cells: (a) the ectoderm covering the outside of the gastrula and (b) the endoderm lining the archenteron. The mesoderm buds off from the inner tip of the archenteron, and its cells migrate to form a third layer of cells between the ectoderm and endoderm.

The archenteron is the primitive gut, and the blastopore is situated in what will be the region of the anus in the adult sea star. This latter fact is important from an evolutionary viewpoint because it marks a major fork in the evolution of *coelomates*, which have body cavities completely lined with mesoderm. The phyla Echinodermata (for example, sea star) and Chordata (for example, frogs, humans) are called *deuterostomes* because the blastopore marks the region of the anus in the adult. Phyla Mollusca (for example, snails, clams, octopuses), Annelida (for example, earthworms), and Arthropoda (for example, lobsters, insects) are called *protostomes* because the blastopore marks the region of the mouth in the adult.

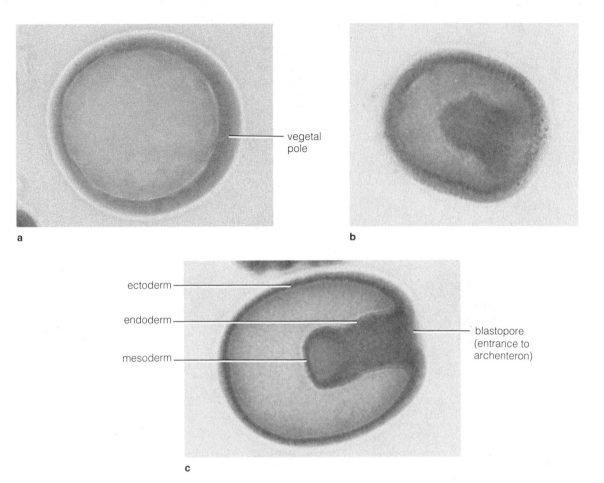

vegetal pole

ectoderm

endoderm

mesoderm

blastopore (entrance to archenteron)

Figure 37-7 Gastrulation in the sea star (320 ×). (**a**) late blastula; (**b**) early gastrula; (**c**) late gastrula. (Photos by D. Morton.)

B. Frog

Gastrulation in the frog is affected by the large amount of yolk in the vegetal hemisphere. Because the pigmented cells of the animal pole divide faster, they partially overgrow the yolk-laden cells of the vegetal pole.

Examine living or preserved specimens, as well as supplemental models of an early and a late gastrula as described in section I.B for earlier developmental stages. Note that gastrulation does not occur simultaneously over the surface of the vegetal pole. It starts at a point just under what will be the anus of the adult frog and continues, forming a crescent (fig. 37-8a) that will close to form a circle around a plug of yolk-laden cells, the *yolk plug* (fig. 37-8b). The initial point of the folding-in is referred to as the *dorsal lip of the blastopore.*

C. Chicken

1. Obtain a prepared slide bearing a whole mount of a chicken gastrula (18 hours incubation) and one of an embryo incubated for 24 hours.

> **CAUTION**
> Do not use the high-power objectives when examining whole mounts of thick material. You will break the coverslip.

2. Observe the gastrula using only the low- and medium-power objectives. Like cleavage, gastrulation in the chicken is influenced by the large amount of yolk. The gastrula does not fold in like a pocket through a blastopore, but rather cells move or migrate into a groove on the blastoderm called the **primitive streak** (fig. 37-9a on p. 495). At one end of the primitive streak, the migrating cells pile up to form *Hensen's node.* This is thought to be equivalent to the dorsal lip of the blastopore.

3. Now examine the slide of the embryo of 24 hours incubation. The three-layered embryo forms in the same axis as the primitive streak but in front of Hensen's node (fig. 37-9b). The three layers from the top of the embryo toward the yolk are ectoderm, mesoderm, and endoderm. One of the first recognizable structures in the embryo is the mesodermal **notochord.**

The notochord induces the formation of the embryo's nervous system from the ectoderm. *Neural folds*

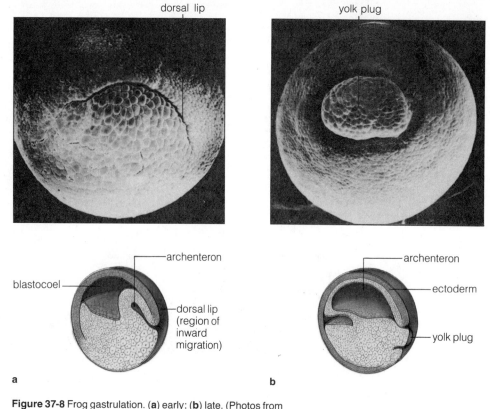

Figure 37-8 Frog gastrulation. (**a**) early; (**b**) late. (Photos from R. Kessel and C. Shih, *Scanning Electron Microscopy in Biology*, Springer-Verlag, 1974. Diagrams after Starr and Taggart, 1989.)

can be seen on either side of the notochord. With time, the neural folds will fuse like a zipper, anterior to posterior, and in doing so form the **neural tube** (fig. 37-10 on p. 496).

Likewise, the head of the embryo has lifted off the surface of the yolk and in doing so has formed the anterior portion of the digestive tract, the *foregut*.

Also prominent are two rows of **somites** on either side of the notochord. These are segmental condensations of mesoderm that will later develop into the skeleton, the skeletal muscles of the trunk, and the dermis of the skin. How many pairs of somites are present in your embryo?

---

4. If your instructor has fertile hen's eggs incubated for 18 and 24 hours, you will be instructed how to examine this material under the dissection microscope.

D. Human

Examine the models illustrating human gastrulation. Human gastrulation follows much the same scenario as that of the chicken. However, before gastrulation a cavity forms between the inner cell mass and the trophoblast. This is the cavity of the amnion (fig. 37-11). Also, cells from the inner cell mass grow downward, along the inner surface of the trophoblast, fuse,

and form the yolk sac. Between the amniotic and yolk sac cavities is the embryonic disk. The **embryonic disk** in mammals is the equivalent of the blastoderm in the chicken. Thus, gastrulation commences in the embryonic disk with the formation of a primitive streak.

Implantation of the embryo in the uterus occurs at the same time as the formation of the amnion and yolk sac, about eight days after fertilization of the ovum in the upper oviduct. The outer layer of the trophoblast erodes away the maternal tissues so that the blastocyst can sink into the wall of the uterus. The inner layer of the trophoblast forms the chorion. By this time, the ectoderm of the amnion and chorion as well as the endoderm of the yolk sac are coated with mesoderm derived from the inner cell mass.

III. Extraembryonic Membranes

The **amnion, yolk sac,** and **chorion** are three of the four **extraembryonic membranes.** The fourth is the **allantois.**

MATERIALS

Per student pair:
- dissection microscope
- fertile hen's eggs incubated for 5 days (optional)

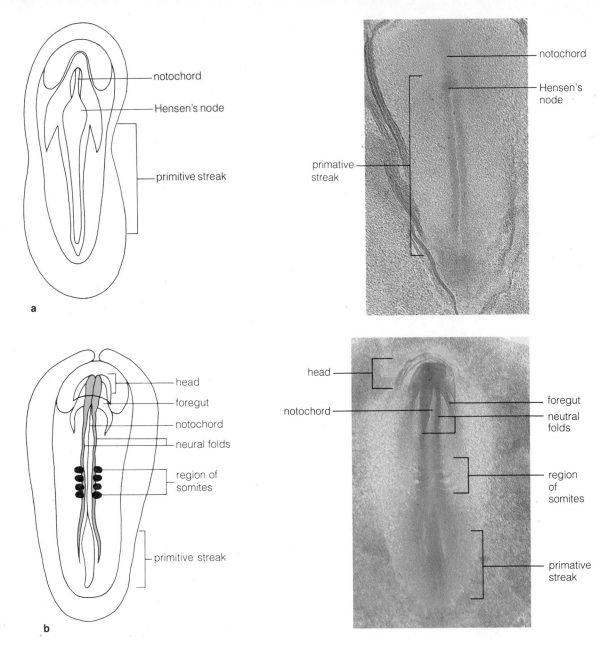

Figure 37-9 Gastrulation in the chicken. (**a**) gastrulation (18 hours of incubation) (28 ×); (**b**) early embryo (24 hours of incubation) (16 ×). The embryos have been removed from the surface of the yolk. (Photos courtesy Ripon Microslides, Inc.)

Per student group (bench or table):

- preserved pig fetus and placenta with injected vessels
- models of human intrauterine development

PROCEDURE

A. Chicken

The amnion in the chicken forms from four folds of ectoderm and mesoderm: one in front of the developing embryo, one behind it, and one on each side. These in time fuse over the developing embryo's back.

The outer wall of the fused folds becomes the chorion; the inner wall becomes the amnion. The chorion forms a sac that surrounds the developing embryo and the other extraembryonic membranes. The amnion forms the fluid-filled **amniotic cavity,** in which the developing embryo is suspended.

The yolk sac is formed when endoderm, accompanied by mesoderm containing a rich network of blood vessels, spreads over the yolk. The yolk sac serves as the digestive organ for the developing embryo. The endodermal cells secrete enzymes that digest the yolk. The end products of digestion diffuse into the blood vessels and are carried into the developing embryo.

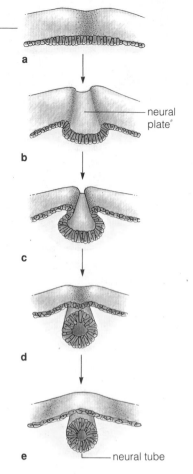

Figure 37-10 Formation of neural tube. (After Starr and Taggart, 1989.)

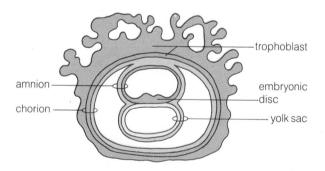

Figure 37-11 Transverse section through a human embryo as it appears after implantation in the uterine wall.

The floor of the hindgut folds out like a pocket to form the allantois, which is lined on the inside by endoderm and covered on the outside by mesoderm. The mesoderm forms a rich network of blood vessels. The cavity of the allantois functions as a dump for excretory wastes.

The yolk sac and allantois both function as embryonic respiratory organs.

If your instructor has fertile hen's eggs incubated for about five days, you will be instructed how to ex-amine extraembryonic membranes under the dissection microscope.

The extraembryonic membranes of the chicken are illustrated in figure 37-12.

B. Mammals

The formation of the extraembryonic membranes in mammals is quite similar to that of the chicken. One difference is the earlier formation of the amnion and allantois. Because there is little yolk in the zygote of most mammals, the yolk sac is generally smaller.

After implantation, the trophoblast and the maternal tissues of the uterus start to form the **placenta.** When complete, the placenta brings the blood of the mother and the embryo very close to each other but does not allow them to mix. Diffusion across this thin barrier allows the placenta to function as the digestive, respiratory, and excretory organs of the developing embryo.

The mesoderm of the allantois forms the **umbilical cord** and the *umbilical arteries* and *umbilical vein* contained therein. The mesoderm of the allantois also directs the formation of connecting blood vessels in the placenta. By the time the embryo's circulatory system is established, a circuit of vessels to and from the placenta is complete.

Once the organs and basic body shape of an embryo are established, the embryo is called a **fetus.** This transition occurs about one-third of the way through the time spent in the uterus. This time spent in the uterus is called the *gestation period*. Look at the models of intrauterine (within the uterus) development in humans and identify the structures labeled in figure 37-13.

Examine the preserved pig fetus and its placenta (fig. 37-14 on p. 498). The umbilical arteries have been injected with different colored latex to help you identify the maternal and fetal blood vessels. Identify the fetus, umbilical cord, amnion, fetal vessels, and maternal vessels.

IV. Organogenesis

Organogenesis is the fifth stage of development. As the name suggests, it is during this period that the developing animal's organs and adult form are achieved.

MATERIALS

Per student:
- compound microscope
- prepared slides with whole mounts of chicken embryos at 48 and 72 hours of incubation

Per student group:
- fertile hen's eggs incubated for 48 and 72 hours and longer (optional)
- dissection microscope

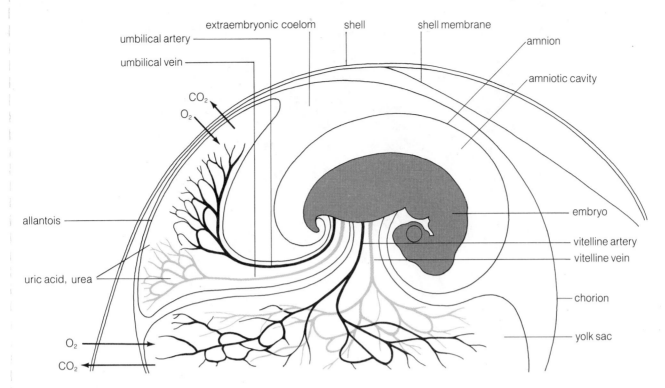

Figure 37-12 Extraembryonic membranes of the chicken. (After B.M. Patten, National Sigma Xi Lecture, reprinted in *American Scientist,* 39:225–243, 1951. Used by permission.)

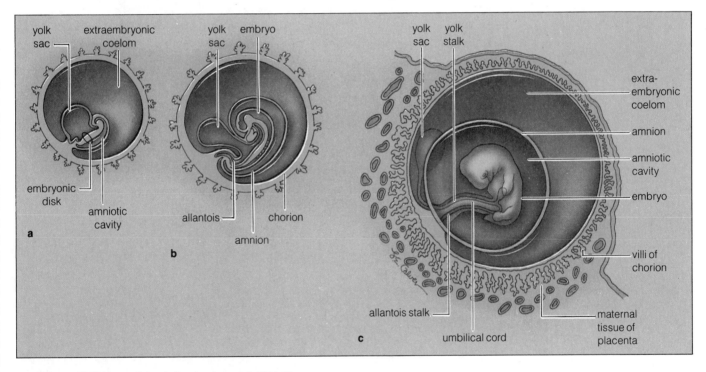

Figure 37-13 Human intrauterine development. (After Starr, 1991.)

ANIMAL DEVELOPMENT: CLEAVAGE, GASTRULATION, AND LATE DEVELOPMENT

497

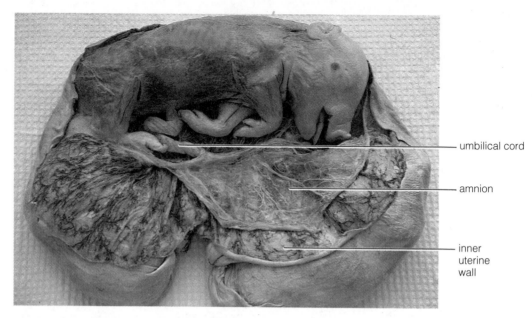

umbilical cord

amnion

inner
uterine
wall

Figure 37-14 Extraembryonic membranes of the pig. Fetal vessels have been injected with yellow latex. The maternal vessels have been injected with red latex for arteries and blue latex for veins. (Specimen courtesy Ward's Natural Science Establishment, Inc.; photo by D. Morton.)

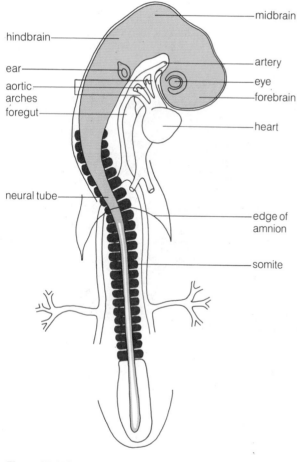

midbrain

hindbrain

ear

aortic
arches

foregut

artery

eye

forebrain

heart

neural tube

edge of
amnion

somite

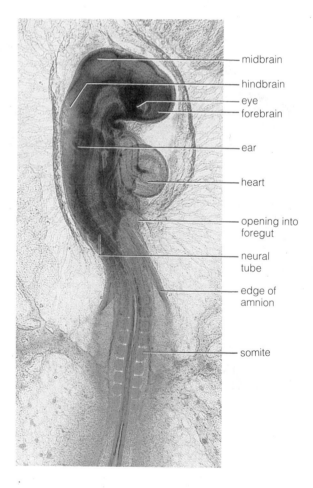

midbrain

hindbrain

eye
forebrain

ear

heart

opening into
foregut

neural
tube

edge of
amnion

somite

Figure 37-15 Development of the chicken — 48 hours of incubation (18 ×). (Photo courtesy Ripon Microslides, Inc.)

PROCEDURE

1. Obtain and examine under the low-power lens a slide of a whole mount of a chicken embryo incubated for 48 hours (fig. 37-15). At this time, development of the front end of the neural tube has produced the *forebrain*, *midbrain*, and *hindbrain*. The head of the embryo has turned to the right, and growth of the brain has caused it to bend over itself. The *primitive eye* can now be seen forming on both sides of the forebrain. The *primitive ear* has just formed on both sides of the hindbrain.

The first portion of the circulatory system to develop is the heart. The *primitive heart* can be seen bulging to the embryo's right. Three pairs of *aortic arches* are present as well as other blood vessels.

Has the number of somites increased compared with those of the chicken embryo incubated for 24 hours?

(yes or no) _____

Note that the front half of the embryo is enclosed by the folds that form the amnion and chorion. Only the amnion, especially its lower edge, can be seen.

2. Now examine the slide of a chicken embryo incubated for 72 hours with the dissection microscope (fig. 37-16). By this stage the embryo is lying on its right side, and the rate of development is accelerating. New features include another pair of *aortic arches*, the *wing buds*, *hindlimb buds*, and a *tail bud*. The heart has established contact with the *vitelline arteries* and *vein*, which connect the embryo with its source of nourishment, the yolk in the yolk sac.

3. If your instructor has fertile hen's eggs incubated for 48 hours, 72 hours, and longer, you will be instructed how to examine this material under the dissection microscope.

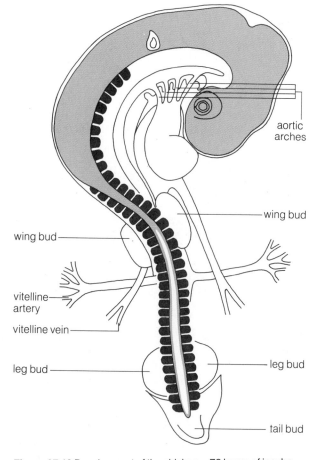

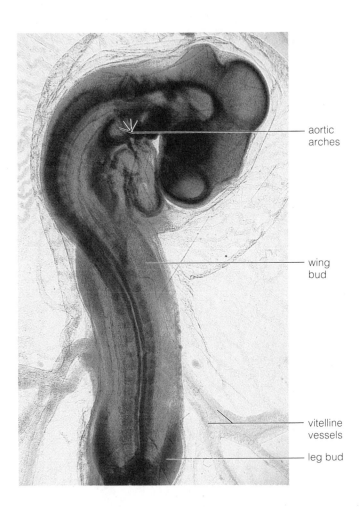

Figure 37-16 Development of the chicken — 72 hours of incubation (14x). (Photo courtesy Ripon Microslides, Inc.)

_____ 1. Choose the correct chronological order for the first five stages of development listed below: (1) fertilization, (2) cleavage, (3) organogenesis, (4) gametogenesis, (5) gastrulation. (a) 1,2,3,4,5; (b) 1,4,2,5,3; (c) 2,1,4,5,3; (d) 4,1,2,5,3.

_____ 2. The three primary germ layers are formed during (a) fertilization, (b) cleavage, (c) organogenesis, (d) gastrulation.

_____ 3. Muscle tissues, connective tissues, and the epithelia lining the urinary and reproductive systems arise from (a) mesoderm, (b) endoderm, (c) ectoderm, (d) two of the above.

_____ 4. Cleavage is complete and unequal in the (a) sea star, (b) frog, (c) chicken, (d) human.

_____ 5. The cavity within a blastula is called the (a) archenteron, (b) blastocoel, (c) blastocyst, (d) blastopore.

_____ 6. The entrance into the cavity formed during gastrulation in the sea star and frog is called the (a) archenteron, (b) blastocoel, (c) blastocyst, (d) blastopore.

_____ 7. The blastocyst consists of the (a) inner cell mass, (b) trophoblast, (c) placenta, (d) a and b.

_____ 8. In which of the following species do all of the cells derived from the zygote contribute to the body of the embryo? (a) frog, (b) chicken, (c) human, (d) b and c.

_____ 9. How many extraembryonic membranes are there? (a) one, (b) two, (c) three, (d) four.

_____ 10. The majority of mammals derive most of their nourishment during development from (a) the yolk sac, (b) the placenta, (c) the fluid in the amnionic cavity, (d) none of the above.

E X E R C I S E   3 7

Animal Development: Cleavage, Gastrulation, and Late Development

P O S T - L A B   Q U E S T I O N S

1. Define:

   a. *cleavage*

   b. *gastrulation*

   c. *organogenesis*

2. Identify the following sea star developmental stages (185×) (photos by D. Morton).

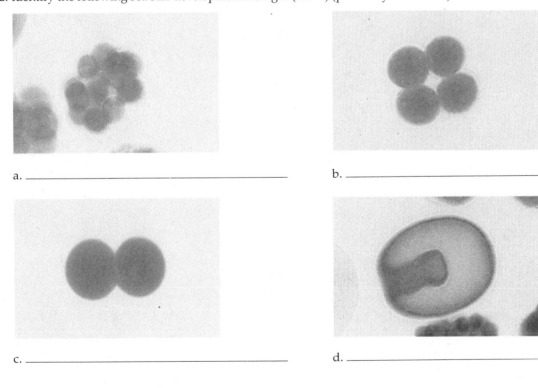

a. _____   b. _____

c. _____   d. _____

3. How do the amount and distribution of yolk in animal zygotes affect cleavage?

4. In what ways is gastrulation in humans more like that of the chicken than that of the sea star?

5. List the primary germ layers formed by gastrulation. What tissues will they form in the adult?

6. List the extraembryonic membranes of a chicken embryo and give their functions.

7. Describe the formation of the placenta in mammals. What are its functions?

Ecology:

Living Organisms in Their Environment

OBJECTIVES

After completing this exercise you will be able to:

1. define *ecology, population, habitat, community, ecosystem, producer, trophic level, primary consumer, secondary consumer, decomposer, herbivore, carnivore, omnivore;*

2. construct a food web;

3. determine the trophic level to which an organism belongs;

4. identify the three basic types of survivorship curves and describe the trends exhibited by each.

INTRODUCTION

"Ecology." The word probably brings to mind the activities of groups trying to create a clean environment, or lobbying the government to prevent loss of natural areas, or protesting the policy and/or actions of some other groups. To the biologist, this is *not* what ecology as a science is all about. Rather, these activities are environmental concerns that may have grown out of ecological studies.

The word *ecology* is derived from the Greek *oikos,* which means "house." Broadly defined, **ecology** is the study of the interactions between living organisms and their environment. Individuals, whether they are bacteria or humans, do not exist in a vacuum. Rather, they interact with one another in complex ways. As such, ecology is an extremely diverse and complex study. There are so many aspects of ecology that it is difficult to pick one, or even a few, to represent the entire science.

Let's look at the levels of organization considered in an ecological study.

A **population** of organisms is a group of individuals *of the same species* occupying a given area at a given time. For example, you and your colleagues in the laboratory at the moment are a population; you are all *Homo sapiens.* Your population consists of humans in biology class right now. The place you are occupying, the lab room, is your **habitat.** Of course, you are sharing your habitat with other organisms; unseen bacteria and possibly insects are present in the room as well. Maybe there are plants decorating the lab room; if so, there are other populations in your habitat.

A **community** consists of all those populations of *all* species occupying a given area at a given time. To use our simple example, our laboratory community consists of humans, bacteria, insects, and plants.

Our community exists in a physical and chemical environment. The combination of the community and its environment comprise an **ecosystem.** Our laboratory ecosystem consists of humans, bacteria, insects, plants, an atmosphere, lab benches, walls, floors, light, heat, and so on. As you see, an ecosystem has both living and nonliving components.

In this exercise, we will introduce some of the principles of ecology. We will study three concepts to illustrate how ecological studies are performed: food webs, survivorship curves, and population density and distribution.

I. Food Webs

Let's consider the community of organisms in any environment. Each population plays a different role within the structure of the community. Each community consists of producers, consumers, and decomposers.

Producers are autotrophic organisms, such as most plants and a few microorganisms. They utilize the energy of the sun or chemical energy to synthesize organic (carbon-containing) compounds from inorganic compounds.

Consumers are heterotrophic organisms. In ecological studies, consumers are usually classified into feeding levels (**trophic levels**) by what they eat.

• **Primary consumers** are **herbivores,** animals that eat plant material.

• **Secondary consumers** are organisms that eat primary consumers or other secondary consumers. These organisms are **carnivores.** (Parasites — organisms that utilize living tissue for their nourishment — might be considered primary or secondary consumers depending on whether they are feeding on a producer or another consumer.)

• **Decomposers** include fungi, most bacteria, and some protistans that break down organic material into smaller molecules, which are then recycled into the ecosystem.

Sometimes feeding strategies cross trophic levels.

• **Omnivores** are organisms that eat either producers or consumers.

How would you classify yourself with respect to your feeding strategy?

Table 38-1 A Simplified Food Web

Trophic Level	Organism(s)
Decomposers	
Secondary consumers (parasites)	
Secondary consumers (carnivores)	
Omnivores	
Primary consumers (herbivores)	
Producers	

PROCEDURE

Below you will find a list of organisms in a forest community and their source(s) of energy. In the blank beside each, list the *most specific* trophic level from the descriptions presented above.

• Human (black raspberries, hickory nuts, deer, rabbits) _____

• Black raspberry (sun) _____

• Deer (plants) _____

• Bear (black raspberries, deer) _____

• Coyote (deer, rabbits, black raspberries) _____

• Nematode (living hickory tree roots) _____

• Bacterial species 1 (black raspberries) _____

• Bacterial species 2 (deer) _____

• Bacterial species 3 (dead hickory trees, dead black raspberries, dead humans, dead black bears, dead deer)

_____

• Weasels (young rabbits)

_____

• Mosquito (blood of living humans, deer, and bears)

_____

• Hickory tree (sun) _____

• Cyanobacterium (sun) _____

• Fungal species 1 (living black raspberries) _____

• Fungal species 2 (dead hickory trees, dead black raspberries) _____

• Rabbit (black raspberries) _____

Now, in table 38-1, construct a **food web** by placing the names of the organisms in their appropriate trophic levels. Then connect arrows to the organisms to complete the food web.

II. The Flow of Energy Through an Ecosystem

Ecologists have learned through careful measurement that the amount of energy captured decreases at each succeeding trophic level. Of the total amount of the energy of the sun falling on a green plant, the plant is capable of capturing only about 1% of the energy. That is to say that only a small fraction of the light energy falling on the surface of a leaf is converted to the chemical energy of ATP. A primary consumer may only capture about 10% of the energy stored in a green plant. A secondary consumer may capture only 10% of the energy stored in the body of the animal that *it* eats.

How much of the solar energy that was captured by plants has been captured by the secondary consumer that feeds on a herbivore?

_____

Suppose another secondary consumer eats the first (herbivore-eating) secondary consumer. Assuming a similar flow of energy, how much of the sun's original energy does this secondary consumer gain?

_____

_____

Suppose an omnivore can obtain all the nutritional requirements necessary for life by either eating plant material or animal material. From an energetics stand-point, by which route will the greater amount of sun's energy be captured?

_____

_____

By what different routes could this organism obtain equal amounts of the sun's energy?

_____

_____

_____

III. Survivorship*

Within a population, some individuals die very young while others live into old age. To a large extent, the *pattern* of survivorship is species-dependent. Generally, three patterns of survivorship have been identified. These three have been summarized by *survivorship curves*, graphs that indicate the pattern of mortality (death) in a population.

Humans in highly developed countries with good health-care services are characterized by a Type I curve, in which there is high survivorship until some age, then high mortality. The insurance industry has generated information to determine risk groups. The premiums they charge are based upon the risk group to which an individual belongs.

While survivorship curves for humans are relatively easy to generate, information about other species is more difficult to determine. It can be quite a trick to simply determine the age of an individual plant or animal, not to mention watching an entire population over a period of years. However, the principle of determining survivorship can be demonstrated in the laboratory using nonliving objects.

In this exercise we will sudy the populations of dice and soap bubbles, using them as models of real populations to construct survivorship curves. We will subject these populations to different kinds of stress to determine the effects upon survivorship curves.

MATERIALS

Per student:
• 15-cm ruler

_____

Adapted from an exercise courtesy J. Shepherd, Mercer University.

Per student group (4):
• bucket containing 50 dice
• soap bubble solution and wand
• survivorship frame (see figure 38-1)
• stopwatch or digital watch

Per lab room:
• overhead transparency of figs. 38-2 and 38-3
• 1 set of red, blue, black, and green pens (Sharpie® or similar marker)
• overhead projector
• projection screen

PROCEDURE

A. Dice Survivorship

Work in a group of four for this experiment. One person should be assigned to dump the dice, another to record data, while the other two count.

Population 1.

1. Empty the bucket of fifty dice onto the floor.

2. Assume that all individuals that come up as 1s die of heart disease. All others survive.

3. Pick up all the 1s, set them aside (in the cemetery), and count the number of individuals who have survived. Record the number of *survivors* in this generation (Generation 1) in table 38-2 on p. 506.

4. Return the survivors to the bucket.

5. Dump the survivors onto the floor again and remove the deaths (1s) that occurred during this second generation.

6. Count and record the number of survivors.

7. Continue this process until all the dice have died from heart disease.

Population 2.

1. Start again with a full bucket of fifty dice. Assume that 1s die of heart disease and 2s die of cancer. Proceed as described for Population 1, recording numbers in table 38-2, until all dice are dead.

2. Now determine the percentage of survivors for each generation with the following formula:

$$\text{percentage surviving} = \frac{\text{number surviving}}{50} \times 100\%$$

B. Soap Bubble Survivorship

Work in groups of four for this experiment. One person will blow bubbles, a second group member will serve as the timer, a third observes survivorship, and the fourth records data in table 38-3 on p. 506.

Three different populations of soap bubbles will be formed based upon actions of group members, with each group being assigned one population to work with during this exercise:

Table 38-2 Dice Population Data

Generation	Population 1 (heart disease only)		Population 2 (cancer & heart disease)	
	Number Surviving	Percentage Surviving	Number Surviving	Percentage Surviving
0	50	100%	50	100%
1				
2				
3				
4				
5				
6				
7				
8				
9				
10				
11				
12				
13				
14				
15				
16				
17				
18				
19				
20				
21				
22				
23				
24				
25				

Table 38-3 Soap Bubble Population Data

Age at Death (seconds)	Put a Check for Each Bubble Dying	Total Number Dying at This Age	Number Surviving to This Age	Percentage Surviving to This Age
0			50	100%
1				
2				
3				
4				
5				
6				
7				
8				
9				
10				
11				
12				
13				
14				
15				
16				
17				
18				
19				
20				
21				
22				
23				
24				
25				
26				
27				
28				
29				
30 +				

Population 1: Once the bubble leaves the wand, group members wave, blow, or fan in an effort to keep the bubble in the air and prevent it from breaking (dying).

Population 2: Group members do nothing to interfere with the bubbles or keep them up in the air.

Population 3: This group uses a wand mounted on a wooden frame (fig. 38-1). The group member blowing bubbles tries to blow the bubbles through the opening in the frame. Bubbles that break rather than passing through the frame are timed and included in the data. Bubbles that fall without passing through or breaking on the frame are ignored (not counted in the data). *Do not attempt to manipulate the frame in any way so as to increase the chances that the bubbles will pass through it.*

1. Practice blowing bubbles for a few minutes until they can be generated with the single end of the wand.

2. Once the bubble is free of the wand, the timer should start the watch. When the bubble bursts, the timer notes the time and puts a check mark next to the appropriate age at death in table 38-3.

3. Obtain data on fifty bubbles.

4. Summarize your data as follows:

a. Count the number of checks (the number of bubbles dying) at each age. Record the number in the column marked "Total Number Dying at This Age."

b. By subtracting the number dying at each age from 50, determine and record the number surviving at each age. For example, if five bubbles broke (died) at age 1 second, then 50 − 5 = 45 survived at least 1 second.

c. Calculate the percentage surviving at each age. Since at birth (moment the bubble left the wand) fifty bubbles were alive, 100% were alive at age 0. Use the following formula:

$$\text{percentage surviving to this age} = \frac{\text{number surviving}}{50} \times 100\%$$

d. Plot the percentage surviving on the graphs in figures 38-2 and 38-3. (See section C.)

C. Plotting Survivorship Curves

1. Use figure 38-2 to make an *arithmetic plot* for survivorship of the dice and soap bubble populations. Note that the horizontal lines are all the same distance apart on this graph.

Use open circles to plot the percentage of dice surviving at each age (throw of the dice). Use closed circles to plot the percentage of soap bubbles surviving at each age, transposing data from the "Percentage Surviving" columns of tables 38-2 and 38-3. Connect the points from each population by drawing straight lines with the aid of a ruler.

2. We will use figure 38-3 to make a *logarithmic plot* of survivorship. Note that on this *semilog* paper, the horizontal lines become closer together toward the top of the page.

Perhaps you've never used semilog paper. Examine figure 38-3 more closely. Note that the scale on the vertical axis runs from 1 at the bottom to 9 near the middle, then from 1 to 9 near the top; there is a 1 at the topmost line. The lines are spaced according to the *logarithms* of the numbers.

The 1 at the top represents 100%, the 9 below it 90%, and so on. This means that the 1 in the middle of the page is 10%, while the 9 below it represents 9%; the 8, 8%; and so forth until the 1 at the bottom, which represents 1%.

Plot the survivorship of the dice and soap bubbles that your group generated, transferring the numbers from the "Percentage Surviving" columns from tables 38-2 and 38-3. Use open circles to plot the percentage of dice surviving and closed circles for the percentage

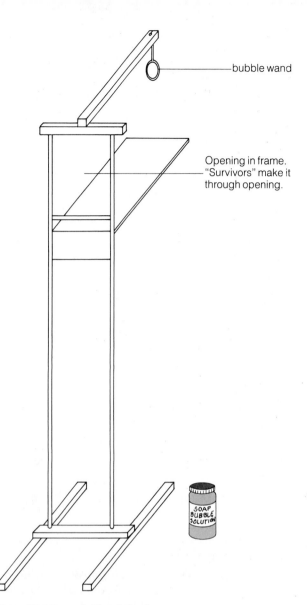

Figure 38-1 Frame for Population 3.

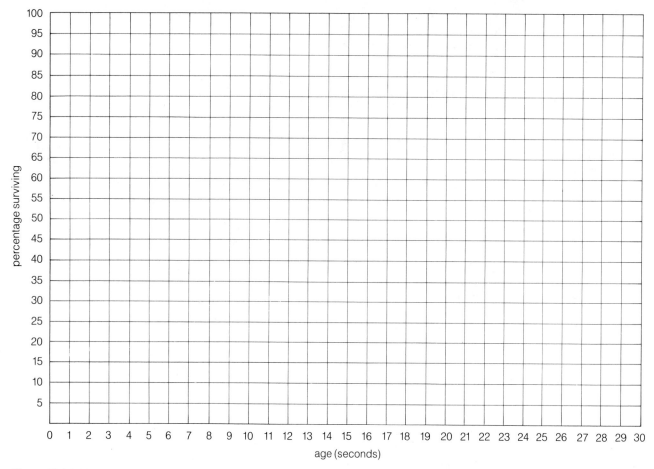

Figure 38-2 Arithmetic plot of survivorship.

of soap bubbles surviving. Use a ruler to connect the points, forming a curve for each.

After you have made your own plots, copy the plot from your population onto the overhead transparency your instructor has for this purpose, using the pens provided. Different colored pens are available so that each population will have a separate color.

D. Interpretation of the Survivorship Curves

Examine first the plots from the dice populations. In the first population (heart disease only), a constant one-sixth, or 17%, of the population dies at each age. In the second population, two-sixths, or 33%, dies at each age. As you see, on the *arithmetic plot* these data form a smooth curve, while on the *logarithmic plot* they form a straight line.

Both types of plots provide useful information. A straight line on a logarithmic plot indicates the death rate is *constant*. In the arithmetic plot it is easier to see that more individuals die at a young age than at older ages.

In natural populations, three basic trends of survivorship affecting population size have been identified. These are represented in figure 38-4 and summarized in table 38-4.

Now examine the survivorship curves for the soap bubble population.

How do they compare with the Type I, II, and III curves in figure 38-4?

_____

_____

Do any of the soap bubble populations show a constant death rate for at least part of their lifespan? If so, which?

_____

_____

How did the treatments that Populations 1, 2, and 3 were subjected to affect the shape of the curves?

_____

_____

_____

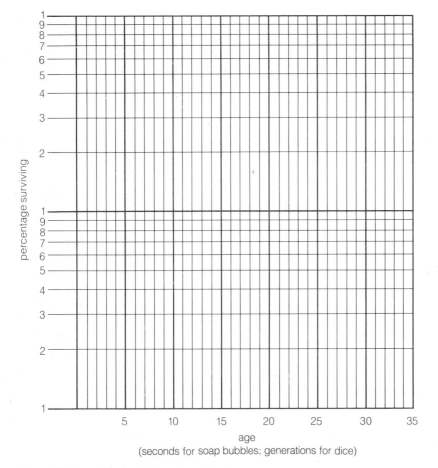

Figure 38-3 Logarithmic plot of survivorship.

Table 38-4 Survivorship Curves	
Type	**Trend**
I	Low mortality early in life, most deaths occurring in a narrow time span at maturity.
II	Rate of mortality fairly constant at all ages.
III	High mortality early in life.

OPTIONAL

IV. Exercise: Stream Ecology

Your instructor may provide you with an exercise dealing with the collection and analysis of stream ecology data.

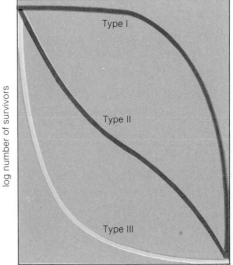

age (percent of lifespan)

Figure 38-4 Three generalized types of survivorship curves. (After Starr and Taggart, 1989.)

____ 1. The Greek word *oikos* means (a) environment, (b) habitat, (c) house, (d) community.

____ 2. A population includes (a) all organisms of the same species in one place at a given moment, (b) the organisms and their physical environment, (c) the physical environment in which an organism exists, (d) all organisms within the environment at a given moment.

____ 3. The place that an organism resides is its (a) community, (b) habitat, (c) population, (d) all of the above.

____ 4. An ecosystem (a) includes populations but not habitats, (b) is the same as *oikos*, (c) consists of a community and the physical and chemical environment, (d) is none of the above.

____ 5. Which of the following is *not* a consumer? (a) bacterium, (b) human, (c) bean plant, (d) fungus.

____ 6. The trophic level that an organism belongs to (a) is determined by what it uses as an energy source, (b) determines the population to which it belongs, (c) can be thought of as its feeding level, (d) a and c above.

____ 7. A decomposer (a) is exemplified by a fungus, (b) is the same as a parasite, (c) feeds on living organic material, (d) feeds on primary consumers only.

____ 8. An herbivore (a) eats plant material, (b) eats primary producers, (c) would be exemplified by a deer, (d) all of the above.

____ 9. Which term would best describe a human being? (a) omnivore, (b) herbivore, (c) carnivore, (d) parasite.

____ 10. A survivorship curve shows the (a) number of individuals surviving to a particular age, (b) cause of death of an individual, (c) place an organism exists in its environment, (d) trophic level in which an organism is found.

EXERCISE 38

Ecology: Living Organisms in Their Environment

POST-LAB QUESTIONS

1. Define the terms *herbivore, carnivore, omnivore.*

2. Define the terms *primary producer, primary consumer, secondary consumer.*

3. Characterize yourself, using the terms listed in questions 1 and 2. Justify your answer.

4. Which type of survivorship curve describes a population of organisms that produces a very large number of offspring, most of which die at a very early age, only a few surviving to old age? Give an example of a population of this type.

5. Would you expect a population in which most members survive for a long time to produce few or many offspring? Which would be most advantageous to the population as a whole?

6. Suppose a human population exhibits a Type III survival curve. What would you expect to happen to the curve over time if a dramatic improvement in medical technology takes place?

7. What would you expect to happen to a population whose birth rate is about equal to the death rate?

8. Examine the food web you constructed in table 38-1. Suppose there were a severe outbreak of rabies among the coyote population, resulting in the death of the entire population. Draw a graph and explain the impact of the loss of this predator on the rabbit, deer, and raspberry populations.

9. You probably have read newspaper articles describing the so-called ozone hole over Antarctica and health-care cautions that destruction of the earth's protective ozone layer will allow harmful ultraviolet (UV) light to reach the surface. Already, a significant increase in skin cancer is being noticed, a disease known to be caused by excessive exposure to UV light (including that produced by tanning lights). Perhaps not so well known is the damage UV light causes to the photosynthetic machinery of green plants and protists.

   a. Describe and explain what would happen over time to the populations of organisms of your food web (table 38-1) if ozone depletion becomes massive.

   b. What causes depletion of atmospheric ozone?

   c. Suggest what *you* can do to minimize the damage to our ozone layer.

10. Returning to the survivorship model you created using dice: You found that the chance of dying from heart disease is one-sixth for each die, indicating that survivorship is essentially the same for each age group. Relate what happened in your model with a realistic projection showing at what ages most humans die of heart disease.

Measurement Conversions

Metric to American Standard	American Standard to Metric
Length	**Length**
1 mm = 0.039 inch	1 inch = 2.54 cm
1 cm = 0.394 inch	1 foot = 0.305 m
1 m = 3.28 feet	1 yard = 0.914 m
1 m = 1.09 yards	1 mile = 1.61 km
1 km = 0.622 mile	
Volume	**Volume**
1 mL = 0.0338 fluid ounce	1 fluid ounce = 29.6 mL
1 L = 4.23 cups	1 cup = 237 mL
1 L = 2.11 pints	1 pint = 0.474 L
1 L = 1.06 quarts	1 quart = 0.947 L
1 L = 0.264 gallon	1 gallon = 3.79 L
Mass	**Mass**
1 mg = 0.0000353 ounce	1 ounce = 28.3 gm
1 gm = 0.0353 ounce	1 pound = 0.454 kg
1 kg = 2.21 pounds	

Terms of Orientation in and Around the Animal Body

A. Body Shapes

1. **Symmetry.** The body can be divided into almost identical halves.

2. **Asymmetry.** The body cannot be divided into almost identical halves (for example, many sponges).

3. **Radial symmetry.** The body is shaped like a cylinder (for example, sea anemone) or wheel (for example, sea star).

4. **Bilateral symmetry.** The body is shaped like ours in that it can be divided into halves by only one symmetrical plane (midsagittal).

B. Directions in the Body

1. **Dorsal.** At or toward the back surface of the body.

2. **Ventral.** At or toward the belly surface of the body.

3. **Anterior.** At or toward the head of the body — ventral surface of humans.

4. **Posterior.** At or toward the tail or rear end of the body — dorsal surface of humans.

5. **Medial.** At or near the midline of a body. The prefix *mid-* is often used in combination with other terms (for example, midventral).

6. **Lateral.** Away from the midline of a body.

7. **Superior.** Over or placed above some point of reference — toward the head of humans.

8. **Inferior.** Under or placed below some point of reference — away from the head of humans.

9. **Proximal.** Close to some point of reference or close to a point of attachment of an appendage to the trunk of the body.

10. **Distal.** Away from some point of reference or away from a point of attachment of an appendage to the trunk of the body.

11. **Longitudinal.** Parallel to the midline of a body.

12. **Axis.** An imaginary line around which a body or structure can rotate. The midline or *longitudinal axis* is the central axis of a symmetrical body or structure.

13. **Axial.** Placed at or along an axis.

14. **Radial.** Arranged symmetrically around an axis like the spokes of a wheel.

C. Planes of the Body

1. **Sagittal.** Passes vertically to the ground and divides the body into right and left sides. The *midsagital* or *median plane* passes through the longitudinal axis and divides the body into right and left halves.

2. **Frontal.** Passes at right angles to the sagittal plane and divides the body into dorsal and ventral parts.

3. **Transverse.** Passes from side to side at right angles to both the sagittal and frontal planes and divides the body into anterior and posterior parts — superior and inferior parts of humans. This plane of section is often referred to as a cross section.

Illustration References

Abramoff, P., and R. G. Thomson. 1982. *Laboratory Outlines in Biology III*. New York: W. H. Freeman.

Boolootian, R. A., and K. A. Stiles Trust. 1981. *College Zoology*. Tenth Edition. New York: Macmillan.

Case, C. L., and T. R. Johnson. 1984. *Experiments in Microbiology*. Menlo Park, California: Benjamin/Cummings.

Fowler, I. 1984. *Human Anatomy*. Belmont, California: Wadsworth.

Gilbert, S. G. 1966. *Pictorial Anatomy of the Fetal Pig*. Second Edition. Seattle, Washington: University of Washington Press.

Glase, J. C., et al. 1975. *Investigative Biology*. Ithaca, New York.

Hickman, C. P. 1961. *Integrated Principles of Zoology*. Second Edition. St. Louis, Missouri: C. V. Mosby.

Hickman, C. P., et al. 1978. *Biology of Animals*. Second Edition. St. Louis, Missouri: C. V. Mosby.

Jensen, W. A., et al. 1979. *Biology*. Belmont, California: Wadsworth.

Kessel, R. G., and R. H. Kardon. 1979. *Tissues and Organs*. New York: W. H. Freeman.

Kessel, R. G., and C. Y. Shih. 1974. *Scanning Electron Microscopy in Biology*. New York: Springer-Verlag.

Lytle, C. F., and J. E. Wodsedalek, 1984. *General Zoology Laboratory Guide*. Complete Version. Ninth Edition. Dubuque, Iowa: Wm. C. Brown.

Patten, B. M. 1951. *American Scientist* 39: 225–243.

Scagel, R. F., et al. 1982. *Nonvascular Plants*. Belmont, California: Wadsworth.

Sheetz, M., et al. 1976. *The Journal of Cell Biology* 70: 193.

Shih, C. Y., and R. G. Kessel. 1982. *Living Images*. Boston: Science Books International/Jones and Bartlett Publishers.

Stanier, R., et al. 1986. *The Microbial World*. Fifth Edition. Englewood Cliffs, New Jersey: Prentice-Hall.

Starr, C., and R. Taggart. 1984. *Biology*. Third Edition. Belmont, California: Wadsworth.

Starr, C., and R. Taggart. 1987. *Biology*. Fourth Edition. Belmont, California: Wadsworth.

Starr, C., and R. Taggart. 1989. *Biology*. Fifth Edition. Belmont, California: Wadsworth.

Starr, C. 1991. *Biology: Concepts & Applications*. Belmont, California: Wadsworth.

Steucek, G. L., et al. 1985. *American Biology Teacher* 471: 96–99.

Storer, T., et al. 1979. *General Zoology*. New York: McGraw-Hill.

Villee, C. A., et al. 1973. *General Zoology*. Fourth Edition. Philadelphia: W. B. Saunders.

Weller, H., and R. Wiley. 1985. *Basic Human Physiology*. Boston: PWS Publishers.

Wischnitzer, S. 1979. *Atlas and Dissection Guide for Comparative Anatomy*. Third Edition. New York: W. H. Freeman.

Wolfe, S. L. 1985. *Cell Ultrastructure*. Belmont, California: Wadsworth.